THE ECONOMY AS AN EVOLVING COMPLEX SYSTEM II

THE ECONOMY AS AN EVOLVING COMPLEX SYSTEM II

Editors

W. Brian Arthur
Santa Fe Institute
Santa Fe, New Mexico

Steven N. Durlauf
Department of Economics
University of Wisconsin at Madison
Santa Fe Institute
Santa Fe, New Mexico

David A. Lane
University of Modena
Italy

Proceedings Volume XXVII

Santa Fe Institute
Studies in the Sciences of Complexity

Addison-Wesley
The Advanced Book Program
Reading, Massachusetts

Publisher: David Goehring
Executive Editor: Jeff Robbins
Production Manager: Pat Jalbert-Levine

Director of Publications, Santa Fe Institute: Ronda K. Butler-Villa
Production Manager, Santa Fe Institute: Della L. Ulibarri
Publication Assistant, Santa Fe Institute: Marylee Thomson

This volume was typeset using TEXtures on a Macintosh IIsi computer.
Camera-ready output from a Hewlett Packard Laser Jet 4M Printer.

Addison-Wesley is an imprint of Addison Wesley Longman, Inc.

ISBN 0-201-95988-7 (Hardcover)
ISBN 0-201-32823-2 (Paperback)

1 2 3 4 5 6 7 8 9-MA-0100999897
First printing, September 1997

Find us on the World Wide Web at
http://www.aw.com/gb/

About the Santa Fe Institute

The *Santa Fe Institute* (SFI) is a private, independent, multidisciplinary research and education center, founded in 1984. Since its founding, SFI has devoted itself to creating a new kind of scientific research community, pursuing emerging science. Operating as a small, visiting institution, SFI seeks to catalyze new collaborative, multidisciplinary projects that break down the barriers between the traditional disciplines, to spread its ideas and methodologies to other individuals, and to encourage the practical applications of its results.

All titles from the *Santa Fe Institute Studies in the Sciences of Complexity* series will carry this imprint which is based on a Mimbres pottery design (circa A.D. 950–1150), drawn by Betsy Jones. The design was selected because the radiating feathers are evocative of the outreach of the Santa Fe Institute Program to many disciplines and institutions.

Santa Fe Institute
Studies in the Sciences of Complexity

Proceedings Volumes

Vol.	Editors	Title
I	D. Pines	Emerging Syntheses in Science, 1988
II	A. S. Perelson	Theoretical Immunology, Part One, 1988
III	A. S. Perelson	Theoretical Immunology, Part Two, 1988
IV	G. D. Doolen et al.	Lattice Gas Methods for Partial Differential Equations, 1990
V	P. W. Anderson, K. Arrow, & D. Pines	The Economy as an Evolving Complex System, 1988
VI	C. G. Langton	Artificial Life: Proceedings of an Interdisciplinary Workshop on the Synthesis and Simulation of Living Systems, 1988
VII	G. I. Bell & T. G. Marr	Computers and DNA, 1990
VIII	W. H. Zurek	Complexity, Entropy, and the Physics of Information, 1990
IX	A. S. Perelson & S. A. Kauffman	Molecular Evolution on Rugged Landscapes: Proteins, RNA and the Immune System, 1990
X	C. G. Langton et al.	Artificial Life II, 1992
XI	J. A. Hawkins & M. Gell-Mann	Evolution of Human Languages, 1992
XII	M. Casdagli & S. Eubank	Nonlinear Modeling and Forecasting, 1992
XIII	J. E. Mittenthal & A. B. Baskin	The Principles of Organization in Organisms, 1992
XIV	D. Friedman & J. Rust	The Double Auction Market: Institutions, Theories, and Evidence, 1993
XV	A. S. Weigend & N. A. Gershenfeld	Time Series Prediction: Forecasting the Future and Understanding the Past, 1994
XVI	G. Gumerman & M. Gell-Mann	Understanding Complexity in the Prehistoric Southwest, 1994
XVII	C. G. Langton	Artificial Life III, 1994
XVIII	G. Kramer	Auditory Display, 1994
XIX	G. Cowan, D. Pines, & D. Meltzer	Complexity: Metaphors, Models, and Reality, 1994
XX	D. H. Wolpert	The Mathematics of Generalization, 1995
XXI	P. E. Cladis & P. Palffy-Muhoray	Spatio-Temporal Patterns in Nonequilibrium Complex Systems, 1995
XXII	H. Morowitz & J. L. Singer	The Mind, The Brain, and Complex Adaptive Systems, 1995
XXIII	B. Julesz & I. Kovács	Maturational Windows and Adult Cortical Plasticity, 1995
XXIV	J. A. Tainter & B. B. Tainter	Evolving Complexity and Environmental Risk in the Prehistoric Southwest, 1996
XXV	J. Rundle, D. Turcotte, & W. Klein	Reduction and Predictability of Natural Disasters, 1996
XXVI	R. K. Belew & M. Mitchell	Adaptive Individuals in Evolving Populations: Models and Algorithms, 1996
XXVII	W. B. Arthur, S. Durlauf, & D. Lane	The Economy as an Evolving Complex Systems II, 1997

Lectures Volumes

Vol.	Editor	Title
I	D. L. Stein	Lectures in the Sciences of Complexity, 1989
II	E. Jen	1989 Lectures in Complex Systems, 1990
III	L. Nadel & D. L. Stein	1990 Lectures in Complex Systems, 1991
IV	L. Nadel & D. L. Stein	1991 Lectures in Complex Systems, 1992
V	L. Nadel & D. L. Stein	1992 Lectures in Complex Systems, 1993
VI	L. Nadel & D. L. Stein	1993 Lectures in Complex Systems, 1995

Lecture Notes Volumes

Vol.	Author	Title
I	J. Hertz, A. Krogh, & R. Palmer	Introduction to the Theory of Neural Computation, 1991
II	G. Weisbuch	Complex Systems Dynamics, 1991
III	W. D. Stein & F. J. Varela	Thinking About Biology, 1993
IV	J. M. Epstein	Nonlinear Dynamics, Mathematical Biology, and Social Science, 1997
V	H. F. Nijhout, L. Nadel, & D. L. Stein	Pattern Formation in the Physical and Biological Sciences, 1997

Reference Volumes

Vol.	Author	Title
I	A. Wuensche & M. Lesser	The Global Dynamics of Cellular Automata: Attraction Fields of One-Dimensional Cellular Automata, 1992

Contributors to This Volume

Anderson, Philip W., Joseph Henry Laboratories of Physics, Badwin Hall, Princeton University, Princeton, NJ 08544

Arthur, W. B., Santa Fe Institute, 1399 Hyde Park Road, Santa Fe, NM 87501

Blume, Lawrence E., Department of Economics, Uris Hall, Cornell University, Ithaca, NY 14853

Brock, William A., Department of Economics, University of Wisconsin at Madison, Madison, WI 53706

Darley, V. M., Division of Applied Sciences, Harvard University, Cambridge, MA 02138

Durlauf, Steven, Department of Economics, University of Wisconsin at Madison, Madison, WI 53706 and Santa Fe Institute, 1399 Hyde Park Road, Santa Fe, NM, 87501

Geanakoplos, John, Cowles Foundation, Yale University, 30 Hillhouse Avenue, New Haven, CT 06520

Holland, John H., Department of Computer Science and Engineering, University of Michigan, Ann Arbor, MI 48109 and Santa Fe Institute, 1399 Hyde Park Road, Santa Fe, NM 87501

Ioannides, Yannis M., Department of Economics, Tufts University, Medford, MA 02155

Kauffman, Stuart A., Santa Fe Institute, 1399 Hyde Park Road, Santa Fe, NM 87501

Kirman, Alan P., G.R.E.Q.A.M., E.H.E.S.S. and Université d'Aix-Marseille III, Institut Universitaire de France, 2 Rue de la Charite, 13002 Marseille, FRANCE

Kollman, Ken, Department of Political Science and Center for Political Studies, University of Michigan, Ann Arbor, MI 48109

Krugman, Paul, Department of Economics, Stanford University, Stanford, CA 95305

Lane, David, Department of Political Economy, University of Modena, ITALY

LeBaron, Blake, Department of Economics, University of Wisconsin, Madison, WI 53706 and Santa Fe Institute, 1399 Hyde Park Road, Santa Fe, NM 87501

Leijonhufvud, Axel, Center for Computable Economics, and Department of Economics, UCLA, 405 Hilgard Avenue, Los Angeles, CA 90095

Lindgren, Kristian, Institute of Physical Resource Theory, Chalmers University of Technology and Göteborg University, S-412 96 Göteborg, SWEDEN

Manski, Charles F., Department of Economics, University of Wisconsin at Madison, Madison, WI 53706

Maxfield, Robert, Department of Engineering and Economic Systems, Stanford University, Stanford, CA 95305

Miller, John H., Department of Social and Decision Sciences, Carnegie Mellon University, Pittsburgh, PA 15213

North, Douglass C., Department of Economics, Washington University, St. Louis, MO 63130-4899

Padgett, John F., Department of Political Science, University of Chicago, Chicago, IL 60637

Page, Scott, Division of Humanities and Social Sciences, California Institute of Technology 228-77, Pasadena, CA 91125

Palmer, Richard, Department of Physics, Duke University, Durham, NC 27706 and Santa Fe Institute, 1399 Hyde Park Road, Santa Fe, NM 87501

Shubik, Martin, Yale University, Cowles Foundation for Research in Economics, Department of Economics, P.O. Box 208281, New Haven, CT 06520-8281

Tayler, Paul, Department of Computer Science, Brunel University, London, UK

Tesfatsion, Leigh, Department of Economics, Heady Hall 260, Iowa State University, Ames, IA 50011-1070

Contents

Acknowledgment

The conference at which these papers were presented was sponsored by Legg Mason, whose support we gratefully acknowledge. Over the years the Santa Fe Institute's Economics Program has benefited from the generosity of Citicorp, Coopers & Lybrand, The John D. and Catherine T. MacArthur Foundation, McKinsey and Company, the Russell Sage Foundation, and SFI's core support. We thank Eric Beinhocker, Caren Grown, Win Farrell, Dick Foster, Henry Lichstein, Bill Miller, John Reed, and Eric Wanner, not only for their organizations' financial support but for the moral and intellectual support they have provided. Their many insights and suggestions over the years have greatly bolstered the program. We also thank the members of SFI's Business Network, and the many researchers who have taken part in the program. George Cowan took a chance early on that an economics program at the Institute would be a success. We thank him for his temerity.

One of the pleasures of working at the Santa Fe Institute is the exemplary staff support. In particular we thank Ginger Richardson, the staff Director of Programs, and Andi Sutherland, who organized the conference this book is based on. We are very grateful to the very able publications people at SFI, especially Marylee Thomson and Della Ulibarri.

Philip Anderson and Kenneth Arrow have been guiding lights of the SFI Economics Program since its inception. Their intellectual and personal contributions are too long to enumerate. With respect and admiration, this book is dedicated to them.

W. Brian Arthur, Steven N. Durlauf, and David A. Lane

THE ECONOMY AS AN EVOLVING COMPLEX SYSTEM II

W. B. Arthur,* S. N. Durlauf, and D. Lane†**
*Citibank Professor, Santa Fe Institute, 1399 Hyde Park Road, Santa Fe, NM 87501
**Department of Economics, University of Wisconsin at Madison, 53706 and Santa Fe Institute, 1399 Hyde Park Road, Santa Fe, NM, 87501
†Department of Political Economy, University of Modena, ITALY

Introduction

PROCESS AND EMERGENCE IN THE ECONOMY

In September 1987, twenty people came together at the Santa Fe Institute to talk about "the economy as an evolving, complex system." Ten were theoretical economists, invited by Kenneth J. Arrow, and ten were physicists, biologists, and computer scientists, invited by Philip W. Anderson. The meeting was motivated by the hope that new ideas bubbling in the natural sciences, loosely tied together under the rubric of "the sciences of complexity," might stimulate new ways of thinking about economic problems. For ten days, economists and natural scientists took turns talking about their respective worlds and methodologies. While physicists grappled with general equilibrium analysis and noncooperative game theory, economists tried to make sense of spin glass models, Boolean networks, and genetic algorithms.

The meeting left two legacies. The first was a volume of essays, *The Economy as an Evolving Complex System*, edited by Arrow, Anderson, and David Pines. The other was the founding, in 1988, of the Economics Program at the Santa Fe Institute, the Institute's first resident research program. The Program's mission was to encourage the understanding of economic phenomena from a complexity perspective, which involved the development of theory as well as tools for modeling and for empirical analysis. To this end, since 1988, the Program has brought researchers to Santa Fe, sponsored research projects, held several workshops each year, and published several dozen working papers. And, since 1994, it has held an annual summer school for economics graduate students.

This volume, *The Economy as an Evolving Complex System II*, represents the proceedings of an August 1996 workshop sponsored by the SFI Economics Program. The intention of this workshop was to take stock, to ask: What has the complexity perspective contributed to economics in the past decade? In contrast to the 1987 workshop, almost all of the presentations addressed economic problems, and most participants were economists by training. In addition, while some of the work presented was conceived or carried out at the Institute, some of the participants had no previous relation with SFI—research related to the complexity perspective is under active development now in a number of different institutes and university departments.

But just what *is* the complexity perspective in economics? That is not an easy question to answer. Its meaning is still very much under construction, and, in fact, the present volume is intended to contribute to that construction process. Indeed, the authors of the essays in this volume by no means share a single, coherent vision of the meaning and significance of complexity in economics. What we will find instead is a family resemblance, based upon a set of interrelated themes that together constitute the current meaning of the complexity perspective in economics.

Several of these themes, already active subjects of research by economists in the mid-1980s, are well described in the earlier *The Economy as an Evolving Complex System*: In particular, applications of nonlinear dynamics to economic theory and data analysis, surveyed in the 1987 meeting by Michele Boldrin and William Brock; and the theory of positive feedback and its associated phenomenology of path dependence and lock-in, discussed by W. Brian Arthur. Research related to both these themes has flourished since 1987, both in and outside the SFI Economics Program. While chaos has been displaced from its place in 1987 at center stage of the interest in nonlinear dynamics, in the last decade economists have made substantial progress in identifying patterns of nonlinearity in financial time series and in proposing models that both offer explanations for these patterns and help to analyze and, to some extent, predict the series in which they are displayed. Brock surveys both these developments in his chapter in this volume, while positive feedback plays a central role in the models analyzed by Lane (on information contagion), Durlauf (on inequality) and Krugman (on economic geography), and lurk just under the surface of the phenomena described by North (development) and Leijonhufvud (high inflation).

Looking back over the developments in the past decade and the papers produced by the program, we believe that a coherent perspective—sometimes called the "Santa Fe approach"—has emerged within economics. We will call this the complexity perspective, or Santa Fe perspective, or occasionally the process-and-emergence perspective. Before we describe this, we first sketch the two conceptions of the economy that underlie standard, neoclassical economics (and indeed most of the presentations by economic theorists at the earlier 1987 meeting). We can call these conceptions the "equilibrium" and "dynamical systems" approaches. In the equilibrium approach, the problem of interest is to derive, from the rational choices of individual optimizers, aggregate-level "states of the economy" (prices in general equilibrium analysis, a set of strategy assignments in game theory with associated payoffs) that satisfy some aggregate-level consistency condition (market-clearing, Nash equilibrium), and to examine the properties of these aggregate-level states. In the dynamical systems approach, the state of the economy is represented by a set of variables, and a system of difference equations or differential equations describes how these variables change over time. The problem is to examine the resulting trajectories, mapped over the state space. However, the equilibrium approach does not describe the *mechanism* whereby the state of the economy changes over time—nor indeed how an equilibrium comes into being.[1] And the dynamical system approach generally fails to accommodate the distinction between *agent*- and *aggregate*-levels (except by obscuring it through the device of "representative agents"). Neither accounts for the emergence of new kinds of relevant state variables, much less new entities, new patterns, new structures.[2]

To describe the complexity approach, we begin by pointing out six features of the economy that together present difficulties for the traditional mathematics used in economics:[3]

DISPERSED INTERACTION. What happens in the economy is determined by the interaction of many dispersed, possibly heterogeneous, agents acting in parallel. The action of any given agent depends upon the anticipated actions of a limited number of other agents and on the aggregate state these agents cocreate.

[1]Since an a priori intertemporal equilibrium hardly counts as a mechanism.

[2]Norman Packard's contribution to the 1987 meeting addresses just this problem with respect to the dynamical systems approach. As he points out, "if the set of relevant variables changes with time, then the state space is itself changing with time, which is not commensurate with a conventional dynamical systems model."

[3]John Holland's paper at the 1987 meeting beautifully—and presciently—frames these features. For an early description of the Santa Fe approach, see also the program's March 1989 newsletter, "Emergent Structures."

NO GLOBAL CONTROLLER. No global entity controls interactions. Instead, controls are provided by mechanisms of competition and coordination among agents. Economic actions are mediated by legal institutions, assigned roles, and shifting associations. Nor is there a universal competitor—a single agent that can exploit all opportunities in the economy.

CROSS-CUTTING HIERARCHICAL ORGANIZATION. The economy has many levels of organization and interaction. Units at any given level—behaviors, actions, strategies, products—typically serve as "building blocks" for constructing units at the next higher level. The overall organization is more than hierarchical, with many sorts of tangled interactions (associations, channels of communication) across levels.

CONTINUAL ADAPTATION . Behaviors, actions, strategies, and products are revised continually as the individual agents accumulate experience—the system constantly adapts.

PERPETUAL NOVELTY. Niches are continually created by new markets, new technologies, new behaviors, new institutions. The very act of filling a niche may provide new niches. The result is ongoing, perpetual novelty.

OUT-OF-EQUILIBRIUM DYNAMICS. Because new niches, new potentials, new possibilities, are continually created, the economy operates far from any optimum or global equilibrium. Improvements are always possible and indeed occur regularly.

Systems with these properties have come to be called *adaptive nonlinear networks* (the term is John Holland's[5]). There are many such in nature and society: nervous systems, immune systems, ecologies, as well as economies. An essential element of adaptive nonlinear networks is that they do not act simply in terms of stimulus and response. Instead they anticipate. In particular, economic agents form expectations—they build up models of the economy and act on the basis of predictions generated by these models. These anticipative models need neither be explicit, nor coherent, nor even mutually consistent.

Because of the difficulties outlined above, the mathematical tools economists customarily use, which exploit linearity, fixed points, and systems of differential equations, cannot provide a deep understanding of adaptive nonlinear networks. Instead, what is needed are new classes of combinatorial mathematics and population-level stochastic processes, in conjunction with computer modeling. These mathematical and computational techniques are in their infancy. But they emphasize the *discovery* of structure and the *processes* through which structure *emerges* across different levels of organization.

This conception of the economy as an adaptive nonlinear network—as an evolving, complex system—has profound implications for the foundations of economic theory and for the way in which theoretical problems are cast and solved. We interpret these implications as follows:

COGNITIVE FOUNDATIONS. Neoclassical economic theory has a unitary cognitive foundation: economic agents are rational optimizers. This means that (in the usual interpretation) agents evaluate uncertainty probabilistically, revise their evaluations in the light of new information via Bayesian updating, and choose the course of action that maximizes their expected utility. As glosses on this unitary foundation, agents are generally assumed to have common knowledge about each other and rational expectations about the world they inhabit (and of course cocreate). In contrast, the Santa Fe viewpoint is pluralistic. Following modern cognitive theory, we posit no single, dominant mode of cognitive processing. Rather, we see agents as having to cognitively structure the problems they face—as having to "make sense" of their problems—as much as solve them. And they have to do this with cognitive resources that are limited. To "make sense," to learn, and to adapt, agents use variety of distributed cognitive processes. The very categories agents use to convert information about the world into action emerge from experience, and these categories or cognitive props need not fit together coherently in order to generate effective actions. Agents therefore inhabit a world that they must cognitively interpret—one that is complicated by the presence and actions of other agents and that is ever changing. It follows that agents generally do not optimize in the standard sense, not because they are constrained by finite memory or processing capability, but because the very concept of an optimal course of action often cannot be defined. It further follows that the deductive rationality of neoclassical economic agents occupies at best a marginal position in guiding effective action in the world. And it follows that any "common knowledge" agents might have about one another must be attained from concrete, specified cognitive processes operating on experiences obtained through concrete interactions. Common knowledge cannot simply be assumed into existence.

STRUCTURAL FOUNDATIONS. In general equilibrium analysis, agents do not interact with one another directly, but only through impersonal markets. By contrast, in game theory all players interact with all other players, with outcomes specified by the game's payoff matrix. So interaction structures are simple and often extreme—one-with-all or all-with-all. Moreover, the internal structure of the agents themselves is abstracted away. [4] In contrast, from a complexity perspective, structure matters. First, network-based structures become important. All economic action involves interactions among agents, so economic functionality is both constrained and carried by networks defined by recurring patterns of interaction among agents. These network structures are characterized by relatively sparse ties. Second, economic action is structured by emergent social roles and by socially supported procedures—that is,

[4]Except in principal-agent theory or transaction-costs economics, where a simple hierarchical structure is supposed to obtain.

by institutions. Third, economic entities have a recursive structure: they are themselves comprised of entities. The resulting "level" structure of entities and their associated action processes is not strictly hierarchical, in that component entities may be part of more than one higher-level entity, and entities at multiple levels of organization may interact. Thus, reciprocal causation operates between different levels of organization—while action processes at a given level of organization may sometimes by viewed as autonomous, they are nonetheless constrained by action patterns and entity structures at other levels. And they may even give rise to new patterns and entities at both higher and lower levels. From the Santa Fe perspective, the fundamental principle of organization is the idea that units at one level combine to produce units at the next higher level. [5]

WHAT COUNTS AS A PROBLEM AND AS A SOLUTION. It should be clear by now that exclusively posing economic problems as multiagent optimization exercises makes little sense from the viewpoint we are outlining—a viewpoint that puts emphasis on process, not just outcome. In particular, it asks how new "things" arise in the world—cognitive things, like "internal models"; physical things, like "new technologies"; social things, like new kinds of economic "units." And it is clear that if we posit a world of perpetual novelty, then outcomes cannot correspond to steady-state equilibria, whether Walrasian, Nash, or dynamic-systems-theoretical. The only descriptions that can matter in such a world are about transient phenomena—about process and about emergent structures. What then can we know about the economy from a process-and-emergence viewpoint, and how can we come to know it? Studying process and emergence in the economy has spawned a growth industry in the production of what are now generally called "agent-based models." And what counts as a solution in an agent-based model is currently under negotiation. Many of the papers in this volume—including those by Arthur et al., Darley and Kauffman, Shubik, Lindgren, Kollman et al., Kirman, and Tesfatsion—address this issue, explicitly or implicitly. We can characterize these as seeking emergent structures arising in interaction processes, in which the interacting entities anticipate the future through cognitive procedures that themselves involve interactions taking place in multilevel structures.

A description of an approach to economics, however, is not a research program. To build a research program around a process-and-emergence perspective, two things have to happen. First, concrete economic problems have to be identified for which the approach may provide new insights. A number of candidates are offered in this volume: artifact innovation (Lane and Maxfield), the evolution of trading networks (Ioannides, Kirman, and Tesfatsion), money (Shubik), the origin and spatial distribution of cities (Krugman), asset pricing (Arthur et al. and

[5] We need not commit ourselves to what constitutes economic "units" and "levels." This will vary from problem context to problem context.

Brock), high inflation (Leijonhufvud) persistent differences in income between different neighborhoods or countries (Durlauf). Second, cognitive and structural foundations for modeling these problems have to be constructed and methods developed for relating theories based on these foundations to observable phenomena (Manski). Here, while substantial progress has been made since 1987, the program is far from complete.

The essays in this volume describe a series of parallel explorations of the central themes of process and emergence in an interactive world—of how to study systems capable of generating perpetual novelty. These explorations do not form a coherent whole. They are sometimes complementary, sometimes even partially contradictory. But what could be more appropriate to the Santa Fe perspective, with its emphasis on distributed processes, emergence, and self-organization? Here are our interpretations of the research directions that seem to be emerging from this process:

COGNITION. The central cognitive issues raised in this volume are ones of interpretation. As Shubik puts it, "the interpretation of data is critical. It is not what the numbers are, but what they mean." How do agents render their world comprehensible enough so that "information" has meaning? The two papers by Arthur, Holland, LeBaron, Palmer, and Tayler and by Darley and Kauffman consider this. They explore problems in which a group of agents take actions whose effects depend on what the other agents do. The agents base their actions on expectations they generate about how other agents will behave. Where do these expectations come from? Both papers reject common knowledge or common expectations as a starting point. Indeed, Arthur et al. argue that common beliefs cannot be deduced. Because agents must derive their expectations from an imagined future that is the aggregate result of other agents' expectations, there is a self-reference of expectations that leads to deductive indeterminacy. Rather, both papers suppose that each agent has access to a variety of "interpretative devices" that single out particular elements in the world as meaningful and suggest useful actions on the basis of the "information" these elements convey. Agents keep track of how useful these devices turn out to be, discarding ones that produce bad advice and tinkering to improve those that work. In this view, economic action arises from an evolving ecology of interpretive devices that interact with one another through the medium of the agents that use them to generate their expectations.

Arthur et al. build a theory of asset pricing upon such a view. Agents—investors—act as market statisticians. They continually generate expectational models—interpretations of what moves prices in the market—and test these by trading. They discard and replace models if not successful. Expectations in the market therefore become endogenous—they continually change and adapt to a market that they create together. The Arthur et al. market settles into a rich psychology, in which speculative bubbles, technical trading, and persistence of volatility emerge. The homogeneous rational expectations of the standard literature become a special case—possible in theory but unlikely to emerge in practice. Brock presents a variant of this approach, allowing agents to switch between a limited number of expectational models. His model is simpler than that of Arthur et al., but he achieves analytical results, which he relates to a variety of stylized facts about financial times series, many of which have been uncovered through the application of nonlinear analysis over the past decade.

In the world of Darley and Kauffman, agents are arrayed on a lattice, and they try to predict the behavior of their lattice neighbors. They generate their predictions via an autoregressive model, and they can individually tune the number of parameters in the model and the length of the time series they use to estimate model parameters. Agents can change parameter number or history length by steps of length 1 each period, if by doing so they would have generated better predictions in the previous period. This induces a coevolutionary "interpretative dynamics," which does not settle down to a stable regime of precise, coordinated mutual expectations. In particular, when the system approaches a "stable rational-expectations state," it tends to break down into a disordered state. They use their results to argue against conventional notions of rationality, with infinite foresight horizons and unlimited deductive capability.

In his paper on high inflation, Leijonhufvud poses the same problem as Darley and Kauffman: Where should we locate agent cognition, between the extremes of "infinite-horizon optimization" and "myopic adaptation"? Leijonhufvud argues that the answer to this question is context dependent. He claims that in situations of institutional break-down like high inflation, agent cognition shifts toward the "short memory/short foresight adaptive mode." The causative relation between institutional and cognitive shifts becomes reciprocal. With the shrinking of foresight horizons, markets for long-term loans (where long-term can mean over 15 days) disappear. And as inflation accelerates, units of accounting lose meaning. Budgets cannot be drawn in meaningful ways, the executive arm of government becomes no longer fiscally accountable to parliament, and local governments become unaccountable to national governments. Mechanisms of social and economic control erode. Ministers lose control over their bureaucracies, shareholders over corporate management.

The idea that "interpretative devices" such as explicit forcasting models and technical-trading rules play a central role in agent cognition fits with a more general set of ideas in cognitive science, summarized in Clark.[2] This work rejects the notion that cognition is all "in the head." Rather, interpretive aids such as autoregressive models, computers, languages, or even navigational tools (as in Hutchins[6]) and institutions provide a "scaffolding," an external structure on which much of task of interpreting the world is off-loaded. Clark[2] argues that the distinctive hallmark of in-the-head cognition is "fast pattern completion," which bears little relation to the neoclassical economist's deductive rationality. In this volume, North takes up this theme, describing some of the ways in which institutions scaffold interpretations of what constitutes possible and appropriate action for economic agents.

Lane and Maxfield consider the problem of interpretation from a different perspective. They are particularly interested in what they call attributions of functionality: interpretations about what an artifact does. They argue that new attributions of functionality arise in the context of particular kinds of agent relationships, where agents can differ in their interpretations. As a consequence, cognition has an unavoidable social dimension. What interpretations are possible depend on who interacts with whom, about what. They also argue that new functionality attributions cannot be foreseen outside the particular generative relationships in which they arise. This unforeseeability has profound consequences for what constitutes "rational" action in situations of rapid change in the structure of agent-artifact space.

All the papers mentioned so far take as fundamental the importance of cognition for economic theory. But the opposite point of view can also be legitimately defended from a process-and-emergence perspective. According to this argument, overrating cognition is just another error deriving from methodological individualism, the very bedrock of standard economic theory. *How* individual agents decide what to do may not matter very much. What happens as a result of their actions may depend much more on the interaction structure through which they act—who interacts with whom, according to which rules. Blume makes this point in the introduction to his paper on population games, which, as he puts it, provide a class of models that shift attention "from the fine points of individual-level decision theory to dynamics of agent interaction." Padgett makes a similar claim, though for a different reason. He is interested in formulating a theory of the firm as a locus of transformative "work," and he argues that "work" may be represented by "an orchestrated sequence of actions and reactions, the sequence of which produces some collective result (intended or not)." Hence, studying the structure of coordinated action-reaction sequences may provide insight into the organization of economic activity, without bringing "cognition" into the story at all. Padgett's paper is inspired by recent work in chemistry and biology (by Eigen and Schuster[3] and by Fontana and Buss,[4] among others) that are considered exemplars of the complexity perspective in these fields.

STRUCTURE. Most human interactions, even those taking place in "economic" contexts, have a primarily social character: talking with friends, asking advice from knowledgeable acquaintances, working together with colleagues, living next to neighbors. Recurring patterns of such social interactions bind agents together into networks.[6] According to standard economic theory, what agents do depends on their values and available information. But standard theory typically ignores where values and information come from. It treats agents' values and information as exogenous and autonomous. In reality, agents learn from each other, and their values may be influenced by others' values and actions. These processes of learning

[6] There is a voluminous sociological literature on interaction networks. Recent entry points include Noria and Eccles,[7] particularly the essay by Granovetter entitled "Problems of Explanation in Economic Sociology," and the methodological survey of Wasserman and Faust.[8]

and influencing happen through the social interaction networks in which agents are embedded, and they may have important economic consequences. For example, one of the models presented in Durlauf's paper implies that value relationships among neighbors can induce persistent income inequalities between neighborhoods. Lane examines a model in which information flowing between agents in a network determines the market shares of two competing products. Kirman's paper reviews a number of models that derive economic consequences from interaction networks.

Ioannides, Kirman, and Tesfatsion consider the problems of how networks emerge from initially random patterns of dyadic interaction and what kinds of structure the resulting networks exhibit. Ioannides studies mathematical models based on controlled random fields, while Tesfatsion works in the context of a particular agent-based model, in which the "agents" are strategies that play Prisoner's Dilemma with one another. Ioannides and Tesfatsion are both primarily interested in networks involving explicitly economic interactions, in particular trade. Their motivating idea, long recognized among sociologists (for example, Baker[1]), is that markets actually function by means of networks of traders, and what happens in markets may reflect the structure of these networks, which in turn may depend on how the networks emerge.

Local interactions can give rise to large-scale spatial structures. This phenomenon is investigated by several of the papers in this volume. Lindgren's contribution is particularly interesting in this regard. Like Tesfatsion, he works with an agent-based model in which the agents code strategies for playing two-person games. In both Lindgren's and Tesfatsion's models, agents adapt their strategies over time in response to their past success in playing against other agents. Unlike Tesfatsion's agents, who meet randomly and decide whether or not to interact, Lindgren's agents only interact with neighbors in a prespecified interaction network. Lindgren studies the emergence of spatiotemporal structure in agent space—metastable ecologies of strategies that maintain themselves for many agent-generations against "invasion" by new strategy types or "competing" ecologies at their spatial borders. In particular, he compares the structures that arise in a lattice network, in which each agent interacts with only a few other agents, with those that arise in a fully connected network, in which each agent interacts with all other agents. He finds that the former "give rise to a stable coexistence between strategies that would otherwise be outcompeted. These spatiotemporal structures may take the form of spiral waves, irregular waves, spatiotemporal chaos, frozen patchy patterns, and various geometrical configurations." Though Lindgren's model is not explicitly economic, the contrast he draws between an agent space in which interactions are structured by (relatively sparse) social networks and an agent space in which all interactions are possible (as is the case, at least in principle, with the impersonal markets featured in general equilibrium analysis) is suggestive. Padgett's paper offers a similar contrast, in a quite different context.

Both Durlauf and Krugman explore the emergence of geographical segregation. In their models, agents may change location—that is, change their position in a social structure defined by neighbor ties. In these models (especially Durlauf's), there are many types of agents, and the question is under what circumstances, and through what mechanisms, do aggregate-level "neighborhoods" arise, each consisting predominantly (or even exclusively) of one agent type. Thus, agents' choices, conditioned by current network structure (the agent's neighbors and the neighbors at the sites to which the agent can move), change that structure; over time, from the changing local network structure, an aggregate-level pattern of segregated neighborhoods emerges.

Kollman, Miller, and Page explore a related theme in their work on political platforms and institutions in multiple jurisdictions. In their agent-based model, agents may relocate between jurisdictions. They show that when there are more than three jurisdictions, two-party competition outperforms democratic referenda. The opposite is the case when there is only one jurisdiction and, hence, no agent mobility. They also find that two-party competition results in more agent moves than does democratic referenda.

Manski reminds us that while theory is all very well, understanding of real phenomena is just as important. He distinguishes between three kinds of causal explanation for the often observed empirical fact that "persons belonging to the same group tend to behave similarly." One is the one we have been describing above: the behavioral similarities may arise through network interaction effects. But there are two other possible explanations: *contextual*, in which the behavior may depend on exogenous characteristics of the group (like socioeconomic composition); or *correlated effects*, in which the behavior may be due to similar *individual* characteristics of members of the group. Manski shows, among other results, that a researcher who uses the popular linear-in-means model to analyze his data and "observes equilibrium outcomes and the composition of reference groups cannot empirically distinguish" endogenous interactions from these alternative explanations. One moral is that nonlinear effects require nonlinear inferential techniques.

In the essays of North, Shubik, and Leijonhufvud, the focus shifts to another kind of social structure, the institution. North's essay focuses on institutions and economic growth, Shubik's on financial institutions, and Leijonhufvud's on high-inflation phenomenology. All three authors agree in defining institutions as "the rules of the game," without which economic action is unthinkable. They use the word "institution" in at least three senses: as the "rules" themselves (for example, bankruptcy laws); as the entities endowed with the social and political power to promulgate rules (for example, governments and courts); and as the socially legitimized constructions that instantiate rules and through which economic agents act (for example, fiat money and markets). In whichever sense institutions are construed, the three authors agree that they cannot be adequately understood from a purely economic, purely political, or purely social point of view. Economics, politics, and society are inextricably mixed in the processes whereby institutions come into being. And they change and determine economic, political, and social action. North also insists that institutions have a cognitive dimension through the aggregate-level "belief systems" that sustain them and determine the directions in which they change.

North takes up the question of the emergence of institutions from a functionalist perspective: institutions are brought into being "in order to reduce uncertainty," that is, to make agents' worlds predictable enough to afford recognizable opportunities for effective action. In particular, modern economies depend upon institutions that provide low transaction costs in impersonal markets.

Shubik takes a different approach. His analysis starts from his notion of strategic market games. These are "fully defined process models" that specify actions "for all points in the set of feasible outcomes." He shows how, in the context of constructing a strategic market game for an exchange economy using fiat money, the full specification requirement leads to the logical necessity of certain kinds of rules that Shubik identifies with financial institutions. Geanakoplos' paper makes a similar point to Shubik's. Financial instruments represent promises, he argues. What happens if someone cannot or will not honor a promise? Shubik already introduced the logical necessity of one institution, bankruptcy law, to deal with defaults. Geanakoplos introduces another, collateral. He shows that, in equilibrium, collateral as an institution has institutional implications—missing markets.

Finally, in his note concluding the volume, Philip Anderson provides a physicist's perspective on a point that Fernand Braudel argues is a central lesson from the history of long-term socioeconomic change. Averages and assumptions of agent homogeneity can be very deceptive in complex systems. And processes of change are generally driven by the inhabitants of the extreme tails of some relevant distribution. Hence, an interesting theoretical question from the Santa Fe perspective is: How do distributions with extreme tails arise, and why are they so ubiquitous and so important?

WHAT COUNTS AS A PROBLEM AND AS A SOLUTION. While the papers here have much to say on cognition and structure, they contain much less discussion on what constitutes a problem and solution from this new viewpoint. Perhaps this is because it is premature to talk about methods for generating and assessing understanding when what is to be understood is still under discussion. While a few of the papers completely avoid mathematics, most of the papers do present mathematical models—whether based on statistical mechanics, strategic market games, random graphs, population games, stochastic dynamics, or agent-based computations. Yet sometimes the mathematical models the authors use leave important questions unanswered. For example, in what way do equilibrium calculations provide insight into emergence? This troublesome question is not addressed in any of the papers, even those in which models are presented from which equilibria are calculated—and insight into emergence is claimed to result. Blume raises two related issues in his discussion of population games: whether the asymptotic equilibrium selection theorems featured in the theory happen "soon enough" to be economically interesting; and whether the invariance of the "global environment" determined by the game and interaction model is compatible with an underlying economic reality in which rules of the game undergo endogenous change. It will not be easy to resolve the inherent tension between traditional mathematical tools and phenomena that may exhibit perpetual novelty.

As we mentioned previously, several of the papers introduce less traditional, agent-based models. Kollman, Miller, and Page discuss both advantages and difficulties associated with this set of techniques. They end up expressing cautious optimism about their future usefulness. Tesfatsion casts her own paper as an illustration of what she calls "the alife approach for economics, as well as the hurdles that remain to be cleared." Perhaps the best recommendation we can make to the reader with respect to the epistemological problems associated with the process-and-emergence perspective is simple. Read the papers, and see what you find convincing.

REFERENCES

1. Baker W. "The Social Structure of a National Securities Market." *Amer. J. Sociol.* **89** (1984): 775–811.
2. Clark, A. *Being There: Putting Brain, Body, and World Together Again.* Cambridge, MA: MIT Press, 1997.
3. Eigen, M., and P. Schuster. *The Hypercycle.* Berlin: Springer Verlag, 1979.
4. Fontana, W., and L. Buss. "The Arrival of the Fittest: Toward a Theory of Biological Organization." *Bull. Math. Biol.* **56** (1994): 1–64
5. Holland, J. H. "The Global Economy as an Adaptive Process." In *The Economy as an Evolving Complex System*, edited by P. W. Anderson, K. J. Arrow, and D. Pines, 117–124. Santa Fe Institute Studies in the Sciences of Complexity, Proc. Vol. V. Redwood City, CA: Addison-Wesley, 1988.
6. Hutchins, E. *Cognition in the Wild.* Cambridge, MA: MIT Press, 1995.
7. Noria, N., and R. Eccles (Eds.) *Networks and Organizations: Structure, Form, and Action.* Cambridge, MA: Harvard Business School Press, 1992.
8. Wasserman, W., and K. Faust. *Social Network Analysis: Methods and Applications.* Cambridge, MA: Cambridge University Press, 1994.

W. Brian Arthur,† John H. Holland,‡ Blake LeBaron,* Richard Palmer,* and Paul Tayler**
†Citibank Professor, Santa Fe Institute, 1399 Hyde Park Road, Santa Fe, NM 87501
‡Professor of Computer Science and Engineering, University of Michigan, Ann Arbor, MI 48109 and Santa Fe Institute, 1399 Hyde Park Road, Santa Fe, NM 87501
*Associate Professor of Economics, University of Wisconsin, Madison, WI 53706 and Santa Fe Institute, 1399 Hyde Park Road, Santa Fe, NM 87501
*Professor of Physics, Duke University, Durham, NC 27706 and Santa Fe Institute, 1399 Hyde Park Road, Santa Fe, NM 87501
**Department of Computer Science, Brunel University, London

Asset Pricing Under Endogenous Expectations in an Artificial Stock Market

We propose a theory of asset pricing based on heterogeneous agents who continually adapt their expectations to the market that these expectations aggregatively create. And we explore the implications of this theory computationally using our Santa Fe artificial stock market.[1]

Asset markets, we argue, have a recursive nature in that agents' expectations are formed on the basis of their anticipations of other agents' expectations, which precludes expectations being formed by deductive means. Instead, traders continually hypothesize—continually explore—expectational models, buy or sell on the basis of those that perform best, and confirm or discard these according to their performance. Thus, individual beliefs or expectations become endogenous to the market, and constantly compete within an ecology of others' beliefs or expectations. The ecology of beliefs coevolves over time.

Computer experiments with this endogenous-expectations market explain one of the more striking puzzles in finance: that market traders often believe in such concepts as technical trading, "market psychology," and bandwagon effects, while

[1]For a less formal discussion of the ideas in this paper see Arthur.[3]

academic theorists believe in market efficiency and a lack of speculative opportunities. Both views, we show, are correct, but within different regimes. Within a regime where investors explore alternative expectational models at a low rate, the market settles into the rational-expectations equilibrium of the efficient-market literature. Within a regime where the rate of exploration of alternative expectations is higher, the market self-organizes into a complex pattern. It acquires a rich psychology, technical trading emerges, temporary bubbles and crashes occur, and asset prices and trading volume show statistical features—in particular, GARCH behavior—characteristic of actual market data.

1. INTRODUCTION

Academic theorists and market traders tend to view financial markets in strikingly different ways. Standard (efficient-market) financial theory assumes identical investors who share rational expectations of an asset's future price, and who instantaneously and rationally discount all market information into this price.[2] It follows that no opportunities are left open for consistent speculative profit, that technical trading (using patterns in past prices to forecast future ones) cannot be profitable except by luck, that temporary price overreactions—bubbles and crashes—reflect rational changes in assets' valuations rather than sudden shifts in investor sentiment. It follows too that trading volume is low or zero, and that indices of trading volume and price volatility are not serially correlated in any way. The market, in this standard theoretical view, is rational, mechanistic, and efficient. Traders, by contrast, often see markets as offering speculative opportunities. Many believe that technical trading is profitable,[3] that something definable as a "market psychology" exists, and that herd effects unrelated to market news can cause bubbles and crashes. Some traders and financial writers even see the market itself as possessing its own moods and personality, sometimes describing the market as "nervous" or "sluggish" or "jittery." The market in this view is psychological, organic, and imperfectly efficient. From the academic viewpoint, traders with such beliefs—embarrassingly the very agents assumed rational by the theory—are irrational and superstitious. From the traders' viewpoint, the standard academic theory is unrealistic and not borne out by their own perceptions.[4]

While few academics would be willing to assert that the market has a personality or experiences moods, the standard economic view has in recent years

[2] For the classic statement see Lucas,[34] or Diba and Grossman.[16]

[3] For evidence see Frankel and Froot.[19]

[4] To quote one of the most successful traders, George Soros[47]: "this [efficient market theory] interpretation of the way financial markets operate is severely distorted. . . .It may seem strange that a patently false theory should gain such widespread acceptance."

begun to change. The crash of 1987 damaged economists' beliefs that sudden price changes reflect rational adjustments to news in the market: several studies failed to find significant correlation between the crash and market information issued at the time (e.g., Cutler et al.[12]). Trading volume and price volatility in real markets are large—not zero or small, respectively, as the standard theory would predict[32,44,45]—and both show significant autocorrelation.[7,21] Stock returns also contain small, but significant serial correlations.[18,33,39,48] Certain technical-trading rules produce statistically significant, if modest, long-run profits.[10] And it has long been known that when investors apply full rationality to the market, they lack incentives both to trade and to gather information.[23,24,36] By now, enough statistical evidence has accumulated to question efficient-market theories and to show that the traders' viewpoint cannot be entirely dismissed. As a result, the modern finance literature has been searching for alternative theories that can explain these market realities.

One promising modern alternative, the noise-trader approach, observes that when there are "noise traders" in the market—investors who possess expectations different from those of the rational-expectations traders—technical-trading strategies such as trend chasing may become rational. For example, if noise traders believe that an upswing in a stock's price will persist, rational traders can exploit this by buying into the uptrend, thereby exacerbating the trend. In this way positive-feedback trading strategies—and other technical-trading strategies—can be seen as rational, as long as there are nonrational traders in the market to prime these strategies.[13,14,15,46] This "behavioral" noise-trader literature moves some way toward justifying the traders' view. But it is built on two less-than-realistic assumptions: the existence of unintelligent noise traders who do not learn over time that their forecasts are erroneous; and the existence of rational players who possess, by some unspecified means, full knowledge of both the noise traders' expectations and their own class's. Neither assumption is likely to hold up in real markets. Suppose for a moment an actual market with minimally intelligent noise traders. Over time, in all likelihood, some would discover their errors and begin to formulate more intelligent (or at least different) expectations. This would change the market, which means that the perfectly intelligent players would need to readjust *their* expectations. But there is no reason these latter would know the new expectations of the noise-trader deviants; they would have to derive their expectations by some means such as guessing or observation of the market. As the rational players changed, the market would change again. And so the noise traders might again further deviate, forcing further readjustments for the rational traders. Actual noise-trader markets, assumed stationary in theory, would start to unravel; and the perfectly rational traders would be left at each turn guessing the changed expectations by observing the market.

Thus, noise-trader theories, while they explain much, are not robust. But in questioning such theories we are led to an interesting sequence of thought. Suppose we were to assume "rational," but nonidentical, agents who do not find themselves in a market with rational expectations, or with publicly known expectations. Suppose we allowed each agent continually to observe the market with an eye to discovering profitable expectations. Suppose further we allowed each agent to adopt these when discovered and to discard the less profitable as time progressed. In this situation, agents' expectations would become endogenous—individually adapted to the current state of the market—and they would cocreate the market they were designed to exploit. How would such a market work? How would it act to price assets? Would it converge to a rational-expectations equilibrium—or would it uphold the traders' viewpoint?

In this chapter we propose a theory of asset pricing that assumes fully heterogeneous agents whose expectations continually adapt to the market these expectations aggregatively create. We argue that under heterogeneity, expectations have a recursive character: agents have to form their expectations from their anticipations of other agents' expectations, and this self-reference precludes expectations being formed by deductive means. So, in the absence of being able to deduce expectations, agents—no matter how rational—are forced to hypothesize them. Agents, therefore, continually form individual, hypothetical, expectational models or "theories of the market," test these, and trade on the ones that predict best. From time to time they drop hypotheses that perform badly, and introduce new ones to test. Prices are driven endogenously by these induced expectations. Individuals' expectations, therefore, evolve and "compete" in a market formed by others' expectations. In other words, agents' expectations coevolve in a world they cocreate.

The natural question is whether these heterogeneous expectations coevolve into homogeneous rational-expectations beliefs, upholding the efficient-market theory, or whether richer individual and collective behavior emerges, upholding the traders' viewpoint and explaining the empirical market phenomena mentioned above. We answer this not analytically—our model, with its fully heterogeneous expectations, is too complicated to allow analytical solutions—but computationally. To investigate price dynamics, investment strategies, and market statistics in our endogenous-expectations market, we perform carefully controlled experiments within a computer-based market we have constructed, the SFI Artificial Stock Market.[5]

The picture of the market that results from our experiments, surprisingly, confirms both the efficient-market academic view and the traders' view. But each is valid under different circumstances—in different regimes. In both circumstances, we initiate our traders with heterogeneous beliefs clustered randomly in an interval near homogeneous rational expectations. We find that if our agents very slowly adapt their forecasts to new observations of the market's behavior, the market converges to a rational-expectations regime. Here "mutant" expectations cannot get a profitable footing; and technical trading, bubbles, crashes, and autocorrelative behavior do not emerge. Trading volume remains low. The efficient-market theory prevails.

If, on the other hand, we allow the traders to adapt to new market observations at a more realistic rate, heterogeneous beliefs persist, and the market self-organizes

[5]For an earlier report on the SFI artificial stock market, see Palmer et al.[38]

into a complex regime. A rich "market psychology"—a rich set of expectations—becomes observable. Technical trading emerges as a profitable activity, and temporary bubbles and crashes occur from time to time. Trading volume is high, with times of quiescence alternating with times of intense market activity. The price time series shows persistence in volatility, the characteristic GARCH signature of price series from actual financial markets. And it shows persistence in trading volume. And over the period of our experiments, at least, individual behavior evolves continually and does not settle down. In this regime, the traders' view is upheld.

In what follows, we discuss first the rationale for our endogenous-expectations approach to market behavior; and introduce the idea of collections of conditional expectational hypotheses or "predictors" to implement this. We next set up the computational model that will form the basic framework. We are then in a position to carry out and describe the computer experiments with the model. Two final sections discuss the results of the experiments, compare our findings with other modern approaches in the literature, and summarize our conclusions.

2. WHY INDUCTIVE REASONING?

Before proceeding, we show that once we introduce heterogeneity of agents, deductive reasoning on the part of agents fails. We argue that in the absence of deductive reasoning, agents must resort to *inductive* reasoning, which is both natural and realistic in financial markets.

A. FORMING EXPECTATIONS BY DEDUCTIVE REASONING: AN INDETERMINACY

We make our point about the indeterminacy of deductive logic on the part of agents using a simple arbitrage pricing model, avoiding technical details that will be spelled out later. (This pricing model is a special case of our model in section 3, assuming risk coefficient λ arbitrarily close to 0, and gaussian expectational distributions.) Consider a market with a single security that provides a stochastic payoff or dividend sequence $\{d_t\}$, with a risk-free outside asset that pays a constant r units per period. Each agent i may form individual expectations of next period's dividend and price, $E_i[d_{t+1}|I_t]$ and $E_i[p_{t+1}|I_t]$, with conditional variance of these combined expectations, $\sigma^2_{i,t}$, given current market information I_t. Assuming perfect arbitrage, the market for the asset clears at the equilibrium price:

$$p_t = \beta \sum_j w_{j,t}(E_j[d_{t+1}|I_t] + E_j[p_{t+1}|I_t])\,. \tag{1}$$

In other words, the security's price p_t is bid to a value that reflects the current (weighted) average of individuals' market expectations, discounted by the factor

$\beta = 1/(1+r)$, with weights $w_{j,t} = (1/\sigma_{j,t}^2)/\sum_k 1/\sigma_{k,t}^2$ the relative "confidence" placed in agent j's forecast.

Now, assuming intelligent investors, the key question is how the individual dividend and price expectations $E_i[d_{t+1}|I_t]$ and $E_i[p_{t+1}|I_t]$, respectively, might be formed. The standard argument that such expectations can be formed rationally (i.e., using deductive logic) goes as follows. Assume *homogeneous* investors who (i) use the available information I_t identically in forming their dividend expectations, and (ii) know that others use the same expectations. Assume further that the agents (iii) are perfectly rational (can make arbitrarily difficult logical inferences), (iv) know that price each time will be formed by arbitrage as in Eq. (1), and (v) that (iii) and (iv) are common knowledge. Then, expectations of future dividends $E_i[d_{t+k}|I_t]$ are by definition known, shared, and identical. And homogeneity allows us to drop the agent subscript and set the weights to $1/N$. It is then a standard exercise (see Diba and Grossman[16]) to show that by setting up the arbitrage, Eq. (1), for future times $t+k$, taking expectations across it, and substituting backward repeatedly for $E[p_{t+k}|I_t]$, agents can iteratively solve for the current price as[6]

$$p_t = \beta^k \sum_{k=1}^{\infty} E[d_{t+k}|I_t]\,. \tag{2}$$

If the dividend expectations are unbiased, dividend forecasts will be upheld on average by the market and, therefore, the price sequence will be in rational-expectations equilibrium. Thus, the price fluctuates as the information $\{I_t\}$ fluctuates over time, and it reflects "correct" or "fundamental" value, so that speculative profits are not consistently available. Of course, rational-expectations models in the literature are typically more elaborate than this. But the point so far is that if we are willing to adopt the above assumptions—which depend heavily on homogeneity—asset pricing becomes deductively determinate, in the sense that agents can, in principle at least, logically derive the current price.

Assume now, more realistically, that traders are intelligent but heterogeneous—each may differ from the others. Now, the available shared information I_t consists of past prices, past dividends, trading volumes, economic indicators, rumors, news, and the like. These are merely qualitative information plus data sequences, and *there may be many different, perfectly defensible statistical ways*, based on different assumptions and different error criteria, to use them to predict future dividends.[1,30] Thus, there is no objectively laid down, expectational model that differing agents can coordinate upon, and so there is no objective means for one agent to know other agents' expectations of future dividends. This is sufficient to bring indeterminacy to the asset price in Eq. (1). But worse, the heterogeneous price expectations

[6]The second, constant-exponential-growth solution is normally ruled out by an appropriate transversality condition.

$E_i[p_{t+1}|I_t]$ are also indeterminate. For suppose agent i attempts rationally to deduce this expectation, he may take expectations across the market clearing Eq. (1) for time $t+1$:

$$E_i[p_{t+1}|I_t] = \beta E_i\left[\sum_j \{w_{j,t+1}\left(E_j\left[d_{t+2}|I_t\right] + E_j\left[p_{t+2}|I_t\right]\right)\}\,|I_t\right]. \tag{3}$$

This requires that agent i, in forming his expectation of price, take into account his expectations of *others'* expectations of dividends and price (and relative market weights) two periods hence. To eliminate, in like manner, the price expectation $E_j[p_{t+2}|I_t]$ requires a further iteration. But this leads agents into taking into account their expectations of others' expectations of others' expectations of future dividends and prices at period $t+3$—literally, as in Keynes'[27] phrase, taking into account "what average opinion expects the average opinion to be."

Now, under homogeneity these expectations of others' expectations collapsed into single, shared, objectively determined expectations. Under heterogeneity, however, not only is there no objective means by which others' dividend expectations can be known, but attempts to eliminate the other unknowns, the price expectations, merely lead to the repeated iteration of subjective expectations of subjective expectations (or, equivalently, subjective priors on others' subjective priors)—an infinite regress in subjectivity. Further, this regress may lead to instability: If investor i believes that others believe future prices will increase, he may revise his expectations to expect upward-moving prices. If he believes that others believe a reversion to lower values is likely, he may revise his expectations to expect a reversion. We can, therefore, easily imagine swings and swift transitions in investors' beliefs, based on little more than ephemera—hints and perceived hints of others' beliefs about others' beliefs.

Under heterogeneity then, deductive logic leads to expectations that are not determinable. Notice the argument here depends in no way on agents having limits to their reasoning powers. It merely says that given differences in agent expectations, there is no logical means by which to arrive at expectations. And so, perfect rationality in the market can not be well defined. Infinitely intelligent agents cannot form expectations in a determinate way.

B. FORMING EXPECTATIONS BY INDUCTIVE REASONING

If heterogeneous agents cannot deduce their expectations, how then do they form expectations? They may observe market data, they may contemplate the nature of the market and of their fellow investors. They may derive expectational models by sophisticated, subjective reasoning. But in the end all such models will be—can only be—hypotheses. There is no objective way to verify them, except by observing their performance in practice. Thus, agents, in facing the problem of choosing appropriate predictive models, face the same problem that statisticians face when choosing appropriate predictive models given a specific data set, but no objective means by which to choose a functional form. (Of course, the situation here is made more difficult by the fact that the expectational models investors choose affect the price sequence, so that our statisticians' very choices of model affect their data and so their choices of model.)

In what follows then, we assume that each agent acts as a market "statistician."[7] Each continually creates multiple "market hypotheses"—subjective, expectational models—of what moves the market price and dividend. And each simultaneously tests several such models. Some of these will perform well in predicting market movements. These will gain the agent's confidence and be retained and acted upon in buying and selling decisions. Others will perform badly. They will be dropped. Still others will be generated from time to time and tested for accuracy in the market. As it becomes clear which expectational models predict well, and as poorly predicting ones are replaced by better ones, the agent learns and adapts. This type of behavior—coming up with appropriate hypothetical models to act upon, strengthening confidence in those that are validated, and discarding those that are not—is called *inductive reasoning*.[8] It makes excellent sense where problems are ill defined. It is, in microscale, the scientific method. Agents who act by using inductive reasoning we will call inductively rational.[9]

Each inductively rational agent generates multiple expectational models that "compete" for use within his or her mind, and survive or are changed on the basis of their predictive ability. The agents' hypotheses and expectations adapt to the current pattern of prices and dividends; and the pattern of prices changes to reflect the current hypotheses and expectations of the agents. We see immediately that the market possesses a *psychology*. We define this as the collection of market hypotheses, or expectational models or mental beliefs, that are being acted upon at a given time.

If there were some attractor inherent in the price-and-expectation-formation process, this market psychology might converge to a stable unchanging set of heterogeneous (or homogeneous) beliefs. Such a set would be statistically validated, and would, therefore, constitute a rational-expectations equilibrium. We investigate whether the market converges to such an equilibrium below.

[7]The phrase is Tom Sargent's.[42] Sargent argues similarly, within a macroeconomic context, that to form expectations agents need to act as market statisticians.

[8]For earlier versions of induction applied to asset pricing and to decision problems, see Arthur[1,2] (the *El Farol* problem), and Sargent.[42] For accounts of inductive reasoning in the psychological and adaptation literature, see Holland et al.,[25] Rumelhart,[41] and Schank and Abelson.[43]

[9]In the sense that they use available market data to learn—and switch among—appropriate expectational models. Perfect inductive rationality, of course, is indeterminate. Learning agents can be arbitrarily intelligent, but without knowing others' learning methods cannot tell a priori that *their* learning methods are maximally efficient. They can only discover the efficacy of their methods by testing them against data.

3. A MARKET WITH INDUCED EXPECTATIONS

A. THE MODEL

We now set up a simple model of an asset market along the lines of Bray[9] or Grossman and Stiglitz.[24] The model will be neoclassical in structure, but will depart from standard models by assuming heterogeneous agents who form their expectations inductively by the process outlined above.

Consider a market in which N heterogeneous agents decide on their desired asset composition between a risky stock paying a stochastic dividend, and a risk-free bond. These agents formulate their expectations separately, but are identical in other respects. They possess a constant absolute risk aversion (CARA) utility function, $U(c) = -\exp(-\lambda c)$. They communicate neither their expectations nor their buying or selling intentions to each other. Time is discrete and is indexed by t; the horizon is indefinite. The risk-free bond is in infinite supply and pays a constant interest rate r. The stock is issued in N units, and pays a dividend, d_t, which follows a given exogenous stochastic process $\{d_t\}$ not known to the agents.

The dividend process, thus far, is arbitrary. In the experiments we carry out below, we specialize it to an AR(1) process

$$d_t = \overline{d} + \rho(d_{t-1} - \overline{d}) + \varepsilon_t \,, \tag{4}$$

where ε_t is gaussian, i.i.d., and has zero mean, and variance σ_e^2.

Each agent attempts, at each period, to optimize his allocation between the risk-free asset and the stock. Assume for the moment that agent i's predictions at time t of the next period's price and dividend are normally distributed with (conditional) mean and variance, $E_{i,t}[p_{t+1} + d_{t+1}]$, and $\sigma^2_{t,i,p+d}$. (We say presently how such expectations are arrived at.) It is well known that under CARA utility and gaussian distributions for forecasts, agent i's demand, $x_{i,t}$, for holding shares of the risky asset is given by:

$$x_{i,t} = \frac{E_{i,t}(p_{t+1} + d_{t+1} - p(1+r))}{\lambda \sigma^2_{i,t,p+d}} \,, \tag{5}$$

where p_t is the price of the risky asset at t, and λ is the degree of relative risk aversion.

Total demand must equal the number of shares issued:

$$\sum_{i=1}^{N} x_{i,t} = N \,, \tag{6}$$

which closes the model and determines the clearing price p—the current market price—in Eq. (5) above.

It is useful to be clear on timing in the market. At the start of time period t, the current dividend d_t is posted, and observed by all agents. Agents then use this information and general information on the state of the market (which includes the historical dividend sequence $\{\ldots d_{t-2}, d_{t-1}, d_t\}$ and price sequence $\{\ldots p_{t-2}, p_{t-1}\}$) to form their expectations of the next period's price and dividend $E_{i,t}(p_{t+1} + d_{t+1})$. They then calculate their desired holdings and pass their demand parameters to the specialist who declares a price p_t that clears the market. At the start of the next period the new dividend d_{t+1} is revealed, and the accuracies of the predictors active at time t are updated. The sequence repeats.

B. MODELING THE FORMATION OF EXPECTATIONS

At this point we have a simple, neoclassical, two-asset market. We now break from tradition by allowing our agents to form their expectations individually and inductively. One obvious way to do this would be to posit a set of individual-agent expectational models which share the same functional form, and whose parameters are updated differently by each agent (by least squares, say) over time, starting from different priors. We reject this in favor of a different approach that better reflects the process of induction outlined in section 2 above. We assume each agent, at any time, possesses a multiplicity of linear forecasting models—hypotheses about the direction of the market, or "theories of the market"—and uses those that are both best suited to the current state of the market and have recently proved most reliable. Agents then learn, not by updating parameters, but by discovering which of their hypotheses "prove out" best, and by developing new ones from time to time, via the genetic algorithm. This structure will offer several desirable properties: It will avoid biases introduced by a fixed, shared functional form. It will allow the individuality of expectations to emerge over time (rather than be built in only to a priori beliefs). And it will better mirror actual cognitive reasoning, in which different agents might well "cognize" different patterns and arrive at different forecasts from the same market data.

In the expectational part of the model, at each period, the time series of current and past prices and dividends are summarized by an array or information set of J market descriptors. And agents' subjective expectational models are represented by sets of *predictors*. Each predictor is a *condition/forecast* rule (similar to a Holland classifier which is a condition/action rule) that contains both a market condition that may at times be fulfilled by the current state of the market and a forecasting formula for next period's price and dividend. Each agent possesses M such individual predictors—holds M hypotheses of the market in mind simultaneously—and uses the most accurate of those that are *active* (matched by the current state of the market). In this way, each agent has the ability to "recognize" different sets of states of the market, and bring to bear appropriate forecasts, given these market patterns.

It may clarify matters to show briefly how we implement this expectational system on the computer. (Further details are in Appendix A.) Suppose we summarize the state of the market by $J = 13$ bits. The fifth bit might correspond to "the price has risen the last 3 periods," and the tenth bit to "the price is larger than 16 times dividend divided by r," with 1 signaling the occurrence of the described state, and 0 its absence or nonoccurrence. Now, the condition part of all predictors corresponds to these market descriptors, and thus, also consists of a 13-bit array, each position of which is filled with a 0, or 1, or # ("don't care"). A condition array matches or "recognizes" the current market state if all its 0's and 1's match the corresponding bits for the market state with the #'s matching either a 1 or a 0. Thus, the condition (####1########) "recognizes" market states in which the price has risen in the last 3 periods. The condition (#########0###) recognizes states where the current price is not larger than 16 times dividend divided by r. The forecasting part of each predictor is an array of parameters that triggers a corresponding forecasting expression. In our experiments, all forecasts use a linear combination of price and dividend, $E(p_{t+1} + d_{t+1}) = a(p_t + d_t) + b$. Each predictor then stores specific values of a and b. Therefore, the full predictor (####1####0###)/(0.96,0) can be interpreted as "*if* the price has risen in the last 3 periods, and if the price is not larger than 16 times dividend divided by r, *then* forecast next period's price plus dividend as 96% of this period's." This predictor would recognize—would be activated by—the market state (0110100100011) but would not respond to the state (0110111011001).

Predictors that can recognize many states of the market have few 1's and 0's. Those more particularized have more 1's and 0's. In practice, we include for each agent a default predictor consisting of all #'s. The genetic algorithm creates new predictors by "mutating" the values in the predictor array, or by "recombination"—combining part of one predictor array with the complementary part of another.

The expectational system then works at each time with each agent observing the current state of the market, and noticing which of his predictors match this state. He forecasts next period's price and dividend by combining statistically the linear forecast of the H most accurate of these active predictors, and given this expectation and its variance, uses Eq. (5) to calculate desired stock holdings and to generate an appropriate bid or offer. Once the market clears, the next period's price and dividend are revealed and the accuracies of the active predictors are updated.

As noted above, learning in this expectational system takes place in two ways. It happens rapidly as agents learn which of their predictors are accurate and worth acting upon, and which should be ignored. And it happens on a slower time scale as the genetic algorithm from time to time discards nonperforming predictors and creates new ones. Of course these new, untested predictors do not create disruptions—they will be acted upon only if they prove accurate. This avoids brittleness and provides what machine-learning theorists call "gracefulness" in the learning process.

We can now discern several advantages of this multibit, multipredictor architecture. One is that this expectational architecture allows the market to have potentially different dynamics—a different character—under different states or circumstances. Because predictors are pattern-recognizing expectational models, and so can "recognize" these different states, agents can "remember" what happened before in given states and activate appropriate forecasts. This enables agents to make swift *gestalt*-like transitions in forecasting behavior should the market change.

Second, the design avoids bias from the choice of a particular functional form for expectations. Although the forecasting part of our predictors is linear, the multiplicity of predictors conditioned upon the many combinations of market conditions yield collectively at any time and for any agent a nonlinear forecasting expression in the form of a piecewise linear, noncontinuous forecasting function whose domain is the market state space, and whose accuracy is tuned to different regions of this space. (Forecasting is, of course, limited by the choice of the binary descriptors that represent market conditions.)

Third, learning is concentrated where it is needed. For example, $J = 12$ descriptors produces predictors that can distinguish more than four thousand different states of the market. Yet, only a handful of these states might occur often. Predictor conditions that recognize states that do not occur often will be used less often, their accuracy will be updated less often and, other things being equal, their precision will be lower. They are, therefore, less likely to survive in the competition among predictors. Predictors will, therefore, cluster in the more visited parts of the market state space, which is exactly what we want.

Finally, the descriptor bits can be organized into classes or information sets which summarize fundamentals, such as price-dividend ratios or technical-trading indicators, such as price trend movements. The design allows us to track exactly which information—which descriptor bits—the agents are using or ignoring, something of crucial importance if we want to test for the "emergence" of technical trading. This organization of the information also allows the possibility setting up different agent "types" who have access to different information sets. (In this chapter, all agents see all market information equally.)

A neural net could also supply several of these desirable qualities. However, it would be less transparent than our predictor system, which we can easily monitor to observe which information agents are individually and collectively using at each time.

4. COMPUTER EXPERIMENTS: THE EMERGENCE OF TWO MARKET REGIMES

A. EXPERIMENTAL DESIGN

We now explore computationally the behavior of our endogenous-expectations market in a series of experiments. We retain the same model parameters throughout these experiments, so that we can make comparisons of the market outcomes using the model under identical conditions with only controlled changes. Each experiment is run for 250,000 periods to allow asymptotic behavior to emerge if it is present; and it is run 25 times under different random seeds to collect cross-sectional statistics.

We specialize the model described in the previous section by choosing parameter values, and, where necessary, functional forms. We use $N = 25$ agents, who each have $M = 100$ predictors, which are conditioned on $J = 12$ market descriptors. The dividend follows the AR(1) process in Eq. (4), with autoregressive parameter ρ set to 0.95, yielding a process close to a random walk, yet persistent.

The 12 binary descriptors that summarize the state of the market are the following:

1–6 Current price × interest rate/dividend $> 0.25, 0.5, 0.75, 0.875, 1.0, 1.125$
7–10 Current price > 5-period moving average of past prices (MA), 10-period MA, 100-period MA, 500-period MA
11 Always on (1)
12 Always off (0)

The first six binary descriptors—the first six bits—reflect the current price in relation to current dividend, and thus, indicate whether the stock is above or below fundamental value at the current price. We will call these "fundamental" bits. Bits 7–10 are "technical-trading" bits that indicate whether a trend in the price is under way. They will be ignored if useless, and acted upon if technical-analysis trend following emerges. The final two bits, constrained to be 0 or 1 at all times, serve as experimental controls. They convey no useful market information, but can tell us the degree to which agents act upon useless information at any time. We say a bit is "set" if it is 0 or 1, and predictors are selected randomly for recombination, other things equal, with slightly lower probabilities the higher their specificity—that is, the more set bits they contain (see Appendix A). This introduces a weak drift toward the all-# configuration, and ensures that the information represented by a particular bit is used only if agents find it genuinely useful in prediction. This market information design allows us to speak of "emergence." For example, it can be said that technical trading has emerged if bits 7–10 become set significantly more often, statistically, than the control bits.

We assume that forecasts are formed by each predictor j storing values for the parameters a_j, b_j, in the linear combination of price and dividend, $E_j[p_{t+1} +$

$d_{t+1}|I_t] = a_j(p_t + d_t) + b_j$. Each predictor also stores a current estimate of its forecast variance. (See Appendix A.)

Before we conduct experiments, we run two diagnostic tests on our computer-based version of the model. In the first, we test to see whether the model can replicate the rational-expectations equilibrium (r.e.e.) of standard theory. We do this by calculating analytically the homogeneous rational-expectations equilibrium (h.r.e.e.) values for the forecasting parameters a and b (see Appendix A), then running the computation with all predictors "clamped" to these calculated h.r.e.e. parameters. We find indeed that such predictions are upheld—that the model indeed reproduces the h.r.e.e.—which assures us that the computerized model, with its expectations, demand functions, aggregation, market clearing, and timing sequence, is working correctly. In the second test, we show the agents a given dividend sequence and a calculated h.r.e.e. price series that corresponds to it, and test whether they individually learn the correct forecasting parameters. They do, though with some variation due to the agents' continual exploration of expectational space, which assures us that our agents are learning properly.

B. THE EXPERIMENTS

We now run two sets of fundamental experiments with the computerized model, corresponding respectively to slow and medium rates of exploration by agents of alternative expectations. The two sets give rise to two different *regimes*—two different sets of characteristic behaviors of the market. In the slow-learning-rate experiments, the genetic algorithm is invoked every 1,000 periods on average, predictors are crossed over with probablity 0.3, and the predictors' accuracy-updating parameter θ is set to 1/150. In the medium-exploration-rate experiments, the genetic algorithm is invoked every 250 periods on average, crossover occurs with probability 0.1, and the predictors' accuracy-updating parameter θ is set to 1/75.[10] Otherwise, we keep the model parameters the same in both sets of experiments, and in both we start the agents with expectational parameters selected randomly from a uniform distribution of values centered on the calculated homogeneous rational-expectations ones. (See Appendix A.) In the slow-exploration-rate experiments, no non-r.e.e. expectations can get a footing: the market enters an evolutionarily stable, rational-expectations regime. In the medium-exploration-rate experiments, we find that the market enters a complex regime in which psychological behavior emerges, there are significant deviations from the r.e.e. benchmark, and statistical "signatures" of real financial markets are observed.

We now describe these two sets of experiments and the two regimes or phases of the market they induce.

[10]At the time of writing, we have discovered that the two regimes emerge, and the results are materially the same, if we vary *only* the rate of invocation of the genetic algorithm.

THE RATIONAL-EXPECTATIONS REGIME. As stated, in this set of experiments, agents continually explore in prediction space, but under low rates. The market price, in these experiments, converges rapidly to the homogeneous rational-expectations value adjusted for risk, even though the agents start with nonrational expectations. In other words, homogeneous rational expectations are an attractor for a market with endogenous, inductive expectations.[11] This is not surprising. If some agents forecast differently than the h.r.e.e. value, then the fact that most other agents are using something close to the h.r.e.e. value, will return a market-clearing price that corrects these deviant expectations: There is a natural, if weak, attraction to h.r.e.e. The equilibrium within this regime differs in two ways from the standard, theoretical, rational-expectations equilibrium. First, the equilibrium is neither assumed nor arrived at by deductive means. Our agents instead arrive inductively at a homogeneity that overlaps that of the homogeneous, theoretical rational expectations. Second, the equilibrium is a stochastic one. Agents continually explore alternatives, albeit at low rates. This testing of alternative explorations, small as it is, induces some "thermal noise" into the system. As we would expect, in this regime, agents' holdings remain highly homogeneous, trading volume remains low (reflecting only variations in forecasts due to mutation and recombination) and bubbles, crashes, and technical trading do not emerge. We can say that in this regime the efficient-market theory and its implications are upheld.

THE COMPLEX OR RICH PSYCHOLOGICAL REGIME. We now allow a more realistic level of exploration in belief space. In these experiments, as we see in Figure 1, the price series still appears to be nearly identical to the price in the rational-expectations regime. (It is lower because of risk attributable to the higher variance caused by increased exploration.)

On closer inspection of the results, however, we find that complex patterns have formed in the collection of beliefs, and that the market displays characteristics that differ materially from those in the rational-expectations regime. For example, when we magnify the difference between the two price series, we see systematic evidence of temporary price bubbles and crashes (Figure 2). We call this new set of market behaviors the rich-psychological, or complex, regime.

This appearance of bubbles and crashes suggests that technical trading, in the form of buying or selling into trends, has emerged in the market. We can check this rigorously by examining the information the agents condition their forecasts upon.

[11]Within a simpler model, Blume and Easley[5] prove analytically the evolutionary stability of r.e.e.

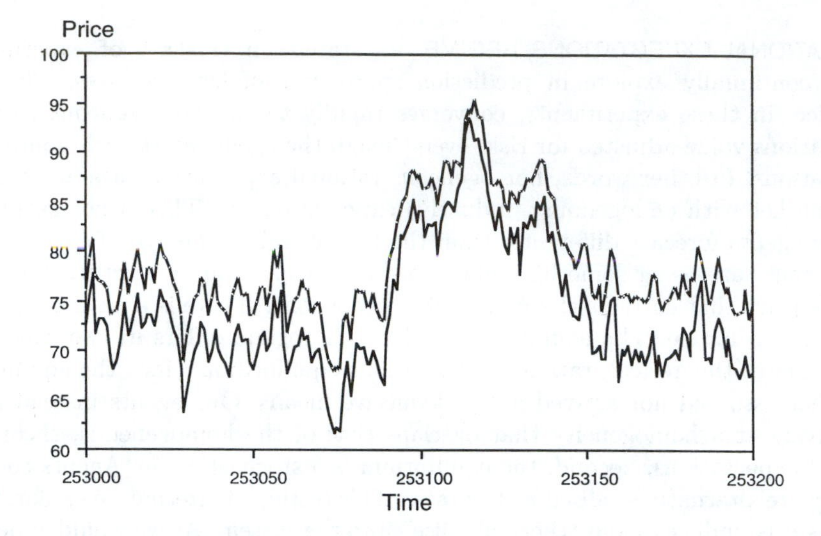

FIGURE 1 Rational-expectations price vs. price in the rich psychological regime. The two price series are generated on the same random dividend series. The upper is the homogeneous r.e.e. price, the lower is the price in the complex regime. The higher variance in the latter case causes the lower price through risk aversion.

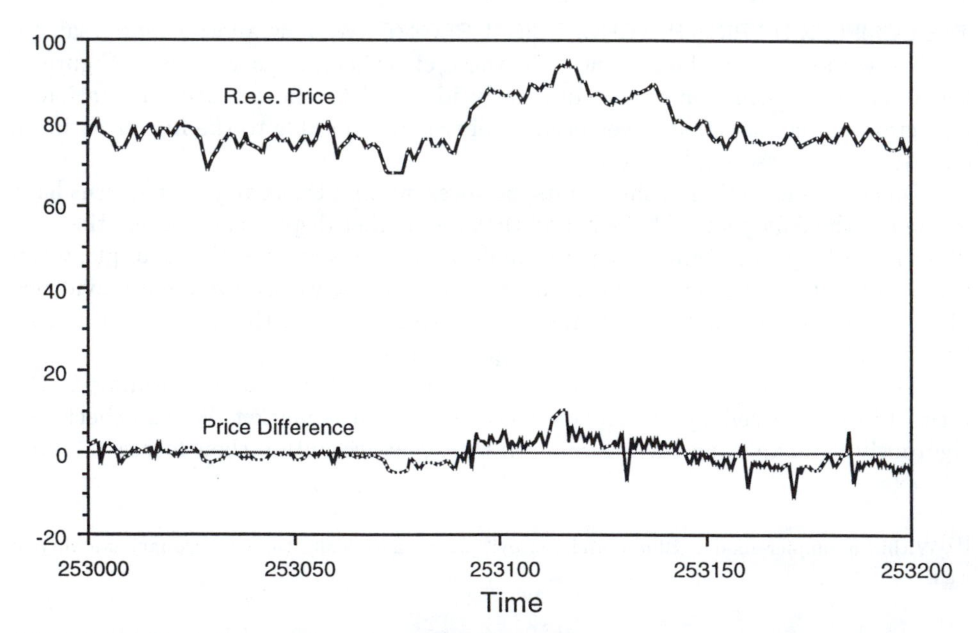

FIGURE 2 Deviations of the price series in the complex regime from fundamental value. The bottom graph shows the difference between the two price series in Figure 1 (with the complex series rescaled to match the r.e.e. one and the difference between the two doubled for ease of observation). The upper series is the h.r.e.e. price.

Figure 3 shows the number of technical-trading bits that are used (are 1's or 0's) in the population of predictors as it evolves over time. In both sets of experiments, technical-trading bits are initially seeded randomly in the predictor population. In the rational-expectations regime, however, technical-trading bits provide no useful information and fall off as useless predictors are discarded. But in the complex regime, they bootstrap in the population, reaching a steady-state value by 150,000 periods. Technical trading, once it emerges, remains.[12]

Price statistics in the complex regime differ from those in the rational-expectations regime, mainly in that kurtosis is evident in the complex case (Table 1) and that volume of shares traded (per 10,000 periods) is about 300% larger in the complex case, reflecting the degree to which the agents remain heterogeneous in their expectations as the market evolves. We note that fat tails and high volume are also characteristic of price data from actual financial markets.

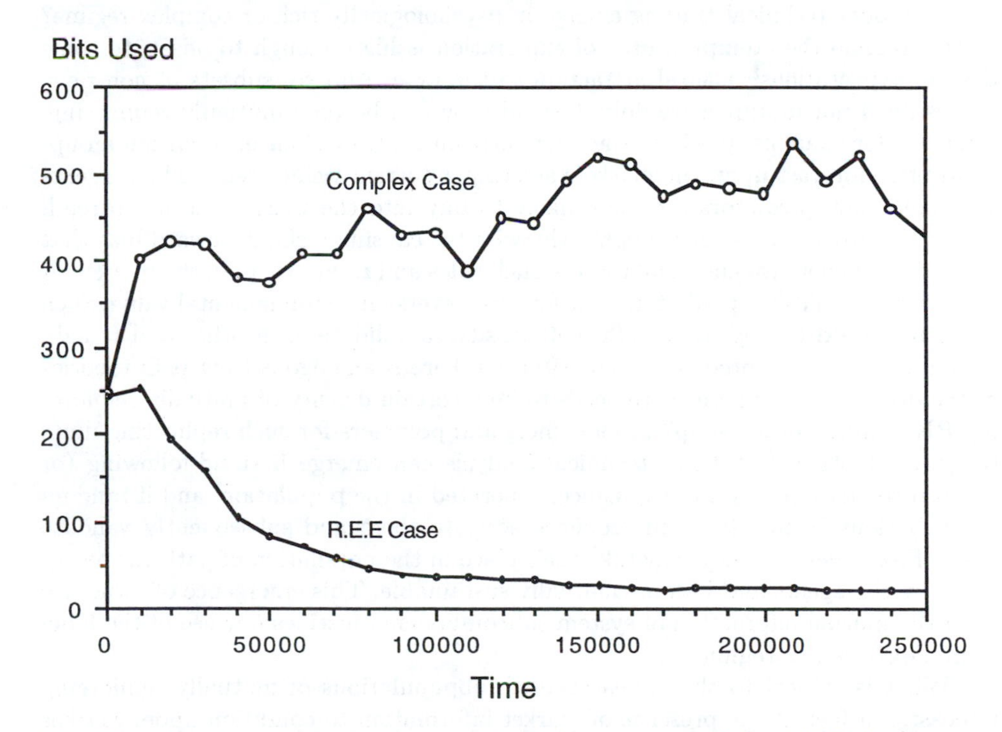

FIGURE 3 Number of technical-trading bits that become set as the market evolves, (median over 25 experiments in the two regimes).

[12] When we run these experiments informally to 1,000,000 periods, we see no signs that technical-trading bits disappear.

TABLE 1 Returns and volume statistics (medians) in the two regimes collected for 25 experiments after 250,000 periods.

	Mean	Std. Dev.	Skewness	Kurtosis[1]	Vol. traded
R.e.e. Regime	0.000	2.1002	0.0131	0.0497	2,460.9
Complex Regime	0.000	2.1007	0.0204	0.3429	7,783.8

[1] Kurtosis numbers are excess kurtosis (i.e., kurtosis -3).

How does technical trading emerge in psychologically rich or complex regime? In this regime the "temperature" of exploration is high enough to offset, to some degree, expectations' natural attraction to the r.e.e. And so, subsets of non-r.e.e. beliefs need not disappear rapidly. Instead they can become mutually reinforcing. Suppose, for example, predictors appear early on that, by chance, condition an upward price forecast upon the markets showing a current rising trend. Then, agents who hold such predictors are more likely to buy into the market on an uptrend, raising the price over what it might otherwise be, causing a slight upward bias that might be sufficient to lend validation to such rules and retain them in the market. A similar story holds for predictors that forecast reversion to fundamental value. Such predictors need to appear in sufficient density to validate each other and remain in the population of predictors. The situation here is analogous to that in theories of the origin of life, where there needs to be a certain density of mutually reinforcing RNA units in the "soup" of monomers and polymers for such replicating units to gain a footing.[17,26] Thus, technical analysis can emerge if trend-following (or mean-reversion) beliefs are, by chance, generated in the population, and if random perturbations in the dividend sequence activate them and subsequently validate them. From then on, they may take their place in the population of patterns recognized by the agents and become mutually sustainable. This emergence of structure from the mutual interaction of system subcomponents justifies our use of the label "complex" for this regime.

What is critical to the appearance of subpopulations of mutually reinforcing forecasts, in fact, is the presence of market information to condition upon. Market states act as "sunspot-like" signals that allow predictors to coordinate upon a direction they associate with that signal. (Of course, these are not classic sunspots that convey no real information.) Such coordination or mutuality can remain in the market once it establishes itself by chance. We can say the ability of market states to act as signals primes the mutuality that causes complex behavior. There is no need to assume a separate class of noise traders for this purpose. We can test this signaling conjecture in further experiments where we "turn off" the condition part of all predictors (by filling them with nonreplaceable #'s). Now forecasts cannot differentiate among states of the market, and market states cannot act as signals. We find, consistent with our conjecture that signaling drives the observed patterns, that the complex regime does not emerge. As a further test of the significance of technical-trading signals, we regress the current price on the previous periods plus the technical indicator (price > 500-period moving average). In the rational-expectations regime, the technical indicator is of course not significant. In the complex regime, the trend indicator *is* significant (with t-value of 5.1 for the mean of the sample of 25 experiments), showing that the indicator does indeed carry useful market information. The corresponding test on actual financial data shows a similar result.[10]

One of the striking characteristics of actual financial markets is that both their price volatility and trading volume show persistence or autocorrelation. And volatility and volume show significant cross-correlation. In other words, both volume and volatility remain high or low for periods of random length, and they are interrelated. Our inductive market also shows persistence in volatility or GARCH behavior in the complex regime (see Figure 4), with the Chi-square statistic in the Engle GARCH Test significant at the 95% level.[13] It also shows persistence in trading volume (see Figure 5), as well as significant cross-correlation between trading volume and volatility (see Figure 6). The figures include corresponding correlations for the often-used market standard, IBM stock. (Note that because our time period and actual market days do not necessarily match, we should expect no exact overlap. But qualitatively, persistence in our market and IBM's is similar.) These correlations are not explained by the standard model, where theoretically they are zero.

Why financial markets—and our inductive market—show these empirical "signatures" remains an open question. We conjecture a simple evolutionary explanation. Both in real markets and in our artificial market, agents are constantly exploring and testing new expectations. Once in a while, randomly, more successful expectations will be discovered. Such expectations will change the market, and trigger further changes in expectations, so that small and large "avalanches" of change will cascade through the system. (Of course, on this very short time-lag scale, these avalanches occur not through the genetic algorithm, but by agents changing their active predictors.) Changes then manifest in the form of increased volatility and increased volume. One way to test this conjecture is to see whether autocorrelations increase as the predictor accuracy-updating parameter θ in Eq. (7) in Appendix A is increased. The larger θ is, the faster individual agents "switch" among their

[13]Autocorrelated volatility is often fitted with a Generalized Autoregressive Conditional Heteroscedastic time series. Hence, the GARCH label. See Bollerslev et al.[7] and Goodhart and O'Hara.[21]

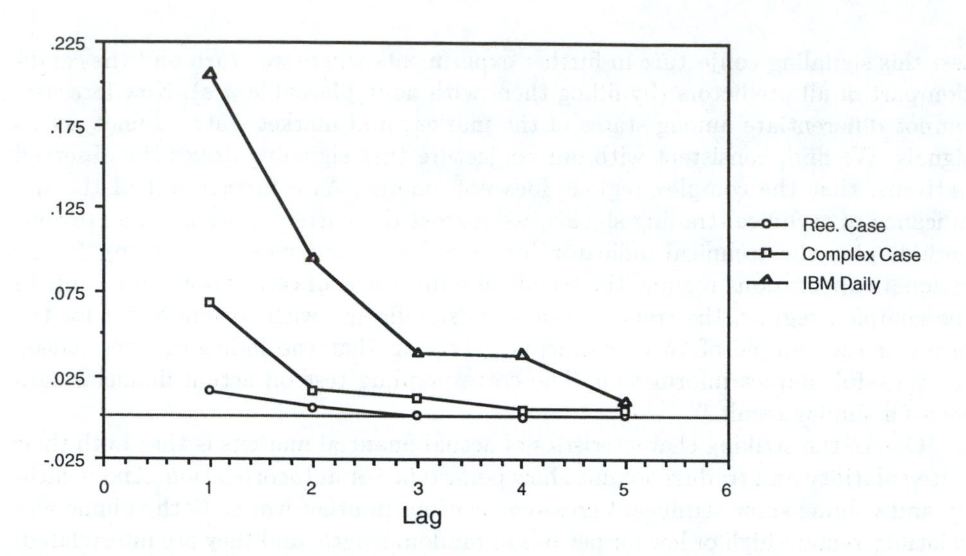

FIGURE 4 Autocorrelation of volatility in rational-expectations and complex regimes, and in IBM daily returns.

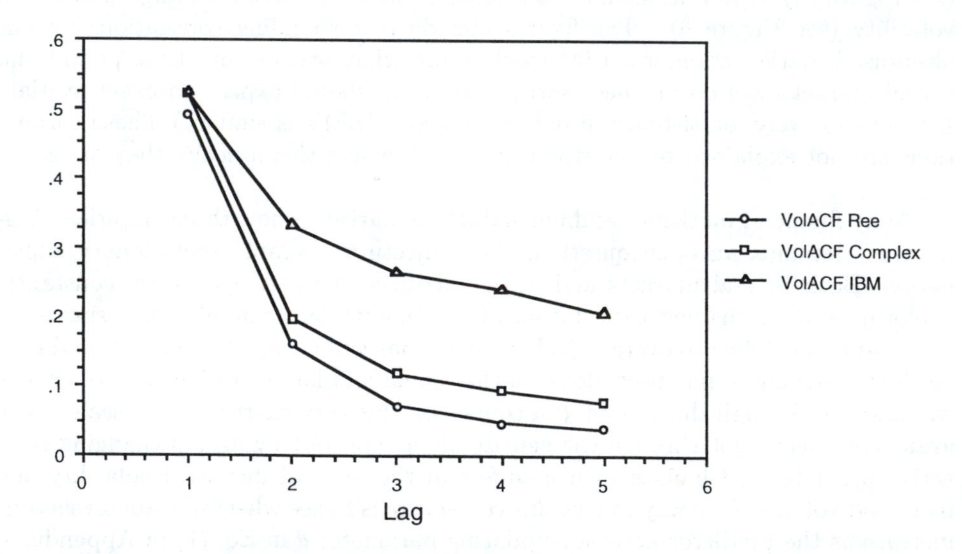

FIGURE 5 Autocorrelation of trading volume in the rational-expectations and complex regimes, and in IBM daily returns.

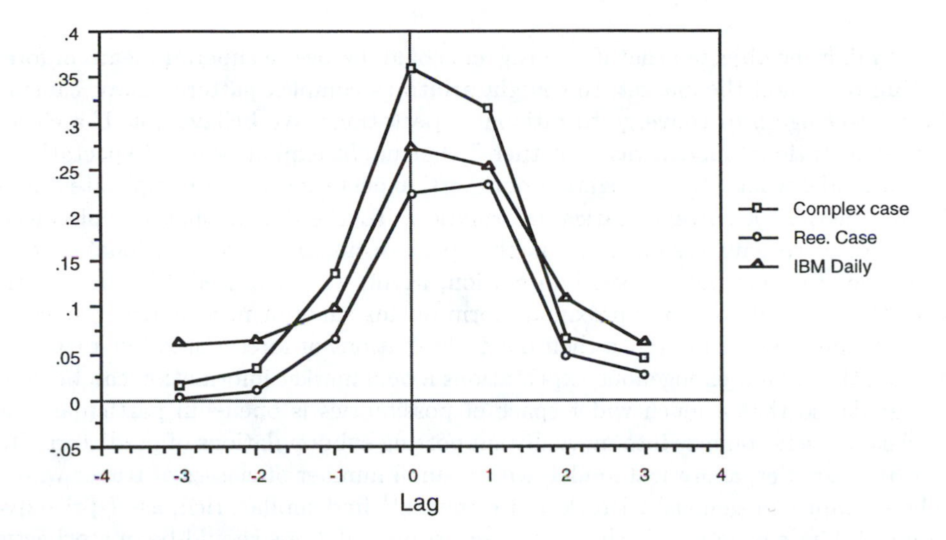

FIGURE 6 Cross-correlation of trading volume with volatility, in the rational-expectations and complex regimes, and in IBM daily returns.

predictors. Thus, the more such switches should cascade. Experiments confirm that autocorrelations indeed increase with θ. Such cascades of switching in time are absorbed by the market, and die away. Hence, our evolutionary market exhibits periods of turbulence followed by periods of quiescence, as do actual markets.[14]

5. DISCUSSION

To what extent is the existence of the complex regime an artifact of design assumptions in our model? We find experimentally by varying both the model's parameters and the expectational-learning mechanism, that the complex regime and the qualitative phenomena associated with it are robust. These are not an artifact of some deficiency in the model.[15]

[14]For a discussion of volatility clustering in a different model, see Youssefmir and Huberman[50]; and also Grannan and Swindle.[22]

[15]One design choice might make a difference. We have evaluated the usefulness of expectational beliefs by their accuracy rather than by the profit they produce. In practice, these alternatives may produce different outcomes. For example, buying into a price rise on the basis of expectations may yield a different result if validated by profit instead of by accuracy of forecast when "slippage" is present, that is, when traders on the other side of the market are hard to find. We believe, but have not proved, that the two criteria lead to the same qualitative results.

It might be objected that if some agents could discover a superior means of forecasting to exploit the market, this might arbitrage complex patterns away, causing the market again to converge to rational expectations. We believe not. If a clever metaexpectational model was "out there" that might exploit others' expectations, such a model would, by aggregation of others' expectations, be a complicated nonlinear function of current market information. To the degree that the piecewise linear form we have assumed covers the space of nonlinear expectational models conditioned on current market information, agents would indeed, via the genetic algorithm, pick up on an approximate form of this superior metamodel. The complex regime owes its existence then not to limitations of forecasting, but rather to the fact that in our endogenous-expectations model market information can be used as signals, so that a much wider space of possibilities is open—in particular, the market can self-organize into mutually supporting subpopulations of predictors. (In fact, in a simpler, analytical model, with a small number of classes of trader whose beliefs adapt endogenously, Brock and Hommes[11] find similar, rich, asset-price dynamics.) There is no reason these emergent subpopulations should be in stochastic equilibrium. Indeed, agents may mutually adapt their expectations forever, so that the market explores its way through this large space, and is nonstationary. In some early exploratory experiments, we "froze" successful agents' expectations, then reinjected these agents with their previously successful expectations much later. The reintroduced agents proved less successful than average, indicating that the market had evolved and was nonstationary.

It might be also objected that by our use of condition bits in the predictors, we have built technical trading into our model. And so it is no surprise that it appears in the complex regime. But actually, only the possibility of technical trading is built in, not its use. The use of market descriptors is selected against in the model. Thus, market signals must be of value to be used, and technical trading emerges only because such market signals induce mutually supporting expectations that condition themselves on these market signals.

If the market has a well-defined psychology in our model, does it also experience "moods"? Obviously not. But, notice we assume that agents entertain more than one market hypothesis. Thus, we can imagine circumstances of a prolonged "bull-market" uptrend to a level well above fundamental value in which the market state activates predictors that indicate the uptrend will continue, and simultaneously other predictors that predict a rapid downward correction. Such combinations, which occur easily in both our market and actual markets, could well be described as "nervous."

What about trade, and the motivation to trade in our market? In the rational-expectations literature, the deductively rational agents have no motivation to trade, even where they differ in beliefs. Assuming other agents have access to different information sets, each agent in a prebidding arrangement arrives at identical beliefs. Our inductively rational agents (who do not communicate directly), by contrast, do not necessarily converge in beliefs. They thus retain a motivation to trade, betting ultimately on their powers as market statisticians. It might appear that, because our agents have equal abilities as statisticians, they are irrational to trade at all. But although their *abilities* are the same, their luck in finding good predictors diverges over time. And at each period, the accuracy of their predictors is fully accounted for in their allocations between the risk-free and risky asset. Given that agents can only act as market statisticians, their trading behavior is rational.

Our endogenous-expectation theory fits with two other modern approaches. Our model generalizes the learning models of Bray and others[8,42] which also assume endogenous updating of expectations. But while the Bray models assume homogeneous updating from a shared nonrational forecast, our approach assumes heterogeneous agents who can discover expectations that might exploit any patterns present. Our evolutionary approach also has strong affinities with the evolutionary models of Blume and Easley.[5,6] These assume populations of expectational (or more correctly, investment) rules that compete for survival in the market in a given population of rules, and that sometimes adapt. But the concern in this literature is the selective survival of different, competing, rule types, not the emergence of mutually supportive subpopulations that give rise to complex phenomena, nor the role of market signals in this emergence.

Our inductively rational market, of course, leaves out many details of realism. In actual financial markets, investors do not perfectly optimize portfolios, nor is full market clearing achieved each period. Indeed, except for the formation of expectations, our market is simple and neoclassical. Our object, however, is not market realism. Rather it is to show that given the inevitable inductive nature of expectations when heterogeneity is present, rich psychological behavior emerges—even under neoclassical conditions. We need not, as in other studies,[20,28] assume sharing of information nor sharing of expectations nor herd effects to elicit these phenomena. Nor do we need to invoke "behaviorism" or other forms of irrationality.[49] Herding tendencies and quasi-rational behavior may be present in actual markets, but they are not necessary to our findings.

6. CONCLUSION

In asset markets, agents' forecasts create the world that agents are trying to forecast. Thus, asset markets have a reflexive nature in that prices are generated by traders' expectations, but these expectations are formed on the basis of anticipations of *others'* expectations.[16] This reflexivity, or self-referential character of expectations, precludes expectations being formed by deductive means, so that perfect rationality ceases to be well defined. Thus, agents can only treat their expectations as hypotheses: they act inductively, generating individual expectational models that

[16]This point was also made by Soros[47] whose term *reflexivity* we adopt.

they constantly introduce, test, act upon, discard. The market becomes driven by expectations that adapt endogenously to the ecology these expectations cocreate.

Experiments with a computerized version of this endogenous-expectations market explain one of the more striking puzzles in finance: Standard theory tends to see markets as efficient, with no rationale for herd effects, and no possibility of systematic speculative profit, whereas traders tend to view the market as exhibiting a "psychology," bandwagon effects, and opportunities for speculative profit. Recently the traders' view has been justified by invoking behavioral assumptions, such as the existence of noise traders. We show, without behavioral assumptions, that both views can be correct. A market of inductively rational traders can exist in two different regimes: Under a low enough rate of exploration of alternative forecasts, the market settles into a simple regime which corresponds to the rational-expectations equilibrium of the efficient-market literature. Under a more realistic rate of exploration of alternative forecasts, the market self-organizes into a complex regime in which rich psychological behavior emerges. Technical trading appears, as do temporary bubbles and crashes. And prices show statistical features—in particular, GARCH behavior—characteristic of actual market data. These phenomena arise when individual expectations that involve trend following or mean reversion become mutually reinforcing in the population of expectations, and when market indicators become used as signaling devices that coordinate these sets of mutually reinforcing beliefs.

Our endogenous-expectations market shows that heterogeneity of beliefs, deviations from fundamental trading, and persistence in time series can be maintained indefinitely in actual markets with inductively rational traders. We conjecture that actual financial markets lie within the complex regime.

APPENDICES

APPENDIX A: DETAILS OF THE MARKET'S ARCHITECTURE

MODEL PARAMETERS. Throughout the experiments we set the interest rate r to 0.1, and agents' risk-aversion parameter λ to 0.5. The parameters of the dividend process in Eq. (4) are $\rho = 0.95$, $\bar{d} = 10$, $r = 0.1$, $\sigma_\epsilon^2 = 0.0743$. (This error variance value is selected to yield a combined price-plus-dividend variance of 4.0 in the h.r.e.e.)

PREDICTOR ACCURACY. The accuracy, or precision, of agent i's jth predictor is updated each time the predictor is active, and is recorded as the inverse of the moving average of squared forecast error:

$$e_{t,i,j}^2 = (1-\theta)e_{t-1,i,j}^2 + \theta[(p_{t+1} + d_{t+1}) - E_{t,i,j}(p_{t+1} + d_{t+1})]^2 \tag{7}$$

with $\theta = 1/75$ in the complex regime, and $1/150$ in the rational-expectations regime.

This accuracy is used in three places. First, if multiple predictors are active, only the most accurate is used. Second, it is part of the fitness measure for selecting predictors for recombination in the genetic algorithm. This fitness measure is defined as

$$f_{t,i,j} = M - e_{t,i,j}^2 - Cs \tag{8}$$

where M is a constant; s is specificity, the number of bits that are set (not #) in the predictor's condition array; and $C = 0.005$ is a cost levied for specificity. The value of M is irrelevant, given tournament rankings.

Third, agents use the error variance of their current predictor for the forecast variance in the demand Eq. (5). (We keep this latter variance fixed between genetic algorithm implementations, updating it to its current value in Eq. (7) at each invocation.)

INITIAL EXPECTATIONS. We initialize agents' expectations in both regimes by drawing the forecasting parameters from a uniform distribution of values centered upon the h.r.e.e. ones. We select a to be uniform $(0.7, 1.2)$ and b to be uniform $(-10, 19.002)$. The variance of all new predictors is initialized in all cases to the h.r.e.e. value of 4.0.

THE GENETIC ALGORITHM. New predictors are generated by updating each agent's predictor set at random intervals, on average every 250 periods or 1,000 periods, depending on the regime, asynchronously across agents. The worst performing (least accurate) 20% of the agent's 100 predictors are dropped, and are replaced by new ones, using uniform crossover and mutation. The agents are initialized by seeding them with random predictors: condition bits are set to 0 or 1 with probability 0.1, otherwise to #. This avoids bias in choosing predictors at the outset, and allows intelligent behavior to bootstrap itself up as the artificial agents generate predictive models that perform better. For the bitstrings, these procedures are standard genetic algorithm procedures for mutation and crossover (uniform crossover is used which chooses a bit at random from each of the two parents). The forecasting parameter vectors are mutated by adding random variables to each individual component. And they are crossed over component-wise, or by taking linear combinations of the two vectors, or by selecting one or the other complete vector. Each of these procedures is performed with equal probability. Crossover on a predictor is performed with probability 0.3 or 0.1 in the rational-expectations and complex regimes, respectively. Individual bits are mutated with probability 0.03. New predictors are brought into the predictor set with variance set to the average of their parents. If a bit has been changed, the new predictor's variance is set to the average of that of all predictors. If this new variance is lower than the variance of the current default predictor less an absolute deviation, its variance is set to the median of the predictors' variance. This procedure gives new predictors a reasonable chance of becoming used.

MARKET CLEARING. The price is adjusted each period by directly solving Eqs. (5) and (6) for p, which entails passing agents' forecasting parameters to the clearing equation. In actual markets, of course, the price is adjusted by a specialist who may not have access to agents' demand functions. But we note that actual specialists, either from experience or from their "books," have a keen feel for the demand function in their markets, and use little inventory to balance day-to-day demand. Alternatively, our market-clearing mechanism simulates an auction in which the specialist declares different prices and agents continually resubmit bids until a price is reached that clears the market.

CALCULATION OF THE HOMOGENEOUS RATIONAL-EXPECTATIONS EQUILIBRIUM. We calculate the homogeneous r.e.e. for the case where the market price is a linear function of the dividend $p_t = fd_t + g$ which corresponds to the structure of our forecasts. We can then calculate f and g from the market conditions at equilibrium. A homogenous equilibrium demands that all agents hold 1 share, so that, from Eq. (5)

$$E_t(p_{t+1} + d_{t+1}) - (1+r)p_t = \lambda\sigma^2_{p+d}\,. \tag{9}$$

From the dividend process Eq. (4) and the linear form for the price, we can calculate $\sigma^2_{p+d} = (1+f)\sigma^2_e$ and $E_t(p_{t+1} + d_{t+1})$ as

$$E_t(p_{t+1} + d_{t+1}) = (1+f)\left[(1-\rho)\bar{d} + \rho d_t\right] + g\,.$$

Noting that the right side of Eq. (9) is constant, we can then solve for f and g as

$$f = \frac{\rho}{1+r-\rho}\,,$$
$$g = \frac{(1+f)\left[(1-\rho)\bar{d} - \lambda\sigma^2_e\right]}{r}\,.$$

Therefore, the expression:

$$E_t(p_{t+1} + d_{t+1}) = (1+r)p_t + \frac{\lambda(2+r)\sigma^2_e}{1+r-\rho} \tag{10}$$

is the h.r.e.e. forecast we seek.

APPENDIX B: THE SANTA FE ARTIFICIAL STOCK MARKET

The Santa Fe Artificial Stock Market has existed since 1989 in various designs (see Palmer et al.[38] for a description of an earlier version). Since then a number of other artificial markets have appeared: e.g., Beltratti and Margarita,[4] Marengo and Tordjman,[35] and Rieck.[40] The Santa Fe Market is a computer-based model that can be altered, experimented with, and studied in a rigorously controlled way. Most of the artificial market's features are malleable and can be changed to carry out different experiments. Thus, the artificial market is a framework or template that can be specialized to focus on particular questions of interest in finance: for example, the effects of different agents having access to different information sets or predictive behaviors; or of a transaction tax on trading volume; or of different market-making mechanisms.

The framework allows other classes of utility functions, such as constant relative risk aversion. It allows a specialist or market maker, with temporary imbalances in fulfilled bids and offers, made up by changes in an inventory held by the specialist. It allows a number of alternative random processes for $\{d_t\}$. And it allows for the evolutionary selection of agents via wealth.

The market runs on a NeXTStep computational platform, but is currently being ported to the Swarm platform. For availability of code, and for further information, readers should contact Blake LeBaron or Richard Palmer.

ACKNOWLEDGMENTS

We are grateful to Kenneth Arrow, Eric Beinhocker, Larry Blume, Buz Brock, John Casti, Steven Durlauf, David Easley, David Lane, Ramon Marimon, Tom Sargent, and Martin Shubik for discussions of the arguments in this chapter, and of the design of the artificial market. All errors are our own.

REFERENCES

1. Arthur, W. B. "On Learning and Adaptation in the Economy." Working Paper 92-07-038, Santa Fe Institute, Santa Fe, NM, 1992.
2. Arthur, W. B. "Inductive Behavior and Bounded Rationality." *Amer. Econ. Rev.* **84** (May 1994): 406–411.
3. Arthur, W. B. "Complexity in Economic and Financial Markets." *Complexity* **1** (1995): 20–25.
4. Beltratti, A., and S. Margarita. "Simulating an Artificial Adaptive Stock Market." Mimeo, Turin University, 1992.
5. Blume, L., and D. Easley. "Learning to Be Rational." *J. Econ. Theory* **26** (1982): 340–351.
6. Blume, L., and D. Easley. "Evolution and Market Behavior." *J. Econ. Theory* **58** (1990): 9–40.
7. Bollerslev, T., R. Y. Chou, N. Jayaraman, and K. F. Kroner. "ARCH Modeling in Finance: A Review of the Theory and Empirical Evidence." *J. Econometrics* **52** (1990): 5–60.
8. Bossaerts, P. "The Econometrics of Learning in Financial Markets." *Econometric Theory* **11** (1995): 151–189.
9. Bray, M. "Learning, Estimation, and Stability of Rational Expectations." *J. Econ. Theory* **26** (1982): 318–339.
10. Brock, W., J. Lakonishok, and B. LeBaron. "Simple Technical Trading Rules and the Stochastic Properties of Stock Returns." Working Paper 91-01-006, Santa Fe Institute, Santa Fe, NM, 1991.
11. Brock, W., and C. H. Hommes. "Models of Complexity in Economics and Finance." Mimeo, Department of Economics, University of Wisconsin, Madison, 1996.
12. Cutler, D. M., J. M. Poterba, and L. H. Summers. "What Moves Stock Prices?" *J. Portfolio Mgmt.* **15** (1989): 4–12.
13. De Long, J. B., A. Shleifer, L. H. Summers, and R. J. Waldmann. "Noise Trader Risk in Financial Markets." *J. Pol. Econ.* **98** (1990): 703–738.

14. De Long, J. B., A. Shleifer, L. H. Summers, and R. J. Waldmann. "Positive Feedback Strategies and Destabilizing Rational Speculation." *J. Fin.* **45** (1990): 379–395.
15. De Long, J. B., A. Shleifer, L. H. Summers, and R. J. Waldmann. "The Survival of Noise Traders in Financial Markets." *J. Bus.* **64** (1991): 1–18.
16. Diba, B. T., and H. I. Grossman. "The Theory of Rational Bubbles in Stock Prices." *Econ. Jour.* **98** (1988): 746–754.
17. Eigen, M., and P. Schuster. *The Hypercycle: A Principle of Natural Self-Organization.* New York: Springer, 1979.
18. Fama, E. F., and K. R. French. "Permanent and Temporary Components of Stock Market Prices." *J. Pol. Econ.* **96** (1988): 246–273.
19. Frankel, J. A., and K. A. Froot. "Chartists, Fundamentalists, and Trading in the Foreign Exchange Market." *AEA Papers & Proc.* **80** (1990): 181–185.
20. Friedman, D., and M. Aoki. "Inefficient Information Aggregation as a Source of Asset Price Bubbles." *Bull. Econ. Res.* **44** (1992): 251–279.
21. Goodhart, C. A. E., and M. O'Hara. "High Frequency Data in Financial Markets: Issues and Applications." Unpublished manuscript, London School of Economics, 1995.
22. Grannan, E. R., and G. H. Swindle. "Contrarians and Volatility Clustering." Working Paper 94-03-010, Santa Fe Institute, Santa Fe, NM, 1994.
23. Grossman, S. J. "On the Efficiency of Competitive Stock Markets Where Traders Have Diverse Information." *J. Fin.* **31** (1976): 573–585.
24. Grossman, S. J., and J. Stiglitz. "On the Impossibility of Informationally Efficient Markets." *Amer. Econ. Rev.* **70** (1980): 393–408.
25. Holland, J. H., K. J. Holyoak, R. E. Nisbett, and P. R. Thagard. *Induction.* Cambridge, MA: MIT Press, 1986.
26. Kauffman, S. *The Origin of Order: Self-Organization and Selection in Evolution.* New York: Oxford University Press, 1993.
27. Keynes, J. M. *General Theory of Employment, Interest, and Money.* London: Macmillan, 1936.
28. Kirman, A. "Epidemics of Opinion and Speculative Bubbles in Financial Markets." In *Money and Financial Markets*, edited by M. Taylor. London: Macmillan, 1991.
29. Kleidon, A. W. "Anomalies in Financial Economics: Blueprint for Change?" *J. Bus.* **59** (1986): 285–316 (supplement).
30. Kurz, M. "On the Structure and Diversity of Rational Beliefs." *Econ. Theory* **4** (1994): 877–900.
31. Kurz, M. "Asset Prices with Rational Beliefs." Working Paper 96-003, Economics Department, Stanford University, 1995.
32. Leroy, S. F., and R. D. Porter. "Stock Price Volatility: Tests Based on Implied Variance Bounds." *Econometrica* **49** (1981): 97–113.
33. Lo, A. W., and C. MacKinlay. "Stock Prices Do Not Follow Random Walks: Evidence from a Simple Specification Test." *Rev. Fin. Stud.* **1** (1988): 41–66.

34. Lucas, R. E. "Asset Prices in an Exchange Economy." *Econometrica* **46** (1978): 1429–1445.
35. Marengo, L., and H. Tordjman. "Speculation, Heterogeneity, and Learning: A Model of Exchange Rate Dynamics." Working Paper WP-95-17, IIASA, 1995.
36. Milgrom, P., and N. Stokey. "Information, Trade, and Common Knowledge." *J. Econ. Theory* **26** (1982): 17–27.
37. Nelson, R., and S. Winter. *An Evolutionary Theory of Economic Change.* Cambridge, MA: Harvard University Press/Bellknap, 1982.
38. Palmer, R. G., W. B. Arthur, J. H. Holland, B. LeBaron, and P. Tayler. "Artificial Economic Life: A Simple Model of a Stockmarket." *Physica D* **75** (1994): 264–274.
39. Poterba, J. M., and L. H. Summers. "Mean Reversion in Stock Prices: Evidence and Implications." *J. Fin. Econ.* **22** (1988): 27–59.
40. Rieck, C. "Evolutionary Simulation of Asset Trading Strategies." In *Many-Agent Simulation and Artificial Life*, edited by E. Hillenbrand and J. Stender. Washington, DC: IOS Press, 1994.
41. Rumelhart, D. E. *Human Information Processing.* New York: Wiley, 1977.
42. Sargent, T. J. *Bounded Rationality in Macroeconomics.* New York: Oxford University Press, 1993.
43. Schank, R., and R. P. Abelson. *Scripts, Plans, Goals, and Understanding.* Hillsdale, NJ: Erlbaum, 1977.
44. Shiller, R. J. "Do Stock Prices Move Too Much to Be Justified by Subsequent Changes in Dividends?" *Amer. Econ. Rev.* **71** (1981): 421–436.
45. Shiller, R. J. *Market Volatility.* Cambridge, MA: MIT Press, 1989.
46. Shleifer, A., and L. H. Summers. "The Noise Trader Approach to Finance." *J. Econ. Perspectives* **4** (1990): 19–33.
47. Soros, G. *The Theory of Reflexivity.* New York: Soros Fund Management, 1994.
48. Summers, L. H. "Does the Stock Market Rationally Reflect Fundamental Values?" *J. Fin.* **46** (1986): 591–601.
49. Thaler, R. H. *Advances in Behavioral Finance.* New York: Russell Sage Foundation, 1993.
50. Youssefmir, M., and B. Huberman. "Clustered Volatility in Multiagent Dynamics." Working Paper 95-05-051, Santa Fe Institute, Santa Fe, NM, 1995.

V. M. Darley* and S. A. Kauffman**
*Division of Engineering and Applied Sciences, Harvard University, Cambridge, MA 02138; E-mail: darley@fas.harvard.edu
**Santa Fe Institute, 1399 Hyde Park Road, Santa Fe, NM 87501; E-mail: stu@santafe.edu

Natural Rationality

We propose a method for modeling economic systems in which outcomes depend locally on the predictions that agents make of other agents. We develop population games in which each agent adaptively searches for a good model of its environment. We demonstrate that such systems can exhibit persistent dynamics, cyclic oscillations between low and high complexity states, and other complex, yet endogenous, phenomena. We propose these "adaptively rational" agents as a natural extension of rational expectations, suitable when mutual consistency is not realistic. We discuss the connections between our work and the programs of bounded rationality, evolutionary game theory, and models motivated by statistical mechanics.

1. INTRODUCTION

We seek a definition of *homo economicus* which relaxes the assumption that globally optimal, mutually consistent strategies are selected by a system of autonomous agents. We believe the development, dynamics, stability, and change of this optimality should be modeled, not graven in stone. We accept Nash's[19] judgment that:

> "It is unnecessary to assume that the participants have full knowledge of the total structure of the game, or the ability and inclination to go through any complex reasoning processes. But the participants are supposed to accumulate empirical information on the relative advantages of the various pure strategies at their disposal."

Ours is part of a growing body of research studying the complex relationship between population dynamics and Nash equilibria. In fact, an explicit population in modeling games was first introduced in Schelling's[18] model of segregation in housing markets. Such models often go further than Nash's admission that "actually, of course, we can only expect some sort of approximate equilibrium, since the information, its utilization, and the stability of the average frequencies will be imperfect." With an explicit population we can address when Nash's "approximate equilibrium" is only metastable or unstable (and, therefore, of lesser relevance), and model the actual dynamics in such cases. We adopt this approach and, for instance, we show that even for simple cooperative games, equilibrium is not assured: punctuated equilibria, metastability, limit cycles, and other emergent properties may all be observed.

Sargent[17] discusses a large variety of boundedly rational macroeconomic models, and neatly summarizes the basic philosophy underlying that program of study:

> "Rational expectations imposes two requirements on economic models: individual rationality, and mutual consistency of perceptions about the environment...I interpret a proposal to build models with 'boundedly rational agents' as a call to retreat from the second piece of rational expectations (mutual consistency of perceptions) by expelling rational agents from our model environments and replacing them with 'artificially intelligent' agents who behave like econometricians. These 'econometricians' theorize, estimate, and adapt in attempting to learn about probability distributions which, under rational expectations they already know."

We take a jointly boundedly rational, population-based modeling approach and present a class of evolutionary games based upon the following assumptions:

- large numbers of agents interact locally in a multidimensional lattice;
- they use predictive models from a class specified by:

  the past history of observations to use for predictions;
  the complexity of model to construct using that data.

- in their interactions, the agents form predictions of one another and base their behaviors upon those predictions, in an attempt to optimize their utility—a function of the forthcoming and historical behavior of the system;
- they change their choice of model according to past predictive success via a given decision rule (the evolutionary learning process).

So our model considers local interactions only, and deals with games in which there are a continuum of possible responses. See Blume[5] for a related approach to $K \times K$ local interaction games, with an asymptotic analysis using techniques from statistical physics (which we shall discuss in section 5), and Ellison[10] for results comparing local with global interactions. In particular, Ellison highlights the importance of transients and rates of convergence—if the rate of convergence of a system is slow compared with its expected unperturbed lifetime, then the history-dependent transient is more important than the equilibrium state. He proves the expected result that local interactions can give rise to much (e.g., 10^{100} times) faster convergence than global interactions.

Lane[13] argues for the importance of such transients, and that the concept of a metastable state, which may contain the seeds of its own destruction, may be more important than pure stability. Our current work supports this stress on the importance of transients; indeed the concept of a metastable state is crucial to an understanding of the systems we consider.

We begin in section 2 with some simple adaptive agents. Our agents are not Bayesian optimizers, rather they make predictions of other agents using their predictive model. They then optimize their utility in a myopic fashion based upon those predictions. They also adjust their predictive models over time, picking the best allowed adjustment that exists (details will be given in section 3). We show how these simple adaptively optimizing dynamics can drive agents playing such games away from steady state equilibria.

Given that simple equilibria are unstable, what is then possible? We expect a very interesting dynamic, based upon a refinement of the argument given in Kauffman.[11] We introduce our results in section 4.1, but give an intuitive description here. In order to predict one another's behavior optimally, complex adaptive agents will build optimally complex and, hence, boundedly rational models of one another. Such adaptive agents might well coevolve to the "edge of chaos" (the boundary region between a disordered and ordered world state), via the following framework:

1. Given finite data, models that can optimize the capacity to generalize accurately must be of optimal, intermediate complexity (neither overly complex nor overly simple models can generalize effectively).
2. When adaptive agents make models of one another as part of their mutual ongoing behavior, the eventual failure of any finite, approximate model of another's behavior drives substitution of a "nearby," optimally complex model of the other's behavior which now appears to be the best fit to the other's behavior. The point is that, given any finite set of data, multiple models of about the

same complexity will fit the data roughly as well. As the data stream evolves, overlapping patches of the data are optimally fit by nearby models drawn from a set of models of the same complexity.

3. Adaptive agents may persistently alter their models of one another's behavior. Since a change in agent behavior follows from a change in model, such agents must coevolve with one another using changing models of one another's behavior.
4. Presumably such coevolving behavior can be chaotic, ordered, or at the edge of chaos. The ordered extreme in which changes to models are rare (so models are mutually consistent) corresponds closely to a rational expectations state, whilst the chaotic extreme corresponds to rapidly changing models of the agents. At the edge of chaos, models of one another would be poised, tending to change, unleashing avalanches of changes throughout the system of interacting agents. An appearance of punctuated equilibria seems likely.
5. A qualitative argument suggests that in a persistent attempt to optimize prediction about the behavior of other agents, adaptive agents will alter their finite optimally complex models of one another so that the entire system approaches the edge of chaos.

We expect the following picture: if the dynamics are very stable and mutually consistent, then each agent has an abundance of reliable data about the behavior of the other agents. Given more data, each agent naturally attempts to improve his capacity to generalize about the other agents' behavior by constructing a *more precise model* of the others' actions. In our systems, this model is more sensitive to small alterations in other agents' behavior. Thus, as agents adopt more precise models to predict better, the coevolving system of agents tends to be driven from the ordered regime toward the chaotic regime. Conversely in the chaotic regime, each agent has very limited *reliable* data about the other agents' behavior. Thus in order to optimize the capacity to generalize, each agent is driven to build a *less precise model* of the other agents' behavior. These less precise models are less sensitive to the vagaries of others' behavior, and so the system is driven from the chaotic regime toward the ordered regime. So we expect an attracting state on or around the edge of chaos. The precise form of this attractor is discussed in section 4.2.2. Furthermore, we expect model precision and model complexity to be dynamically anticorrelated. This is because a stabilizing world will be better predicted by a simpler model and vice versa.

Thus we expect that the natural definition of *homo economics* which we seek is one in which agent complexity, information use, forward-planning horizons, and recursive modeling depth (my model of your model of my model of...) are dynamically constrained within a finite "bubble" of activity, on average poised between the ordered and chaotic regimes of behavior. This is the first step within that search, in which we address model complexity, precision, and information-use.

After introducing our model and results, we present a mean-field analysis of the aforementioned state, and some extensions of that analysis based upon dynamical considerations. Finally, we discuss extensions to our work and add a few general conclusions.

2. ADAPTIVE AGENTS

Given a system of interacting agents, "rational expectations" (RE) is taken to mean that the conditional expectation of the future each agent has is consistent with the future generated by actions based upon those very expectations. "Perfectly rational agents" act so as to maximize their expected utility under any given information constraints. Assumptions such as these have been used to build a rigorous mathematical formalism for economics. Now, most such rational expectations models implicitly or explicitly consider either

1. the representative agent framework, in which the entire economy is modeled with one aggregate agent;
2. only interactions between pairs of agents, or an agent and an averaged whole.

This research forms part of a growing body that seeks to redefine the assumptions of economics by modeling uncertainty, limited information, and bounded rationality. A simple example will suffice to make our point. Given a system with a large number of possible Nash equilibria, one requires additional parameters indexing agents' beliefs to select a particular rational expectations state.[1] If one lets agents adaptively select amongst those parameters and also among a larger class of beliefs, of which rational expectations is but a subset, then what will the agents do? Will they evolve to a particular rational expectations equilibrium (possibly path dependent)?

If the answer is "no," they must pick something quite different. It is within this context that we seek a definition of the natural rationality of agents whose beliefs are not arbitrarily constrained by assumptions such as rational expectations.

Our models can be considered as one of a string of possible relaxations of RE, moving from agents who adjust a fixed model in an attempt to learn the correct distribution (which is assumed to be of that form), to agents who adjust both model and model parameters, to populations of agents whose models predict *each other* and in which, therefore, the dynamics are completely endogenous.

We shall situate our models by describing some of the links in this chain, then in section 3 we define our model using all of these ingredients.

Bray[6] has studied a simple model of adaptation to a rational expectations state. In this model the equilibrium price p_t for a single commodity is given by:

$$p_t = a + bp^e_{t+1} + u_t \,. \tag{1}$$

[1]See Kreps[12] for related discussion of equilibrium selection.

This is a market clearing condition with p^e_{t+1} the price that market participants expect at time $t+1$. Now the steady-state rational expectations solution is $p^e_{t+1} = \beta \quad \forall t$, where $\beta = a/(1-b)$. Bray addresses the question of what happens if we relax rational expectations and assume people form the expectation p^e_{t+1} by an average over past prices p_t. Bray's model deals directly with a least-squares learning process of a simple model, which effectively implies an equation of motion for $\beta_t = p^e_{t+1}$ (the expected price at time $t+1$), of the form:

$$\beta_t = \beta_{t-1} + \frac{1}{t}(p_{t-1} - \beta_{t-1}) . \tag{2}$$

Bray shows that a representative agent learning in a system whose dynamics are governed by Eqs. (1)–(2) will converge to the rational expectations equilibrium with probability one iff $b < 1$. The models we use are functionally more complex than Eq. (2), but they do not use information from the entire past. Applied to Bray's representative agent, our model is:

$$\beta_t = \alpha_0 + \sum_{i=1}^{c} \alpha_i p_{t-i} \tag{3}$$

for some positive integer c, where we will write α as a shorthand for $(\alpha_0, \alpha_1, \ldots, \alpha_c)$. This class of models can approximate a wide variety of functional forms, and its range of dynamics includes complex oscillatory time series. This begins to approach the natural goal of replacing the global expectation of a representative agent, with multiple expectations and predictions formed by a small group of agents.

Model (3) is quite arbitrary as it stands. However, if the agent optimizes α as a function of empirical observations, then it can build an accurate predictive model over time, and the model is then quite appropriate—especially for the case in which agents do not have prior knowledge of the expected equilibrium state (e.g., steady state, oscillatory, or more complex dynamics may all be encompassed within the same framework). Hence, the dynamics of the price p_t are driven by the representative agent's behavior in the following way:

$$p_t = a + b\left(\alpha_0 + \sum_{i=1}^{c} \alpha_i p_{t-i}\right) + u_t \tag{4}$$

where

$$\alpha = \arg\min \frac{1}{T} \sum_{i=1}^{T} \left(p_{t-i} - \alpha_0 - \sum_{j=1}^{c} \alpha_j p_{t-i-j}\right)^2 , \tag{5}$$

where we have introduced another parameter T, the history (number of lags) over which the agent optimizes the mean square error in the predictions of its model. Now Eq. (4) is a simple linear rule, and Eq. (5) is an easy linear programming problem. However, the combination of the two into an update rule of the form $p_t = F(p_{t-1}, \ldots, p_{t-T})$ is highly nonlinear, and will generally possess a more rich dynamical structure—which in our models leads to instability.

If we consider the simplest nontrivial case, $c = 1$, then $\alpha = (\alpha_0, \alpha_1)$ and:

$$p_t = a + b\frac{1}{\mathrm{Var}(p_{t-1})}\left(\langle p_t\rangle\langle p_{t-1}p_{t-1}\rangle - \langle p_{t-1}\rangle\langle p_t p_{t-1}\rangle - p_{t-1}.(\langle p_t\rangle\langle p_{t-1}\rangle - \langle p_t p_{t-1}\rangle)\right) \tag{6}$$

where

$$\langle p_{t'}\rangle = \frac{1}{T}\sum_{i=1}^{T} p_{t'-i},\ \langle p_{t'}p_{t''}\rangle = \frac{1}{T}\sum_{i=1}^{T} p_{t'-i}p_{t''-i}$$

and $\mathrm{Var}(p_{t'})$ is the variance of p_t over the range $t'-1, t'-2, \ldots, t'-T$.

Intuitively for large T, this system will be more stable, although stability is always tempered by the value of b. Figure 1 shows how the stability of this system varies as a function of T and b. This is a short-time-scale approximation to the true phase portrait, which is independent of initial conditions in the presence of noise. In the figure, the "instability" is a measure of how long the system spends away from the unique equilibrium point. The "least squares optimization" adaptive approach is one ingredient of the class of adaptive models we shall introduce in section 3. There are many other examples in the literature, from Bayesian learning to genetic algorithms, neural networks, and classifier systems. Sometimes such systems use a representative agent, sometimes a population of agents with heterogenous beliefs.[2]

More formally, for the simple case above, without loss of generality set $a = 0$, and assume we are near equilibrium, so $\langle p_t\rangle \approx 0$. In fact let us assume $\langle p_t\rangle = O(1/T)$. Then

$$p_t = bp_{t-1}\frac{\langle p_t p_{t-1}\rangle}{p_{t-1}^2} + O\left(\frac{1}{T}\right) + \text{noise}\,. \tag{7}$$

Approximate solutions to this equation are of the form $p_t = e^{zt} + O(1/T)$, where $z = 1/2\log b$. Therefore, provided T is large, so the $O(1/T)$ terms are actually small, the update Eq. (6) will converge to the unique fixed point $a/(1-b)$ iff $0 \le b < 1$. For smaller T we anticipate values of b close to, but smaller than 1 to be unstable. This is precisely the picture calculated empirically in Figure 1.

Finally, we should note that stabilizing simple conditions on α could be imposed ($\sum \alpha_i = 1$ is an obvious candidate). But in the short term such conditions would lead to a worsening of the agent's predictive ability. It is only if we wish to consider very foresightful agents who plan far into the future (and do not discount that future too rapidly) that such conditions might be realistic. We do not pursue that line of research here.

[2]In the latter case, Arifovic[2] has shown the convergence to a rational expectations equilibrium in Bray's model may take place even when $b \not< 1$.

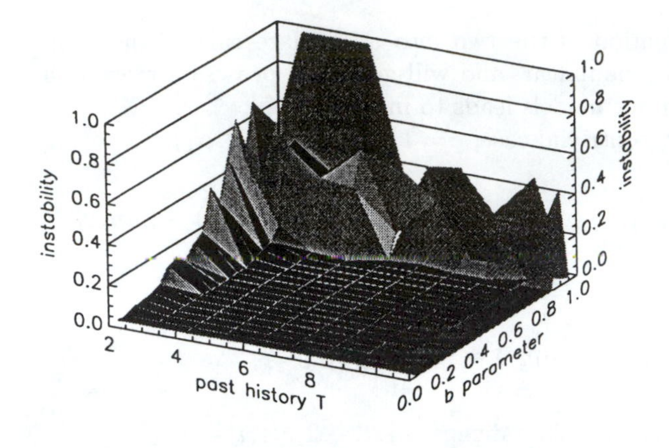

FIGURE 1 A phase portrait for update rule Eq. (6).

3. MODEL DEFINITION

We use the ingredients introduced earlier to model predictive economic systems[3] in which each agent bases its own behavior upon predictions of a small group of other agents. Allowing the agents to change their beliefs over time, we observe these systems in computer simulations, and perform approximate theoretical analyses, to understand the nature of the system's attractor. The attractor may be a static equilibrium, or an attracting dynamic, and the conditions under which a given system evolves to one state or another must be determined. The predictive techniques used are similar to Eqs. (4) and (5), yet even with such simple models, if we reject artificial stabilizing constraints on agent models, very diverse dynamics are observed and convergence to static equilibrium is not to be expected.

We shall present a formal specification of our population models, including the predictive models the agents use, and the manner in which they update those models, and how their model predictions dictate their behavior. This includes discussion of locality of neighborhood, risk-awareness, heterogeneity, cooperation versus competition, and the "stabilized" and "forward" predictive scenarios we study.

The model we shall study for the rest of this paper may be formalized as follows:

[3]Our results may also be applied under some circumstances to biological contexts, in which individual bacteria, say, interact and compete and cooperate in the production of mutually useful chemicals and the absorption of resources from the environment.

DEFINITION. A predictive system $\mathcal{P}$ is a sextuplet $\langle N, L, A, B, M, g\rangle$, where:

$$\begin{aligned}
&N \text{ is the number of agents in the system,}\\
&L \text{ is a regular lattice with exactly } N \text{ sites,}\\
&A = \{a_1, a_2, \ldots, a_N\} \text{ is the set of agents,}\\
&B \text{ is a Euclidean space of possible behaviors,}\\
&M \text{ is the space of possible predictive models,}\\
&u : B \times B^{|\mathcal{N}|} \to \mathbf{R} \text{ is the utility.}
\end{aligned}$$

Here, the lattice L prescribes a function $\mathcal{N} : A \to A^r \subseteq 2^A$ which describes an agent's neighbors. The neighborhood size r is constant, and we shall often write it as $|\mathcal{N}|$. The neighborhood relation will be reflexive, but not transitive, and we shall use multidimensional square lattices. Each agent has the same utility function. Throughout this paper we shall take B to be the set of reals $[0, 1000]$. Agents who try to exceed these limits are constrained to act within them.

The entire state of the system at any time t maybe specified by the time series of behaviors $b_t(a) = (b_t, b_{t-1}, b_{t-2}, \ldots, 0)$ each agent a has generated, and the current model $\mu_t(a) \in M$ each agent uses. It is assumed the time series of behaviors are common knowledge but the agents' models are not. For simulation purposes a short artificial past $\{b_t; t < 0\}$ is generated randomly to initialize the system. We assume the information necessary for the evaluation of u, i.e., the past time series, is common knowledge. This is a first-order decision strategy.

The utility is used in two ways in our systems: (i) via forward expectations and maximization to dictate an agent's behavior; (ii) retrospectively to give the quality of an agent's model.

Both of these uses deal with local interactions and local information. This, and our wish to investigate global dynamics and coordination problems, is why we restrict u to be a function from the agent's own behavior B and the local neighborhood of behaviors $B^{|\mathcal{N}|}$ at a single time period (it is myopic). Thus, we only use point expectations.

3.1 UTILITY FUNCTIONS

Our agents act via a best-response dynamic. Given predictions of their local environment, they maximize their utility function (all agents have the same function, although they will obviously use that function on a different information set). There is a unique best response for the utility functions we consider, of which there are three general types (phrased in terms of best-response production quantities):

COORDINATION. $b_{\text{br}} = 1/\mathcal{N}(\sum_j b_j)$; each agent must try to output precisely the average of the quantities that its neighbors output (so the utility function is $-|b - 1/\mathcal{N}(\sum_j b_j)|$).

SUBSTITUTION. $b_{\text{br}} = 1000 - (\sum_j b_j)$; each agent acts so that the sum of its production and that of its neighbors is a fixed constant. Clearly each agent is taking part in a particular market of fixed size.

COORDINATION WITH PREFERENCES. $b_{\text{br}} = \lambda D + (1-\lambda) \cdot 1/\mathcal{N}(\sum_j b_j)$, so each agent must output a quantity biased between a fixed value and the average of its neighbors. The fixed value becomes the average for the whole system, and becomes a dominant attractor, with attraction given by the level of bias. A good interpretation of this is: each agent has a private valuation D, with a certain level of preference λ. Its action is a weighted sum of its personal preferences (to carry out D) and the influence of its neighbors (to coordinate). This kind of strategy is more naturally considered in the realm of social interaction and discrete choice (see Brock[7] for such a model with local interactions).

This last game may also be considered a "dominant Nash equilibrium" case of the coordination game. It is mathematically identical to the game in Bray's study of adaptive price formation (with $\lambda = b$ and $a = (1-\lambda)D$).

These three game types are special cases of the following best-response rule:

$$b_{\text{br}} = \mu + \lambda(\langle b \rangle - \mu) \tag{8}$$

This is a simple linear combination of the average of the locally observed behaviors ($\langle b \rangle = 1/\mathcal{N}(\sum_j b_j)$), with a single fixed point $b = \mu$ except for the case $\lambda = 1$ for which there are infinitely many fixed points, independent of μ. The utility function implied by Eq. (8) is $u(b, \{b_j\}) = -|b - \mu - \lambda(1/\mathcal{N}(\sum_j b_j - \mu))|$. According to the value of λ we classify the best-response rule into the following categories:

$\lambda > 1$	unstable coordination game
$\lambda = 1$	basic coordination game
$0 < \lambda < 1$	dominant Nash/coordination with preferences game
$\lambda = 0$	zero game
$-1 < \lambda < 0$	dominant substitution game
$\lambda = -1$	exact substitution game
$\lambda < -1$	unstable substitution game

We primarily study coordination games with $0 < \lambda \leq 1$, but we also address our results to substitution games. These utility functions imply assumptions of risk-neutrality; a more general payoff function would include observed predictive errors over some history, weighted by both their variance and mean.[4] A payoff function that heavily penalizes the variance in errors in preference to the mean is known as "risk averse." A risk-neutral payoff seeks only to minimize the mean. The payoff function we use is risk-neutral and blind to the past.

3.2 PREDICTIVE MODELS

At date t, agent a_i is completely specified by its predictive model $\mu_t(a)$ and its past time series $b_t(a)$, where the space of predictive models satisfies the following properties:

DEFINITION. A predictive model $\mu \in M$ defines a triplet T, c, and p, where:

$T = T(\mu) \in \mathbf{N}$, the length of history to use as data for predictions;

$c = c(\mu) \in \mathbf{N}$, the complexity of the predictive model;

$p = p(\mu) : B^{T(1+|\mathcal{N}|)} \longrightarrow B^{|\mathcal{N}|}$, a mapping from local historical observations to future predictions;

so that μ allows an agent to form a prediction of the subsequent behavior of its neighbors, based upon their past behavior for the previous T consecutive time-steps.[5] Clearly both T and c constrain the choice of p.

We use the predictor, α, as introduced earlier, except that it operates endogenously now (there is no global p_t). The basic model specifies how to calculate α to predict an expected behavior $b_{t+1,e}$ from its preceding c behaviors: b_t, $b_{t-1}, \ldots, b_{t-c+1}$, via the following linear recurrence relation:

$$b_{t+1,e} = \alpha_0^* + \sum_{t'=1}^{c} \alpha_{t'}^* b_{t+1-t'} \,. \tag{9}$$

In order to calculate α^*, we must minimize the total error over the history of length T. For computational reasons the standard technique is to minimize the least squares error over some past[6]:

$$\text{error}_j^*(T, c) = \sum_{t=0}^{T-c-1} \left\{ b_t^j - \alpha_0^* - \sum_{t'=1}^{c} \alpha_{t'}^* b_{t-t'}^j \right\}^2 . \tag{10}$$

[4] In which case it may be necessary to separate the decision-making u from the "payoff" utility u.

[5] Other examples of the complexity could be the number of Fourier modes or the size of a correlation matrix used to fit the series.

[6] However, we also consider random initialization of the α followed by random adjustment over time in which better predictive models are retained (this is known as "hill-climbing"), for which our experimental results are unchanged.

The sums here are given for the jth agent, a_j, with time series $\{b_t^j\}$. There are a large number of choices to consider when extending such a predictive model to a population in which local neighbors are modeled. The basic choices concern whether we believe different time series are related; whether our neighbors should be modeled individually or together, and whether an agent should include its own time series in its model estimation. It transpires that such choices are usually not important, so we will fix on the case in which each agent has a single model with $1+c|\mathcal{N}|$ coefficients, which are calculated with a simultaneous regression on all the neighbors' time series. This model is then used to predict the average $\langle b_t \rangle$ directly. We now calculate the optimal $\alpha^* = \{\alpha_0^*\} \cup \{\alpha_{j,1}^*, \ldots, \alpha_{j,c}^*; j \in \mathcal{N}_i\}$, so that the prediction is given by:

$$b_{t+1}^{\text{pred}} = \alpha_0^* + \sum_{j \in \mathcal{N}_i} \sum_{t'=1}^{c} \alpha_{j,t'}^* b_{j,t+1-t'} \tag{11}$$

and the coefficients α^* are given by a least-squares minimization over the set of neighboring lagged time series of length T. Further details on the more general cases in which agents may have separate models of each neighbor or use their own time series as regressors are given in Darley.[9] These scenarios are portrayed in Figure 2.

The techniques used to perform the minimizations are singular-value decomposition (details to be given[9]). Let us note, however, that numerical stability factors[16] mean that it is actually better to solve the minimization problem rather than differentiating the above equations and solving the subsequent equalities exactly.

Under any of these techniques, once neighboring predictions are calculated, the best-response rule/utility function dictate the agent's own behavior.

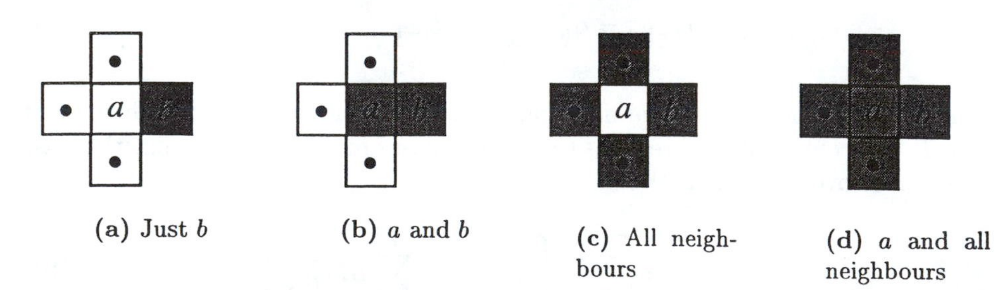

(a) Just b (b) a and b (c) All neighbours (d) a and all neighbours

FIGURE 2 For agent a to predict a given neighbor b using local information it can form a predictive model using any of the four data-sets shown. Furthermore, for (c) and (d) the agent may either use one or $|\mathcal{N}|$ models.

3.3 DYNAMICAL UPDATE RULE

The system's dynamics stem from the ability of the agents to tune their predictive models, based upon differences between observed and expected behaviors, in an attempt to optimize their own behavior with respect to the changing environment. We do not impose any exogenous shocks on our models. This procedure can be summarized by the following sequence of actions, carried out at every time-step, in parallel, by every agent a_i:

Prediction($i, \mathcal{P}$) at time t:

i. Calculate private predictions $b^j_{t+1,e} \in B$ giving the expected behavior of all other agents ($j \neq i$) at time $t+1$.

ii. Find $b^* = \arg\max_{b \in B} u(b, \{b^j_{t+1,e}\})$, agent a_i's predicted optimal response.

iii. Carry out action b^*.

iv. Observe actual behaviors b^j_{t+1} and, using g, calculate agent a_i's utility $u^* = u(b^*, \{b^j_{t+1}\})$.

v. If $\exists \mu'_i$ with $T' = T \pm 1$ or $c' = c \pm 1$ s.t. $u(b^*_{\mu'_i}, \{b^j_{t+1}\}) > u^*$ then pick the *best* such new model μ'_i. This is the model update rule.

Step (v) is the "predictive model update rule," which dictates the main dynamical properties of our systems. Each agent compares its utility under its current model, μ, with the utility it would have had under a perturbed model, μ'. If any of the alternative models would have been better, the *best* such model is selected. Under our perturbation scheme, movement in the space of predictive models, M, is local, by steps of length 1 in either the T or c direction. Hence, discrete changes to an agent's model may be considered as moves in the two-dimensional discrete lattice of history-complexity pairs, and agents can be thought of as performing local search in this space as they seek the best model of the world around them.

As remarked earlier, the payoff function is blind to the past, and the only effect of a particularly poor prediction is presumably a change in one's predictive model to a better pair of parameters. The only level of evolution in the system is the survival of certain models (and, therefore, certain parameter values). As a consequence we do not directly investigate phenomena pertaining to the *accumulation* of utility (rule (v) only compares instantaneous utilities).

Finally, we shall note two facts: first, the sequence of update steps (i)–(v) is completely deterministic, with no exogenous perturbations or shocks; second, the entire model is evolving in synchrony. This differs from the approach of Blume[5] who considers "strategy revision opportunities" independently exponentially distributed, such that the probability of two or more agents updating simultaneously is zero.

3.3.1 HETEROGENEITY AND THE USE OF INFORMATION. Our agents are heterogenous because of two facts: that they may choose different models (or groups of models); and that the information upon which they base their predictions may be different.

Given some slight variation in the world (originally brought about at least by nonuniform initial conditions), each agent will operate with u upon a slightly different set of behaviors giving rise to a variety of actions. We shall investigate whether this initial heterogeneity grows, diminishes, or even vanishes with the evolution of the system.

It is worth pointing out that this research does not currently differentiate fully between the two sources of heterogeneity (models and information): at a first approximation the information used to model a given neighbor (its behavior time series) is common to all agents modeling that neighbor, so models are the only source of local heterogeneity. However, the manner in which correlations between agents are modeled is constrained to utilize slightly different neighborhood information sources. This is because we currently constrain agents to use local information only to make their predictions. This means the four agents who make models of a given common neighbor are all constrained to use different information sets to construct their models if they wish to model correlations.

Further research will allow agents to use nonlocal information. These more sophisticated models will enable us to pinpoint more accurately the assumptions from which heterogeneous nonstationary worlds can be derived.

3.4 WHAT TO PREDICT?

Finally, given a class of predictive models, an economic strategy, and a set of neighborhood relations, we must decide what the agents should predict and the manner in which that determines their own behaviors. We shall consider two scenarios:

1. the *forward predictive scenario* in which each agent predicts what its neighbors will do tomorrow and uses the utility u to determine its own behavior. This is the natural, obvious predictive method.
2. the *stabilized predictive scenario* in which each agent ignores the just realized predictions. The dynamics are as follows: all agents make predictions, adjust their models based upon the success of those predictions, but then those predictions are forgotten and ignored. The agents generate a new set of predictions and repeat. Hence, the models are updated as above, but the behaviors and predictions undergo nothing more than a process of iterated refinement.

The first scenario is the natural one. We also study the second because it is, effectively, a stabilized version of the first. Rather than predicting forward in time, the agent effectively re-predicts what it should have just done, and then carries out that action (the process could be considered one of perpetual refinement of action, not unlike the simple adaptive processes considered earlier).

We will find that the dynamics which arise from these different scenarios can be quite different in character. Intuitively scenario 1 may lead to excessively unstable dynamics, as the agents' forward predictions diverge from one another. A more sophisticated predictive model may be required to follow the dynamics. Scenario 2 on the other hand may be too stable, with predictions from one date to the next hardly varying at all.

A third and fourth scenario, which we leave to future research, are the following:

3. each agent predicts using a discounted sum of expected future utility based upon neighbor predictions over a given planning horizon (and seeks to optimize the length of that horizon);
4. agents build a hierarchy of meta-models of each other—my model of your model of my model of,. . .and optimize the cut-off height of this hierarchy.

We shall present our results for scenario 2 first, in section 4.1, which illustrates the coherently organized coupling between model-update events. Then in section 4.2 we give our results for scenario 1 in rather more detail.

4. OBSERVATION AND SIMULATION

A system such as ours has a large number of possible rational expectations states. There are clearly an infinite number of possible Nash equilibria (using the coordination strategy at least), and many more periodic equilibrium states are possible. One important consideration when given multiple equilibria is to try and understand the problem of equilibrium selection. The parameters that characterize the choice of equilibrium index beliefs the agents have (individually and collectively) about their world. One common use of adaptive models is in equilibrium selection in just such a scenario. From our perspective we would like to understand more than just selection; we would like to know what happens when models are not forced to be stable. This implies that our agents do not have as a priori (and rather ad hoc) beliefs that the world is heading inexorably to a simple static equilibrium (this is implicit in Bray's model and our extensions to it). Our agents will naturally pick from a class of models (which includes a set implying stationarity) so as to maximize their immediate gain.

So one question we must address is: Is a static (coordinated) equilibrium selected for? If not, then we will concern ourselves with understanding and explaining whatever nonstationary dynamics are observed. In particular we attempt to formulate a categorization of the *natural* rationality of agents whose mutual interactions form their world. We shall compare the results of our analysis and experimentation with the intuitive ideas introduced earlier.

In order to initiate the simulations, agents are given an artificial randomized past time series, and history and complexity parameters. All the results are robust to changes in the initialization technique (Gaussian, sinusoidal, and uniformly random initializations have all been tested.) We now present our experimental results for the two predictive scenarios.

4.1 OBSERVATIONS OF THE STABILIZED SCENARIO

The following results are all for predictive scenario 2. As remarked earlier, the stabilized scenario allows us to investigate feedback and propagation of information between models in a more restricted stable world. For this scenario, there is little difference in agent *behavior* dynamics between the small and large-population cases (the behaviors are simple and stable across the system). However, the large case shows interesting spatial order in the space of agent *models*, so we shall consider that case exclusively.

We observe two regimes of behavior, which we label "coherent" and "random," each preceded by a short transitory phase. These two regimes are qualitatively different.inxxpredictive models, agent scenarios

During the first, coherent regime predictive errors decrease exponentially fast, whilst variance in agent behavior (system heterogeneity) collapses exponentially fast onto the system mean. These are diffusive spatial dynamics in which any behavioral heterogeneity disperses rapidly. Such dynamics can be generated by a wide class of models in which agents try to imitate each other using simple adaptive models; such models are presented elsewhere.[9] During the second, random regime predictive errors and system heterogeneity have reached a lower bound at which they remain. The occurrence of these regimes is explained below.

Model update dynamics are more interesting: define an "avalanche" to be the number of consecutive time-steps an agent spends adjusting its predictive parameters c, T. Then in the coherent regime, we observe a power-law relationship between avalanche frequency and size: $f = c/l^k$, where $k = 0.97 \pm 0.07$. Hence, the model-update process has organized model change into a *critical* state in which large, long-duration avalanches may occur. This requires coordination to build up endogenously across significant sub-groups of the population of agents, so we refer to this state as "self-organized." In the random regime the avalanche frequency distribution develops into an exponential fall-off: $f = p^{l-1}(1-p)$, where $p = 0.323 \pm 0.005$. Furthermore, an examination of spatial (rather than temporal) avalanches in the lattice gives the same *power-law then exponential* result. In that sense the system is scale-invariant in both space and time.

We should point out that there is a large *practical* difference between a power-law and an exponential fall off of avalanche size. For the exponential case, large avalanches effectively *never* occur, whereas in the coherent regime we have much data for disturbances of size nearing 100, for instance. This is an important distinction.

The reason for the dramatic change between regimes is as follows: in the first regime behavior variance and predictive errors are both converging to zero exponentially fast (the former at least is expected for a diffusive dynamic). Once the differences between models' predictions are less than that discernible under the numerical representation used, numerical rounding errors contribute more than model difference. At that point model selection becomes a random process. This hypothesis has been confirmed using test experiments of varying floating point accuracy. The curve fit $p^{l-1}(1-p)$ in the random regime is just the expected number of consecutive moves in a random walk with probability p of moving.

The random regime is, therefore, of lesser importance, and the self-organized critical behavior can be considered the predominant characteristic of the internal dynamics of the stabilized scenario. This critical behavior exists in a wide variety of observables, indicating that the system does truly self-tune to a boundary intermediate between order and disorder. Although coherence is eventually lost due to the overwhelming stability we have imposed, leading to a degenerate dynamic, it is clear that basic coevolutionary forces between the agents' models have profound influence upon the global dynamics, and the macroscopic behavior can be captured with a relatively simpler picture.

An extension of the above observations can be found if each agent models each of its neighbors using a totally separate model. We still find the same two regimes (and avalanche characteristics) as before, but now we get exponential convergence to a *nonuniform* state. Each agent has a different behavior, but such behaviors are coordinated so the agents are still in a high-utility state. So, whereas the old system converged to a system-wide fixed mean, zero variance; the new system converges to a system-wide fixed mean, but nonzero variance.

This raises a basic issue: persistent local diversity requires at least the capability of modeling that diversity—an agent which uses a single model for each of its neighbors (or, more generally, for the entire information set it observes) believes the system's equilibrium states are much simpler than the agent with multiple models who has no such presupposition. In systems such as ours in which the dynamics are endogenous, and agents' beliefs are reflected in those dynamics, it is important to take those beliefs into consideration.

4.2 OBSERVATIONS OF THE FORWARD PREDICTIVE SCENARIO

The following results are all for predictive scenario 1. This is a more natural predictive situation, in which we can expect both models and behaviors to exhibit interesting dynamics.

The most clear difference between this and the stabilized scenario is in observations of the agents' behaviors. They no longer always settle down over time. There are two generic cases:

SMALL SYSTEM OR STABILIZING MODELS—agents coordinate across the system on relatively simple strategies. The variance in behavior across the system is very low, and its mean follows a simple monotonic path.

LARGE SYSTEM—an interesting interplay between between periods of coordinated behavior, and periods of disordered rapidly changing behaviors is observed.

Figure 3 shows the basic behavior-space dynamics for large systems. Comparing this with Figure 6, we can see that the coordinated time-phases are precisely linked with periods of very low variance in agent behavior, whereas the uncoordinated periods show very high variance levels. Hence, the agents *all* converge to a particular selection, retain that behavior for some time, with only quite minor fluctuations, and finally the coordination breaks down and the agents pass through a disordered bubble of activity, selecting any of a great range of behaviors before settling upon a new metastable state.

Consider the coordination and substitution games we introduced. We can summarize our results for these very succinctly: the "unstable" variants are indeed unstable because best-response dynamics drive the system away from equilibrium; dominant-Nash games are identical to coordination games (which are the specific case with no domination, $\lambda = 1$) in the short term, but have more robust equilibria in the long term. Their short-term dynamics and the general dynamics of the coordination game exhibit interesting punctuated equilibria, cyclic oscillations,. . .which we shall analyze in detail below.

We shall discuss the case with small populations first, before considering the large-population case.

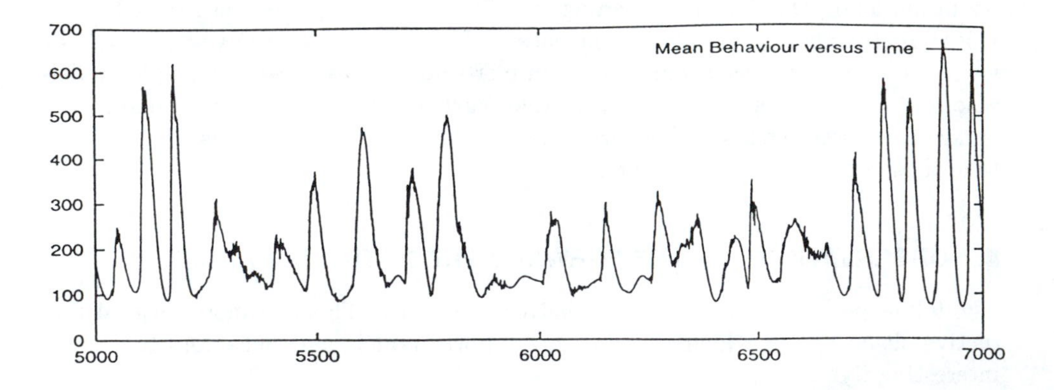

FIGURE 3 Mean agent behavior in a 10×10 world. The maxima/minima represent almost completely homogenous behavior; in between behavior, is highly disordered.

4.2.1 SMALL POPULATIONS. The large-scale dynamics described above are sufficiently complex that it is insightful in this case to consider small populations in which we can better see how the agents influence each other, and in which there is less room for variation, and fewer degrees of freedom to go unstable.

The first thing to learn from Figure 4 is that the agents can learn to coordinate on particular behaviors, and that that coordination can take place in a relatively rapid, damped oscillatory form. Plots (b) and (c) show that coordination need not be on a static equilibrium, if the agents have a sufficiently large percentage of the total information available, i.e., if each agent is modeling a reasonably large fraction of the total number of agents, then global coordination on a t-dependent path may be achieved. Notice that, since the agents are constrained to the range $[0, 1000]$, this will cause a temporary instability on any path which hits the bounds.

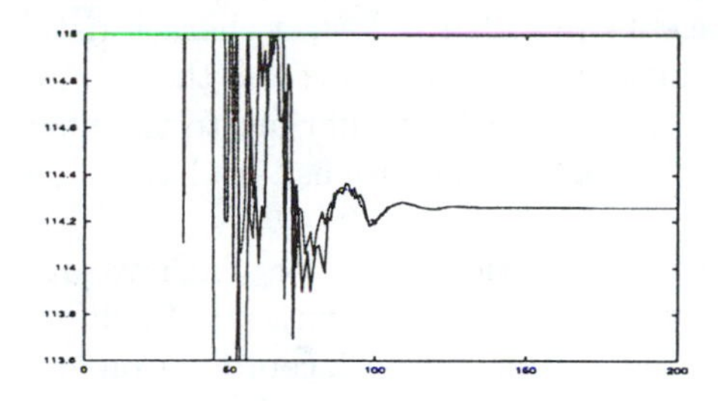

(a) 2 agents, unit neighbourhood — the models become consistent

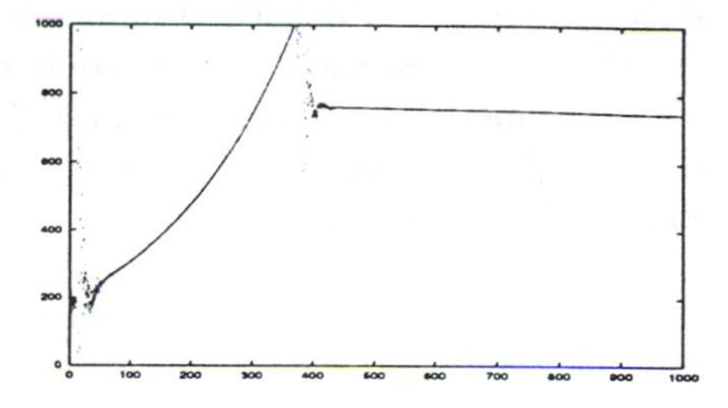

(b) 2×2 agents, $\mathcal{N} = 4$; the models become consistent

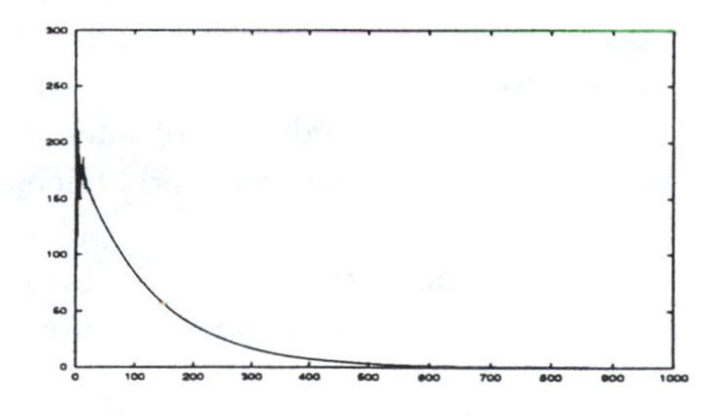

(c) 4×4 agents, $\mathcal{N} = 4$; again consistent models develop (very quickly in this case)

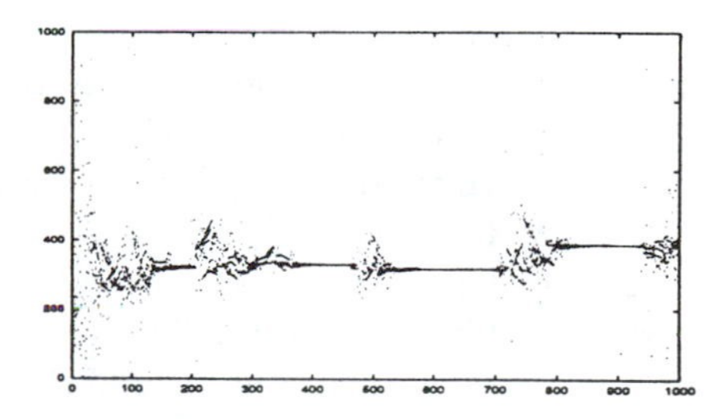

(d) 10 agents, unit neighbourhood. Meta-stability and punctuated equilibria are now observed generically.

FIGURE 4 Agent behaviors in small populations. The horizontal axes are time; vertical is behavior. Each agent's behavior at each time step is plotted as a dot, which are connected in the first figure.

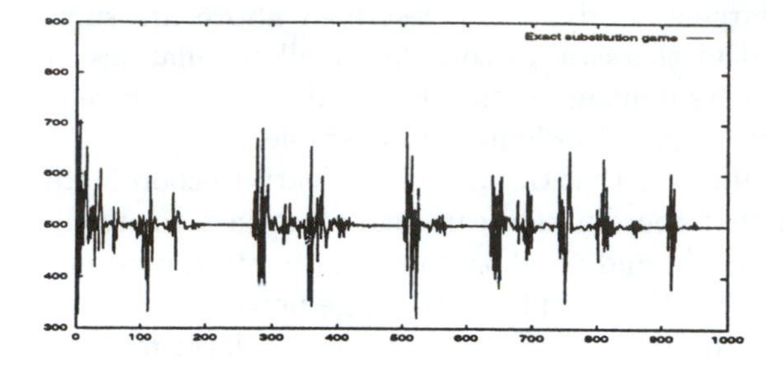

FIGURE 5 Behavior of a sample agent taken from a ring of 10 playing the exact substitution game.

The exact paths in Figures 4(a), (b), and (c) are completely history-dependent, although the range of possible qualitative dynamics is rather small. When we consider systems in which the proportion of agents modeled is low, a characteristic dynamic emerges. Equilibrium is only possible on a constant value, and those equilibria are metastable. The system will leap from one such equilibrium to the next via disordered transitions. This *qualitative* dynamic is very robust, and the only historical influence is in the actual equilibrium values.

These results are all identical for the exact substitution game, for which we give a sample plot in Figure 5. The stabler variants of these games, with $|\lambda| < 1$ damp down to the unique equilibrium after a λ-dependent time, for both coordination and substitution games. For large populations this still occurs, but only after significant transient periods.

4.2.2 LARGE POPULATIONS. The only characteristic simple coordinated states for larger populations are static equilibria. These occur with great regularity, and are highly uniform across the entire system. Figure 6 shows how the variance of the agents' behaviors drops almost to zero at these times—a spatial plot of such a state is shown in Figure 7(a). Such states are only metastable, however, and hence, oscillatory dynamics are observed.[7]

All the large systems we consider show a punctuated equilibrium dynamic. Successive equilibria are destabilized endogenously, leading to wild fluctuations before the agents settle to another equilibrium.

Furthermore, these dynamics are not transient for the exact coordination or substitution game. However, interestingly enough, given some small level of dominance in the game ($|\lambda| = 0.9$, say), the dynamics do become transient after sufficiently long periods of time. The agents eventually correlate their behaviors and models sufficiently that they reach of state of small fluctuations about a static equilibrium. The surprising dynamics of history and complexity by which this may be achieved are shown in Figure 9. Both the early punctuated equilibrium regime, and

[7]The interested reader, with fast Internet access, should point their web browser at `http://www.fas.harvard.edu/~darley/Vince-Thesis.html` for some some movies of the two-dimensional evolution of such a system.

the transition last for significant periods of time. Note, however, that external perturbations can destabilize that system, so that the *natural* dynamics may be either the punctuated or static equilibrium depending upon the natural frequency with which exogenous perturbations occur. As remarked in our introduction, it is always important to remember that distinction when analyzing systems with long transients, since that will dictate which of the possible behaviors is actually observed.

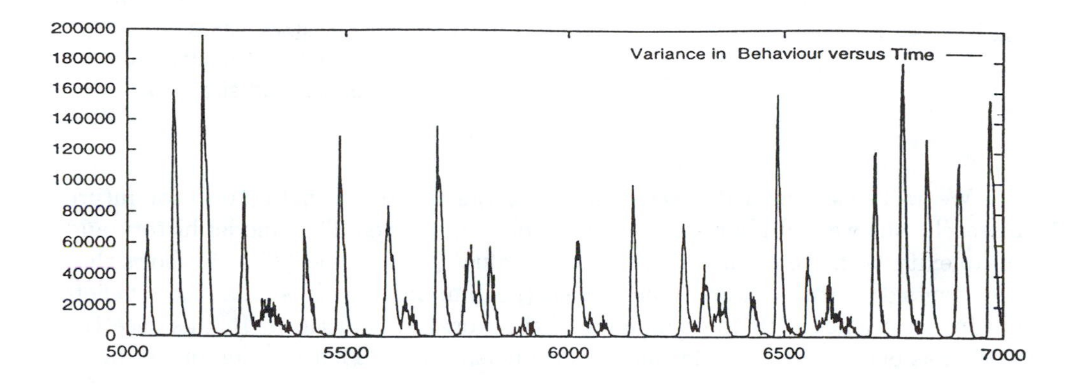

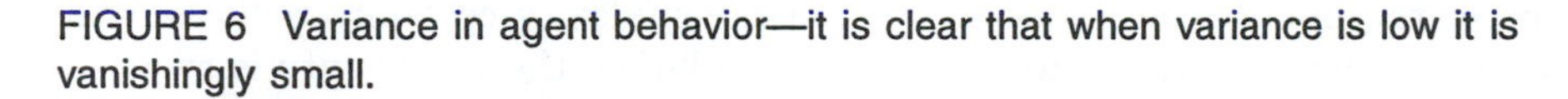

FIGURE 6 Variance in agent behavior—it is clear that when variance is low it is vanishingly small.

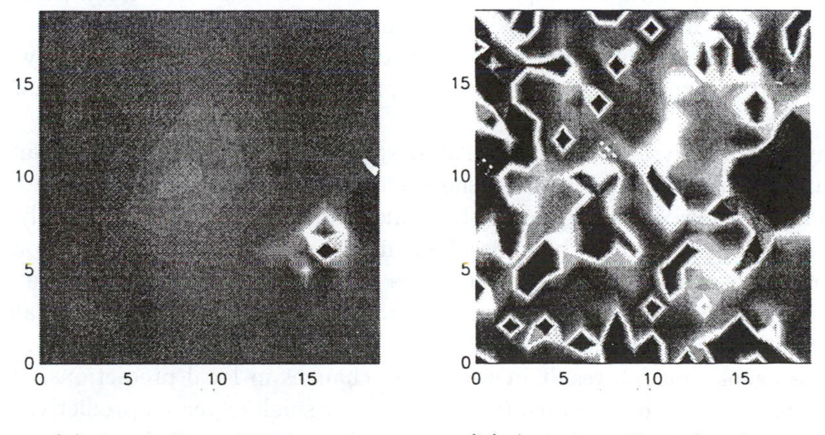

(a) A coordinated regime (b) An uncoordinated regime

FIGURE 7 Spatial distribution of behaviors for the forward predictive scenario. This is for a 20×20 square lattice of agents, with smoothing applied for representational purposes.

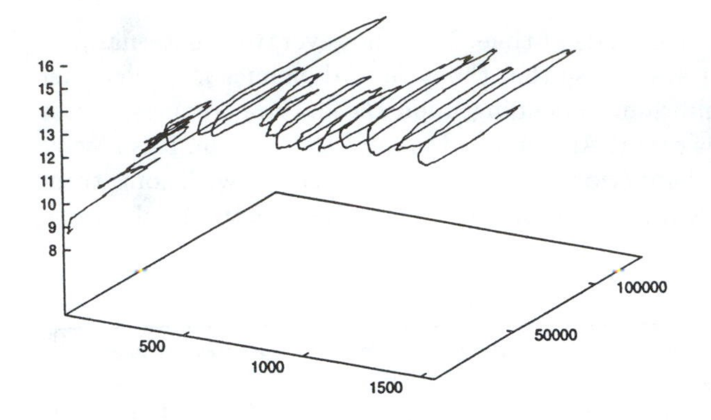

FIGURE 8 Limit cycle behavior in $(t, \sigma_b^2, \langle c \rangle)$ space—the vertical axis is mean model complexity, the left-right axis is time, and the depth axis is variance in agent behavior (a measure of system heterogeneity). This is for a system with 900 agents in a toroidal 2-d lattice with a neighborhood size of 4.

We will postpone a detailed analysis of the transition in Figure 9 for future research, but we will point out a number of simple facts. That model history and complexity both eventually evolve to small values ($\langle c \rangle \approx 2$ and $\langle T \rangle \approx 5$) shows that the agents do learn to change their models (i.e., implicit beliefs) so that those beliefs are in accordance with the dynamics of the world around them. In this case the world has become very simple, and the agents learn that. This is an important point which is not so readily discernible in the presence of more unstable dynamics—for instance in the limit-cycle, Figure 8, the mean complexity certainly reduces with system heterogeneity, but the world destabilizes before the complexity is reduced too far. A possible intuitive explanation for why the reduction takes so long in Figure 9 is that more complex models or models with longer histories are only discarded because of subtle effects of over-fitting the data. However, this is work in progress, and we expect to give more precise, rigorous reasons in a future analytical work.

Why does the system spontaneously destabilize? These instabilities start at a point, and propagate rapidly across the system.[8] They are caused by the following dynamic: a locally simple world state allows agents to build successively more precise models of the local dynamics (i.e., modeling the exact shape of local oscillations, with small expected error). Once an agent begins modeling the shape of the dynamics sufficiently closely, rather than some generic average, small "extraneous" influences from not-so-nearby agents cause neighboring changes, which under the agent's model, result in very large changes in local predictions. Within our systems, a very precise model (i.e., with a very small expected predictive error) is much more sensitive to local perturbations than a more generic average model. Hence, these "extraneous" perturbations will cause an agent with a precise model

[8]Again the interested reader is referred to the movies which have been made available on the internet to help gain an intuitive feel for these models.

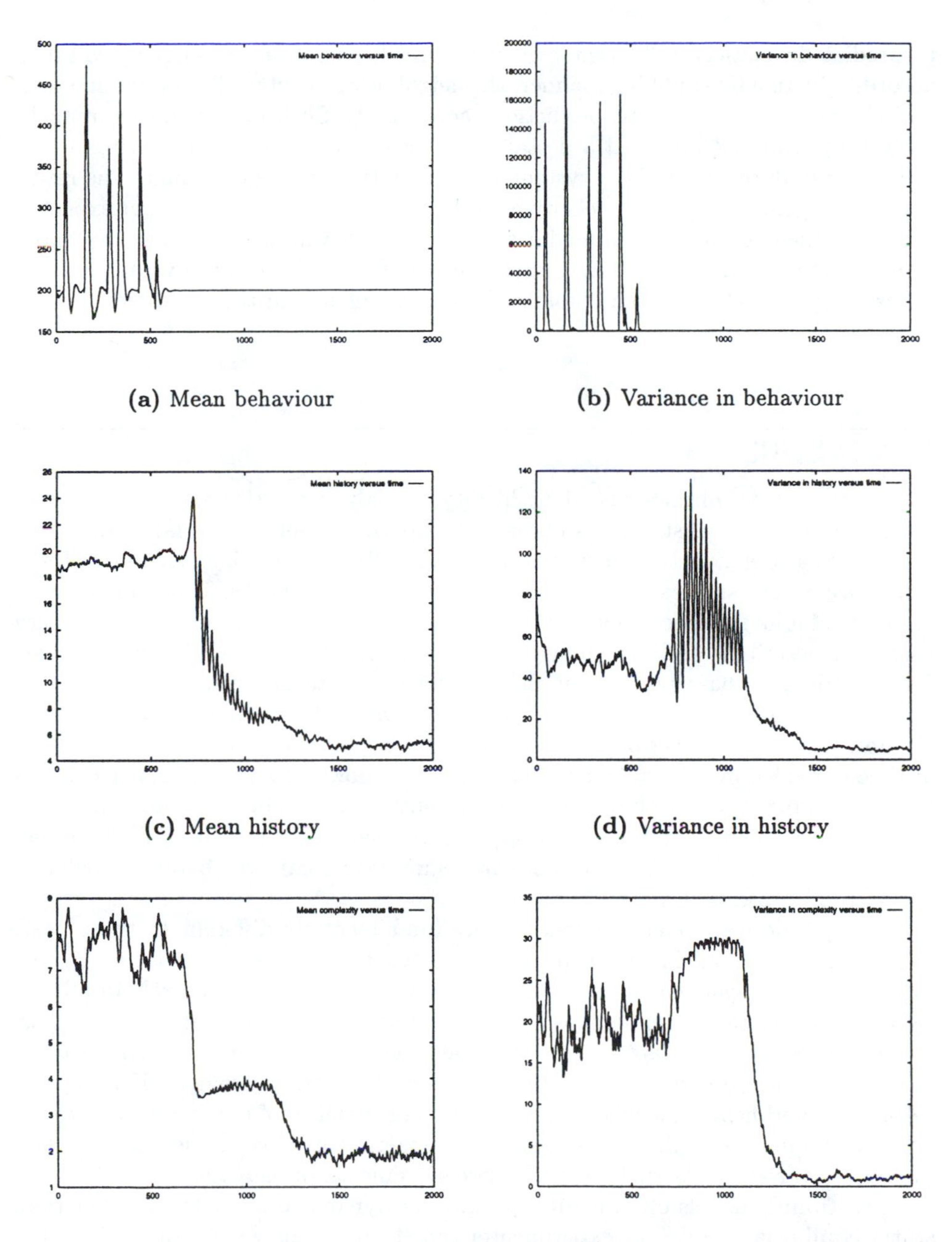

(a) Mean behaviour (b) Variance in behaviour

(c) Mean history (d) Variance in history

(e) Mean complexity (f) Variance in complexity

FIGURE 9 Long transients for a dominant Nash game, followed by a particular transition to a stable equilibrium. The equilibrium persists for at least 10,000 time steps. The form of the transition is in fact typical for this type of system.

to predict dramatically different behaviors, and hence, its own behavior will leap accordingly. In a self-fulfilling manner, the neighboring agents will observe that leap and change their subsequent predictions accordingly. Such large changes immediately propagate outward and a destabilized wave is observed, propagating across the entire uniform state of the system, and disrupting that equilibrium. The result is that all agents will pick models with more degrees of freedom but larger expected errors. In the extremely disordered case, these models will predict a basic weighted average, since their observations are too chaotic for precision to be valuable. Such "averaging" models will then stabilize the system again and the cycle repeats.

5. ANALYSIS

Condensed matter physics has a long history of analyzing lattice models in general. These will often be abstractions of ferro- and para-magnetic materials, taking the form of Ising spin models (and numerous variants). In general, for such models, there exist parameters, such as the temperature and external magnetic field, which have a direct, profound influence upon the given system's dynamics. These models are very useful for describing and explaining experimental observations of physical systems. For example, a phase portrait of the equilibrium state under different values of the temperature and external field can be determined experimentally. In such a portrait, different regimes of behavior are noted, separated by phase changes which are accompanied by critical transitions (exhibiting power-law scaling in correlations between changes to variables) and discontinuities in certain observable variables (e.g., lattice magnetization). By slowly varying the temperature or field we can move the system through these different regimes, and can watch the behavior of the system's observables.

The predictive models we consider are fundamentally different to these basic spin models, in that there are no such obvious parameters to be tuned by the experimenter. In particular, the temperature, which plays a crucial role in all statistical physics, has no immediate and clear interpretation (or role) in our models. Although we could perhaps observe an analogue of the temperature or at least of the entropy of the system, those values emerge from the system itself—they are self-tuned, and hence, the whole system has the capability of moving itself through different regimes and phase changes. As an experimenter, we do not have control over these variables. As such, we call these systems *self-organizing.*

Self-tuning models often exhibit interesting dynamical attractors, rather than static equilibria. Unlike an experimenter-tuned model, in which for a given parameter set a given equilibrium (or equivalence class of equilibria) is determined (provided there are no energy barriers preventing it from being reached in the time-scales under consideration), a self-tuning system relies upon its own dynamics to dictate the environment in which those very dynamics operate. Provided the system is not overweeningly stable, it can only really settle down to a dynamic equilibrium, a coherent mode of *active* behavior which is self-reinforcing. The inherent dynamicity of the system will normally destabilize any potential static equilibria.

Such phenomena are observed in many realms in the real world, and an interesting paradigm for their study has developed from the work of Per Bak.[3]

5.1 MEAN-FIELD APPROACH

A standard approach from which to begin an analysis of such lattice systems is the mean-field approximation. We can apply this whenever the system's constituent parts interact in a primarily local fashion. We approximate by considering that each agent is only really interacting with a hypothetical "mean field," the average of the forces (i.e., field) generated by the actions of the agents in the rest of the system, perturbed slightly by its own actions (or that of a small cluster of agents).

5.1.1 APPROXIMATING ASSUMPTIONS. We shall make certain assumptions designed to contain a minimal specification of what we consider to be the essential properties of our models. We do not yet attempt to derive these approximations from the system's precise microdynamics:

1. That one can derive from each agent a real number representing the sophistication of the agent's model "σ." Moreover, that each sophistication, denoted $s(\sigma)$, labels an equivalence class of models. Within each class there is variation in model, given in this simple case by the directional degree(s) of freedom of σ. As such, we shall assume the set Σ of all possible σ is a vector space over the reals. Σ will be chosen as a succinct characterization of the space of possible models.
   Formally then, for every agent, a, we have a corresponding model given by σ_a, where we define the usual inner product over Σ:
   $$\langle \sigma_a, \sigma_{a'} \rangle \to \mathbf{R}\,. \tag{12}$$
   So that the "sophistication" of a given agent's model is: $s(\sigma_a) = \sqrt[+]{\langle \sigma_a, \sigma_a \rangle}$. We shall often write $|\sigma_a|^2$ for $s(\sigma_a)^2$. Small elements of Σ correspond to simpler models, and the directional degrees of freedom represent different types of behavior. This defines a sophistication equivalence relation over Σ:
   $$\sigma_a \sim \sigma_{a'} \Leftrightarrow s(\sigma_a) = s(\sigma_{a'})\,. \tag{13}$$
   Two models (and, therefore, agents) are in the same equivalence class iff the models have the same sophistication.
2. From the observable behaviors in the world, we can derive a vector field, such that an agent may be considered to be in a local behavior field β', which is

some externally imposed behavior pattern β (assumed to be very small) plus a local field provided by the neighboring agents.

$$\beta' = \beta'(\beta, \text{ local agents}). \tag{14}$$

3. There is a global difficulty of prediction, $\gamma \in \mathbf{R}^+$, derived from observations of the character and complexity of time series behavior observed in the predictive system.

$$\gamma = \gamma(\{b(a)\}_{a \in A}). \tag{15}$$

5.1.2 MEAN-FIELD ASSUMPTIONS. Given the above assumptions, it is reasonable to assume that the average effective model σ_m in the field β' will be given by the following law:

$$\sigma_m \propto \frac{\text{behavior field } \beta'}{\text{difficulty of prediction}}. \tag{16}$$

This states the average agent's model is proportional to the local behavior field and inversely proportional to the difficulty of prediction in the system as a whole. We assume that if prediction is hard the agents will tend to pick simpler models.

The mean-field assumption is that the behavior field, due to neighboring agents, is a function of the average model σ_m. We shall consider σ_m to be small so we can expand β' in a power-series about β. Note that β is the sole source of exogenous effects; other terms are purely endogenous. The first term is given by:

$$\beta' = \beta + l\sigma_m. \tag{17}$$

This relates the observed behavior field β' to the character of the exogenous forcing and average local model. Note that it implicitly addresses the interactions from the point of view of a single individual. It can be extended to consider small clusters of individuals, but the qualitative consequences of the results are not changed. Now, writing γ for the difficulty of prediction,

$$\begin{aligned} \sigma_m &= \frac{c(\beta + l\sigma_m)}{\gamma} \\ &= \frac{c\beta}{\gamma - lc} = \frac{c\beta}{\gamma - \gamma_c}, \end{aligned}$$

where c, l are constants and $\gamma_c = lc$. This solution is only valid for $\gamma > \gamma_c$, otherwise σ_m points in the opposite direction to β, which is not meaningful. For $\gamma < \gamma_c$ we must expand β' to third order (second order terms are not present for reasons of symmetry) in σ_m:

$$\beta' = \beta + (l - b\langle\sigma_m, \sigma_m\rangle)\sigma_m \tag{18}$$
$$\beta' = \beta + (l - b|\sigma_m|^2)\sigma_m \tag{19}$$

and we find:

$$\sigma_m \propto \beta^{\frac{1}{3}} \text{ for } \gamma = \gamma_c . \tag{20}$$

For $\gamma < \gamma_c$,

$$\sigma_m^2 = \frac{(\gamma_c - \gamma)}{cb} . \tag{21}$$

Following assumption 1, since σ_m is not just a real number, this implies that the mean sophistication $s(\sigma_m)$ is fixed by Eq. (21), but that the choice of σ_m within that equivalence class is *not* fixed, and a preferred direction will exist.

5.1.3 INTERPRETATION. It is clear from the above analysis that the correlate of temperature in physical systems is the *difficulty of prediction*, and that there are two qualitatively different regimes of behavior, given by $\gamma \gtrless \gamma_c$.

Therefore:

1. $\gamma > \gamma_c$ If the difficulty of prediction is high (the "high temperature" regime), in the absence of external forcing ($\beta = 0$, as in the endogenously generated worlds we study) agents will:

   - Pick simple models (sophistication is proportional to the reciprocal of the difficulty).
   - Since the average of all models is the "zero" model, there will be no preferred choice of σ within any given equivalence class of sophistications. Agents' models of the world are not mutually consistent (the significance of this point will be expanded upon later).

   If there were to be external forcing, signified by β, then a preferred choice of model would exist. Currently this situation has no real interpretation in our predictive systems, $\mathcal{P}$, as they stand. However, were we to impose a certain pattern of cyclical variation in the world (incorporating, for instance, an economic "sunspot" effect (Barnett et al.[4])), we would expect the particular choice of agents' models to reflect the character of that forcing behavior.
2. $\gamma < \gamma_c$ If the difficulty of prediction is low, even without any external forcing, the agents spontaneously break symmetry within their equivalence class of models, σ, to pick out a preferred type of model. This occurs because the directional degree of freedom of the average model σ_m is nonzero. That degree of freedom gives the preferred model.

The interpretation here is that by selecting a preferred model, the symmetry breaking ensures that the agents' models are *mutually consistent.* The system naturally reaches a state exhibiting an important ingredient of what are normally the assumptions of rational expectations.

Of course, this description and analysis only apply directly to a static equilibrium scenario, which we do not expect to arise in these systems. However, one expects a dynamic equilibrium situation to situate those dynamics around any marginally unstable static equilibrium. Indeed the first step in taking the above analysis further is to consider the character of perturbations about the static equilibrium we have so far derived.

There are two important points still to address: that of the assumed difficulty of prediction and that of interpreting the beliefs of agents' models as being mutually consistent or inconsistent.

DIFFICULTY OF PREDICTION. Despite the form of assumption 3, in the analysis which followed, the difficulty γ was treated as a tunable parameter, independent of the agents. In general, it is certainly not independent of the agents, as assumption 3 states, it is a function of the agents' behaviors: $\gamma = \gamma(\{b(a)\}_{a \in A})$. Certainly if the agents behave in a reasonably varied fashion it will be easier to discern their underlying models and predict, than if all agents behave similarly (given some underlying level of noise and that we wish to consider relative ability of predictions). So the difficulty of prediction will be an emergent observable, tuned by the dynamics of the system. Its dynamics over time will depend upon the following forces:

i. the relative ability of a successful versus unsuccessful agent will be greater in a higher variance world;
ii. agents with fancy models will be more susceptible to subtle changes in the behavior of the world, i.e., a smaller discrepancy is capable of disproving their models;
iii. if the systems are sufficiently dynamic in the sense that agents must adjust their models over time in order to do relatively well in the world, then nonstationarity is maintained.
iv. if the world is *too* static or simple, an agent which encourages more varied behavior in its environment will lead to the potential for relative success to be possible in its neighborhood (following (i)), leading to destabilization of a small region.

CONSISTENCY OF MODEL. The two regimes over γ, are characterized by:

$$\gamma > \gamma_c \Rightarrow \text{ Agents have inconsistent models}$$
$$\gamma < \gamma_c \Rightarrow \text{ Agents have consistent models}$$

What does this actually mean? Consider an equivalence class of models with the same sophistication $s(\sigma)$. Any given choice of σ within that class reflects a *hypothesis of the character of the observed dynamics in the world.*

A simple analogy would be that a given equivalence class selects a fixed number, $s(\sigma)$, of fourier modes (or splines or...) to be used to model a given time series. The set of models within that equivalence class will be the set of all models which use exactly s independent fourier modes. Clearly the models within that class which utilize different fourier modes to describe a series, reflect wildly differing hypotheses about the dynamics of the underlying process generating the given time series.

So, for the former case, the dynamics of the world prevent information of agent's models from transferring through the subsidiary medium of generated behaviors. Agents will pick inconsistent models of the world. In the latter case, information transfer is achieved, and agents will pick consistent models.

So, in the high-difficulty regime, agents' hypotheses about the underlying dynamics of the world are mutually inconsistent (the dynamics of the world prevent information of agent's models from transferring through the subsidiary medium of generated behaviors). In the low-difficulty regime agents will actually select, on average, a single model from a given class (information transfer is achieved). Thus the agents will have a common hypothesis for the world's dynamics.

OBSERVATIONS. An estimate for the concept of difficulty of prediction, upon which the analysis is based, may be obtained from the observed predictive errors. Figure 10 shows how these vary over time, and a very clear correlation between time periods of high-difficulty and uncoordinated behavior can be discerned—compare with Figures 3 and 6.

Although we have, as yet, no statistical estimates, the transitions between coordinated/uncoordinated and small/large errors are very sudden. This lends credence to the concept of an underlying phase transition.

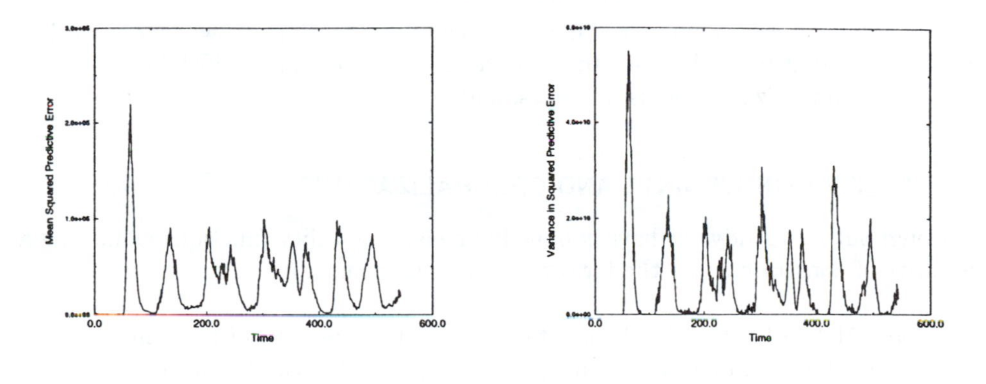

(a) Mean error

(b) Variance in error

FIGURE 10 Averaged predictive errors over a 20×20 world. These can be considered approximate measures of observed predictive difficulty. Errors are uniformly small or uniformly large.

5.2 STATICS AND DYNAMICS

The difficulty of prediction γ is analogous in this exposition to the temperature parameter in magnetic phase transitions, with one crucial difference. It is not a exogenous parameter—it is an endogenous observable, driven by the very dynamics it implicitly tunes. At this point we leave as an open theoretical question whether its emergent dynamics drive the system to the consistent or inconsistent regime, or to fluctuate between the two. The above analysis can only be considered a static stability analysis, and must be extended to derive dynamical results.

Let us first note, however, if models are simple we would expect γ to be small. Hence a large γ encourages a small $|\sigma|$ which in turn produces simpler dynamics and a smaller γ. Conversely, as γ becomes smaller, $\sigma \propto \sqrt{\gamma_c - \gamma}$ increases and so we expect γ to increase. These opposing forces will, therefore, push γ away from extremal values. This is in agreement with the argument presented in the introduction and with empirical observations—we fluctuate between an approximate rational expectations state (in which agents' implicit expectations of the world are in almost perfect agreement), and a highly disordered state.

6. DISCUSSION AND CONCLUSIONS

We shall first discuss a number of limitations and assumptions on which our work is based, and what might be done to alleviate some of the more artificial constraints. We then summarize our goals, progress, and results.

6.1 UTILITY CONSTRAINTS AND GENERALIZATIONS

Throughout this study, we have confined ourselves to utility functions exhibiting a number of constraints (for the forward predictive scenario):

HISTORICAL DEPENDENCY—first note that, given some level of random noise, all the results of our simulations are independent of the initial conditions. In particular, the initial distribution of history and complexity values has no influence upon the long-term dynamics. By independent we mean that the qualitative nature of the system's evolution is not dependent on initial conditions. In the absence of noise, one can find a few pathological initial conditions that are nongeneric (completely uniform starting values for history, complexity, for example).

CONVERGENCE—the distinctions between coordination, substitution, and varying the λ parameter have already been discussed.

CONTINUITY—all the games we use are smooth, such that a small difference in predictions causes only a small difference in behavior. If the space of behaviors (or those behaviors selected by the game) is discretized, then most of the observations and predictions we make are no longer valid. We have not investigated whether this can be alleviated via the choice of a predictive model more suited to discrete predictions.

VARIATION—when making predictions one can use single models or multiple models; predict the average or the individuals. We have already discussed the trade-off between model stability and system size as it effects the system's dynamics.

FREQUENCY—the trade-off between dynamicity of the system and the stabilizing forces of model selection and prediction can be altered by changing the frequency with which the agents can adjust their predictive models (i.e., only carrying out the predictive update step (v) every n th time-step). Our results are very robust against changes to this parameter: even if agents only adjust their model every, say, every 10 time steps, no difference is observed. This is further evidence (though no more than that) of the independence of our results of synchronous versus asynchronous modeling.

TOPOLOGY—the neighborhood size affects the rate of information propagation, and we have noted that system stability increases with neighborhood size (for a fixed number of agents). The type of local neighborhood is not important. If each agent's neighbors are given by a fixed random assignment from the set of N, the results are identical. This accords with our mean-field analysis.

6.2 COOPERATIVE VS. COMPETITIVE GAMES

A significant limitation to our work is the following: the dynamics we have observed in these predictive systems are limited to games which are cooperative in the sense that the best response for a given agent is not excessively bad for its neighbors. Some counter-examples such as the "unstable substitution game" seem of little importance since we understand quite clearly why they are unstable. However, other *nonlinear* games such as the undercut game (in which each agent tries to pick as a behavior a number just a little bit smaller than its neighbors, unless they pick very small numbers in which case it is desirable to pick a very large one) seem to require a more sophisticated modeling and strategic approach than we provide. We have not investigated whether our results can be extended in these directions via a more competitive iterated system, in which multistep forward planning, is incorporated, and nonpure strategies are allowed. Procedures like this allow cooperation to become reasonably stable in the iterated Prisoners' Dilemma, for example.

A direct extrapolation of our results suggests the intuition that such a forward-planning horizon would be evolutionarily bounded in the same manner in which history and complexity are in our current systems.

6.3 PREDICTIVE REGIME CHANGES

Examining the evolution of behaviors through time, as in Figure 3, and in connection with the theoretical results, it is clear that no single, simple model should try to analyze and predict the coordinated and uncoordinated regimes together. A model that seeks to perform better should categorize the data from the past into blocks from each regime. It should use just the data from coordinated periods for predictions when the world is reasonably stable, and the remaining data for the uncoordinated periods. This requires the identification and selection of noncontiguous blocks of data, and hence, a much more sophisticated predictive approach. These kinds of value-judgments must be made when modeling real-world economic/market data—in a healthy economic climate, data affected by political turbulence, depressions, and market crashes will not be used for basic modeling.

Our current research makes no attempt to address these issues. It is an interesting question whether that type of modeling would in fact stabilize our systems.

6.4 THE ACCUMULATION OF PAYOFFS

Ordinarily one associates a payoff function with the gain of something of value (money, energy), which one would expect in any economic market or system of biological survival. However, these predictive systems operate solely via the use of a relative payoff, in which the only use of the payoff is in deciding which model should be used subsequently. The reason for this is that in our models there is no place for the accumulation of payoffs; a more complex class of model would allow the introduction of such a scheme. In particular, it will be possible to study the dynamics of price formation in market systems, an area in which economic theory is notoriously unsuccessful. Work in progress is beginning to achieve an understanding of how predictive agents interact to form prices.

Other extensions to this work, upon which we do not report in detail in this paper, are the following: (i) agents use a probability distribution over an entire space of models rather than a single model, and update that distribution (using a Bayesian rule) according to model success, creating a probability flow over model space. Our results there show that all agents evolve to attach nonzero weight to only a small, finite fraction of the space. In this sense, model *selection* is also bounded; (ii) extensions to Bray's model in which single or multiple agents imitate, learn from, and model each other in simple environments; (iii) forward planning; (iv) hierarchies of models.

6.5 CONCLUSIONS

It has long been observed (for a good discussion, see Anderson[1]) that qualitatively different emergent phenomena arise in large systems, usually accompanied by macroscopic events and correlations brought about by the *accumulation* of what are essentially simple, local microscopic dynamics. The spirit of this paper is that such collective behavior can be harnessed in a dynamic theory of economic systems. Furthermore, we believe that interesting global, dynamical phenomena can arise out of systems in which the agents are relatively simple, with homogenous behavior patterns based upon simple predictive rules. This is no simple case of "complexity begets complexity"; we do not demonstrate that any given complex phenomenon can be recreated with complex, heterogenous underlying rules, rather we wish to show that, given a reasonable, simple class of agents, *they* can generate a wide class of global, punctuated, metastable phenomena.

So, as a first step in that direction, we selected a class of predictive systems, $\mathcal{P}$, designed to relax rational expectations and stationarity assumptions toward a more natural unconstrained dynamic, in which agents' beliefs are allowed to evolve, and hence, select the most advantageous state. The systems were chosen to be as simple as possible, without exogenous shocks, but the games and rules were selected so as to bring about a large amount of *feedback* in the manner in which the interactions accumulate. It is because an individual's environment is no more than its neighbors, that changes in its behavior may cause changes in its neighbors' behaviors which will then both propagate and feedback, generating an interesting dynamic.

6.5.1 NATURAL RATIONALITY. We based our systems, and choice of agents, upon the need to investigate assumptions and questions of local interactions, coordination, optimal use of information, and the resulting dynamics of rationality. We gave heuristic arguments for why dynamics of this form will cause the system to drive itself toward interesting nonstationary dynamic equilibria, intermediate between conceivable ordered and disordered regimes in its character, separated by a critical point.

A mean-field analysis of our systems shows very strong agreement with the qualitative nature of the simulated system's dynamics. There are two regimes: coordinated and uncoordinated, and the system is driven from one to another according to the level of difficulty of prediction in the world (although the mean-field model does not predict the fluctuations in detail), under a self-organizing, endogenous dynamic.

Hence, the system cycles between selecting for a static rational expectations state and breaking down into a disordered state. We studied games with both an infinite number of Nash equilibria and a single equilibrium, expecting that a single equilibrium would stabilize the system so that persistent diversity would not be observed. However, this was shown to be wrong. Punctuated equilibrium may be observed for very long transient periods even in systems with such a single fixed point. Under the influence of noise, such transients can become the predominant dynamic.

The manner in which these dynamics pass from coordinated phases to disordered, uncoordinated bubbles of dynamics is very reminiscent of real-world market phenomena. In that sense, our results show that as systems of autonomous agents grow in size, periods of highly disordered activity (stock market crashes or bubbles?) are to be *expected*, and are not an anomaly. These results hold even in the absence of noise or exogenous perturbations.[9]

There are some connections which can be drawn with business cycles and the exploitation of market niches. For example, if we consider the static state to represent a market in which all the firms have the same strategy, then a firm that offers a slightly different kind of product/service may well be able to exploit the rest of the system's unsophistication. However, our results suggest that such exploitation can *destabilize* the market, leading to a highly disordered regime (bankruptcies, etc.) from which a new standard emerges. It could be argued that the beginnings of such a process can be observed in the airline industry. On a rather different scale, the nature of the "wheel of retailing" (in which firms evolve from innovative, small operations through a more sophisticated growth phase and finally become a larger, stabler more conservative retailer, opening a niche for yet another low-cost, innovative operation to enter the market) could perhaps be analyzed via the same techniques (see McNair[14] and Brown[8] for elaboration on this "wheel").

In summary, our results demonstrate that agents evolve toward a state in which they use a limited selection of models of intermediate history and complexity, unless subjected to a dominant equilibrium when they learn to use simple models. It is *not* advantageous to pick models of ever increasing sophistication, nor to attach nonzero weight to all available information. As such, our agents in their mutual creation of an economic world, do not evolve toward a static rational expectations equilibrium, nor to what one would naïvely consider a perfectly rational state. Rather in their coevolutionary creation of a nonstationary world, they evolve to use a level of rationality which is strictly bounded, constrained within a finite bubble of complexity, information use, model selection. As they fluctuate within those constraints, these agents can still achieve system-wide coordination, but the coordinated states are metastable. The dynamics are driven by the need to create models of greater/lesser precision to match the world's observed dynamics.

We interpret this state as an extension of rationality away from the simple static case to encompass the *natural rationality of nonstationary worlds.* Thus the natural (evolutionarily advantageous) rationality of these systems is a state in which information usage, model sophistication, and the selection of models are all bounded. It is in that sense that the natural rationality of our agents is bounded. Adaptive agents evolve to such a state, mutually creating a system which fluctuates between metastable coordinated and uncoordinated dynamics. This is the first step to understanding what agents *must* do in systems in which their mutual interactions

[9] At a very low level there is some noise caused by rounding and other numerical phenomena.

coevolve to create a nonstationary world, and is, therefore, the first step to an understanding of *homo economicus* in such systems.

7. ACKNOWLEDGMENTS

This research was undertaken with support from Harvard University and the Santa Fe Institute (with core funding from the John D. and Catherine T. MacArthur Foundation, the National Science Foundation, and the U.S. Department of Energy).

This research has benefited enormously from discussions and time at the Santa Fe Institute, in particular with Steve Durlauf, and also with Eric Maskin at Harvard University, and Maja Matarić at the Volen Center for Complex Systems, Brandeis University. Thanks are also due to Ann Bell, James Nicholson, Jan Rivkin, Dave Kane, Phil Auerswald, and Bill Macready.

REFERENCES

1. Anderson, P. W. "More Is Different." *Science* **177(4047)** (1972): 393–396.
2. Arifovic, J. *Learning by Genetic Algorithms in Economic Environments.* Ph.D. Thesis, University of Chicago, 1991.
3. Bak, P., C. Tang, and K. Wiesenfeld. "Self-Organized Criticality: An Explanation of 1/f Noise." *Phys. Rev. Lett.* **59(4)** (1987): 381–384.
4. Barnett, W. A., J. Geweke, and K. Shell, eds. *International Symposium in Economic Theory and Econometrics* (4th: 1987: Austin, TX.), International Symposia in Economic Theory and Econometrics. Cambridge, MA: Cambridge University Press, 1989.
5. Blume, L. "The Statistical Mechanics of Strategic Interaction." *Games & Econ. Behav.* **5** (1993): 387–424.
6. Bray, M. M. "Learning, Estimation, and Stability of Rational Expectations." *J. Econ. Theor.* **26** (1982): 318–339.
7. Brock, W., and S. Durlauf. "Discrete Choice with Social Interactions I: Theory." Technical Report, University of Wisconsin, 1995.
8. Brown, S. "The Wheel of the Wheel of Retailing." *Intl. J. Retail* **3(1)** (1988): 16–37.
9. Darley, Vincent M. "Towards a Theory of Optimising, Autonomous Agents." Ph.D. Thesis, (1997): forthcoming.
10. Ellison, G. "Learning, Local Interaction, and Coordination." *Econometrica* **61(5)** (1993): 1047–1071.
11. Kauffman, Stuart A. *The Origins of Order: Self-Organization and Selection in Evolution.* Oxford: Oxford University Press, 1993.
12. Kreps, D. M. *Game Theory and Economic Modelling.* Oxford: Clarendon Press; New York: Oxford University Press, 1990.
13. Lane, D. "Artificial Worlds in Economics. Parts 1 and 2. *J. Evol. Econ.* **3** (1993): 89–108, 177–197.
14. McNair, M. P. "Significant Trends and Developments in the Post-War Period." In *Competitive Distribution in a Free High Level Economy and Its Implications for the University*, edited by A. B. Smith, 1–25. University of Pittsburgh Press, 1958.
15. Nash, J. *Non-Cooperative Games.* Ph.D. Thesis, Mathematics Department, Princeton University, 1950.
16. Press, W. H., S. A. Teukolsky, W. T. Vetterling, and B. F. Flannery. *Numerical Recipes in C*, 2nd ed. Cambridge: Cambridge Unviersity Press, 1992.
17. Sargent, T. *Bounded Rationality in Macroeconomics.* Oxford: Clarendon Press, 1993.
18. Schelling, T. "Dynamic Models of Segregation." *J. Math. Sociol.* **1** (1971): 143–186.

Steven N. Durlauf
Department of Economics, University of Wisconsin at Madison, Madison, WI 53706; Santa Fe Institute, 1399 Hyde Park Road, Santa Fe, NM 87501

Statistical Mechanics Approaches to Socioeconomic Behavior

This chapter provides a unified framework for interpreting a wide range of interactions models which have appeared in the economics literature. A formalization taken from the statistical mechanics literature is shown to encompass a number of socioeconomic phenomena ranging from out-of-wedlock births to aggregate output to crime. The framework bears a close relationship to econometric models of discrete choice and, therefore, holds the potential for rendering interactions models estimable. A number of new applications of statistical mechanics to socioeconomic problems are suggested.

1. INTRODUCTION

This chapter is designed to provide a unified discussion of the use of statistical mechanics methods[1] in the study of socioeconomic behavior. The use of these methods in the social sciences is still in its infancy. Nevertheless, a growing body of work has

[1]The statistical mechanics models I employ are also referred to as interacting particle system or random fields models.

shown how statistical mechanics and related probability techniques may be used to study the evolution and steady state behavior of heterogeneous populations. Examples of the range of applications of statistical mechanics methods include social pathologies such as out-of-wedlock births and crime (Brock and Durlauf,[20] Glaeser, Sacerdote, and Scheinkman[34]), asset price behavior (Brock[17,18]), expectation formation (Brock and Hommes[21]), business cycles (Bak et al.[7] and Durlauf[25,27]), technology adoption (An and Kiefer[1]), and endogenous preferences (Bell[11]). In addition, Blume[15,16] has shown how these methods can provide insight into the structure of abstract game-theoretic environments.[2]

These disparate phenomena are linked by the possibility that each is determined, at least partially, by direct interactions between economic actors. Put differently, each of these phenomena is a case where the decisions of each individual are influenced by the choices of others with whom he interacts. This interdependence leads to the possibility that polarized behavior can occur at an aggregate level solely due to the collective interdependence in decision-making. This explanation of polarized group behavior may be contrasted with explanations which rely on the presence of highly correlated characteristics among members of a group.

Interactions between economic actors are commonplace in economic models. However, these interactions are typically mediated through markets. What distinguishes the bulk of the interactions in the recent inequality literature is the focus on interdependencies which are direct.[3] Examples of such direct interactions are role model, social norm, and peer group effects.

At first glance, statistical mechanics methods, which underlie the theory of condensed matter, would appear to have little to do with socioeconomic phenomena related to inequality. However, strong metaphorical similarities exist between the two fields of research. The canonical question in statistical mechanics concerns the determinants of magnetization in matter. As a magnetized piece of matter is one in which a substantial majority of the atoms share a common spin (which can either be up or down), magnetization would appear to be an extremely unlikely phenomenon, as it would require the coincidence of many atoms sharing a common property. However, if the probability that one atom has a particular spin is a function of the spins of surrounding atoms, the possibility of collective interdependence renders magnetism understandable. As techniques for the study of economic phenomena, statistical mechanics approaches have proven valuable for studying the

[2] Alternative approaches to the modeling of complex interaction environments include Arthur,[4,5] Ioannides,[35] Kirman,[37] and Krugman.[39]

[3] While markets may not exist to directly mediate these interactions, this does not imply that economic actors do not alter their behavior in order to account for them. For example, as discussed in Bénabou[12,13] and Durlauf,[27,28] the presence of within-neighborhood interactions can play a primary role in determining the composition of neighborhoods.

aggregate behavior of populations facing interdependent binary choices. In particular, statistical mechanics methods hold the promise of providing a general framework for understanding how collective interdependence can lead to the emergence of interesting and rich aggregate behavior.[4]

Of course, any metaphorical similarity between physical and social models of interdependent behavior is of little interest, unless the specific substantive models underlying each can be shown to have similar structures. An important goal of this chapter is to show how statistical mechanics structures naturally arise in a number of socioeconomic environments. Also, it is important to recognize that the potential for interesting aggregate behavior to emerge from individual decisions has been explored in previous economic contexts. Two prominent examples include Becker's[8] work on aggregate demand with zero-intelligence agents and Schelling's[46] analysis of racial segregation.[5] Statistical mechanics approaches should thus be regarded as complementary to disparate strands of previous work.

The rest of this chapter is organized as follows. Section 2 outlines some general issues in modeling binary choices with interactions. Section 3 analyzes binary choice models with global interaction structures. Section 4 analyzes binary choice models with local interactions. Section 5 provides a discussion of limitations and outstanding questions which arise in statistical mechanics models of social behavior. Section 6 contains summary and conclusions.

2. GENERAL CONSIDERATIONS

The statistical mechanics-inspired models of socioeconomic phenomena that typically have been studied focus on environments in which each individual of a group faces a binary choice. Many individual decisions which are relevant to understanding inequality are binary in nature. Standard examples include decisions to have a child out of wedlock, drop out of school, commit a crime. Not only are these decisions of interest in their own right, but they are well-known to affect a broad range of individual socioeconomic outcomes over large time horizons. While the importance of these binary decisions in cross-section and intertemporal inequality is beyond dispute, there is considerable controversy over the role of interactions, and so one purpose of the particular formulation I have chosen is to develop interaction-based models in a way which makes contact with the econometric literature on discrete choice. In addition, this approach provides a method for exploring the interconnections between different approaches to modeling interactions. Brock[17] and Blume[15]

[4] See Crutchfield[24] for a discussion of the meaning of emergent phenomena.

[5] Schelling's model possesses a structure that is quite similar to some of the statistical mechanics models which are discussed below.

originally recognized the connection between discrete choice and statistical mechanics models. The current development follows Brock and Durlauf.[19]

Binary decisions of this type may be formalized as follows. Consider a population of I individuals. Individual i chooses ω_i, whose support is $\{-1, 1\}$. The vector consisting of the choices of all agents in the population is $\underset{\sim}{\omega}$ and the vector of all decisions other than that of agent i is $\underset{\sim}{\omega}_{-i}$.

Individuals are heterogeneous in three different respects. First, agents differ with respect to personal attributes which are characterized by the vector $\underset{\sim}{X}_i$. This vector can include elements ranging from family background to community environment to past behavior. Second, individuals possess distinct expectations concerning the behavior of the population as a whole. Specifically, each agent is associated with a conditional probability measure $\mu_i^e(\underset{\sim}{\omega}_{-i})$. Third, each agent experiences a pair of unobservable random shocks, $\epsilon(\omega_i)$, which influence the payoff of each of the possible choices. The shock $\epsilon(1)$ is distinguished from $\epsilon(-1)$, as certain types of random shocks are only relevant for one of the choices. For example, $\epsilon(1)$ might refer to a shock which reflects mathematical talent and so is only relevant if the person stays in school. Similarly, if $\epsilon(-1)$ is an innovation to the sensitivity of one's fingertips, a virtue in safecracking, then the shock only affects the payoff to dropping out and becoming a criminal.

Taken together, an individual's choice problem may be specified as

$$\max_{\omega_i \in \{-1,1\}} V(\omega_i, \underset{\sim}{X}_i, \mu_i^e(\underset{\sim}{\omega}_{-i}), \epsilon(\omega_i)). \tag{1}$$

At this level of generality, of course, virtually nothing can be said about the properties either of the individual decisions or about the behavior of the population as a whole. Two standard restrictions have been made on the general decision problem Eq. (1) to permit explicit analysis.

First, the decision problem is assumed to be additive in three components,

$$V(\omega_i, \underset{\sim}{X}_i, \mu_i^e(\underset{\sim}{\omega}_{-i}), \epsilon(\omega_i)) = u(\omega_i, \underset{\sim}{X}_i) + S(\omega_i, \underset{\sim}{X}_i, \mu_i^e(\underset{\sim}{\omega}_{-i})) + \epsilon(\omega_i). \tag{2}$$

In this specification, $u(\omega_i, \underset{\sim}{X}_i)$ represents deterministic private utility, $S(\omega_i, \underset{\sim}{X}_i, \mu_i^e(\underset{\sim}{\omega}_{-i}))$ represents deterministic social utility, and $\epsilon(\omega_i)$ represents random private utility. The two private utility components are standard in the economics of discrete choice. Recent work is distinguished by the introduction of social utility considerations.

Second, the form of the social utility and the form of the probability density characterizing random private utility are generally given particular functional forms. (It will become clear that restricting the form of the private utility term has no qualitative effect on the properties of the aggregate population.) The social utility component is formalized by exploiting the intuition that individuals seek to conform in some way to the behavior of others in the population. Formally, a specification which subsumes many specific models may be written as

$$S(\omega_i, \underset{\sim}{X}_i, \mu_i^e(\underset{\sim}{\omega}_{-i})) = -E_i \sum_{j \neq i} \frac{J_j(\underset{\sim}{X}_i)}{2} (\omega_i - \omega_j)^2 . \tag{3}$$

$E_i(\cdot)$ represents the conditional expectation operator associated with agent i's beliefs. The term $J_j(\underset{\sim}{X}_i)/2$ represents the interaction weight which relates i's choice to j's choice and is typically assumed to be nonnegative. Note that this specification can accommodate any assumptions concerning who interacts with whom.[6]

The random utility terms are assumed to be extreme-value distributed, so that

$$\text{Prob}(\epsilon(-1) - \epsilon(1) \leq z) = \frac{1}{1 + \exp(-\beta(\underset{\sim}{X}_i) z)} ; \beta(\underset{\sim}{X}_i) \geq 0 . \tag{4}$$

Anderson, Thisse, and dePalma[2] provide a nice survey of the logistic density and its relationship to interpretations of the random payoff terms. The random terms are assumed to be independent across agents.

These assumptions permit the model to be manipulated into a more transparent form. First, observe that since the support of ω is $\{-1, 1\}$, the first term in the expression can be replaced with $h(\underset{\sim}{X}_i)\omega_i + k(\underset{\sim}{X}_i)$ so long as the functions $h(\underset{\sim}{X}_i)$ and $k(\underset{\sim}{X}_i)$ obey

$$h(\underset{\sim}{X}_i) + k(\underset{\sim}{X}_i) = u(1, \underset{\sim}{X}_i) \tag{5}$$

and

$$-h(\underset{\sim}{X}_i) + k(\underset{\sim}{X}_i) = u(-1, \underset{\sim}{X}_i) \tag{6}$$

since the linear function coincides with the original function on the support of the individual choices.

Second, since $\omega^2 = 1$, social utility can be rewritten as

$$S(\omega_i, \underset{\sim}{X}_i, \mu_i^e(\underset{\sim}{\omega}_{-i})) = \sum_{j \neq i} J_j(\underset{\sim}{X}_i) \cdot (\omega_i E_i(\omega_j) - 1) , \tag{7}$$

which means that social utility is linear in the expected values of the choices in one's reference group.

Now, since the probability of an individual's choice, conditional on his characteristics and expectations, is

$$\begin{gathered} \text{Prob}(\omega_i | \underset{\sim}{X}_i, \mu_i^e(\underset{\sim}{\omega}_{-1})) = \\ \text{Prob}(V(\omega_i, \underset{\sim}{X}_i, \mu_i^e(\underset{\sim}{\omega}_{-i}), \epsilon(\omega_i)) > V(-\omega_i, \underset{\sim}{X}_i, \mu_i^e(\underset{\sim}{\omega}_{-i}), \epsilon(-\omega_i))) , \end{gathered} \tag{8}$$

[6] Of course, $J_j(\underset{\sim}{X}_i)$ could be further generalized. For example, bilateral interactions might depend on the characteristics of both agents, producing interaction weights of the form $J_j(\underset{\sim}{X}_i, \underset{\sim}{X}_j)$.

substituting Eqs. (5), (6), and (7) into Eq. (8) implies

$$\text{Prob}\,(\omega_i|\underset{\sim}{X}_i,\mu_i^e(\underset{\sim}{\omega}_{-i}))\sim\exp\left(\beta(\underset{\sim}{X}_i)h(\underset{\sim}{X}_i)\omega_i+\sum_{j\neq i}\beta(\underset{\sim}{X}_i)J_j(\underset{\sim}{X}_i)\cdot\omega_i E_i(\omega_j)\right).\tag{9}$$

The $k(\underset{\sim}{X}_i)$ terms do not appear in this expression, as they are irrelevant to the comparison of utilities which drive individual choices.

Finally, observe that each of the individual choices is independent once one has conditioned on the set of individual-specific expectations. Hence,

$$\begin{aligned}\text{Prob}\,(\underset{\sim}{\omega}|\underset{\sim}{X}_1,\ldots,\underset{\sim}{X}_I,\mu_1^e(\underset{\sim}{\omega}_{-1}),\ldots,\mu_I^e(\underset{\sim}{\omega}_{-I}))&=\prod_i\text{Prob}(\omega_i|\underset{\sim}{X}_i,\mu_i^e(\underset{\sim}{\omega}_{-i}))\\&\sim\prod_i\exp\left(\beta(\underset{\sim}{X}_i)h(\underset{\sim}{X}_i)\omega_i+\sum_{j\neq i}\beta(\underset{\sim}{X}_i)J_j(\underset{\sim}{X}_i)\cdot\omega_i E_i(\omega_j)\right)\end{aligned}\tag{10}$$

Equation (10) provides a general form for the joint probability measure of individual choices. It has the general form of a Gibbs measure, which is not coincidental. A deep theorem in the statistical mechanics literature, due to Averintsev[6] and Spitzer,[49] is that models of stochastic interactions of the type which have been outlined will generically possess probability measures with Gibbs representations.

Different specializations of the functional forms in these expressions will yield many of the particular statistical mechanics models which have been studied in the literature. These different functional forms differ substantively with respect to the nature of the interaction structure that connects individual decisions.

3. GLOBAL INTERACTIONS

A first class of models has focused on environments in which each individual interacts symmetrically with all other members of the population. I focus on the model studied by Brock and Durlauf[19]; a related example is Aoki.[3]

In this model, each individual is assumed to derive utility from conforming with the average behavior of his reference group. Operationally, this occurs for the special case (relative to the general specification of bilateral interactions)

$$J_j(\underset{\sim}{X}_i)=\frac{J(\underset{\sim}{X}_i)}{I}\tag{11}$$

so that

$$S(\omega_i,\underset{\sim}{X}_i,\mu_i^e(\underset{\sim}{\omega}_{-i}))=J(\underset{\sim}{X}_i)\cdot(\omega_i m_i^e-1)\tag{12}$$

where m_i^e is individual i's expectation of the average population choice level.

Under these assumptions, the joint probability measure characterizing all agents choices obeys

$$\text{Prob}(\underset{\sim}{\omega} \mid \underset{\sim}{X}_1, \ldots, \underset{\sim}{X}_I, \mu_1^e(\underset{\sim}{\omega}_{-1}), \ldots, \mu_I^e(\underset{\sim}{\omega}_{-I})) \sim \prod_i \exp(\beta(\underset{\sim}{X}_i) h(\underset{\sim}{X}_i)\omega_i + \beta(\underset{\sim}{X}_i) J(\underset{\sim}{X}_i) \cdot \omega_i m_i^e) \cdot \tag{13}$$

The large sample behavior of the average choice level in this economy may be analyzed as follows. First, assume that all agents share a common expectation of the average choice level, i.e.,

$$m_i^e = m^e \forall i \,. \tag{14}$$

Second, let $dF_{\underset{\sim}{X}}$ denote the limit of the sequence of empirical probability density functions associated with individual characteristics $\underset{\sim}{X}_i$, where the limit is taken with respect to the population size I. (I assume that such a limit exists.) The strong law of large numbers implies, for any common expected mean, that the sample mean $\bar{m}_I$ of population choices converges with a limit equal to

$$\lim_{I \Rightarrow \infty} \bar{m}_I = \int \tanh(\beta(\underset{\sim}{X}) h(\underset{\sim}{X}) + \beta(\underset{\sim}{X}) J(\underset{\sim}{X}) m^e) dF_{\underset{\sim}{X}} \,. \tag{15}$$

The model is closed by imposing self-consistency in the large economy limit, so that the limit of the sample mean corresponds to the common expected average choice level. A self-consistent equilibrium mean, m^*, is any root of

$$m^* = \int \tanh(\beta(\underset{\sim}{X}) h(\underset{\sim}{X}) + \beta(\underset{\sim}{X}) J(\underset{\sim}{X}) m^*) dF_{\underset{\sim}{X}} \,. \tag{16}$$

When all agents are identical, in that they are associated with the same $\underset{\sim}{X}$, the mean-choice level in the economy will correspond to the roots of

$$m^* = \tanh(\beta h + \beta J m^*) \,. \tag{17}$$

In this special case, the model corresponds to the mean-field approximation of the Curie-Weiss model. Brock's paper[17] is the first instance in which the Curie-Weiss model was given an economic interpretation; the current formulation differs in emphasizing the equivalence between the mean-field approximation of the model and the assumption of a noncooperative interaction environment. The following theorem, taken from Brock and Durlauf,[19] characterizes the number of self-consistent steady states.

THEOREM. 1. Existence of multiple average choice levels in equilibrium.

i. If $\beta J > 1$ and $h = 0$, there exist three roots to Eq. (17). One of these roots is positive, one root is zero, and one root is negative.

ii. If $\beta J > 1$ and $h \neq 0$, there exists a threshold H, (which depends on β and J) such that

   a. for $|\beta h| < H$ there exist three roots to Eq. (17) one of which has the same sign as h, and the others possessing the opposite sign.

   b. for $|\beta h| > H$ there exists a unique root to Eq. (17) with the same sign as h.

Notice that the model exhibits nonlinear behavior with respect to both the parameters βh and βJ. This makes sense intuitively. Conditional on a given private utility difference between the choices 1 and -1, which equals $2h$, there is a level that the conformity effect βJ must reach in order to produce multiple self-consistent mean-choice behavior. Recall that the random utility shocks are i.i.d., so that in absence of the conformity effects, there would be a unique mean whose sign is the same as h. Conditional on a conformity effect $\beta J > 1$, as βh increases in magnitude, any multiplicity will eventually be eliminated. This occurs because eventually the private utility differential between the choices will overcome any tendency for the conformity effect to produce a self-consistent mean with the opposite sign.

This type of model illustrates the complementary nature of the roles of economic fundamentals and social norms in explaining the degree of social pathologies in different neighborhoods. For example, Theorem 1 states that high degrees of conformity can lead to mean-choice levels opposite to that dictated by private utility. To be concrete, even if economic fundamentals, as embodied in h, imply that the average teenager should stay in school, conformity effects can produce an equilibrium in which most teenagers drop out.

It is straightforward to show (Brock and Durlauf[19]) that the equilibrium in which the sign of h is the same as the sign of the mean-choice level produces higher average utility than the equilibrium in which the signs are opposite. Hence, this model illustrates the potential for collectively undesirable behavior, such as high out-of-wedlock birth rates, which is individually optimal. This ranking of average utility provides the appropriate stochastic generalization to the Pareto rankings of equilibria studied in Cooper and John.[23]

4. LOCAL INTERACTIONS

An alternative approach in the study of interactions has focused on the aggregate implications of local interactions. Such models assume that each individual interacts with a strict subset of others in the population. For individual i, this subset is n_i and is referred to as the individual's neighborhood. Hence, for agent i,

$$J_j(\underset{\sim}{X}_i) = 0 \text{ if } j \notin n_i . \tag{18}$$

Typically, the index i has been interpreted so that $|i-j|$ measures distance between individuals. This allows one to construct a neighborhood for agent i by taking all agents within some fixed distance from i. (The distance can vary with direction.) This latter assumption is what renders the interactions local.

I. ENDOGENOUS PREFERENCES

Föllmer[33] introduced statistical mechanics methods to economics by considering an economy with locally interdependent preferences. Specifically, he wished to understand under what circumstances randomness in individual preferences will fail to disappear in the aggregate. Agents in the model are arrayed on the two-dimensional integer lattice Z^2 and possess one of two possible utility functions denoted as U_1 and U_{-1} respectively; agent i's preferences are coded by ω_i so that the first preference type corresponds to $\omega_i = 1$ and the second type to $\omega_i = -1$. The probability that an agent has one utility function is an increasing function of the number of neighbors with the same preferences. In particular, Föllmer assumes that the probability density characterizing the preferences of agents is of the form of the Ising model of statistical mechanics (see Liggett[40] or Spitzer[50]), which means that

$$\begin{gathered}\text{Prob}(\omega_i|\omega_j \forall j \neq i) = \\ \text{Prob}(\omega_i|\omega_j \forall j \text{ such that } |i-j| = 1) \sim \exp\left(J \sum_{|i-j|=1} \omega_i\omega_j\right)\end{gathered} \tag{19}$$

so that the joint probability measure over agent preferences is

$$\text{Prob}(\underset{\sim}{\omega}) \sim \exp\left(J \sum_i \sum_{|i-j|=1} \omega_i\omega_j\right) . \tag{20}$$

As is well-known, there exists a critical value J_c such that if $J < J_c$, then the sample mean produced by a realization of this system will converge to zero (almost surely) when I is infinite; whereas if $J > J_c$, then the sample mean from a realization of this economy will converge (when averaged over larger and larger finite squares each of whose center is a particular agent on the lattice) to one of two possible nonzero values. Hence, Föllmer concluded that for his preference structure, idiosyncratic shocks may have aggregate consequences.

Föllmer's model of endogenous preferences does not directly translate into the discrete choice problem formulated in section 2. The reason for this is that he places a conditional probability structure on agent characteristics in order to study interdependence; in fact there is no choice component to his model.[7] However, there exist two ways in which one can modify Föllmer's model so as both to render its probability structure interpretable within a discrete choice formulation. In each case, one reinterprets the ω_i's as individual choices (of some commodity, for example) in which the utility of a particular choice depends on whether one's neighbors have made the same choice. Further, the model must be made explicitly intertemporal. This intertemporal interpretation of preference interdependence has the feature that it closely preserves the structure of neoclassical utility theory without losing any of Föllmer's qualitative insights; see Becker and Stigler[10] for a defense of this approach to generalizing neoclassical preferences. For clarity of exposition, I reindex relevant variables by time where appropriate.

In one possible modification of Föllmer's model, one moves to a discrete time setting and assumes that the utility of a particular choice is a function of whether others in an agent's neighborhood have made the same choice the previous period.[8] To reformulate Föllmer's model as a discrete choice problem in this fashion, this means that

$$J_j(\underset{\sim}{X}_{i,t}) = 0\,, \tag{21}$$

which eliminates any contemporaneous interactions. Second, by assuming that private utility depends on the past behavior of one's nearest neighbors in a way such that the total number of neighbors making a particular choice at $t-1$ is all that matters with respect to utility at t, the deterministic private utility weight can be expressed as

$$h(\underset{\sim}{X}_{i,t}) = h \cdot \sum_{|i-j|=1} \omega_{j,t-1}\,. \tag{22}$$

This embeds all local interactions in the private utility component. Third, set $\beta(\underset{\sim}{X}_{i,t}) = \beta$, which may be done without loss of generality. Together, this implies that

$$\text{Prob}(\omega_{i,t}|\omega_{j,t-1}\forall j \text{ such that } |i-j|=1) \sim \exp\left(\beta h \sum_{|i-j|=1} \omega_{i,t}\omega_{j,t-1}\right) \tag{23}$$

[7]In fact, I am unaware of any way to explicitly formulate a set of individual decision problems in a noncooperative environment such that Föllmer's conditional probability structure emerges as an equilibrium.

[8]This way of formulating interdependent preferences is strongly defended by Becker.[9]

and

$$\text{Prob}(\underset{\sim}{\omega}_t) \sim \exp\left(\beta h \sum_i \sum_{|i-j|=1} \omega_{i,t}\omega_{j,t-1}\right). \tag{24}$$

The role of idiosyncratic preference shocks in aggregate outcomes, can be reformulated for this model as follows. Do all initial configurations of choices $\underset{\sim}{\omega}_0$ produce the same limiting behavior for the average choice level m_∞? The answer parallels Föllmer's original case. There exists an h_c such that if $h < h_c$, then the mean choice is zero, whereas if $h > h_c$, the mean choice will converge to one of two nonzero values. The probability of converging to any one of these values will depend on $\underset{\sim}{\omega}_0$.

While providing qualitatively similar features to the original Föllmer formulation, the absence of any contemporaneous interactions does create some differences. In particular, the h_c in the dynamic model does not equal the J_c in the static model. This occurs because of the absence of any contemporaneous interactions. An exact equivalence between the static and dynamic models (with respect to parameter values and invariant measures) can be achieved through the following continuous time formulation. Suppose that agents adjust asynchronously in time. In particular, each agent is associated with a separate Poisson process such that at each arrival time for his process, the agent's choice is such that he takes the choices of the nearest neighbors as given. As the probability that any two agents choose at the same time is zero even in the infinite population limit. This model will generate multiple regimes, with $J_c = h_c$.[9]

Föllmer's model illustrates that it is difficult to provide a straightforward interpretation of contemporaneous local interactions in which the interdependence occurs with respect to choices. Alternatively, once can interpret this discussion as demonstrating the sensitivity of local interactions models to the assumptions placed on expectations. To see this, consider a local interaction formulation of social utility which preserves the Ising interaction structure,

$$S(\omega_i, \underset{\sim}{X}_i, \mu_i^e(\underset{\sim}{\omega}_{-i})) = J \sum_{|i-j|=1} E_i(\omega_j). \tag{25}$$

Unlike the global interactions model, it is not clear how to formulate individual expectations. At first glance, it might appear that since the analysis is dealing with nearest neighbors, expectations should exhibit perfect foresight with probability 1, i.e.,

$$E_i(\omega_j) = \omega_j. \tag{26}$$

However, this assumption will not lead to the conditional probability structure (10) for individual choices. The reason is simple. Under perfect foresight, the choices of individuals i and j will be interdependent whenever $|i-j| = 1$, and so each choice will be a function of both $\epsilon_i(\omega_i)$ and $\epsilon_j(\omega_j)$. Therefore, the equilibrium probability

[9] See Liggett[40] for a formal discussion.

will not be the product of a set of densities of i.i.d. extreme-value innovations. The Gibbs representation of the equilibrium probability measure will no longer be valid; no characterization of the equilibrium in this case is known.

An alternative assumption on expectations is that all agents assign the same expectations to each other

$$E_i(\omega_j) = E_k(\omega_l) \forall i, j, k, l \,. \tag{27}$$

In this case, the model can easily be seen to be observationally equivalent to the global interactions model that has already been examined, as Eq. (25) and Eq. (27) combine to yield

$$S(\omega_i, \underset{\sim}{X}_i, \mu_i^e(\underset{\sim}{\omega}_{-i})) = 4JE(\omega)\,, \tag{28}$$

implying that

$$\text{Prob}(\underset{\sim}{\omega}) \sim \exp(4J \sum_i \omega_i E(\omega))\,, \tag{29}$$

which is the same form as Eq. (13). This might appear odd, given the explicit local interaction structure of preferences. In fact, the equivalence is not surprising. When all expectations are identical, and the sample mean is required to equal the population mean, then agents globally interact through the expectations formation process.[10]

While much remains to be understood about these models, the import of this discussion is that intertemporal interactions are at the present time far more tractable than contemporaneous ones in general local interaction environments.

II. GROWTH AND ECONOMIC DEVELOPMENT

A second area where statistical mechanics approaches have been applied is that of cross-country inequality. Durlauf[26] constructs a model based on local technological interactions. A set of infinitely-lived industries is analyzed, each of which maximizes the discounted value of profits. Industries are assumed to be the aggregation of a large number of firms that act identically but noncooperatively. Letting $Y_{i,t}$ denote industry i's output at $t, K_{i,t}$ denote industry i's capital investment at t, and $\mathcal{F}_t$ denote information available at t, discounted expected profits will equal

$$\Pi_{i,t} = E\left(\sum_{j=0}^{\infty} \beta^{t+j}(Y_{i,t+j} - K_{i,t+j})|\mathcal{F}_t\right)\,. \tag{30}$$

[10]Notice that this version of the Ising model is equivalent to the mean-field version of the Curie-Weiss model, indicating how, in noncooperative environments with common expectation assumptions, very different interaction structures may be observationally equivalent.

Each industry can produce output using one of two production techniques. Technique choices are coded so that $\omega_{i,t} = 1$ if technique 1 is chosen, -1 if technique 2 is chosen. Capital fully depreciates after one period of use. Output is produced with a one-period lag, so that investment at t produces output available at $t+1$. The production functions at t depend on the technique choices at $t-1$. Together, these assumptions may be represented by a pair of production functions of the form

$$Y_{i,t+1} = f_1(K_{i,t} - F, \zeta_{i,t}\omega_{j,t-1} \forall j \in \Delta_{k,l}) \tag{31}$$

if technique 1 is chosen, or

$$Y_{i,t+1} = f_{-1}(K_{i,t}, \eta_{i,t}, \omega_{j,t-1} \forall j \in \Delta_{k,l}) \tag{32}$$

if technique 2 is chosen. F is a fixed capital cost which must be paid to employ technique 1 at t. $\zeta_{i,t}$ and $\eta_{i,t}$ are industry-specific productivity shocks, and are assumed to be i.i.d. across industries and time. The term $\Delta_{k,l}$ refers to the interaction range of an industry. For each industry i, $\Delta_{k,l} = \{i-j, \ldots, i+k\}$ so that the spillovers are all taken to be local: past technique choices influence the relative productivity of current techniques. The relative positions of industries with respect to the i index is interpreted as measuring technological similarity.

Finally, the relative productivity of technique 1 at t is enhanced by choices of technique 1 at $t-1$. If $\underset{\sim}{\omega}'$ and $\underset{\sim}{\omega}''$ denote two realizations of $\underset{\sim}{\omega}_{t-1}$ such that $\omega_j' \geq \omega_j'' \forall j \in \Delta_{k,l}$, then

$$\begin{aligned} f_1(K, \zeta_{i,t}, \omega_{j,t-1} &= \omega_j' \forall j \in \Delta_{k,l}) - \\ f_{-1}(K, \eta_{i,t}, \omega_{j,t-1} &= \omega_j' \forall j \in \Delta_{k,l}) \geq \\ f_1(K, \zeta_{i,t}, \omega_{j,t-1} &= \omega_j'' \forall j \in \Delta_{k,l}) - \\ f_{-1}(K, \eta_{i,t}, \omega_{j,t-1} &= \omega_j'' \forall j \in \Delta_{k,l}) \,. \end{aligned} \tag{33}$$

The long-run output dynamics of the model will thus depend on the evolution of technique choices.

The assumptions outlined are sufficient to show that the technique choices in the economy will obey the structure

$$\text{Prob}(\omega_{i,t} | \mathcal{F}_{t-1}) = \text{Prob}(\omega_{i,t} | \omega_{j,t-1} \forall j \in \Delta_{k,l}) \,, \tag{34}$$

which is similar to the dynamic variant of the Ising model discussed above.

Durlauf[26] examines the stability of the choice of technique 1 by assuming that

$$\text{Prob}(\omega_{i,t} = 1 | \omega_{j,t-1} = 1 \forall j \in \Delta_{k,l}) = 1 \,, \tag{35}$$

and studying the conditions under which $\underset{\sim}{\omega}_\infty = \underset{\sim}{1}$ is the unique invariant measure of the system. Using the following bounds on the system of conditional probabilities (34),

$$\Theta_{k,l}^{\min} \leq \text{Prob}(\omega_{i,t} = 1 | \omega_{j,t-1} = -1 \text{ for some } j \in \Delta_{k,l}) \leq \Theta_{k,l}^{\max} \tag{36}$$

the following result is proven.

THEOREM. 2. Uniqueness versus multiplicity of long-run equilibrium as a function of strength of complementarities

For each index set $\Delta_{k,l}$ with at least one of k or l nonzero, there exist numbers $\bar{\Theta}_{k,l}$ and $\underline{\Theta}_{k,l}$, $0 \leq \underline{\Theta}_{k,l} \leq \bar{\Theta}_{k,l} < 1$ such that

A. If $\Theta_{k,l}^{\min} \geq \bar{\Theta}_{k,l}$, then $\text{Prob}(\omega_{i,\infty} = 1 | \underset{\sim}{\omega}_{-1} = -\underset{\sim}{1}) = 1$.

B. If $\Theta_{k,l}^{\max} \leq \underline{\Theta}_{k,l}$, then

   i. $\text{Prob}(\omega_{i,\infty} = 1 | \underset{\sim}{\omega}_{-1} = -\underset{\sim}{1}) < 1$.

   ii. $\text{Prob}(\underset{\sim}{\omega}_{\infty} = \underset{\sim}{1} | \underset{\sim}{\omega}_{-1} = -\underset{\sim}{1}) = 0$.

The main result of the model is that when production decisions are interdependent, then a low level production trap is produced. In terms of the underlying probability structure, the model (and its associated properties) is a generalization of the Stavskaya-Shapiro model, described in Shnirman[47] and Stavskaya and Pyatetskii-Shapiro.[48]

This model can be rewritten in the canonical discrete choice form as follows. For each industry i at t, reinterpret $u(\omega_{i,t}, \cdot, \cdot)$ as the expected discounted profits, which depends on current technique choice.[11] Following Eqs. (5) and (6), $h(\underset{\sim}{X}_{i,t})$ measures the relevant individual- and time-specific deterministic private payoff parameter associated with a technique choice. Further, assume that the only relevant characteristics determining the relative profitability of the two techniques for a given industry are the past technique choices of technologically similar industries,

$$h(\underset{\sim}{X}_{i,t}) = h(\omega_{i-k,t-1}, \ldots, \omega_{i+l,t-1}) . \tag{37}$$

Restrict this function so that it is increasing in all arguments and positive when all previous technique choices were equal to 1,

$$h(1, \ldots, 1) > 0 . \tag{38}$$

Finally, assume that $\beta(\underset{\sim}{X}_{i,t})$ has the property that

$$\beta(\underset{\sim}{X}_{i,t}) = \infty \text{ if } \omega_{i-k,t-1} = \ldots = \omega_{i+l,t-1} = 1 . \tag{39}$$

Subject to these restrictions, there will exist a discrete choice specification which replicates the probabilities in the interacting industries model. This specification reveals an important feature of the particular specification studied in Durlauf[26]: a strong nonlinearity in the case where all influences are positive at $t-1$ versus all other cases. This is not surprising given the unbounded support of the logistic density for finite β combined with the probability 1 assumption on technique choice

[11]Recall that the structure of industries is such that technique choices do not reflect any consequences for future spillover effects, so that an industry always chooses a technique to maximize one-period ahead profits.

under the conditioning in Eq. (35). This suggests that the particular case studied in Durlauf[26] is knife-edge in terms of parameter values under the general discrete choice framework. While this does not affect the theoretical interest of the model, it does suggest that it should be reparameterized if one is to use it in empirical work.

Finally, it is worth noting that when interactions between decisions are all intertemporal, then the assumption of extreme-valued random utility increments can be dropped. The equilibrium properties of the dynamic models in this section can be recomputed under alternative probability densities such as probit which are popular in the discrete choice work. In fact, under the mean-field analysis of global interactions, alternative specifications incorporate probit or other densities as well. In both cases, the large-scale properties of models under alternative error distributions are largely unknown.

5. STOCHASTIC INTERACTION STRUCTURES

A third area of work has focused on cases where the interaction environments are themselves stochastic. Three approaches to this have been taken. The first is to allow for heterogeneity within a group. The second allows for random group formation. The third treats group membership as a choice variable.

I. MULTIPLE AGENT TYPES

Glaeser, Sacerdote, and Scheinkman[34] provide a model of local interactions and crime that fits cleanly into the general discrete choice framework. They consider a model in which individuals are arrayed on a one-dimensional lattice. Each individual has one of three kinds of preferences with reference to the decision to commit a crime. The decision to commit is coded as $\omega_i = 1$. Preference type 1 is such that the agent always decides to commit a crime. Preference type 2 is such that an agent will choose to commit a crime only if the agent to his left does as well. Preference type 3 is such that the agent never commits a crime. Representing the preferences of agents as $U_j(\omega_i, \omega_{i-1})$ where j denotes the preference type,

$$U_1(1,1) > U_1(-1,1); \quad U_1(1,-1) > U_1(-1,-1) \tag{40}$$

$$U_2(1,1) > U_2(-1,1); \quad U_2(-1,-1) > U_2(1,-1) \tag{41}$$

$$U_3(-1,1) > U_3(1,1); \quad U_3(-1,-1) > U_3(1,-1)\,. \tag{42}$$

The distribution of preference types is i.i.d. across agents. The interaction structure described here is a variant of the so-called voting model in statistical mechanics; see Liggett[40] for a detailed description.

From the discrete choice perspective, these preference assumptions can be thought of as doing the following. Each agent possesses a neighborhood consisting of the agent to his left, so that

$$J_j(\underset{\sim}{X}_i) = J(\underset{\sim}{X}_i) \text{ if } j = i-1, 0 \text{ otherwise}. \tag{43}$$

Further, each agent is associated with a latent variable ϕ_i, with support $\{\phi^l, \phi^m, \phi^h\}$. The latent variable is the only personal characteristic which influences individual utility, so that the different preference types can be thought of as induced by ϕ_i. These influences work through the deterministic private and social utility functions according to

$$h(\phi^l) < 0, \;\; J(\phi^l) = 0 \tag{44}$$
$$h(\phi^m) = 0, \;\; J(\phi^m) > 0 \tag{45}$$
$$h(\phi^h) > 0, \;\; J(\phi^h) = 0. \tag{46}$$

The joint probability measure for choices in this model is

$$\text{Prob}(\underset{\sim}{\omega} \,|\, \phi_1, \ldots, \phi_I) \sim \prod_i \exp(\beta h(\phi_i)\omega_i + \beta J(\phi_i) \cdot \omega_i \omega_{i-1}). \tag{47}$$

The unconditional probability measure can be computed once an assumption is made on the form of $dF_{\underset{\sim}{\phi}}$ and thereby allow analysis of cross-sectional patterns. When $\beta = \infty$, the model reduces to the deterministic choice structure of Glaeser, Sacerdote, and Scheinkman.

A nice feature of this model is that by preserving nonoverlapping neighborhoods, the problems created by the contemporaneous determination of choices are avoided in the perfect foresight case. Further, notice that if the interactions are intertemporal and past behavior influences current behavior (as clearly seems natural in this context), then joint behavior will follow

$$\text{Prob}(\underset{\sim}{\omega}_t) \sim \exp\left(\sum_i h_{i,t}\omega_{i,t}\right) \tag{48}$$

where

$$h_{i,t} = h(\omega_{i-1,t-1}, \omega_{i,t-1}, \phi_{i,t}). \tag{49}$$

In this case, the interaction structure is a special case of that studied in Durlauf.[26]

II. RANDOM COMMUNICATION STRUCTURES

Kirman,[37] Ioannides,[35] and Durlauf,[29] have studied economic environments in which the structure of bilateral interactions has important aggregate consequences. In their models, random communication links exist between any pair of agents i and j. Coalitions emerge across any grouping of agents such that a path of direct bilateral communication links can be formed between any pair of members in a coalition. As all agents within a coalition communicate and whereas members of different coalitions do not, this structure can illustrate the role of group membership in phenomena ranging from fluctuations in the prices of a particular good across trading regions to the role of the degree of specialization of labor in explaining business cycles. A rich set of results from random graph theory illustrate how the distribution of coalition sizes will depend sensitively on the probability of the bilateral links. In particular, when the bilateral links are conditionally i.i.d. (in the sense that the probability that any pair of agents is directly linked is independent of whether any other pair is linked), then as $I \Rightarrow \infty$ (1) if the probability of any link is less than $1/I$, then the largest coalition will be of order $\log I$, (2) if the probability equals c/I for $c > 1$, then the largest coalition will be of order I, (3) if the probability is greater than $c \log I/I$ for $c > 1$, then all agents will be members of a common coalition.

Previous papers employing random graph formulations have been interested in the size distribution of coalitions. However, for socioeconomic environments involving networks of contacts or friends, the approach can enrich the exogenous interaction structures which are typically assumed. The discrete choice structure can accommodate random interactions by assigning a probability measure to $J_{i,j}$'s such that

$$J_{i,j} \in \{0,1\}\,, \tag{50}$$

$$J_{i,j} = J_{j,i}\,, \tag{51}$$

$$\text{If } J_{i,l_1} J_{l_1,l_2} \cdots J_{l_m,j} = 1, \text{ then } J_{i,j} = 1\,. \tag{52}$$

Any assumptions about the distribution of the bilateral interactions can be mapped in a straightforward fashion to the $J_{i,j}$ weights subject to these restrictions.

This particular structure is related to the Mattis model in statistical mechanics, which is described in Fischer and Hertz.[32] In the Mattis model, the individual weights are defined by

$$J_{i,j} = J\xi_i\xi_j \tag{53}$$

where ξ_i has support $\{1,-1\}$ and is distributed i.i.d. across i. Using the transformation

$$\zeta_{i,j} = \frac{\xi_i\xi_j + 1}{2}\,, \tag{54}$$

then the Mattis model will produce a random graph structure and associated interactions weights with the formula

$$J_{i,j} = J\zeta_{i,j} \, . \tag{55}$$

Notice that in this formulation, the bilateral links between any two agents will no longer be conditionally independent, as they are for the random graph structures which have been studied in the economics literature. Dependence is a natural assumption if agents with common attributes are more likely to communicate with each other.

III. SELF-ORGANIZING NEIGHBORHOOD COMPOSITION

Bénabou[12,13] and Durlauf[27,28] have emphasized the importance of endogeneity in neighborhood structure. This class of models embodies interactions of the type which have already been surveyed, with an essential difference. While agents interact noncooperatively within a neighborhood, they choose neighborhoods in recognition of these interactions.

Evolving neighborhood composition can be accommodated in the statistical mechanics approach through the $J_j(\underset{\sim}{X}_{i,t})$. This can be seen in two steps. First, suppose that at time t, each agent is assigned to a neighborhood n, $n = 1 \ldots N$. By allowing the interaction weights to depend on whether one is in a common or different neighborhood as other agents, one can replicate the interactive structure of the endogenous neighborhoods models. The endogenous neighborhood model will be complete once a neighborhood assignment rule is specified. Since such a rule will presumably depend upon the attributes of all agents in the economy, this implies that

$$n_{i,t} = \Lambda(\underset{\sim}{X}_{1,t}, \ldots, \underset{\sim}{X}_{I,t}) \, . \tag{56}$$

Now, the specification of this function is far from trivial, as it must embody factors such as house price or rental differences which endogenously determine the assignment of individuals. What is relevant to the current discussion is that there is no reason in principle that the statistical mechanics approach cannot accommodate a rich neighborhood structure.

Different specifications of the neighborhood determination will correspond to different models in the literature. For example, if

$$J_j(\underset{\sim}{X}_{i,t}) = 1 \text{ if } n_{i,t} = n_{j,t}, \; 0 \text{ otherwise} \; , \tag{57}$$

then general form (10) will produce a Brock-Durlauf[19] model specification for each of the neighborhoods.

While Bénabou and Durlauf have emphasized the role of economic segregation in neighborhood formation, observe that neighborhoods can be given alternative definitions. For example, by treating neighborhood membership as determined by ethnicity, one can use the model to study differences in ethnic group behavior; more generally, individual interaction weights can depend on a multiplicity of factors which reflect the different communities or reference groups which characterize individuals. Alternatively, one can allow for random interactions in the way discussed above, where the individual probabilities are determined by individual attributes. This can capture some of the ideas on social connections analyzed in Montgomery[45] and Chwe.[22]

The introduction of heterogeneous and time dependent social utility weights $J_j(\underset{\sim}{X}_{i,t})$ in the random interaction and endogenous neighborhood models links the analysis of socioeconomic interactions with the frontier of research in statistical mechanics. Specifically, statistical mechanics has in the last two decades focused on the behavior of systems in which the configuration of elements obeys the canonical probability structure

$$\text{Prob}(\underset{\sim}{\omega}) \sim \exp\left(\sum_i \sum_j J_{i,j}\omega_i\omega_j\right), \tag{58}$$

when the $J_{i,j}$ terms are a function of something other than $|i-j|$, this structure is known as an anisotropic ferromagnet. When $J_{i,j} < 0$ for some i and j, the system is known as a spin glass. See Fischer and Hertz[32] and Mézard, Parisi, and Virasoro[44] for introductions to these models.

Anisotropic ferromagnets and spin glasses can exhibit phase transitions and complex pattern formation. Unlike the models which have been used in economics thus far, these phase transitions can apply to moments of the equilibrium probability measure other than the population mean. Spin glass formulations can (in particular formulations) exhibit phase transitions with respect to the variance of individual choices as well as spatial and intertemporal correlations. These models thus hold the promise of allowing the study of multiple equilibria and multiple steady states in the distribution of social and economic activity. In addition, the spin glass formulation will permit the modeling of interactions within and across neighborhoods. Suppose that each member of a neighborhood assigns a weight to conforming with others in the economy which depends upon the neighborhood in which the individual lives. In this case, there will exist interdependence across the neighborhood mean choices so that the equilibrium mean-choice level of neighborhood n is

$$m_n^* = \int \tanh\left(\beta h(\underset{\sim}{X}) + \beta \sum_r J_{n,r} m^*\right) dF_{n,\underset{\sim}{X}} \quad n = 1 \ldots N \tag{59}$$

where the weights $J_{n,r}$ have been reindexed to reflect interaction weights between and within neighborhoods and $dF_{n,\underset{\sim}{X}}$ refers to the distribution of individual characteristics within neighborhood n. These equations may be used to analyze the consequences of different neighborhood formation rules on the population-wide mean-choice level, thus providing a way to study the aggregate effects of integration and segregation, whose importance in a different context has been studied by Bénabou.[14]

Finally, it is important to recognize that endogenous neighborhood formation represents a class of statistical mechanics models whose properties have yet to be directly explored in the physics or complex systems literatures. Economic models usually will impose $J_{i,j}$ weights which are at least partially determined by individual choices. In fact, it is clear that new forms of phase transition are likely to occur in these systems. The reason is the following. Endogenous neighborhood models typically result in the stratification of neighborhoods by some observable attribute such as income; recent examples include Bénabou,[12,13] Durlauf,[28,29] and Fernandez and Rogerson.[31] These attributes will correlate with the h_i and $J_{i,j}$ terms which distinguish individuals. What this means is that when one considers cross-group inequality, differences can be explained by the presence of multiple equilibrium invariant measures as well as by different characteristics of the populations. One example where this endogeneity matters is the Brock-Durlauf model, where an integrated community mixing agents with high and low $|h_i|$ values can exhibit a unique equilibrium whereas segregated communities can exhibit multiple equilibria for those communities in which members are characterized by low $|h_i|$ values.

6. LIMITATIONS

While the statistical mechanics approach has yielded numerous insights into phenomena ranging from out of wedlock births to crime, the literature suffers from a number of limitations at this time.

First, the use of statistical mechanics methods in social science has exclusively focused on binary choices, which omits an important range of interaction environments. Bénabou[12,14] and Durlauf[28,29] for example, explicitly focus on the level of education investment, which is naturally thought of as continuous. While there is a small literature on n-state spin systems for $n > 2$ (see Yeomans[52] for examples), it not clear that the systems which have been studied are rich enough for social science applications.

Second, there has been relatively little success at this stage in developing rational expectations variants of these models. Dynamic approaches such as Durlauf[25,26,27] have imposed linear technologies precisely in order to avoid the need for firms to forecast future prices. The difficulty with developing rational expectations versions of these models is that the interaction structures embody sufficiently complicated nonconvexities to render the standard fixed point arguments invalid. Important recent work by Blume[16] makes progress on this issue.

Third, there has been little formal econometric work on statistical mechanics models.[12] Manski[41,42] provides a general framework which suggests that even the little empirical work which attempted to uncover interaction effects is flawed by identification problems. Hence, the interaction effects which underlie the theoretical literature, while plausible, are unproven. Brock and Durlauf[19] attempt to address this limitation by providing a general estimation theory for their global interactions framework.

7. CONCLUSION

The goal of this chapter has been to provide a unified perspective on the ways in which economists have employed statistical mechanics to model socioeconomic interactions. My main claim is that the statistical mechanics approach is compatible with good microeconomic reasoning, and thus represents a valuable additional tool for a range of research questions. This claim has been made on the basis of unifying a number of applications in a discrete choice framework. This framework encompasses a range of approaches in the literature, and shares a common form with the logistic likelihood function. This common form holds the promise that the theoretical insights which have been generated by the statistical mechanics approach can be matched by empirical evidence.

ACKNOWLEDGMENTS

I thank William Brock for innumerable conversations on the issues raised in this chapter as well as Lawrence Blume, Kim-Sau Chung, Brian Krauth, Susan Nelson, Richard Palmer, and conference participants at the National Bureau of Economic Research and Santa Fe Institute for comments on an earlier draft. Financial support from the National Science Foundation, the John D. and Catherine T. MacArthur Foundation, and the Santa Fe Institute is greatly appreciated.

[12]A recent exception is Topa.[51] This is not to say that statistical mechanics models have not been used to interpret empirical findings. For example, Glaeser, Sacerdote, and Scheinkman[34] show how their local interactions model can explain wide discrepancies between crime rates in cities.

REFERENCES

1. An, M., and N. Kiefer. "Local Externalities and Societal Adoption of Technologies." *J. Evol. Econ.* **5** (1995): 103–117.
2. Anderson, S., A. de Palma, and J.-F. Thisse. *Discrete Choice Theory of Product Differentiation.* Cambridge, MA: MIT Press, 1992.
3. Aoki, M. "Economic Fluctuations with Interactive Agents: Dynamic and Stochastic Externalities." *Japanese Econ. Rev.* **46** (1995): 148–165.
4. Arthur, W. B. "Urban Systems and Historical Path Dependence." In *Urban Systems and Infrastructure*, edited by R. Herman and J. Ausubel. Washington D.C.: National Academy of Sciences/National Academy of Engineering, 1987.
5. Arthur, W. B. "Increasing Returns, Competing Technologies and Lock-In by Historical Small Events: The Dynamics of Allocation Under Increasing Returns to Scale." *Econ. J.* **99** (1989): 116–131.
6. Averintsev, M. "On a Method of Describing Discrete Parameter Fields." *Problemy Peredachi Informatsii* **6** (1970): 100–109.
7. Bak, P., K. Chen, J. Scheinkman, and M. Woodford. "Aggregate Fluctuations from Independent Sectoral Shocks: Self-Organized Criticality in a Model of Production and Inventory Dynamics." *Ricerche Economiche* **47** (1993): 3–30.
8. Becker, G. "Irrational Behavior and Economic Theory." *J. Pol. Econ.* **70** (1962): 1–13.
9. Becker, G. *Accounting for Tastes.* Cambridge, MA: Harvard University Press, 1996.
10. Becker, G., and G. Stigler. "De Gustibus Non Est Disputandum." *Amer. Econ. Rev.* **67** (1977): 76–90.
11. Bell, A. "Dynamically Interdependent Preferences in a General Equilibrium Environment." Mimeo, Department of Economics, Vanderbilt University, 1995.
12. Bénabou, R. "Workings of a City: Location, Education, and Production." *Quart. J. Econ.* **CVIII** (1993): 619–652.
13. Bénabou, R. "Equity and Efficiency in Human Capital Investment: The Local Connection." *Rev. Econ. Studies* **62** (1996): 237–264.
14. Bénabou, R. "Heterogeneity, Stratification, and Growth: Macroeconomic Implications of Community Structure and School Finance." *Amer. Econ. Rev.* **86** (1996): 584–609.
15. Blume, L. "The Statistical Mechanics of Strategic Interaction." *Games & Econ. Behav.r* **5** (1993): 387–424.
16. Blume, L. "Population Games." Mimeo, Department of Economics, Cornell University, Ithaca, NY, 1996.
17. Brock, W. "Pathways to Randomness in the Economy: Emergent Nonlinearity and Chaos in Economics and Finance." *Estudios Economicos* **8(1)** (1993):

3–55, and Social Systems Research Institute Reprint #410, Department of Economics, University of Wisconsin, Madison, WI, 1993.
18. Brock, W. "Asset Price Behavior in Complex Environments." Mimeo, University of Wisconsin, Madison, WI, 1995.
19. Brock, W., and S. Durlauf. "Discrete Choice with Social Interactions I: Theory." Mimeo, University of Wisconsin, Madison, WI, 1995.
20. Brock, W., and S. Durlauf. "Discrete Choice with Social Interactions II: Econometrics." Mimeo in progress, University of Wisconsin, Madison, WI, 1996.
21. Brock, W., and C. Hommes. "Rational Routes to Randomness." Mimeo, Department of Economics, University of Wisconsin, Madison, WI, 1995.
22. Chwe, M. "Structure and Strategy in Collective Action: Communication and Coordination in Social Networks." Mimeo, Department of Economics, University of Chicago, 1996.
23. Cooper, R., and A. John. "Coordinating Coordination Failures in Keynesian Models." *Quart. J. Econ.* **103** (1988): 441–463.
24. Crutchfield, J. "Is Anything Ever New? Considering Emergence." In *Complexity: Metaphors, Models, and Reality*, edited by G. Cowan, D. Pines, and D. Meltzer. Santa Fe Studies in the Sciences of Complexity, Proc. Vol. XIX, Reading, MA: Addison-Wesley, 1994.
25. Durlauf, S. "Multiple Equilibria and Persistence in Aggregate Fluctuations." *Amer. Econ. Rev.* **81** (1991): 70–74.
26. Durlauf, S. "Nonergodic Economic Growth." *Rev. Econ. Studies* **60** (1993): 349–366.
27. Durlauf, S. "Path Dependence in Aggregate Output." *Ind. & Corp. Change* **1** (1994): 149–172.
28. Durlauf, S. "A Theory of Persistent Income Inequality." *J. Econ. Growth* **1** (1996): 75–93.
29. Durlauf, S. "Neighborhood Feedbacks, Endogenous Stratification, and Income Inequality." In *Dynamic Disequilibrium Modelling: Proceedings of the Ninth International Symposium on Economic Theory and Econometrics*, edited by W. Barnett, G. Gandolfo, and C. Hillinger. Cambridge, MA: Cambridge University Press, 1996.
30. Durlauf, S. "An Incomplete Markets Model of Business Cycles." *Comp. & Math. Org. Theory* **2** (1996): 191–212.
31. Fernandez, R., and R. Rogerson. "Income Distribution, Communities, and the Quality of Public Education." *Quart. J. Econ.* **111** (1996): 135–164.
32. Fischer, K., and J. Hertz. *Spin Glasses.* Cambridge, MA: Cambridge University Press, 1991.
33. Föllmer, H. "Random Economies with Many Interacting Agents." *J. Math. Econ.* **1** (1974): 51–62.
34. Glaeser, E., B. Sacerdote, and J. Scheinkman. "Crime and Social Interactions." *Quart. J. Econ.* **CXI** (1996): 507–548.
35. Ioannides, Y. "Trading Uncertainty and Market Structure." *Intl. Econ. Rev.* **31** (1980): 619–638.

36. Kindermann, R., and J. L. Snell. *Markov Random Fields and Their Applications.* Providence: American Mathematical Society, 1980.
37. Kirman, A. "Communication in Markets: A Suggested Approach." *Econ. Lett.* **12** (1983): 1–5.
38. Kollman, K., J. Miller, and S. Page. "Adaptive Parties in Spatial Elections." *Am. Pol. Sci. Rev.* **86** (1992): 929–937.
39. Krugman, P. *The Self-Organizing Economy.* Oxford: Basil Blackwell, 1996.
40. Liggett, T. *Interacting Particle Systems.* New York: Springer-Verlag, 1985.
41. Manski, C. "Identification of Endogenous Social Effects: The Reflection Problem." *Rev. Econ. Studies* **60** (1993): 531–542.
42. Manski, C. "Identification Problems in the Social Sciences." In *Sociological Theory*, edited by P. Marsden, vol. 23. Cambridge: Basil Blackwell, 1993.
43. Manski, C. "Identification of Anonymous Endogenous Social Interactions." Mimeo, University of Wisconsin, Madison, WI, 1995.
44. Mézard, M., G. Parisi, and M. Virasoro. *Spin Glass Theory and Beyond.* Singapore: World Scientific, 1987.
45. Montgomery, J. "Social Networks and Labor Market Outcomes: Towards a Dynamic Analysis." *Amer. Econ. Rev.* **81** (1991): 1408–1418.
46. Schelling, T. "Dynamic Models of Segregation." *J. Math. Sociol.* **1** (1971): 143–186.
47. Shnirman, N. G. "On the Ergodicity of a Markov Chain." *Problems in Information Theory* **20** (1968): 115–124.
48. Stavskaya, O. N., and I. I. Pyatetskii-Shapiro. "Homogeneous Networks of Spontaneous Active Elements." *Problems in Information Theory* **20** (1968): 91–106.
49. Spitzer, F. "Markov Random Fields and Gibbs Ensembles." *Am. Math. Monthly* **78** (1971): 142–154.
50. Spitzer, F. *Random Fields and Interacting Particle Systems.* Providence: American Mathematical Society (reprinted lecture notes), 1971.
51. Topa, G. "Social Interactions, Local Spillovers, and Unemployment." Mimeo, University of Chicago, IL, 1996.
52. Yeomans, J. *Statistical Mechanics of Phase Transitions.* Oxford: Oxford University Press, 1992.

David Lane
Department of Political Economy, University of Modena, ITALY; E-mail: lane@unimo.it

Is What Is Good For Each Best For All? Learning From Others In The Information Contagion Model

Suppose you are thinking about seeing a movie. Which one should you see? Even if you keep up with the reviews in the newspapers and magazines you read, you will probably also try to find out which movies your friends are seeing—and what they thought of the ones they saw. So which movie you decide to see will depend on what you learn from other people, people who already went through the same choice process in which you are currently engaged. And after you watch your chosen movie, other people will ask *you* what film you saw and what you thought of it, so your experience will help inform their choices and hence their experiences as well.

Now suppose you are a regular reader of *Variety,* and you follow the fortunes of the summer releases from the major studios. One of the films takes off like a rocket, but then loses momentum. Nonetheless, it continues to draw and ends up making a nice profit. Another starts more slowly but builds up an audience at an ever-increasing rate and works its way to a respectable position on the all-time earners' list. Most of the others fizzle away, failing to recover their production and distribution costs.

The preceding paragraphs describe the same process at two different levels: the *individual-level* decision process and the *aggregate-level* market-share allocation process. The first paragraph highlighted a particular feature of the individual-level process: people learn from other people, what they learn affects what they do—and then what they do affects what others learn. Thus, the process that results in allocation of market-share to the competing films is characterized by an informational feedback, that derives from the fact that individuals learn from the experience of others.

In this chapter, I discuss a model designed to isolate the effects of this informational feedback on the market-share allocation process, at least with respect to a particular kind of choice problem and a particular underlying social structure. In the model, agents choose sequentially between two competing products, each of which is characterized by a number that measures how well the product performs (that is, its "intrinsic value"). From each of their informants, agents learn which product the informant selected and an estimate of the product's performance characteristic. The agents' social space has a simple structure: agents randomly sample their informants from the pool of agents who have already made their choice.[1]

The chapter focuses on some subtle and surprising ways in which agent characteristics and connectivity structure affect the market-shares of the competing products. In particular, I describe two examples of properties that seem desirable at the individual level, but which turn out to have undesirable effects at the aggregate level: what is good for each is, in a certain sense, bad for all.

The chapter proceeds as follows. In section 1, I present the information contagion model, first introduced in Arthur and Lane[3], and consider what it might mean for an agent to act rationally in the context of this model. In section 2, I show that giving agents access to more information does not necessarily lead to a better outcome at the aggregate level: *increasing* the number of informants for each agent can *decrease* the proportion of agents that end up adopting the better product. In section 3, I compare two different procedures for integrating data obtained from informants into a choice between the two products. One of these procedures uses Bayesian updating and maximizes expected utility; the other uses a rule-of-thumb based on an insufficient statistic to estimate the true performance characteristics. It turns out that the rule-of-thumb leads to an asymptotic market-share of 100% for the superior product, as the number of agents goes to infinity, no matter how small the actual difference between the products. In contrast, the procedure based on Bayesian updating and expected utility maximization can result in substantial market-share for the inferior product.

[1]In fact, social learning is channeled by pre-existing social networks: you get your information about films from friends or at least acquaintances, not random film-goers. Few papers in the economic social learning literature try to deal with this fact. An exception is Ahn and Kiefer,[1] whose agents base their adoption decisions on information obtained from their neighbors on a d-dimensional lattice.

1. THE INFORMATION CONTAGION MODEL

In the information contagion model, a countable set of homogeneous agents chooses sequentially between two products, A and B. The relative value of the two products to agents depends on their respective "performance characteristics," real numbers c_A and c_B, respectively. If agents knew the value of these numbers, they would choose the product associated with the larger one.[2]

All agents have access to the same public information about the two products, which does not change throughout the allocation process. The process begins with an initial "seed" of r A-adopters and s B-adopters. Successive agents supplement the public information by information obtained from n randomly sampled[3] previous adopters. They translate their information into a choice between A and B by means of a decision rule D. Here and elsewhere, agents are homogeneous: they differ only with respect to the information they obtain from their sampled informants.

From each sampled informant, an agent learns two things: which product the informant adopted and the value of that product's performance characteristic perturbed by a standard normal observation error. That is, from informant i, agent j learns the values of two random variables, X_{ji} and Y_{ji}. X_{ji} takes values in $\{A, B\}$; Y_{ji} is a normal random variable, with mean $c_{X_{ji}}$ and variance 1. Given $\{X_{j1}, \ldots, X_{jn}\}$, $Y_{j1}, \ldots, Y_{jn}$ are independent. Finally, $Y_1, Y_2, \ldots$ are independent random vectors, given $\{X_1, X_2, \ldots\}$.

A particular information contagion model is specified by two real parameters c_A and c_B; three integer parameters r, s, and n; and a decision function D from $\{\{A, B\} \times R\}^n \rightarrow \{A, B\}$.[4] For any such specification, it can be shown by a simple modification of arguments in Arthur and Lane,[3] based on results from Hill, Lane, and Sudderth,[12] that the proportion of adopters of product A converges with probability 1. Moreover, the support of its limiting distribution is $\{x : x \text{ in } [0,1], f(x) = x \text{ and } f'(x) \leq 1\}$, where f is given by

$$f(x) = \sum_{i=0}^{n} p(k) \binom{n}{k} x^k (1-x)^{n-k}$$

[2]The prices of the two products do not enter explicitly into the model. If we suppose that these prices are known and fixed throughout the market-share allocation process, then we can suppose that the performance characteristics are measured in a way that adjusts for price differences.

[3]Asymptotically, it does not matter whether the sampling is with or without replacement. Obviously, sampling without replacement is more consistent with the underlying interpretative narrative. However, when I calculate market-allocation distributions after a finite number of adoptions in the next section, I will use sampling with replacement. Among other advantages, I do not then have to worry about whether n is greater than $r + s$!

[4]Since the agents are homogeneous and share the same public information, the effect of that information on their decision is just incorporated into the form of D. Agents' decisions can differ only on the basis of the private information obtained from sampled informants.

and $p(k)$ is the probability that an agent who samples k previous purchasers of A—and hence, $(n-k)$ previous purchasers of B—will choose product A; $p(k)$ depends on c_A, c_B, n, and D, but not r and s.

NOTE: In the rest of the chapter, assertions about the limiting distribution for the number of adopters of each product are based upon calculating p and then finding the roots of $f(x)-x$.

RATIONALITY IN THE INFORMATION CONTAGION MODEL

A conventionally rational agent in the information contagion model would begin by coding the publicly available information about product performance into a prior distribution for (c_A, c_B). After he observed the value of (X, Y) from his informants, he would compute his posterior distribution for (c_A, c_B):

$$\text{post}(c_A, c_B) \propto \text{Lik}_{(x,y)}(c_A, c_B)\ \text{prior}(c_A, c_B),$$

where

$$\begin{aligned}\text{Lik}_{(X,Y)}(c_A, c_B) &= P[(X,Y)|(c_A, c_B)] \\ &= P[Y|X,(c_A, c_B)]P[X|(c_A, c_B)].\end{aligned}$$

Finally, he would compute his expected utility for $c_A(c_B)$ using his marginal posterior distribution for $c_A(c_B)$, and then would choose the product with the higher expected utility.

Assuming that the agent (correctly) models the components of Y as independent random variables each equal to the appropriate performance characteristic perturbed by a standard normal error, the first factor in the likelihood function is

$$P[y|x,(c_A, c_B)] = \prod_{i=1}^{n} \varphi(y_i - c_{x_i}),$$

where φ is the standard normal density function.

The second likelihood factor, $P[X|(c_A, c_B)]$, is more problematic. Let n_A be the number of A-adopters among an agent's informants. By exchangeability, $P[X|(c_A, c_B)]$ is proportional to $P[n_A|(c_A, c_B)]$ as a function of (c_A, c_B). How might our agent generate a probability distribution for n_A, given c_A and c_B? Let R and S represent the number of agents who have already adopted A and B, respectively. The distribution of n_A, given R and S, is either hypergeometric or binomial, depending on whether we suppose sampling is with or without replacement. So thinking about the distribution for n_A given c_A and c_B is equivalent to thinking about the distribution of R and S given c_A and c_B—and that means modeling the market-share allocation process. To do that, agents have to make assumptions about the other agents' sources of information and about the way in which they process that information to arrive at their product choices. How might we imagine that the agents do this in such a way that they all arrive at the same answer—that is, that agents' sources of information and decision processes become common knowledge? There are only two possibilities: either this common knowledge is *innate,* or it is a product of *social learning,* since knowledge *about* others must be learned through interaction *with* others. The idea that common knowledge of decision processes is innate is implausible; that common knowledge of others' information sources is innate is impossible. So to suppose that our agents share a model of the market-allocation process is to assume the successful accomplishment of a particularly complicated feat of social learning, much more complicated than the one the information contagion model is designed to study.

Basing a model that purports to study the effects of social learning in a restricted context on an assumption that presupposes we know its effects in a grander one, seems inadmissibly circular to me.[5] If, on the other hand, we abandon the idea that the agents have common knowledge of each other's information sources and decision processes, but that each has his own private way of modeling these things, then how can we expect the agents to model the market-allocation process at all, depending as it does on their beliefs about others' beliefs about others' beliefs. . .?[6]

These considerations lead me to suppose that our rational agent gives up the attempt to model how n_A depends on c_A and c_B. That is, I suppose that he regards his knowledge about the initial conditions r and s, and about the other agents' information sources and decision processes, as sufficiently vague that he just assigns

[5] But apparently not to others. Banerjee,[5] Bikhchandani et al.,[7] Lee,[17] and Smith and Sorenson,[19] among others, discuss models in which agents observe the actions of every preceding adopter, so for them the problem of modeling something equivalent to n_A given c_A and c_B does not arise. Still, these authors have to make some strong common-knowledge assumptions in order for their agents to extract information from the decisions other agents make: that all agents know the private information sources (not the information itself) of all the other agents, that they all model these sources in the same way, and that they all update available information Bayesianally. In the setting of these authors' social-learning models, these common knowledge assumptions are sufficient to generate rational individual-level decisions that sequentially generate the market allocation process. Smith and Sorenson[19] are particularly attentive to how the various underlying assumptions drive the results that they and the other authors who have contributed to this literature have obtained.
Banerjee and Fudenberg[6] (BF), on the other hand, suppose that agents obtain information only from a sample of previous adopters as in the information contagion model. As a result, they need to make more stringent rationality and common-knowledge assumptions than the authors described in the previous paragraph. In their model, agents carry out their Bayesian updating by first computing a Bayesian equilibrium distribution, supposing the "rationality" of all previous adopters, and then conditioning on this distribution as a datum. BF agents, then, not only have common knowledge of agents' information sources and Bayesianity, but of each other's "desire" to be in an equilibrium state! (Of course, BF merely assume that their agents "are" in the equilibrium; I can interpret this existential claim in process terms only through an assumption the agents make that that is where they all *want* to be.) BF's Bayesian equilibrium, by the way, can only be calculated under the assumption that there are an uncountable number of agents. BF's is a rather strong, not to say bizarre, attribution of common knowledge.

[6] Arthur[2] argues that social learning that results in common knowledge of agent rationality is in fact impossible in circumstances like this.

a noninformative uniform distribution to n_A for all possible values c_A and c_B.[7] If he does so, his likelihood function reduces to

$$\text{Lik}_{(X,Y)}(c_A, c_B) = P[Y|X, (c_A, c_B)] .$$

I think that agents who do Bayesian updating with this likelihood function and then maximize the resulting expected utility are as rational as they ought to be in the information contagion context. I will call such agents "almost-rational."

I now describe a three-parameter family of decision rules for almost-rational agents, introduced in Arthur and Lane.[3] I will refer to these rules, and to the agents who use them, as "Arthur-Lane Bayesian optimizers" (hereafter ALBO):

ALBO PRIOR DISTRIBUTION: c_A and c_B are independent and normally distributed. If we suppose, as I will henceforth, that the public information does not distinguish between the two products, then the distributions for c_A and c_B have common mean, μ, and variance, σ^2.

ALBO LIKELIHOOD FUNCTION: Given X and the product performance characteristics c_A and c_B, the Y's are independent normal random variables with mean the appropriate (as determined by X) performance characteristic and variance equal to 1.

ALBO UTILITY FUNCTION: ALBOs use a utility function in the constant-risk-aversion family, parameterized by the nonnegative constant λ as follows: [8]

$$u(c) = \begin{cases} -e^{-2\lambda c} & \text{if } \lambda > 0 , \\ c & \text{if } \lambda = 0 . \end{cases}$$

[7] If we suppose agents have countably additive prior distributions for r and s, these distributions for n_A could only be coherent (in the sense of Heath and Sudderth[10]) if agents believed that other agents chose each product in a way that disregards the information in Y about product quality, and if their prior distribution for (r, s) is symmetric. However, I suspect that the distributions might be coherent no matter what assumptions agents made about each other, if their opinions about r and s were sufficiently vague to be expressed in the form of finitely, but not countably additive, symmetric prior; or if their assumptions about other agents' knowledge and decision processes were sufficiently diffuse.

[8] When λ equals 0, agents are risk-neutral; the greater the value of λ, the more risk averse are the agents.

ALBO DECISION RULE: The expected utility for product $i (i = A$ or $B)$ is

$$E(U_i) = -\exp[2\lambda(\lambda\sigma_i^2 - \mu_i)],$$

where μ_i is the mean of the agent's posterior distribution for c_A, and σ_i^2 is the variance of this distribution. Thus, the ALBO decision rule is: choose A if

$$\mu_A - \lambda\sigma_A^2 > \mu_B - \lambda\sigma_B^2,$$

and if the inequality is reversed, choose B.

Let n_i represent the number of i adopters a particular agent samples and $\bar{Y}_i$ the average value of the observations obtained from the sampled agents. Then

$$\mu_i - \lambda\sigma_i^2 = \frac{1}{n_i + \sigma^{-2}}(n_i\bar{Y}_i + \sigma^{-2}\mu - \lambda).$$

To analyze the market-share allocation process with ALBO agents, we need to calculate the appropriate $p(k)$, the probability that an agent will adopt A if he samples k previous A adopters and $n - k$ B adopters:

$$\begin{aligned} p(k) &= P(E(U_A) > E(U_B) | n_A = k) \\ &= P\left(k\bar{Y}_A + \sigma^{-2}\mu - \lambda > \frac{k + \sigma^{-2}}{n - k + \sigma^{-2}}[(n-k)\bar{Y}_B + \sigma^{-2}\mu - \lambda]\right). \end{aligned}$$

Hence,

$$p(k) = \Phi\left[\frac{(2k-n)[\sigma^{-2}(c_B - \mu) + \lambda] + k(n - k + \sigma^{-2})(c_A - c_B)}{\sqrt{k(\sigma^{-2} + n - k)^2 + (n-k)(\sigma^{-2} + k)^2}}\right],$$

where Φ is the standard normal cdf.

2. MORE CAN BE LESS: INFORMATION INVERSION AND PATH DEPENDENCE

MORE INFORMATION IS GOOD AT THE INDIVIDUAL LEVEL...

Suppose you ask an agent in the information contagion model how many informants he would like to have. Clearly, his answer should depend on how much each interview cost him, in time and trouble as well as money. But suppose that observations are completely costless. Then, from a purely statistical point of view, the agent would like to have as many informants as possible: after all, the more information he gets about each product, the better he can estimate c_A and c_B—and the more certain he can be about which is really the better product, *no matter what the values of* c_A *and* c_B *actually are.* Moreover, for an ALBO agent, additional (costless) information is always desirable on decision-theoretic grounds: his expected value for the expected utility of the product he would choose after an additional observation is always greater than the expected utility of the product he would choose on the basis of his current information.[9]

Thus, if we asked each of an infinite sequence of ALBO agents whether he would rather interview, say, 3 or 20 informants, all of them would opt for 20.

BUT NOT NECESSARILY AT THE AGGREGATE LEVEL

Now suppose we have two societies of ALBO agents, in one each agent interviews 3 informants and in the other, 20. According to the arguments in the previous paragraph, agents in the second society on average feel better about their choices than do their counterparts in the first society—and with justice, since their choices are indeed better founded. Surely, then, a higher proportion of the agents in the second society than in the first will end up adopting the better product?

Not necessarily: Figure 1 shows some typical results.[10] In the comparisons summarized there, product B is slightly superior to product $A(c_B - c_A = .1)$, the agents' prior distribution is well-calibrated for $B(\mu = c_B)$, and the prior variance equals 1. The four curves in Figure 1 show how the proportion of B-adopters varies as a function of the number of informants n, for four different values of the agents' risk aversion parameter λ.

The bottom curve in Figure 1, corresponding to the case in which agents are risk-neutral ($\lambda = 0$), is strictly increasing: as we might expect, the proportion of agents adopting the superior product increases with the number of informants per agent. But for larger values of λ,[11] this monotonicity no longer holds. Rather, "information inversions" occur—for some range of values of n, *more* information available to individual agents results in *fewer* adoptions of the superior product. When $\lambda = 1$, the proportion adopting the superior product decreases as the number

[9]The argument goes as follows: Let I_i represent the information an agent obtains from his first i informants: $I_i = \{(X_1, Y_1), (X_2, Y_2), \ldots (X_i, Y_i)\}$. Further, let A_i and B_i represent the expected utility the agent calculates for products A and B, respectively, on the basis of I_i, and let $U_i = \max(A_i, B_i)$. Thus, U_i is the value to the agent of his currently preferred product. It is easy to show that $\{A_j\}$and $\{B_j\}$ are martingales with respect to the σ-fields generated by $\{I_j\}$, so $\{U_j\}$ is a submartingale: $E(U_{i+1}|I_i) \geq U_i$. Since the support of Y_{i+1} is unbounded above and below, an agent will change his mind about which product he prefers on the basis of an additional observation with positive probability. Thus, the above inequality is strict.

[10]Typical in the sense that I have found information inversions corresponding to every value of the paremeters $d = c_A - c_B$, μ, and σ^2 that I have investigated, for some range of values of λ. As the $\lambda = 0$ curve in Figure 1 indicates, inversion is not generic—at least in this case there was no inversion for $n < 1000$, the largest value I calculated. (That does not rule out the possibility of an inversion somewhere: I have found examples where an inversion occurs for the first time at a very large value of n, of the order of 1000!) In any event, inversion seems to be a very general phenomenon in the information contagion model.

[11]Here, for $\lambda \geq 0.8$.

of informants increases from 3 to 9. Thereafter, it increases. But the increase is very slow; in fact, the proportion adopting the superior product with 20 informants per agent (62.63%) is actually a little lower than it is with 3 (63.97%).

As λ increases, the information inversion becomes much more dramatic. For $\lambda = 1.6$, with 4 informants per agent, the superior product takes more than 96% market-share, while with *twenty*-4 per agent it gets only 67.9%. As the number of informants increases beyond 24, the proportion adopting the superior product

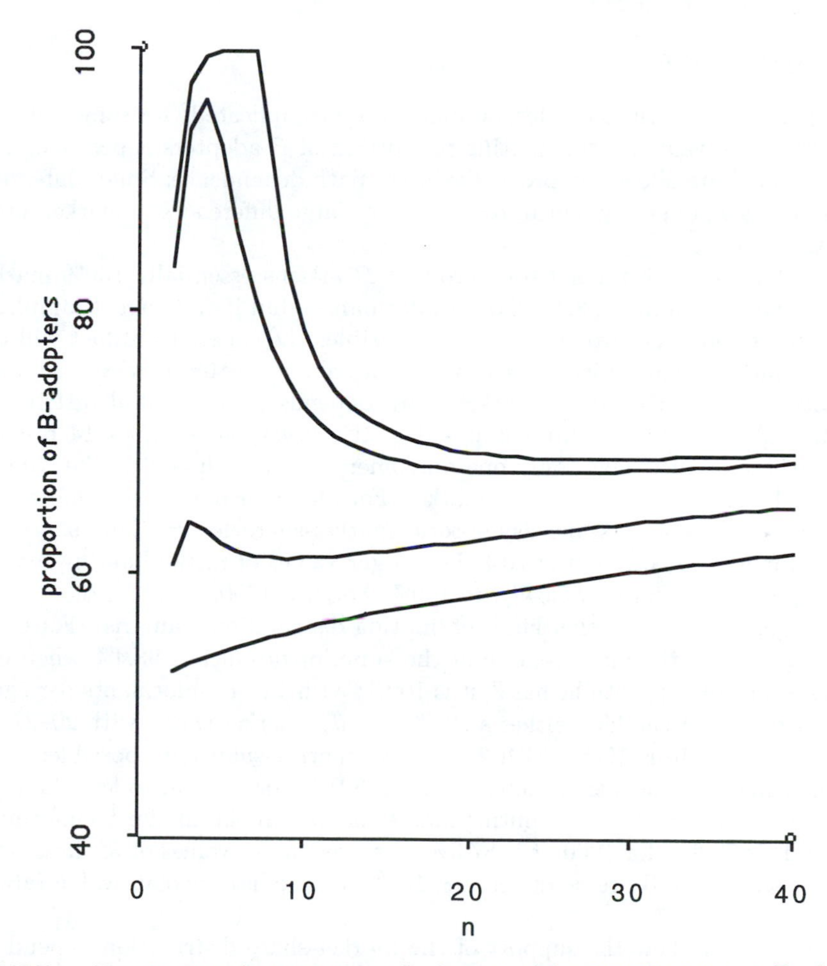

FIGURE 1 The asymptotic proportion of adopters of the superior product B: $d = -0.1$; $\mu = c_B, \sigma^2 = 1$. Each of the four curves corresponds to a different value of λ: for bottom curve, $\lambda = 0$, and for the others, in ascending order, $\lambda = 1, 1.6, 1.7$.

increases, but very slowly: when each agent interviews *1000* previous adopters, the superior product's asymptotic market-share is only 87.6%, more than 6% smaller than it is when each agent has only 3 informants! When $\lambda = 1.7$, the market-share of the superior product is 99.98% with 7 informants per agent, only 69.47% with 20, and hits bottom (68.98%) with 27. With 1000 informants per agent, the superior product obtains 87.7% of the market, about the same as with 2 informants and nearly 10% less than with 3.

PATH DEPENDENCE

For λ larger than 1.7, the situation becomes more complicated: for some values of n, more than one value for the limiting proportion of B-adopters is possible. That is, the market-share allocation process exhibits path dependence. Small differences in who samples whom early on can result in very large differences in market-shares later in the process.

When λ equals 1.8, the superior product B attains essentially 100% market-share[12] if every agent has 7, 8, 9, 10 or 11 informants—but if every agent samples 12 or 13 previous adopters, two outcomes are possible: the superior product will take the whole market asymptotically, or it will attain about an 80% market-share! The probability that B takes 100% market-share depends on the initial distribution (r, s); any value between 0 and 1 is possible. If every agent samples 14 previous adopters, there is once again only one outcome: this time, however, the superior product will claim only 76.6% of the market. For all larger numbers of informants, there is also only one possible asymptotic market-share for B. This share is a minimum for $n = 32$, where it is 70%. For larger values of n, the limiting market-share increases, as usual very slowly, to 100%. For $n = 1000$, it is 87.8%.

For larger values of λ, other kinds of limiting distributions can arise. For example, if λ equals 2.2, the market-share of the superior product is 98.4% when each agent has 2 informants; with he has 3, it is 100%; with 4 to 19 informants per agent, the superior product can have either a 100% or a *0%* market-share; with 20–22, the only possibility again is 100%; with 23–35, two market-shares are possible: one is 100%, the other decreases as n increases, from 82.7% for $n = 23$, to less than 75% for $n = 35$; thereafter, only one limiting market-share can obtain: for 50 informants per agent, it is 73.8%; for 1000, it is 87.3%. For even larger values of λ, three possible market-shares for B are possible: 0%, 100%, or one intermediate value (always greater than 50%).

Figure 2 shows how the support of the market-share distribution depends on λ and n for a set of parameters that differs from the one considered above only in the difference between the prior mean μ and c_B (here the difference is 2, for the example of Figure 1 it is 0).

[12]To be precise, greater than 99.99999%.

Five different types of market-share distributions appear in the figure.[13] In the first type, which occurs in an irregular-shaped region in the lower left-hand corner, the superior product dominates the market: its limiting market-share is essentially 100%.[14] In the second type, one of the products will attain essentially 100% market-share, but either product can dominate. This type of market-share allocation always obtains for sufficiently large values of the risk aversion parameter λ, for any fixed values of the other model parameters. In the third type of market-structure, there are three possible limiting market-share distributions: essentially 100% market domination by either one of the two products, or stable market-sharing at some intermediate level (> 50% for the superior product). For the particular parameter values used here, the intermediate solution gives between 70% and 85% market-share to the superior product, for values of λ less than 1.5 and n less than 100. In the fourth type of market-structure, either the superior product dominates or there is a single intermediate level of stable market-sharing (here again the market-share of the superior product is between 70% and 85%). In the fifth type of market-structure, only stable market-sharing at one possible intermediate level is possible. This type of market structure obtains for sufficiently large values of n, when the values of the other parameters are fixed at any particular set of values.

When the values of all other parameters are fixed and the number of informants n goes to infinity, the market-share of the superior product increases to 100%. Only in this doubly-asymptotic[15] sense, is it true that the "best possible" outcome at the aggregate level can be achieved by getting more and more information at the individual level.

For a class of decision-rule closely related to ALBO, the diffuse ALBO rules, even this weak asymptotic aggregate-level optimality property for the individual-level adage "more information is better" fails. Diffuse ALBO rules employ the same updating and decision processes as ALBO rules, but with respect to the standard uniform improper prior for μ. This prior distribution yields a proper posterior for $c_A(c_B)$ as long as there is at least one $A-(B-)$adopter among the agent's informants. If every informant turns out to have adopted the same product, a diffuse ALBO agent adopts the same product, regardless of what he learned about its performance. The diffuse ALBO rule corresponding to just two informants per agent has "optimal" aggregate-level effects: the limiting proportion of agents adopting the

[13]I think, but have not yet been able to prove, that these are the only types of market-share distributions that can arise with any parameterizations of the information contagion model, with one exception. Some models can lead to an absolutely continuous distribution on [0,1]: for example, if each agent adopts the product that his only informant adopted; or if the two products are in fact equal and ALBO agents with well-calibrated priors sample exactly two previous adopters.

[14]See footnote 12.

[15]In the number of adopters and in the number of informants per adopter.

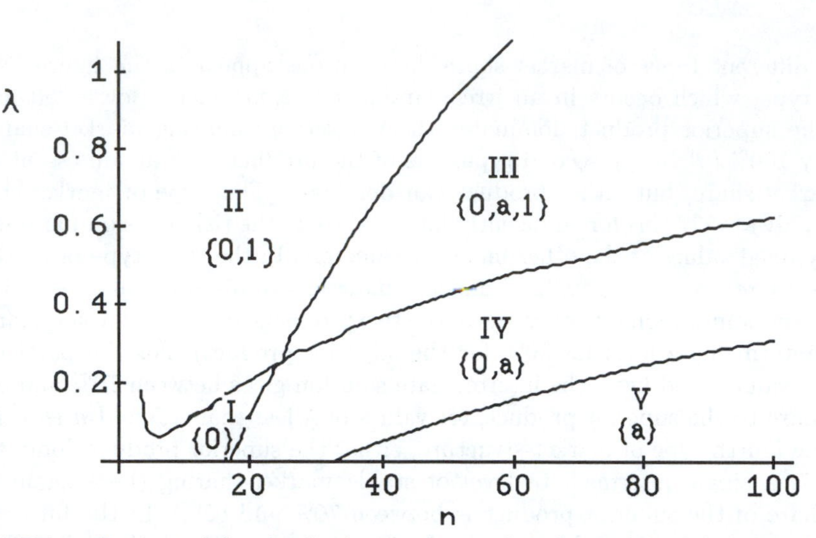

FIGURE 2 Support of the distribution for the market-share of the inferior product: $d = -0.1$, $\mu = c_B + 2$, $\sigma^2 = 1$.

superior product is 100%—actually, not just essentially—no matter how small the difference between the two products.[16] This level of aggregate-level "performance" is only approached, never attained, as n goes to infinity.

FINITE NUMBER OF ADOPTERS

All the results discussed above are asymptotic in the sense that they refer to an infinite sequence of adopters. Since the proportion of agents who adopt each product converges with probability one, asymptotic market-shares are easy to describe: as we have seen, the limiting market-share is often unique, and when it is not (except in very special cases), its distribution has at most 3 points. The situation after a finite number of adoptions—say 100—is much more difficult to summarize succinctly. First, the market-share of the superior product after 100 adoptions is a random variable that can take on any of 101 possible values. Second, the dependence of this distribution on the parameters r and s cannot be ignored. Under these circumstances, it is reasonable to ask whether—and how—the "more is less" information inversion phenomenon shows up for finite-sized agent societies.

[16] If the two products have equal performance characteristics, the limiting share of product A may be any number between 0 and 1. The market-share allocation process is a Polya urn process, so the a priori distribution for A's limiting market-share is $\beta(r, s)$.

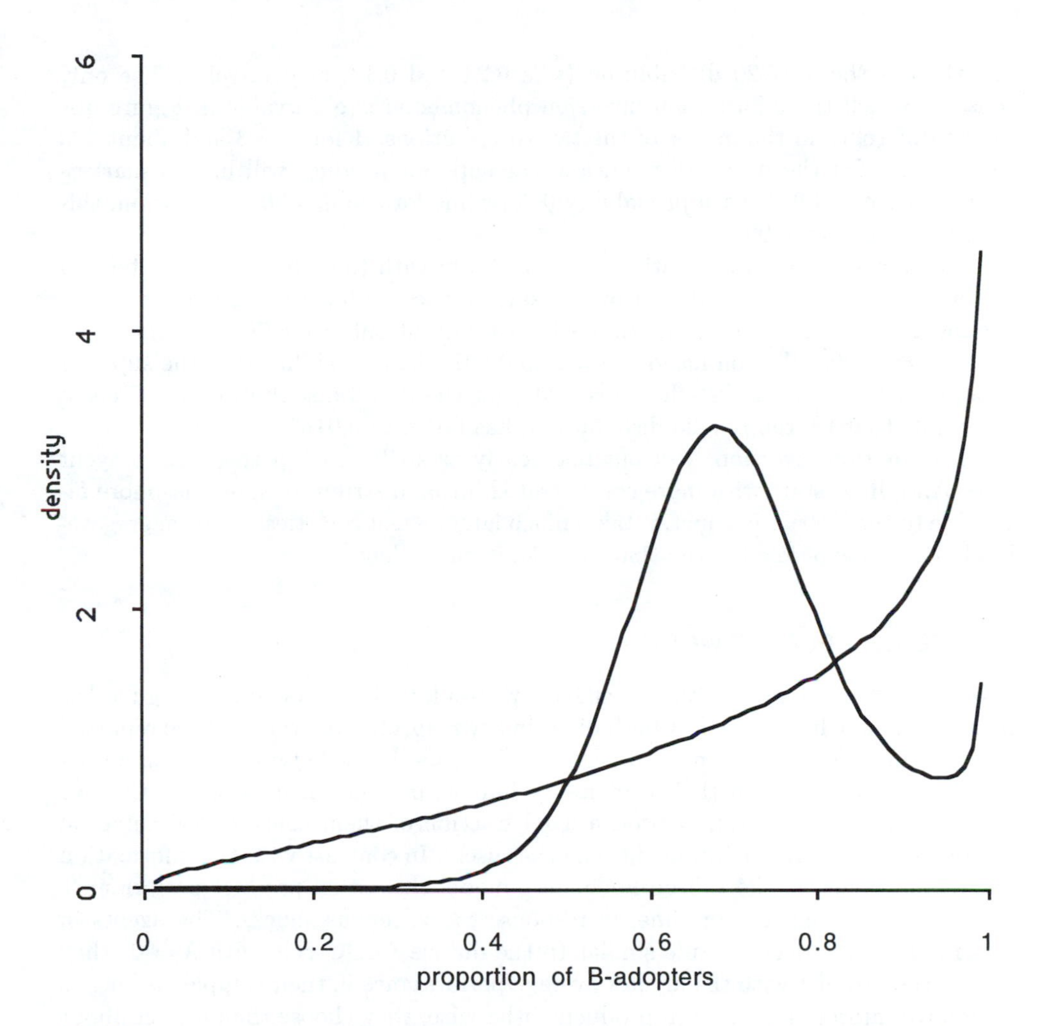

FIGURE 3 Distribution of the proportion of adopters of the superior product B after 100 adoptions: $d = -0.1$, $\mu = c_B$, $\sigma^2 = 1$, $\lambda = 1.6$, $r = 1$, $s = 2$. For the density with mode at 1, $n = 3$; for the other density, $n = 20$.

The answer to "whether" is: of course, but not uniformly, in r and s. Figure 3 shows the distribution for the proportion of adopters of the superior product B, with the same values for $d, \mu - c_B$, and σ^2 as in the Figure 1 example, λ equal to 1.6, and two different values of n: 3 and 20. The asymptotic market-shares of the superior product in these two cases are 93.88% and 68.04%, respectively. In the example illustrated in Figure 3, the initial proportion of B-adopters is $2/3$. The means of the two distributions are nearly identical: 0.712 and 0.707 for $n = 3$ and $n = 20$, respectively. Not surprisingly, the $n = 3$ distribution is much more spread out than is the $n = 20$ distribution (sd's 0.24 and 0.13, respectively). The only sense in which the information inversion phenomenon is observable in Figure 3 is in the difference in the modes of the two distributions: 1 for $n = 3$ and about 2/3 for $n = 20$. For the $n = 3$ distribution, the superior product will have a market-share in excess of 90% with probability 0.28, while for the $n = 20$ distribution, this probability is only 0.0915.

The market-share distributions change slowly with the number of adopters, of course in the direction predicted by the asymptotic results. After 5000 adoptions, the mean of the $n = 3$ distribution has increased, but only to 0.75, while the mean of the $n = 20$ distribution has decreased to 0.697; the probability that the superior product will have a market-share over 90% for the $n = 3$ distribution has risen to 0.3, while that for the $n = 20$ distribution, has fallen to 0.0185.

Information inversion, yes; but not nearly as striking as in the infinite agent case. And, if we start with more concentrated initial distributions, or ones more favorable to the inferior product, it takes much larger agent societies for the aggregate-level advantage of less information to make itself evident.

A NOTE ON RELATED WORK

Two other "more information/less efficiency" results arise in social-learning models in the economic literature, but both differ in key respects from those I have reported here. In the model of Ellison and Fudenberg,[9] agents choose between two competing products. Each agent in the uncountably infinite population uses one of the two products, and in each time period a fixed fraction of them consider switching on the basis of information obtained from other users. In contrast with the information contagion model, in the Ellison-Fudenberg model the actual product performance characteristics change over time, in response to exogenous shocks. The agents in this model use a decision rule similar to the diffuse ALBO rule with $\lambda = 0$: they choose the product with the highest average performance in their sample, as long as they have information on both products; otherwise, they choose the product about which they have information.

Ellison and Fudenberg find that when the number of informants per agent is small, the proportion adopting the superior product converges to 1, which does not happen when the number of informants is larger. This Ellison-Fudenberg version of information inversion is driven by their model's assumed "shock" to the products' performance characteristics: a currently unpopular product can be "resurrected" in a period in which its performance is particularly high only if many agents find out about this good performance, which happens with higher probability the more informants each agent has. Thus, more informants per agent guarantee that neither product is eliminated. More subtle relations between individual and aggregate level do not arise in the Ellison-Fudenberg model, because their assumption of an uncountable number of agents lets the strong law of large numbers sweep the delicate probabilistic effects of feedback under the rug.

In Banerjee,[5] heterogeneous agents encounter an investment opportunity only by hearing about others who chose to invest. The yield of the investment depends on a state of the world unknown to the agents. For some of the agents, the investment is profitable no matter what the state of the world, but for others it is not. Agents update their state of the world distribution rationally when they learn that someone has invested, and then decide whether or not to invest themselves. The probability that an agent will learn about an investment in a particular time interval is an increasing function of the number of agents who have invested. The key to deciding whether to invest is the time at which one first learns about the investment.

Banerjee's process begins when a certain fraction x of agents learn about the investment opportunity *and* the true state of the world; this is a one-time revelation, not repeated for later investors who must infer the state of the world as described above. Surprisingly, the aggregate-level efficiency (measured as the fraction of those who should invest that do, as a function of the true state of the world) does not increase with x, the amount of "hard information" injected into the system. To me, this result is interesting and counter-intuitive, but the phenomenon is different from what happens in the information contagion model, since in Banerjee's model the relevant "more information" comes in at the aggregate not the individual level.

If we range further afield than social-learning models, we can encounter a number of results that bear a certain resemblance to what I found in the information contagion context. The general issue is: whether (and if so when) more information at the microlevel of a multilevel system result in degraded higher-level system performance? Many examples of this phenomenon come to mind. Think of stochastic Hopfield networks (see Hertz, Krogh, and Palmer,[11] Chap. 2): their performance as associative memories is enhanced when each unit responds to its input signals from other units *perturbed by noise*. Here, I interpret "less noise" as "more information": thus, more information at the individual unit level, about the state of other units, can result in poorer aggregate- or system-level performance.[17] Or the "complexity catastrophe" that happens in the NK model, for sufficiently large values of the connectivity parameter K (Kauffman[13]). It would be interesting to speculate about whether there are some common underlying processes linking these different phenomena.

2. RATIONALITY, SUFFICIENCY, AND EFFICIENCY: MAX DOMINATES ALBO

In section 1, I argued that ALBO agents were as rational as we have a right to expect agents to be in the information contagion context. In section 2, I showed that

[17] In Hopfield networks, too much noise always results in poor performance, while in the information contagion context very little—but not *no*—information at the individual level can in some circumstances result in very good aggregate-level results.

the aggregate-level "efficiency"—the proportion of agents that adopts the superior product—of ALBO rules depends in a complicated way on the model parameters. In particular, the market-share of the superior product can be substantially less than 100%, at least when the real difference between the products' performance characteristics is not too large.[18] In this section, I introduce another decision rule, which I call the Max rule. The Max rule has the remarkable property that it always leads to 100% market-share for the superior product, no matter how small the difference between the two products. Thus, from a social efficiency perspective, it dominates ALBO rules, however attractive the Bayesian updating and maximizing expected utility prescriptions may be at the individual level. Moreover, the Max rule violates one of the central canons of statistical inference and decision theory: it is based on a function of the data (X, Y) that is not sufficient for the unknown parameters c_A and c_B!

The Max rule was "discovered" in the course of an experiment carried out by Narduzzo and Warglien.[18] This experiment reproduced the context of the information contagion model, with Venetian business school students as agents. As part of their experiment, Narduzzo and Warglien conducted a protocol analysis, in which subjects explained why they preferred one product to the other. Not surprisingly, none of the subjects claimed to reach their decision by Bayesian optimization. Instead, they typically invoked one or another of four simple rules-of-thumb to account for their choices. One of these rules, by the way, corresponds to the Ellison-Fudenberg rule cited in the last section, the diffuse ALBO rule with $\lambda = 0$. Another, in complete contrast to my rationality argument in section 1, uses only the information in X and completely ignores the information in Y: choose whichever product the majority of your informants chose.[19]

[18]In fact, the proportion of ALBO agents that adopt the superior product can be substantially less than 1 even when the real difference between the products is arbitrarily large. For example, suppose that the agents are well-calibrated for the superior product B (that is, $\mu = c_B$) and, as usual, the public information does not distinguish between the two products, so the agents have the same prior distribution for c_A as for c_B. In such a situation, the market share for the inferior product A need not be essentially 0, no matter how large the actual difference between the two products. The reason is simple: unlike the usual single-agent Bayesian asymptotics, in the information contagion context the effect of the prior distribution does not go away as the number of adopters increases. When the proportion of adopters of the superior product B becomes sufficiently large, an agent with high probability samples only B adopters. Then, his posterior mean for c_A is just the prior mean $\mu = c_B$. Of course, the expected value of the average of his Y-observations is also c_B, so the probability that his posterior mean for c_B is less than his posterior mean for c_A is 1/2. Thus, if he is not very risk averse and the number of his informants is not too large, he will end up adopting A with appreciable probability. For example, if $\lambda = 0$, $n = 5$, and $\sigma^2 = 10$, the inferior product will have a market-share of at least 18%, no matter how low its actual performance characteristic.

[19]Not surprisingly, this "imitation" rule leads to complete market domination by one or the other product. If $r = s$, each product has a probability 1/2 of dominating. See Lane and Vescovini.[15] Kirman[14] provides an interesting analysis of the effects of this rule when $n = 1$, the population of agents is finite, and agents can change their product choices.

The Max rule can be stated as follows: choose the product associated with the highest value observed in the sample. That is, denoting by max the index of $y_{(n)} = \max(y_1, \ldots, y_n)$, the rule is given by $D((x_1, y_1), \ldots, (x_n, y_n)) = x_{\max}$. Note that if all informants adopted the same product, then the Max rule opts for that product, regardless of the magnitude of the observations $y_1, \ldots, y_n$.[20]

The Warglien-Narduzzo experimental subjects who invoked the Max rule justified it with a curious argument. They claimed that they ought to be able to obtain product performance "as good as anyone else," and so they figured that the best guide to how a product would work for them was the best it had worked for the people in their sample who had used it. Narduzzo and Warglien[18] describe this argument as an example of a "well-known bias in human decision-making," which they follow Langer[16] in calling the "illusion of control."

Of course, this justification completely fails to take into account sample-size effects on the distribution of the maximum of a set of i.i.d. random variables. Since the current market leader tends to be over-represented in agents' samples, its maximum observed value will tend to be higher than its competitors', at least as long as the true performance of the two products are about the same. Thus, it seems plausible that in these circumstances a market lead once attained ought to tend to increase. According to this intuition, the Max rule should generate information contagion and thus path-dependent market domination.

This intuition turns out to be completely wrong. In fact, for any $n \geq 2$, the Max rule always leads to a maximally socially efficient outcome:

PROPOSITION: Suppose $n \geq 2$. If the two products, in reality, have identical performance characteristics, the market-share allocation process exhibits path dependence: the limiting distribution for the share of product A is $\beta(r, s)$. Suppose, however, that the two products are different. Then, the better product attains 100% limiting market-share with probability 1.

A proof of this proposition, from Lane and Vescovini,[15] is given in the appendix.

Why ALBO should fail to lead to market domination by the superior product is no mystery. But it is not easy to understand why the Max rule does allow agents collectively to ferret out the superior product. The proof gives no insight into the reason, though it does make clear that the larger is n, the faster will be the convergence to market domination. In fact, I am not sure there is a real "reason" more accessible to intuition than the fact that drives the proof: with the Max rule, the

[20] $p(k)$ can be calculated for the Max rule as follows: $p(n) = 1$ and $p(0) = 0$; for $0 < k < n$, $p(k)$ is just the probability that the maximum of a sample of size k from a $N(c_A, 1)$ distribution will exceed the maximum of an independent sample of size $(n - k)$ from a $N(c_B, 1)$ distribution:

$$p(k) = \int \Phi[y - (c_B - c_A)]^{n-k} k \Phi[y]^{k-1} \phi[y] dy \,,$$

where φ is the standard normal density function.

probability of adopting the inferior product given the current proportion of inferior product adopters is always smaller than that proportion. In fact, the superiority of the Max rule to ALBO rules is not particularly evident in "real time." The associated distributions for the superior product's market-share after a finite number of adoptions, tend to be both more spread-out and more dependent on initial conditions (r and s) than those associated with ALBO rules having the same number of adopters and number of informants per adoption. Nonetheless, these distributions do, sooner or later, inexorably converge to point mass at 100%.

3. SUMMARY

This chapter has focused on a simple instance of a deep and general problem in social science, the problem of micro–macro coordination in a multilevel system. Behavior at the microlevel in the system gives rise to aggregate-level patterns and structures, which then constrain and condition the behavior back at the microlevel. In the information contagion model, the interesting microlevel behavior concerns how and what individual agents choose, and the macrolevel pattern that emerges is the stable structure of market-shares for the two products that results from the aggregation of agent choices. The connection between these two levels lies just in the agents' access to information about the products.

The results described in sections 2 and 3 show that mechanism at the microlevel has a determining and surprising effect at the macrolevel. In particular, what seems good from a microlevel point of view can turn out to have bad effects at the macrolevel, which in turn can materially affect microlevel agents. The chapter highlighted two specific instances of this phenomenon. First, although individual decision-makers should always prefer additional costless information, performance at the aggregate level can decrease as more information is available at the microlevel. Second, despite the fact that individual decision-makers "ought" to maximize their expected utility, a very large population of Max rule followers will achieve a higher level of social efficiency than will a population of Bayesian optimizers.

APPENDIX: PROOFS

1. ALBO AGENTS EXPECT TO GAIN FROM MORE INFORMATION

Let I_i represent the information an agent obtains from his first i informants: $I_i = \{(X_1, Y_1), (X_2, Y_2), \ldots, (X_i, Y_i)\}$. Further, let A_i and B_i represent the expected utility the agent calculates for products A and B, respectively, on the basis of I_i, and let $U_i = \max(A_i, B_i)$. Thus, U_i is the value to the agent of his currently preferred product. I now show that an agent always expects to increase the value of U by obtaining information from an additional informant.

CLAIM 1: $\{A_i\}$ and $\{B_i\}$ are martingales with respect to the σ-fields generated by I_i.

PROOF OF CLAIM: I show that $\{A_i\}$ is a martingale.

$$\begin{aligned} E(A_{i+1}|I_i) &= E(A_{i+1}1_{\{X_{i+1}=A\}}|I_i) + E(A_{i+1}1_{\{X_{i+1}=B\}}|I_i) \\ &= E(A_{i+1}|\{X_{i+1} = A\}, I_i)P(X_{i+1} = A|I_i) + \\ &\quad A_i P(X_{i+1} = B|I_i)\,. \end{aligned}$$

Thus, I need to show that $E(A_{i+1}|\{X_{i+1} = A\}, I_i) = A_i$:

Let n be the number of A-adopters among the first i informants and let $Z_1, \ldots, Z_n$ denote the Y-values obtained from these informants. Let a denote σ^{-2}. Then, given I_i,

$$c_A \sim N\left(\mu_n, \frac{1}{n+a}\right),$$

where

$$\mu_n = \frac{1}{n+a}\left(\sum_{j=1}^{n} Z_i + a\mu\right).$$

Given I_i and $\{X_{i+1} = A\}$,

$$A_i = -\exp\left(-2\lambda\left(\mu_n - \frac{\lambda}{n+a}\right)\right)$$

and

$$A_{i+1} = -\exp\left(-2\lambda\left(\frac{n+a+1}{n+a}A_i + \frac{Y_{i+1}}{n+a+1}\right)\right).$$

Thus,

$$E(A_{i+1}|\{X_{i+1}=A\},I_i) = -\exp\left(-2\lambda\left(\frac{n+a}{n+a+1}\mu_n - \frac{\lambda}{n+a+1}\right)\right) \cdot E\left(\exp\left(-2\lambda\frac{Y_{i+1}}{n+a+1}\right)|\{X_{i+1}=A\},I_i\right).$$

Given I_i and $\{X_{i+1}=A\}$,

$$Y_{i+1} \sim N\left(\mu_n, \frac{n+a+1}{n+a}\right);$$

thus, the second factor above evaluates to

$$\exp\left(-2\frac{\lambda}{n+a+1}\left(\mu_n - \frac{\lambda}{n+a}\right)\right),$$

and so

$$E(A_{i+1}|\{X_{i+1}=A\},I_i) = A_i\,.$$

CLAIM 2: $E(U_{i+1}|I_i) > U_i$ with probability 1: Since $U = \max(A,B), U$ is a submartingale: $E(U_{i+1}|I_i) \geq U_i$. Moreover, since the support of Y_{i+1} is unbounded, the information from the $(i+1)^{st}$ informant will induce the agent to change his mind about which product to obtain with positive probability: thus, the inequality is strict.

2. THE MAX RULE IS MAXIMALLY SOCIALLY EFFICIENT

PROOF Proof: The proposition follows from the two claims below, using the results from Hill, Lane, and Sudderth[12] and Arthur and Lane[3] cited in section 1 of this chapter:

CLAIM 1: If $c_A = c_B$, then $p(k) = k/n$, and hence, $f(x) = x$ for all x in [0,1]: In this case, $p(k)$ is just the probability that the maximum of the sample of the n i.i.d. random variables $Y_1, \ldots, Y_n$ is one of the k of them whose associated X-value is A, then $f(x) = x$, and the market allocation process is a Polya urn scheme, from which the $\beta(r,s)$ limiting distribution follows.

CLAIM 2: If $c_A > c_B$, then $p(k) \geq k/n$ for all k, and for $0 < k < n$ the inequality is strict; hence, $f(x) > x$ for all x in (0,1). We now consider a sample of size k from a $N(c_A, 1)$ distribution and a sample of size $(n-k)$ from a $N(c_B, 1)$. $p(k)$ is the probability that the maximum of the combined sample comes from the first sample, which is clearly a nondecreasing (increasing for $0 < k < n$) function of $c_A - c_B$. That $p(k) \geq k/n$ then follows from claim 1. Since f is just a linear combination of the $p(k)$'s with positive coefficients in (0,1), the inequality $f(x) > x$ follows from the inequalities for each $p(k)$ and the calculation of f in claim 1. The inequality $f(x) > x$, together with the evaluations $f(0) = 0$ and $f(1) = 1$, imply that 1 is the only value in the set $\{x : x \text{ in } [0,1], f(x) = x \text{ and } f'(x) \leq 1\}$. Hence, the limiting market-share of product A is 1 with probability 1.

ACKNOWLEDGMENTS

I would like to thank the Santa Fe Institute and the Italian Ministry of Research for support for this research. I benefited from discussions with Massimo Warglien and Alessandro Narduzzo, and my colleagues at the Santa Fe Institute (especially Brian Arthur and Bob Maxfield) and the University of Modena (especially Andrea Ginzburg). Roberta Vescovini, Giordano Sghedoni, and Raffaella Barozzi contributed invaluable research assistance in the course of their baccalaureate thesis projects.

REFERENCES

1. Ahn, M., and N. Kiefer. "Local Externalities and Societal Adoption of Technologies." *J. Evol. Econ.* **5** (1995): 103–117.
2. Arthur, W. B. "Complexity in Economic and Financial Markets." *Complexity* **1(1)** (1995): 20–25.
3. Arthur, W. B., and D. Lane. "Information Contagion." *Struct. Change & Econ. Dyn.* **4** (1993): 81–104.
4. Banerjee, A. "A Simple Model of Herd Behavior. *Quart. J. Econ.* **107** (1992): 797–817.
5. Banerjee, A. "The Economics of Rumours." *Rev. Econ. Studies* **60** (1993): 309–327.
6. Banerjee, A., and D. Fudenberg. "Word-of-Mouth Learning." preprint, 1995.
7. Bikhchandani, S., D. Hirshleifer, and I. Welch. "A Theory of Fads, Fashion, Custom, and Cultural Change as Informational Cascades." *J. Pol. Econ.* **100** (1992): 992–1026.
8. Ellison, G., and D. Fudenberg. "Rules of Thumb for Social Learning." *J. Pol. Econ.* **101** (1993): 612–643.
9. Ellison, G., and D. Fudenberg. "Word of Mouth Communication and Social Learning." *Quart. J. Econ.* **110** (1995): 93–125.
10. Heath, D., and W. Sudderth. "On Finitely Additive Priors, Coherence, and Extended Admissibility." *Ann. Stat.* **6** (1978): 333–345.
11. Hertz, J. A., A. S. Krogh, and R. G. Palmer, eds. *Introdution to the Theory of Neural Computation.* Santa Fe Institute Studies in the Sciences of Complexity, Lect. Vol. 2. Redwood City, CA: Addison-Wesley, 1991.
12. Hill, B., D. Lane, and W. Sudderth. "A Strong Law for Some Generalized Urn Models." *Ann. Prob.* **6** (1979): 214–226.
13. Kauffman, S. *The Origins of Order: Self-Organization and Selection in Evolution.* New York: Oxford University Press, 1993.

14. Kirman, A. "Ants, Rationality, and Recruitment." *Quart. J. Econ.* **108** (1993): 137–156.
15. Lane, D., and R. Vescovini. "Decision Rules and Market-Share: Aggregation in an Information Contagion Model." *Ind. & Corp. Change* **5** (1996): 127–146.
16. Langer, E. J. "The Illusion of Control." *J. Pers. & Soc. Psych.* **32** (1975): 311–328.
17. Lee, I. "On the Convergence of Informational Cascades." *J. Econ. Theory* **61** (1993): 395–411.
18. Narduzzo, A., and M. Warglien. "Learning from the Experience of Others: An Experiment on Information Contagion." *Ind. & Corp. Change* **5** (1996): 113–126.
19. Smith, L., and P. Sorensen. "Pathological Outcomes of Observational Learning." Working Paper, MIT, Cambridge, MA, 1995.

Yannis M. Ioannides
Department of Economics, Tufts University, Medford, MA 02155; E-mail: yioannid@tufts.edu

Evolution Of Trading Structures

The paper reviews the evolution of trading structures by examining two pertinent strands in the literature on economies with interacting agents, one, works that presume a specified topology of interactions among agents, and two, works that let random mechanisms determine that topology. The paper reviews interactive discrete choice models in isotropic settings and proposes extensions within certain stylized anisotropic settings, which are particularly interesting for economists. In particular, circular patterns of interaction highlight the role of money and credit; tree-type settings depict Walrasian interactions. The paper suggests that the random topology of interaction approach, which has employed random graph theory to study evolution of trading structures, may go beyond analyses of sizes of trading groups and thus exploit the full range of possible topological properties of trading structures.

The paper proposes to integrate those approaches and to exploit their natural complementarities. In the simplest possible version, our synthesis involves individual decisions and expectations, randomness, and nature combining to fix an initial "primordial" topology of interaction. The dynamics of interaction move the economy from then on. The evolution of trading structures depends critically upon multiplicity and stability properties of equilibrium configurations of the interaction model. The paper addresses a number of additional topics, including matching models, spatial aspects of the evolution of trading structures and issues of statistical inference.

1. INTRODUCTION

The notion that economic transactions are driven by the gains that individuals and other decision-making units expect to make through exchanging goods and services lies at the heart of economics. Much of the theoretical structure of modern economics assumes a given physical and institutional infrastructure, through which trades take place. At the same time, modern economics has neglected another equally important prerequisite for trade, namely the ways that agents find out about each other's availability and willingness to engage in transactions.

In the most general setting, the present paper addresses a very broad topic. It looks at spatial, temporal and informational aspects of trading structures, that is, when do transactions take place, to what extent do agents who are geographically separated have to come into physical contact in order to transact, and what information is available to and used by different potential traders?

The casual observer of economic life and activity, whether a historian, geographer, anthropologist or economist, would notice that transactions take place in fairly well-defined time periods, that some economic activity may be evenly distributed over part of the economically-exploited space, such as agriculture, but that most people live in cities or towns, areas of relatively high population density, and most manufacturing also occurs in high-density areas. Economic activity regularly utilizes space. Electricity is produced in specific sites and transported over long distances by means of physically fixed infrastructure, cereals and minerals are transported in barges and ships. Information moves over space in the form of voice, numerical or visual data. The laws of physics, chemistry, and biology define the technologically feasible choices available to the individual members of the economy at any given point in time. Within these constraints, it is the decisions that individuals make which in turn determine how constraints will evolve in the future. The nature and identity of decision makers can change and the infrastructure that defines the environment of trade structures is endogenous.

The Arrow-Debreu model deals with these problems by defining a set of commodities and by taking as given the corresponding set of markets. It is a feature of the Arrow-Debreu model that commodities, defined in terms of physical characteristics, are indexed in terms of points in time, location in space, and contingent events. Evstigneev's extension of the Arrow-Debreu model, which we discuss below, generalizes this notion by endowing the index set with the topology of a directed graph, rather than the tree of events in the Arrow-Debreu model. Many of the questions we raise may be considered as special cases of incomplete markets, where the pertinent contingencies involve the likelihood that agents can or will be connected in order to trade. The paper sets out to exploit the insight that may be gained by considering broad features of economies, which consist of large numbers of agents who can be connected in a variety of ways. We use this approach to integrate our knowledge of trade structures, with a fairly broad view of such structures, and address the largely unresolved issue of the endogeneity of markets and institutions that mediate trades. Research under the rubric of economics of information has addressed aspects of strategic release and processing of information, the transmission of information via market signals, and the search for trading partners. The emergence of the full topology of the modern economy as consequence of preferences and prevailing communication technologies is, however, an important unresolved issue in current research. We adopt an approach for most of the paper which is intended to allow us to understand the *emergent* properties of the economy. We abstract from the vast complexity of the problem by restricting ourselves to graph topologies that roughly describe interconnections between agents and markets.

The importance of the topology of interactions may be underscored by reference to Schelling's theory of residential segregation, which evolves from spatial dependence. The following spatial example, also from Schelling,[100] makes the point succinctly: "If everybody needs 100 watts to read by and a neighbor's bulb is equivalent to half one's own, and everybody has a 60-watt bulb, everybody can read as long as he and both his neighbors have their lights on. Arranged on a circle, everybody will keep his lights on if everybody else does (and nobody will if his neighbors do not); arranged in a line, the people at the ends cannot read anyway and turn their lights off, and the whole thing unravels" (p. 214).

The remainder of the paper is organized as follows. Section 2 introduces the basic intuition of our approach and defines the two broad classes of models. Those which involve a prespecified topology of interaction among agents are discussed first in section 3, and those which do not are discussed in section 4. Both those sections also contain some original syntheses of the existing literature. Some of those new ideas are pursued further in section 6, where a prototype equilibrium model of interactions with endogenous links is developed. A number of other topics, notably including spatial models, issues of statistical inference and matching models, are reviewed in section 5. Section 7 concludes.

2. ECONOMIES WITH INTERACTING AGENTS

Modeling trade links is easiest in a setting with two initially isolated agents where strategic considerations are simple to express.[61] In more complicated settings, intermediation possibilities will affect the likelihood that trade links between two individuals will open up.[73] A moment's reflection suggests that aggregating individual decisions into a pattern of links in a large economy is an inherently very difficult problem to tackle.

Walrasian equilibrium prices are obtained by solving for the zeroes of a system of Walrasian excess demands. Excess demands are defined as additive with respect to the index of agents participating in each Arrow-Debreu market. We are interested in exploring patterns of dependence across agents, which depending upon agents' preferences may imply nonadditive aggregation. When aggregation is additive, it is appropriate to invoke powerful statistical theorems to ensure that small risks in large markets may cause uncertainty, in the guise of individual heterogeneity, to have negligible consequences in-the-large. A large number of agents responding to feedback from one another may cause individual, mutually independent, sources of randomness to reinforce and, thus, result in nonadditive aggregation.

The first demonstration that in complex patterns of trade, feedback to external shocks may be responsible for nonadditive aggregation of individual characteristics is due to B. Jovanovic.[70,71] He shows that individual mutually independent shocks affecting a large number of traders who, because they are interlinked in the form of a circle, are subject to sizable fluctuations in per capita endogenous variables. Such phenomena are, as we shall see, inherent in cyclical patterns of interaction.

The structure of interconnections is crucial for another, relatively unexplored, area of the economics of information, namely how economic news travels in large economies. A prototype investigation here is the work by Allen on the diffusion of information about technological innovation that derives the logistic law from a model based on Markov random fields.[1,2][1] The appropriate concept for random topology of interaction models is that of connectivity, which is discussed in section 4 below.

It may be argued that in an era of nearly ubiquitous communication systems—telephone, fax, electronic mail, beepers, alarms, and superfast communication networks—few communication problems should remain. In practice, much ambiguity and uncertainty remain having to do with coordination, resolution of ambiguities, and failure to provide complete information. The other side of this is that contracts have to be incomplete. Problems of imperfect communication have also been addressed by engineers. It is known that decision-making in decentralized environments in the presence of imperfect information creates delays, and lack of central control gives rise to complex dynamics. Such dynamics have concerned computer scientists.[66][2] It is also known that large communication networks are susceptible to inherent dynamic behavior as lagged responses tend to build up.[12]

Most macroeconomic theorizing in the 1970s and the 1980s rested on some very special models, such as variants of the Samuelson overlapping generations model, which are characterized by stylized restrictions on trading patterns and have been used to motivate the holding of money. The coexistence of a national currency with numerous often geographically restricted credit arrangements (credit cards versus

[1]For tractable treatments of Markov random fields, see Spitzer,[101,102] Preston,[97] and Kindermann and Snell.[78]

[2]The challenges created by interconnected computer systems have been addressed in the popular press.

personal credit versus money) provokes further investigation. It is interesting to ponder general results regarding particular patterns of interaction, defined in terms of time-space events, which combine with appropriate endowments and preferences to give rise to a national currency. It is also interesting to study money (and other assets) as emergent properties of an economy.

2.1 MODELS OF COMMUNICATION IN ECONOMIES WITH INTERACTING AGENTS

Two distinct approaches have developed in the literature on economies with interacting agents since Föllmer[54] formally introduced the term.[3] The older approach, proposed by Kirman,[79] involves the use of random graph theory to model the topology of interconnections as random. Kirman et al.,[82] Ioannides,[68] and Durlauf[39] are the key contributions in this approach. This literature provides a better understanding of trade among a large number of traders when links among traders are random (and thus the number of links per trader varies). Most of this literature has addressed properties of trade that depend only on the *number* of interlinked agents and ignore the actual *topology*, that is, graph topology,[4] of interconnections. The second, and newer, approach is predicated on a given topology of interconnections. Markov random field theory has been employed as a natural modeling device there.

We review the accomplishments of both those approaches. We also propose to synthesize them by introducing the notion of a "grand" compound event of trade outcomes. The likelihood of a particular realization of trade outcomes associated with a given topology of interconnections may be obtained in terms of the probability law that describes the evolution of different patterns of interaction. This concept is new, in the context of this literature, and will be used to analyze the endogenous evolution of patterns of interconnections as an outcome of equilibrium interactions. We shall see that certain stylized anisotropic topologies of interaction lend themselves to particularly interesting economic applications.

3. SPECIFIED TOPOLOGY OF INTERACTION: MARKOV RANDOM FIELDS

The pioneering work in the modern approach to a class of economies with interacting agents is due to Durlauf.[34,35,36,37,38,39] Several of Durlauf's papers share the same analytical core, which involves Markov random field models defined over a

[3] Kirman[81] provides a different review of this literature.

[4] Unless otherwise indicated, the term topology in this paper is meant in the sense of *graph* topology. Other topologies are still pertinent; see Haller.[59,60]

countable infinity of agents located in the two dimensional *lattice* of integers Z^2.[5] We first summarize some basic analytical features of the theory and then turn to its applications for modeling interactions in problems with lattices as well as more general graph topologies.

Let $\omega_{i,j}, \underset{\omega}{\sim}$, and Ω denote, respectively, the activity of each of a continuum of identical agents located in point (i,j) of the lattice Z^2, the vector denoting the joint level of activity of agents in all sites, and the set of all possible states. The vector $\underset{\omega}{\sim}$ is typically referred to as a *configuration*. In a typical application, the set S of all possible values of $\omega_{i,j}$ is assumed to be finite. It is assumed, for simplicity, that the choice set is binary, $S = \{0,1\}$. For example, $\omega_{i,j} = 1$ (0) means the agent produces (does not produce). Alternatively, it may denote that an agent is (is not) informed (cf., Allen[1,2]). The literature proceeds by making assumptions about local interactions and derives global properties of equilibrium distributions. For example, Durlauf assumes that activity by individual ι located in site (i,j) of Z^2 is affected by activity in all neighboring sites, that is $\{\omega_{i-1,j}, \omega_{i+1,j}, \omega_{i,j-1}, \omega_{i,j+1}\}$.

We now briefly outline the extension of the model that allows for more general patterns of interactions. Let $\mathcal{I}$ be a finite set of individuals, $\iota \in \mathcal{I}$, which are identified with the vertices of a graph. Interactions in an economy are defined in terms of a graph $G(V,E)$, where: V is the set of vertices, $V = \{v_1, v_2, \ldots, v_n\}$, a one-to-one map of the set of individuals $\mathcal{I}$ onto itself—the graph is labeled; $n = |V| = |\mathcal{I}|$ is the number of vertices (nodes), which is known as the *order* of the graph; E is a subset of the collection of unordered pairs of vertices; $q = |E|$ is the number of edges, which is known as the *size* of the graph. In a typical application in this literature, nodes denote individuals, and edges denote potential interaction between individuals. It will be useful to define the set of *nearest neighbors* of individual $\iota \in \mathcal{I}$, $\nu(\iota) = \{j \in \mathcal{I} | j \neq \iota, \{\iota, j\} \in E\}$. We define an individual ι's *environment* as the state of all other individuals, formally as a mapping

$$\eta_\iota : \mathcal{I} - \{\iota\} \rightarrow \underbrace{\{0,1\} \times \ldots \times \{0,1\}}_{n-1}.$$

Interactions across individuals are defined in terms of probability distributions for the state of individual ι conditional on her environment, $\pi_\iota(\omega_\iota | \eta_\iota)$. The collection of these distribution functions, $\mathcal{V} = \{\pi_\iota\}_{\iota \in \mathcal{I}}$ is known as a *local specification*. We say that $\mathcal{V}$ is a *nearest neighbor specification* if it implies that the state of individual ι depends only upon the state of her neighbors. A probability measure π_ι is said to define a *Markov random field* if its local specification for ι depends only on knowledge of outcomes for the elements of $\nu(\iota)$, the nearest neighbors of ι. This definition of a *Markov random field* confers a *spatial* property to the underlying stochastic structure, which is more general than the Markov property. We normally

[5]The development of Markov random fields originated in lattices,[32] but was later extended by Preston[97] to general graphs.

impose the assumption that regardless of the state of an individual's neighbors her state is nontrivially random.

Random fields are easier to study in a state of statistical equilibrium. While in most of our own applications we have a finite number of agents in mind, a countable infinity of agents is also possible.[1,2,97] It is an important fact that for infinite graphs there may be more than one measure with the same local characteristics. When this happens the probabilities relating to a fixed finite set will be affected by the knowledge of outcomes arbitrarily far (or, just quite far, if infinity is construed as an approximation for large but finite sets) from that set. Existence of more than one measure with the same local characteristics is known as a *phase transition*.[6]

We say that a measure μ is a *global phase* for model $(G, S, \mathcal{V})$ if μ is *compatible* with $\mathcal{V}$ in the sense that $\mu\{\omega_\iota = s|\eta_\iota\} = \pi_\iota(s|\eta_\iota)$, $s \in S$, where $\pi_\iota(s|\eta_\iota)$ is agent ι's environment. A result from the literature on Markov random fields states that if the set of nodes (individuals) is finite and $\mathcal{V}$ is a strictly positive nearest neighbor specification, then there exists a unique *global phase* which is consistent with the local specification $\mathcal{V}$. The global phase is a Markov random field. The literature has established[58] that every Markov random field is equivalent to a *Gibbs state* for some unique *nearest neighbor potential*. The equivalence between Gibbs states and Markov random fields is responsible for an enormous simplification in the formal treatment of Markov random fields.[7] We proceed with definitions for these terms.

A set of vertices κ in an graph $G(V, E)$, $\kappa \subset V$, is a *clique*, or *simplex*, if every pair of vertices in κ are neighbors. A *potential* $\mathcal{D}$ is a way to assign a number $\mathcal{D}_A(\underset{\sim}{\omega})$, to every subspecification $\underset{\sim}{\omega}_A$ of $\underset{\sim}{\omega} = (\omega_1, \ldots, \omega_\iota, \ldots, \omega_n)$. A potential $\mathcal{D}$ is a *nearest neighbor Gibbs potential* if $\mathcal{D}_A(\underset{\sim}{\omega}) = 0$, whenever A is not a clique. Let $\Pi(\cdot)$ be the probability measure determined on Ω by a nearest neighbor Gibbs potential $\mathcal{D}$, that is

$$\Pi(\underset{\sim}{\omega}) = \frac{1}{\zeta} \exp\left[\sum_{\kappa \subset V} \mathcal{D}_\kappa(\underset{\sim}{\omega})\right], \tag{1}$$

where the sum is taken over all cliques κ on the graph $G(V, E)$ and ζ is a normalizing constant. It turns out that the probability measure for the state of individual ι conditional on her environment η_ι, is given by:

$$\pi_\iota(\omega_\iota|\eta_\iota) = \frac{\exp[\sum_{\kappa \subset V} \mathcal{D}_\kappa(\underset{\sim}{\omega})]}{\sum_{\underset{\sim}{\omega}'} \exp[\sum_{\kappa \subset V} \mathcal{D}_\kappa(\underset{\sim}{\omega}')]}, \tag{2}$$

[6]See Kindermann and Snell,[78] for a simple statement, and Dobrushin[32,33] and Preston[97] for elaboration of conditions for the absence of phase transitions. See also Ellis[42] and Georgii.[53]

[7]In contrast, no such simplification is available if random fields are defined over *directed* graphs, as in the applications pursued by Evstigneev,[47,48,49] Evstigneev and Greenwood,[50] and Evstigneev and Taksar.[51,52]

where $\tilde{\omega}'$ is any configuration which agrees with $\tilde{\omega}$ at all vertices except possibly ι.[8]

The integration of spatial with temporal considerations, which Markov random fields make possible, is a powerful modeling tool. In economic applications, normally it is the long-run equilibrium properties of the model which are of interest and depend upon the parameters of $(G, S, \mathcal{D})$, or alternatively, of $(G, S, \mathcal{V})$. Unless these parameters are endogenous, the model has equilibrium implications, which are quite similar to those of stochastic models indexed on time. For example, planting a rumor may change the system's initial conditions but not its global phase.[1] The dynamic implications of Markov random field models are rather hard to study, unless particular simplified assumptions are made, such assuming isotropic, i.e., homogeneous settings. Allen[1] shows that if $\mathcal{I}$ is assumed to be very large and G is defined as the complete graph—every individual is a neighbor of every other individual, $\nu(\iota) = \mathcal{I} - \{\iota\}, \forall \iota \in \mathcal{I}$—then the Markov random field model of information diffusion in a homogeneous economy may be treated as a birth-and-death model, and implies a logistic growth process for the diffusion of information: the rate of growth of the percentage of people who are informed is proportional to the percentage of people who are uninformed. It is also true that this literature is just starting to deliver completely endogenous equilibria, where the parameters of the the stochastic structure of the random fields are the outcomes of decisions. We return to this below.

The birth-and-death approximation that Allen invokes is a more general property, which serves to illuminate the possibilities afforded by the Markov random field model. Preston[97] proves equivalence between Markov random fields and equilibrium states of time-reversible nearest neighbor birth-death semigroups. A *time-reversible nearest neighbor semigroup* on $\mathcal{P}(V)$, the set of subsets of V, may be used to express the dynamic evolution of changes in configurations, which could potentially serve as the basis for modeling general patterns of trade among a number of agents. One could use a very large state space to describe detailed aspects of trade among a finite number of agents. Pricing, market structure, and many other issues (and, in particular, some of the issues taken up by interesting recent contributions to the literature of economies with interacting agents[9,103]) could be better understood as special cases in this more general approach. We are currently working on such an approach.

3.1 REMARKS

Modeling interactions in terms of Markov random fields requires that the pattern of potential links are the routes through which dependence across sites is transmitted. It is interesting to think of links between agents as emerging endogenously as

[8]This crucial formula follows readily once one recognizes that for any clique κ that does not contain ι, $\mathcal{D}_\kappa(\tilde{\omega}) = \mathcal{D}_\kappa(\tilde{\omega}')$. Thus terms that correspond to cliques that do not contain ι cancel from both the numerator and the denominator of Eq.(2) (Kindermann and Snell,[78] p. 27).

outcomes of individuals' decisions. This paper advocates an approach that would help bridge the distance between the random graph and the Markov random field approaches. Such an approach is outlined in section 6.2 below.

Durlauf has successfully employed the specified topology of interaction approach to a variety of settings. Durlauf[34,35] explores the global, or economy-wide consequences of local coordination failures. Durlauf[38] considers the role of complementarities in economic growth. He shows that local linkages across industries can create sequential complementarities which build up over time to affect aggregate behavior. Also, industries that trade with all other industries can induce takeoff to sustained industrial production and, thus, may be considered as leading sectors, in the sense employed earlier by the economic development literature on the significance of growth through intersectoral linkages. Durlauf's applications involve equilibria which may be nonergodic, and thus allow for different global probability measures for economic activity to be consistent with the same microeconomic characteristics of agents. Durlauf[40,41] deals with an implicit spatial interpretation to study the persistence of income inequality.

As Preston[97] notes, the equivalence between Gibbs states and the equilibrium states of Markov processes defined over graphs prompts the challenge to construct Markov processes whose equilibrium states are Gibbs states with desirable properties, which may be expressed in terms of appropriate potentials. Such a "reverse" approach, especially when combined with the equivalence between Markov random fields and birth-and-death semigroups, suggests a technique of greater applicability than what we have seen to date. Related to this, and virtually unexplored, is the economic interpretation of the phenomenon of phase transitions.

3.2 INTERACTIVE DISCRETE CHOICE

Scholars[9] have been aware of the tantalizing similarity between the Gibbs measure induced by the Gibbs potential and the logit function, which was used by McFadden[92] to construct measures for discrete choice models. Blume[17] and Brock[23] were the first to exploit the analytical significance of this link.[10] Brock[23] articulated what had been lacking in the literature so far, namely a Nash equilibrium concept in a model of the "Manski-McFadden world of interconnected discrete choosers" (p. 20). Brock's approach provides behavioral foundation for *mean field theory,* whereby the Markov random field model may be simplified by replacing direct interaction linkages between pairs of agents by interactions [11] between agents and

[9]Including this author; see Haller and Ioannides,[61] p. 17, fn. 1.

[10] We agree with Brock's claim that "the linkage of [mean field theory] and discrete choice theory presented below appears new to this paper." Brock,[23] p. 20.

[11]It appears that Brock[23] is the first to use the term "interactive" discrete choice. Manski[89] addresses the inference problem posed by the possibility that average behavior in some group influences the behavior of the individuals that comprise the group (the "reflection" problem). There exists great potential for interactive choice models, as Brock and Durlauf point out, and there also exists great number of potential econometric applications that may disentangle *endogenous*

the "mean of the field." This is discussed in more detail further below in section 3.3. The interactive discrete choice model serves to provide a genuine economic interpretation of the local specification introduced above, that is, the family of distribution functions for $\pi_\iota(\omega_\iota|\eta_\iota)$, under the assumption that it constitutes a nearest neighbor specification. Below we use the interactive discrete choice model in preliminary explorations of a number *anisotropic* settings.

Let agent ι who chooses $\omega_\iota, \omega_\iota \in S$, enjoy utility $U(\omega_\iota; \widetilde{\omega}_{\nu(\iota)})$, where $\widetilde{\omega}_{\nu(\iota)}$ denotes the vector containing as elements the decisions made by each of agent ι's nearest neighbors. We assume that an agent's utility function is additively separable from her own decision, ω_ι, and from the impact of her neighbors' decisions upon her own, $\widetilde{\omega}_{\nu(\iota)}$. The specification of an agent's utility function definition of the set of all cliques that include agent ι as a member, $\mathcal{K}(\iota)$. So we have:

$$U_\iota(\omega_\iota; \widetilde{\omega}_{\nu(\iota)}) \equiv u(\omega_\iota) + \omega_\iota \left[\sum_{j \in \mathcal{K}(\iota), j \neq \iota} J_{\iota j} \omega_j \right] + h\omega_\iota + \gamma \epsilon(\omega_\iota) , \tag{3}$$

where γ is a parameter and $\epsilon(\omega_\iota)$ a random variable to be specified shortly below. Agent ι's choice is influenced by those of her nearest neighbors, $j \in \nu(\iota)$, via the vector of interaction effects $J_{\iota j}$.

Following Brock,[23] in view of McFadden,[92] and under the additional assumption that $\epsilon(\omega_\iota)$ is independently and identically type I extreme-value distributed[12] across all alternatives and agents $\iota \in \mathcal{I}$, we may write fairly simple expressions for the choice probabilities. Without loss of generality, let S, the range of the choice variable ω_ι, be binary. Then:

$$\pi_\iota(\omega_\iota = 1 | \widetilde{\omega}_{\nu(\iota)}) = \text{Prob} \left\{ u(1) - u(0) + \left[\sum_{j \in \mathcal{K}(\iota), j \neq \iota} J_{\iota j} \omega_j \right] + h \geq -\gamma[\epsilon(1) - \epsilon(0)] \right\} \tag{4}$$

effects, meaning the propensity of an individual to behave in some way varies with the behavior of the group, from exogenous (contextual) effects, in which behavior varies with the exogenous characteristics of the group, and correlated effects, in which individuals in a group exhibit similar behavior because they have similar individual characteristics or face similar institutional environments. See Ioannides[69] for spatial applications.

[12] If two independent and identically distributed random variables have type I extreme-value distributions, then their difference has a logistic distribution.

which may be written in terms of the logistic integral[13]

$$\pi_\iota(\omega_\iota = 1|\tilde{\omega}_{\nu(\iota)}) = \frac{\exp\left[\beta\left(u(1) - u(0) + [\sum_{j\in\mathcal{K}(\iota),j\neq\iota} J_{\iota j}\omega_j] + h\right)\right]}{1 + \exp\left[\beta\left(u(1) - u(0) + [\sum_{j\in\mathcal{K}(\iota),j\neq\iota} J_{\iota j}\omega_j] + h\right)\right]}, \tag{6}$$

where β, a positive parameter, is an additional behavioral parameter. $\beta = 0$ implies purely random choice. The higher is β the more concentrated is the distribution. While the extreme value assumption for the ϵ's is made for convenience, there are several substantive arguments in its favor. First, the logistic integral is a fairly good approximation to the normal. Second, and much less known, is the fact that the extreme value distribution is the asymptotic distribution, as $n \to \infty$, for $Y_n = \max_{1\leq i\leq n}\{X_1, \ldots, X_n\} - \ell n n$, where $X_1, \ldots, X_n$ are independently and identically distributed random variables with zero mean, drawn from a fairly large class of distributions.[14]

The definition of an agent's utility according to Eq. (3) along with the discrete choice model implied by the assumption that the ϵ's in Eq. (3) are extreme value distributed leads naturally to an interpretation of the underlying Gibbs potential $\mathcal{D}(\tilde{\omega}) \equiv \sum_{\iota\in\mathcal{I}} u(\omega_\iota) + \sum_{\iota\in\mathcal{I}} \omega_\iota[\sum_{j\in\mathcal{K}(\iota),j\neq\iota} J_{\iota j}\omega_j] + h\sum_{\iota\in\mathcal{I}} \omega_\iota$ as a "social utility function" (cf., Brock[23]). To this individual choice system, there corresponds an "aggregate" choice mechanism in terms of the probability that the aggregate state $\tilde{\omega} = (\omega_1, \ldots, \omega_n)$ be chosen. As Brock[23] (p. 19) and Brock and Durlauf[24] note, choices by the n agents are not necessarily social welfare maximizing. Welfare analysis is facilitated by recognizing that expected social utility, defined as the expectation of the maximum, may play the role of expected indirect utility in a discrete choice setting (Manski and McFadden,[90] pp. 198–272).[15] Unfortunately, the convenience of the equivalence between Gibbs measures and the logit model does not carry over in the case of McFadden's generalized extreme value model.[91]

[13]When S is not binary then:

$$\pi_\iota(\omega_\iota|\tilde{\omega}_{\nu(\iota)}) = \frac{\exp\left[\beta\left[u(\omega_\iota) + \omega_\iota\left[\sum_{j\in\mathcal{K}(\iota),j\neq\iota} J_{\iota j}\omega_j\right] + h\omega_\iota\right]\right]}{\sum_{\omega_\iota\in S}\exp\left[\beta\left[u(\omega_\iota) + \omega_\iota\left[\sum_{j\in\mathcal{K}(\iota),j\neq\iota} J_{\iota j}\omega_j\right] + h\omega_\iota\right]\right]}. \tag{5}$$

[14]This class is defined as follows. If $F(x)$ and $f(x)$ denote the probability distribution and probability density functions of the X's, and $(d/dx)((1 - F(x)/f(x)) \to 0$, as $x \to \infty$, then the standardized variable $(Y_n - a_n)/b_n$, with $a_n = F^{-1}(1-1/n)$, $b_n^{-1} = nf(a_n)$, has an extreme value distribution, i.e., its probability distribution function is given by $\exp[-\exp[-y]])$. It is skewed, with a long upper tail, and a mode at 0; in its standard form, its mean is .57722 and its variance $\pi^2/6$ (Cox and Hinkley,[28] p. 473). See also Appendix 1, Pudney,[98] pp. 293–300, and Lerman and Kern.[87]

[15]This function is a convex function of $u(\cdot)$, and its derivative with respect to u yields the respective conditional choice probability. For a recent application, see Rust,[99] p. 136.

3.3 MEAN FIELD THEORY

Brock[23] and Brock and Durlauf[24] attempt to simplify mean field theory and offer some fascinating results. They specify an individual's utility (Eq. (3)) as depending in a number of alternative ways on interaction with others. If it depends on the average decision of all others', then we have: $U_\iota(\omega_\iota; \tilde{\omega}_{\nu(\iota)}) \equiv u(\omega_\iota) + \omega_\iota J \bar{m}_\iota^e$, where $\bar{m}_\iota^e \equiv (|\mathcal{I}| - 1)^{-1} E[\sum_{j \neq \iota} \omega_j]$. In a Nash equilibrium setting, each individual takes others' decisions as given and makes her own decisions subject to randomness that is due to independent and identically distributed draws from an extreme-value distribution. In the simplest possible model, one may take each individual's expectation of the mean to be common across all individuals: $\bar{m}_\iota^e = \bar{m}$, $\forall \iota \in \mathcal{I}$. Once the probability structure has been specified, this condition leads to a fixed point.[16]

Aoki[4,5] is concerned with modeling similar settings where from a single agent's perspective the rest of the environment is construed as a uniform field. In contrast to Brock[23] and Brock and Durlauf,[24] these approaches rest on weaker behavioral motivations. Aoki typically works with the so-called *master equation*, a version of the Chapman-Kolmogorov equation that keeps track of the time evolution of probability distributions in Markov chains. We eschew further discussion here and refer to Aoki.[6]

3.4 STYLIZED ASYMMETRIC PATTERNS OF INTERACTION

We now turn to some globally anisotropic settings, which exhibit some "local" symmetry. Retaining some symmetry not only lends itself naturally to economic interpretation but also confers an advantage in making the model amenable to Nash equilibrium analysis. That assumption is in contrast to the globally symmetric settings underlying mean field theory. We consider the cases alternatively: first, Walrasian star-shaped interaction, and second, completely circular interaction. The case of complete pairwise interactions may serve as a benchmark case and is examined first.

[16]To see that, note that the probability of a social state $\tilde{\omega}$ is given by the product probability $\text{Prob}(\tilde{\omega}|\bar{m}_\iota^e) = \exp[\beta(u(\omega_\iota) + \omega_\iota J \bar{m}_\iota^e)] / \prod_{\iota \in \mathcal{I}} \sum_{\omega_\iota = 0,1} \exp[\beta(u(\omega_\iota) + \omega_\iota J \bar{m}_\iota^e)]$. It follows that the denominator of r.h.s. of this equation may be written as $\sum_{\omega_1=0,1} \cdots \sum_{\omega_n=0,1} \exp[\beta(u(\omega_\iota) + \omega_\iota J \bar{m}_\iota^e)]$, which agrees with Eqs. (1) and (2) above.

3.4.1 COMPLETE PAIRWISE INTERACTION. The isotropic case of complete pairwise interaction is obtained from the general case of Eqs. (1) and (2) by specifying that for each ι, $\nu(\iota) = \mathcal{I} - \{\iota\}$, and $J^{\text{co}} \equiv J_{\iota j}$, $\forall j \in \mathcal{I} - \{\iota\}$. For the discrete choice model generated by Eq. (3) we have:

$$U_\iota(\omega_\iota; \omega_{\mathcal{I}-\{\iota\}}) \equiv u(\omega_\iota) + \omega_\iota J^{\text{co}} \sum_{j \neq \iota} \omega_j + h\omega_\iota + \gamma\epsilon(\omega_\iota) ; \tag{7}$$

$$\text{Prob}\{\omega_\iota | \widetilde{\omega}_{\nu(\iota)}\} = \frac{\exp\left[\beta[u(\omega_\iota) - u(0) + \omega_\iota J^{\text{co}} \sum_{j\neq\iota} \omega_j + h\omega_\iota]\right]}{1 + \exp\left[\beta[u(1) - u(0) + J^{\text{co}} \sum_{j\neq\iota} \omega_j + h]\right]} . \tag{8}$$

The probabilities associated with Nash equilibrium of the interactive discrete choice system may be easier to obtain once we have computed the equilibrium probabilities $\text{Prob}\{\omega_2, \ldots, \omega_n\}$, and so on, of which there 2^{n-1}. Taking advantage of symmetry reduces vastly the number of the relevant unknown probabilities down to n.

Instead of pursuing this further here, we note that the model bears a close resemblance to that of Kauffman[75] (p. 192), the case of the "Grand Ensemble": a given number of nodes n is connected to every other node, and each node is assigned a Boolean function at random from among the maximum possible number of possible logical rules, 2^{2^n}. It is interesting to explore the intuition obtained from the literature related to Kauffman's work, which is discussed in *The Origins of Order.*[75] This model may also be considered as an extension of Allen,[1] which allows us to explore some of the problems Allen takes up in more general settings.

3.5 WALRASIAN STAR-SHAPED INTERACTION

This is obtained from the general case above by treating agent 1 as located at the center of a star, so that for $\forall\iota$, $\iota \neq 1$, $\nu(\iota) = \{1\}$, and $J^{\text{w}} \equiv J_{\iota 1}$, and $J^w \equiv J_{\iota j}$, otherwise. Also, $\nu(1) = \mathcal{I} - \{1\}$, $J^{\text{w}} \equiv J_{1\iota}, \forall\iota \neq 1$. For the discrete choice model generated by Eq. (3) we have:

$$U_\iota(\omega_1; \omega_2, \ldots, \omega_n) \equiv u(\omega_1) + \omega_1 J^{\text{w}} \sum_{\iota=2}^{n} \omega_\iota + h\omega_1 + \gamma\epsilon(\omega_1) ; \tag{9}$$

$$U_\iota(\omega_\iota; \omega_1) \equiv u(\omega_\iota) + \omega_\iota J_1 \omega_1 + h\omega_\iota + \gamma\epsilon(\omega_\iota);\ \iota = 2, \ldots, n. \tag{10}$$

This system implies interactive discrete choice probabilities as follows:

$$\text{Prob}\{\omega_1 | \omega_2, \ldots, \omega_n\} = \frac{\exp\left[\beta[u(\omega_1) - u(0) + \omega_1 J^{\text{w}} \sum_{\iota=2}^{n} \omega_\iota + h\omega_1]\right]}{1 + \exp\left[\beta[u(1) - u(0) + J^{\text{w}} \sum_{\iota=2}^{n} \omega_\iota + h]\right]} . \tag{11}$$

$$\text{Prob}\{\omega_\iota | \omega_1\} = \frac{\exp\left[\beta[u(\omega_\iota) - u(0) + \omega_\iota J_1 \omega_1 + h\omega_\iota]\right]}{1 + \exp\left[\beta[u(1) - u(0) + J_1\omega_1 + h]\right]},\ \forall\iota \neq 1 . \tag{12}$$

The probabilities associated with Nash equilibrium of the interactive discrete choice system may be obtained in terms of the equilibrium probabilities $\text{Prob}\{\omega_2, \ldots, \omega_n\}$. The general case involves 2^{n-1} such probabilities, but taking advantage of symmetry vastly reduces the number of unknowns down to n.

Walrasian star-shaped interaction is seen here as a prototype for trees, which is one of the graph topologies for which threshold properties have been obtained by the random graphs literature. The model may be augmented to allow for branches with different number of nodes and may also serve as a prototype for hierarchical structures. A variety of economic settings may be explored with this model. The extreme-value distribution assumed by the behavioral model fits quite naturally a situation in which agent 1 conducts an auction based on offers by agents $2, \ldots, n$. Alternatively, each of the agents on the periphery may specialize in the production of a differentiated product. The number of agents n may thus reflect the demand for variety, etc.

3.6 CIRCULAR INTERACTION

This is obtained from the general case above by specifying that for each ι, $\nu(\iota) = \{\iota-1, \iota+1\}$, and $J_{\iota,\iota-1}, J_{\iota,\iota+1} \equiv J$, where for symmetry, $\{n+1\} = \{1\}, \{1-1\} = \{n\}$. For the discrete choice model generated by Eq. (3) we have:

$$U_\iota(\omega_\iota; \omega_{\iota-1}, \omega_{\iota+1}) \equiv u(\omega_\iota) + \omega_\iota J[\omega_{\iota-1} + \omega_{\iota+1}] + h\omega_\iota + \gamma\epsilon(\omega_\iota), \tag{13}$$

which implies interactive discrete choice probabilities as follows:

$$\text{Prob}\{\omega_\iota | \omega_{\iota-1}, \omega_{\iota+1}\} = \frac{\exp[u(\omega_\iota) - u(0) + \omega_\iota J[\omega_{\iota-1} + \omega_{\iota+1}] + h\omega_\iota]}{1 + \exp[u(1) - u(0) + J[\omega_{\iota-1} + \omega_{\iota+1}] + h]}. \tag{14}$$

The probabilities associated with Nash equilibrium of the interactive discrete choice system are easy to obtain once we have computed the equilibrium values of the probabilities $\text{Prob}[\omega_1, \ldots, \omega_n]$. The key is to recognize that symmetry implies that

$$\text{Prob}[\omega_1, \ldots, \omega_n] = \text{Prob}[\omega_2, \ldots, \omega_n, \omega_1] = \ldots = \text{Prob}[\omega_n, \omega_1, \ldots, \omega_{n-1}]. \tag{15}$$

Specifically, for the case of $n = 3$ it suffices, based on Eq. (15) to solve for the four unknown probabilities $\text{Prob}[1,1,1]$, $\text{Prob}[1,1,0]$, $\text{Prob}[1,0,0]$, and $\text{Prob}[0,0,0]$. We use the following notation for the respective interactive choice probabilities Eq. (14) in terms of the parameters of the model:

$$\text{Prob}\{1|1,0\} \equiv \frac{\exp[u(1) - u(0) + J + h]}{1 + \exp[u(1) - u(0) + J + h]};$$

$$\text{Prob}\{1|1,1\} \equiv \frac{\exp[u(1) - u(0) + 2J + h]}{1 + \exp[u(1) - u(0) + 2J + h]};$$

$$\text{Prob}\{0|0,0\} \equiv \frac{1}{1 + \exp[u(1) - u(0) + h]}.$$

Symmetry along with the definition of the interactive choice probabilities Eq. (14) imply the following linear system of equations:

$$\begin{aligned}\text{Prob}[1,1,1] + 3\text{Prob}[1,1,0] + 3\text{Prob}[1,0,0] + \text{Prob}[0,0,0] &= 1\,;\\ (\text{Prob}[1,1,1] + \text{Prob}[1,1,0])\text{Prob}\{1|1,1\} &= \text{Prob}\{1,1,1\}\,;\\ (\text{Prob}[1,0,0] + \text{Prob}[0,0,0])\text{Prob}\{0|0,0\} &= \text{Prob}[0,0,0]\,;\\ (\text{Prob}[1,1,0] + \text{Prob}[1,0,0])\text{Prob}\{1|0,1\} &= \text{Prob}[1,1,0]\,.\end{aligned}$$

Let $\mathbf{P}_{ci}(u(1) - u(0), J, h)$, a vector, denote the solution to this system, which is in closed form. An application of this model draws from the classic example (due to Wicksell[107]) of indirect exchange. Because of absence of double coincidence of wants, agents will avail themselves of the opportunity to trade only if circular sequences of trades may be completed. We pursue this further in section 6. We note that the circle model has been addressed by the interacting particle systems literature (Ellis,[42] pp. 190–203). It gives rise to some features which are absent from the Curie-Weiss model, namely a new kind of phase transition described in terms of random waves. This and other features of models of circular interaction are being investigated currently by the author.

3.7 COMPARISONS

The most important differences among the three stylized topologies that we have examined are associated with the fact that the circular and complete graph topologies allow circular trades to be completed. The results reported by Kauffman[75] involving autonomous Boolean networks with randomly assigned Boolean functions are indicative of the sort of results one could obtain by different specifications of preferences. For example, the Boolean functions .OR. and .AND. are conceptually similar to substitutability and complementarity, respectively. Analogous interpretations may be obtained for other Boolean functions. The importance of the preference structure in conjunction with the topology of interconnections may be underscored by the example, due to Schelling,[100] of an extreme case of complementarity, preferences for reading and lighting, which we quoted in the Introduction.

The sharp differences that preferences combined with topology make suggest the potential complexity of outcomes, and great richness in dynamic settings. Further study of these models is currently being pursued by the author.

3.8 THE ARROW-DEBREU-EVSTIGNEEV MODEL

This section attempts to address the challenge of extending the Arrow-Debreu model in order to endow it with a general graph topology. It is critical to note that the Arrow-Debreu model depends on time being naturally ordered. The important extension we introduce below is based on the σ-algebras not being necessarily linearly ordered. It is standard to model uncertainty by means of a set $\mathcal{N}$

of states of the world, indexed by $w \in \Omega$, and a finite number of periods, indexed by $t = 0, 1, \ldots, T$. The information in the economy is exogenously specified and is represented by a sequence of partitions of Ω, $\{F_t \mid t = 0, 1, \ldots, T\}$. At time t an agent knows which event has occurred and which cell of F_t contains the true state. It is standard to assume that information increases through time; F_{t+1} is at least as fine as F_t. It does not imply loss of generality to assume that $F_0 = \Omega$, the universe, and $F_T = \{\omega \mid \omega \in \Omega\}$, the discrete partition, the state of the world is revealed by period T. Let us denote by $\mathcal{F}_t$ the σ-field of events, generated by F_t; $\mathbf{F} = \{F_t; t \in \{0, 1, \ldots, T\}\}$ is the filtration generated by the sequence of partitions F_t. Consumption goods and endowments in terms of goods may be represented as stochastic processes adapted to $\mathcal{F}_t$.

Evstigneev[49] extended the Arrow-Debreu model by invoking the index set used[17] to describe communication links in a large economy. This device possesses a multitude of advantages. It implies as special cases graph topologies which describe patterns of trading links that underlie several popular models in economics as special cases. Moreover, it may be considered as a rigorous extension of the Arrow-Debreu model, where the pattern of trading links has the (graph) topology of a tree, the tree of events. We, therefore, refer to this model as the Arrow-Debreu-Evstigneev model, ADE for short.

A key characteristic of the ADE model is that the *index* set T is endowed with the topology of an *oriented*, that is directed, graph.[18] Let $(\Omega, \mathcal{F}, \mathbf{P})$ be a probability space, as before. It is assumed that to each $t \in T$ there corresponds a set $K(t) \subseteq T$, which contains t and may be interpreted as the set of *descendants* of t, that is the set of agents directly dependent on t. On the graph corresponding to T there is an arrow (arc) leading from t to every element of $K(t)$.

The ADE model explores the notion that the sources of randomness which influence t also influence all agents who descend from her. That is, the σ-algebras $\mathcal{F}_t$ corresponding to each agent, $\mathcal{F}_t \subseteq \mathcal{F}$, satisfy

$$\mathcal{F}_t \subseteq \mathcal{F}_s, \; \forall s \in K(t).$$

This condition, namely that the σ-algebras $\mathcal{F}_t$ do not decrease as we move along the arrows of the graph means that the random factors which influence agent t also influence all agents in $K(t)$. In case of a cycle, the σ-algebras coincide $\mathcal{F}_s = \mathcal{F}'_s$.

The index set T and the σ-algebras do not have to be linearly ordered. In contrast, the counterpart of our index set in a number of standard economic models

[17]See also Evstigneev,[47,48] which solves the problem of optimizing an objective function that roughly corresponds to a theory of teams type problem. The model in Evstigneev,[49] on the other hand, corresponds to a recursive equilibrium-type structure, where economic units may be spatially differentiated and be asymmetrically related to one another.

[18]A generalization to Evstigneev's approach would be to consider $\tilde{T}$ as a *virtual* index set. It would incorporate the set of logical as well technological possibilities that are a prerequisite to actual economic relationships. The actual index set in our setting, T, will be the outcome of individuals' decisions.

has very special, and linearly ordered, structure. In dynamic equilibrium models, it describes the sequential evolution of time. Alternatively, in purely spatial allocation problems T indexes locations in space, sites. Similarly to an agent's descendants we may define the set of of an agent t's *ancestors,* those agents from whom agent t descends. The set of agent t's ancestors is defined as $M(t)$, includes t and consists of the agents $\{s : s \in M(t), \text{if } t \in K(s)\}$.

Evstigneev suggests that time may be introduced explicitly in two alternative ways. One is to think of the elements of the set T as pairs $t = (n, b)$, where $n \in \{1, 2, \dots, N\}$ denotes discrete-valued time within a finite horizon N, and the set $B(n)$, with generic element $b \in B(n)$, is the set of economic units functioning at time t. With each pair, $t = (n, b)$ $(b \in B(n))$ a set $K(n, b) \subseteq T$ is associated such that $(n', b') \in K(n, b)$ implies $n' \geq n$. A second and perhaps simpler way of introducing time is to assume that an integer-valued function $\nu(t)$, $t \in T$, is given on the graph T, and its values $\nu(t)$ are interpreted as moments of functioning of economic units $t \in T$. It is natural to assume here that $\nu(s) \geq \nu(t)$ for $s \in K(t)$. We define the set of agent t's *contemporaries* $C(t)$ as all agents who function at the same point in time.

Evstigneev[49] starts with $Z_t(q)$, the set of preferable programs of agent t under price system $\mathbf{p} : q = (q_s)_{s \in K(t)}$ and proves existence of equilibrium by means of tools from the theory of monotone operators. Evstigneev and Taksar[51,52] offer a sensitivity analysis and an extension for the case of growing economies. Evstigneev's work constitutes a major generalization of the Arrow-Debreu model in a spatial context. Still, like the Arrow-Debreu model, it does not allow for feedback from the anticipation of trade frictions to allocation decisions.

4. MODELS WITH RANDOM TOPOLOGY

The previous section explored interactions across agents, when the topology of interconnections among them is given. Here we study by means of random graph theory how different patterns of interconnections may come about completely randomly. We review the literature that has utilized this theory in models of trading structures. We end the section by taking up a number of economic issues which may be also be analyzed by random graphs but have been paid little attention by the literature.

Random graph models[19] are a natural way to model situations where traders are into contact with other potential trading partners directly as well as indirectly through the partners of their trading partners and so on. There are other good reasons for treating the communication structure of an economy as a random variable.

[19]The seminal work by Erdös and Renyi[45] was followed by a broader set of applications in Erdös and Spencer.[46] See Bollobas[19] and Palmer[96] for the latest sources on this vast, fascinating literature.

It might not be known who communicates with whom, and any two individuals may attempt to contact one another at random, and such attempts may fail stochastically.

Kirman[79] was the first to argue in favor of this approach for studying communication in markets. Kirman et al.,[82] employ it studying coalition formation in large economies. Ioannides[68] uses is it in a model of trading uncertainty, where formation of trading groups rather than consummation of bilateral trades is the object of investigation. Durlauf[39] uses it to study the role of communication frictions in generating coordination failure which in turn leads to aggregate fluctuations. Bernheim and Bagwell[10] invoke random graph theory to assess the number of individuals who are linked through interpersonal transfers.

4.1 RANDOM GRAPH THEORY

Random graph theory introduces randomness by means of two alternative models.[20] Model A is defined in terms of graphs of order n and a probability $p = p(n)$ such that an edge exists between any two vertices i and j from among all possible vertices. The sample space consists of all possible labeled graphs of order n, with each of all $\binom{n}{2}$ possible edges being assigned independently with probability p. Model B is defined in terms of graphs of order n and size q, $0 \leq q \leq \binom{n}{2}$. The sample space consists of all possible labelled graphs of order n and size $q = q(n)$, each occurring with equal probability given by $\binom{\binom{n}{2}}{q}^{-1}$. Model A, B will be referred to as $\mathcal{G}^A_{n,p(n)}$, $\mathcal{G}^B_{n,q(n)}$, respectively.

Random graph theory is concerned with properties of graphs, such as connectedness and others, where the likelihood that they prevail satisfies a threshold property as the order of the graph grows. The literature utilizes results from random graph theory when n tends to infinity without specifying the space within which the graph is imbedded. Almost all of the literature works with homogeneous models, but as Erdös and Renyi[45] speculated, there exists now a bit of literature pertaining to anisotropic models and is discussed below. It is evidence of the richness of this theory that the *emergent*[21] properties of random graphs do appear in homogeneous settings.

In the remainder of this section we discuss certain results from random graph theory which are less known and lend themselves neatly to economic applications. We complete the section with a review of the principal economic applications of random graph theory.

[20] Erdös and Renyi[45] work with Model B but refer to the equivalence between these two models.
[21] I believe that Cohen[26] and Kauffman[75] are the first to refer to the properties of random graphs as emergent.

4.1.1 EMERGENCE OF CYCLES. Of particular interest in the section below where we attempt to explore the actual topology of the random graph is conditions for the emergence of cycles. We know from Erdös and Renyi,[45] the following facts. The number of *cycles* of order k, $k = 3, 4, \ldots,$ contained in $\mathcal{G}^B_{n,cn}$ has a Poisson distribution with mean equal to $(2c)^k/2k$. The number of *isolated* cycles of order $k, k = 3, 4, \ldots$ contained in $\mathcal{G}^B_{n,cn}$ has a Poisson distribution with mean equal to $(2ce^{-2c})^k/2k$. The number of *components* consisting of $k \geq 3$ points and k edges has a Poisson distribution with mean equal to $\left((2ce^{-2c})^k/2k\right)\left(1 + k + k^2/2! + \ldots + k^{k-3}/(k-3)!\right)$. The expected number of *all* cycles contained in $\mathcal{G}^B_{n,cn}$ is equal to $(1/2)\ell n(1/(1-2c)) - c - c^2$, if $c < 1/2$; to $(1/4)\ell nn$, if $c = 1/2$. The probability that $\mathcal{G}^B_{n,cn}$ contains at least one cycle is given by $1 - \sqrt{1-2c}e^{c+c^2}$, $1/2 \geq c$. The total number of points of $\mathcal{G}^B_{n,cn}$ that belong to some cycle is finite and given by $4c^3/(1-2c)$, if $c < 1/2$. The expected number of points of $\mathcal{G}^B_{n,cn}$ which belong to components containing exactly one cycle is given by $(1/2)\sum_{k=3}^{\infty}(2ce^{-2c})^k(1 + k + k^2/2! + \ldots + k^{k-3}/(k-3)!)$, if $c \neq 1/2$, and equal to $(\Gamma(1/3)/12)n^{2/3}$, otherwise. It then follows that for $c < 1/2$ all components of $\mathcal{G}^B_{n,cn}$ are with probability tending to 1, either trees or components containing one cycle. In other words, almost all graphs have no components with more than one cycle. Thus, we are not guaranteed a cycle until $c = 1/2$, or $p(n) \geq 1/n$. Finally, the asymptotic probability of cycles of all orders in $\mathcal{G}^B_{n,n}$ is equal to 1.

4.1.2 CONNECTIVITY AND HAMILTONICITY. A cycle that goes through every node exactly once (a spanning cycle) is known as a *Hamiltonian*. Specifically, we have the following result from Palmer[96], p. 59. In the random graph $\mathcal{G}^A_{n,p(n)}$, with $p(n) = (\ell nn + \ell n[\ell nn] + 2c_n)/n$, the probability of hamiltonicity tends, as $n \to \infty$, to: 0, if $c_n \to -\infty$; $e^{-e^{-2c}}$, if $c_n \to c$; 1, if $c_n \to +\infty$. In contrast, the probability of connectivity in the random graph $\mathcal{G}^A_{n,p(n)}$, with $p(n) = c(\ell nn/n)$ tends to 1, if $c > 1$. This holds in $\mathcal{G}^B_{n,q(n)}$, with $q(n) = (c/2)n\ell nn$. In both cases, the random graph is disconnected if $0 < c < 1$. If $p(n) = (\ell nn + 2c)/n$ the probability of connectivity of $\mathcal{G}^A_{n,p(n)}$ tends to $e^{-e^{-2c}}$. Outside the giant component, there are only isolated agents. Similarly, if $p(n) = (\ell nn + 2c)/n$ the probability of connectivity of $\mathcal{G}^A_{n,p(n)}$ tends to $e^{-e^{-2c}}$. It is thus clear that in order to make hamiltonicity possible, the probability that any two agents be connected must be larger than the value which ensures connectivity at least by a term $(\ell n[\ell nn])/n$, which, of course, decreases with n.

Connectivity is a useful concept in understanding how news travels through a population of agents. For example, if news can travel through direct and indirect contacts, then the probability of connectivity gives the probability that a given piece of news can reach everyone. Hamiltonicity lends itself readily to modeling the emergence of conditions under which fiat money can finance trades, which we take up in section 6 below.

4.1.3 ANISOTROPIC RANDOM GRAPH MODELS. Kovalenko[83] provides a rare example of a random graph model where the edge probabilities are not equal. Specifically, Kovalenko considers random graphs where an edge between nodes i and j may occur with the probability p_{ij} independently of whatever other edges exist. He assumes that the probability tends to 0 as $n \to \infty$ that there are no edges leading out of every node and that there are no edges leading into every node. Under some additional limiting assumptions about the probability structure, he shows that in the limit the random graph behaves as follows: there is a subgraph $\mathcal{A}_1$ of *in-isolated* nodes whose order follows asymptotically a Poisson law; there is a subgraph $\mathcal{A}_2$ of *out-isolated* nodes whose order follows asymptotically a Poisson law; all remaining nodes form a connected subgraph. The orders of $\mathcal{A}_1$ and $\mathcal{A}_2$ are asymptotically independent and their parameters are given in terms of the limit of the probability structure.

4.2 ECONOMIC APPLICATIONS OF RANDOM GRAPHS

Here are some examples of how the economics literature has exploited random graph theory. Kirman[79] works with$\mathcal{G}^A_{n,p(n)}$, where $p(n) \equiv p$, a constant. Then $\mathcal{G}^A_{n,p(n)}$ is strongly connected with probability equal to $\lim_{n\to\infty} : 1 - n(1-p)^{n-1}$, which is, of course, equal to zero. Kirman derives a *Probabilistic Limit Theorem for the Core:* the epsilon core is arbitrarily close to a sequence of replica Walrasian economies. This result is strengthened by Kirman, Oddou, and Weber (1986), where $p(n)$ is a decreasing function of n, which can go to 0 as fast as $1/\sqrt{n}$, for $n \to \infty$. They also prove a probabilistic theorem that all members of a coalition are required to be in direct contact; in that case in particular, $p(n)$ should go to 0 slower than $1/\ell nn$. Ioannides[68] interprets each node as an independently endowed trader and the isolated components of the random graph as trading groups within which trades are allocated according to Walrasian rules.[22] He then considers the evolution of $\mathcal{G}^B_{n,cn}$, where c is a constant, and associates the emergence of the giant component of the random graph, when $n \to \infty$, with elimination of uncertainty for a certain proportion of the economy. The expected number of components of $\mathcal{G}^B_{n,cn}$ divided by the number of traders n is: $1 - c + O(1)/n$, if $c < 1/2$; $1 - c + O(\ell nn)/n$, if $c = 1/2$; and, $(1/2c)\left(x(c) - x^2(c)/2\right)$, if $c > 1/2$, where $x(c)$ is the only solution in $(0,1)$ of the equation $xe^{-x} = 2ce^{-2c}$. That is, it decreases, as c increases, originally linearly and then slower than linearly with c. The random graph contains a unique giant component of order $[1 - x(c)/2c]n$, and the remaining vertices, whose number is roughly equal to $(x(c)/2c)n$, belong to components that are trees of order at most ℓnn. This finding is particularly dramatic if it is presented as a plot of the size of the largest component against c, the ratio of edges to nodes, which has a sigmoidal

[22] Ioannides[67] considers that interlinked traders form mutual funds.

shape with a sharp rise at $c = 1/2$.[23] The random graph $\mathcal{G}^B_{n,cn}$ is of particular interest in that it exhibits an interesting uniformity "in-the-small," in spite of this stark heterogeneity. For any of its points, the probability that it is connected to j other points is given by $(2c)^j e^{-2c}/j!$, $j = 0, 1, \ldots$, i.e., it has a Poisson distribution with mean equal to twice the number of edges per trader. Equivalent statements hold for the random graph $\mathcal{G}^A_{n,2c/n}$.

Durlauf[39] considers the evolution of a set of industries, each consisting of a large number of costlessly communicating firms. When a set of industries establish contact with one another and form a production coalition, whereby all members of a production coalition must be able to communicate directly through a bilateral link with at least one other member of the coalition, output and profits are improved. Durlauf works with $\mathcal{G}^A_{n,p(n)}$ and assumes that the probability of communication between industries i and j, the probability of an edge between nodes i and j, is a continuously increasing function of industry outputs; he also assumes that conditional on industry outputs, the event that industry i communicates with industry j is independent of the event that industry i' communicates with industry j', when $\{i, j\}$ and $\{i', j'\}$ are distinct pairs. Durlauf shows that the following conditions are necessary for existence of aggregate cycles: For all finite n, the probability is strictly positive that two isolated industries be connected when they produce their maximum possible output—that is, not all industries are simultaneously isolated. The probability is strictly less than 1 that two different industries, which are members of the largest possible coalition of size n, be connected directly when they produce their lowest possible output—not all bilateral links will become permanently established. So long as the probability of each bilateral link is between 0 and 1 for all output levels, the model describing aggregate output will be ergodic. The main theorem in Durlauf[39] states that under the assumptions made about the sensitivity of the probability of bilateral communication aggregate output fluctuations will be arbitrarily persistent: high output enhances the process of coalition formation and low output impedes it, so that circumstances of high and low aggregate activity are self-reinforcing. All these papers do take advantage of the fascinating properties of random graphs, but they stop short of making as dramatic a use of the *emergent* properties of communication among large numbers of agents as that by mathematical biologists.[26,75,76]

It is most fortunate that random graph theory has obtained such precise results about trees and cycles which, as we argued earlier in sections 3.5 and 3.6 are of particular relevance for economic applications. Our current research in this area sets out to combine two sets of analytical insights obtained by the two approaches discussed so far, that is, models with specified topology of interaction, discussed in section 3, and models with random topology, discussed in the present one. Our goal is to obtain an integrated model, a grand synthesis, of the evolution of trading structures. This, as we see it, involves two steps: first, agents' decisions

[23]Kauffman[76] uses this model as a metaphor for the origin of an autocatalytic model of chemical reactions that "is. . .almost certainly self-sustaining, alive"(p. 58).

combine with nature to generate a particular structure of interconnections; and second, given a structure of interconnections agents trade. The interplay between those two approaches becomes the basis of the synthesis. Agents' decisions may affect the parameters of the random mechanism of the first step above, and agents' decisions in the second step may affect the further evolution of the economy. Even with static expectations, agents face a trade off: heavy investments in establishing a trade infrastructure makes trade more likely but also costs more. With general structures of preferences, some agents may be better off specializing in the production of certain differentiated goods, whereas other agents may be better off providing intermediation services. Circular trading structures facilitate trade but require, on the average, a greater "thickness" of interconnections, as the discussion in section 4.1.2 immediately above makes clear. Macro-type properties with multiplier effects are more likely to emanate from models with circular trades. Endogeneity of differentiated product varieties may be modeled in tree-type structures.

5. OTHER TOPICS

Spatial aspects of trading structures give rise to issues of spatial complexity and link with the urban economics literature. Our theoretical approach would be complemented by a study of possibilities of statistical inference. Aggregate implications of some of the approaches we have discussed have already been examined extensively by the economies with interacting agents literature. For example, Durlauf[35] derives and tests empirically the model's implications for persistence. Those tests, and other work by Durlauf, work with the aggregate implications of the theory. Techniques of inference that could be aimed at a more "micro" level would be particularly useful. The evolution of trading structures in a spatial context is an area that lends itself to particularly interesting issues of statistical inference. Another would be testing Kauffman's theory of technological change through inference based on the evolution of economic webs.[74] The material below demonstrates that it is fairly straightforward to carry out inference in terms of differences across graphs. Two other issues are taken up in the remainder of this section, namely pricing and matching models.

5.1 SPATIAL MODELS

The modern economy depends critically on differentiated use of space. Such differentiation emerged during very early stages of development, when settlements appeared to take advantage of economies of scale in trading activities relative to a rural hinterland and to exploit other benefits of close human interactions.[11] These processes have been subject of intense investigations by other social scientists but received only scant attention by economists outside the broad area of urban economics, which has almost exclusively focussed on cities.[64,65] Recent work by Paul Krugman on economic geography has revived interest in this area and has given a new impetus by setting a number of new challenges.

Our understanding of the trade infrastructure of the modern economy is unthinkable without an understanding of how cities interact with one another and with their hinterlands. To accomplish this, we must go beyond looking at agents and sites as points in space and bring into the fore their full spatial aspects. However, the graph topology that we have relied upon so far is still useful in understanding a number of key issues.

Modern urban economics has addressed jointly two important questions, namely the role of cities whose primary functions are other than retailing to rural areas and the internal structure of cities. Mills[93] emphasized the scale economies in industrial production associated with the spatial clustering of workers and firms. These economies derive from ease of communication, benefits from matching of workers and firms, and sharing of fixed costs associated with public transportation infrastructure. Cities produce traded and nontraded goods. There is overwhelming evidence that smaller and medium size cities specialize in the production of groups of different goods. As Henderson[65] explains it, "separating industries into different cities allows for a greater degree of scale economy exploitation in each industry relative to a given level of commuting costs and city spatial area" (p. 32).

When economic factors such as scale economies due to specialization and/or urbanization interact with geographical features and historical patterns of settlement the outcome is an uneven spatial distribution of economic activity. Researchers have debated a variety of stylized facts characterizing urban systems in different economies and alleged regularities in city size distributions over time in specific economies as well as across different economies (Henderson,[65] p. 46; Dobkins and Ioannides[31]). In the economics literature, however, there has been more emphasis upon the economic functions of cities and less emphasis upon the structural aspects of systems of cities. Empirical results (Henderson,[65] p. 197) suggest that observed differences in urban concentration across economies reflect the composition of national output, with important agriculture and resource-based activities being associated with lower concentration and greater presence of service-type industry being associated with higher concentration and clustering of activity into multicentered metropolitan areas. Strategic considerations have also been utilized to explain allocation of economic activity over space.[62,63]

Casual observation of the evolution of systems of cities suggests that complexity factors, akin to the ones addressed elsewhere in this paper play an important role. Cities interact in a variety of complex ways, through transportation of goods and movements of people, flows of knowledge, and complex trading arrangements.[85,86] The complexity of real world trading arrangements, especially in their spatial dimensions, closely resembles the theoretical models discussed earlier in the paper. We refrain from pursuing this issue further here, since the paper by Paul Krugman in this volume is dedicated to this topic. We offer an example of how interactions across cities belonging to a system of cities may be highlighted by means of random graph theory.

Let us consider a random graph $\mathcal{G}^B_{n,q(n)}$, where n denotes the number of cities. Let edges represent routes of economic interaction between cities, which in a hypothetical case are placed in the most random way between cities in the system. We are interested in the topology of interconnections across cities as the number of interconnections increases over time. In particular, we are interested in knowing how quickly the isolated parts of the system of cities become integrated into the main interconnected part of the system. If the number of routes per city starts at greater than 1/2 and increases over time, then the giant component grows by absorbing isolated components. It turns out that the probability is equal to $e^{-2k(c_{t_2}-c_{t_1})}$ that an isolated subsystem of k cities (with the topology of a tree) which is present in the system at time t_1, that is, it belongs to $\mathcal{G}^B_{n,c_{t_1}n}$, should still be isolated at time t_2, that is it belongs to $\mathcal{G}^B_{n,c_{t_2}n}$, $c_{t_2} > c_{t_1}$. Thus, the lifetime of an isolated urban subsystem of order k has approximately an exponential distribution with mean value $1/2k$ and is thus independent of the "age" of the tree.

This simple model, which is a direct outgrowth of random graph theory, suggests that fairly rapid thickening of interconnections commonly associated with systems of cities in growing economies over time could result even from a random placement of interconnections among cities. If the number of new links is endogenized, and perhaps related to incentives as perceived by the members of the subsystem of size k, then depending upon the relationship of the number of links with the number of cities we may be able to make different predictions about the properties of the system of cities. Another result from the random graph literature may be helpful in predicting the minimum number of interconnections, as a function of the number of cities that renders the system completely connected. Let the number of interconnections between cities be chosen in such a manner that at each stage every interconnection that has not been chosen has the same probability of being chosen as the next and let the process continue until until all cities become interconnected. Then as $n \to \infty$, in model $\mathcal{G}^B_{n,q(n)}$ $q(n)$ is such that

$$q(n) = \frac{1}{2} n\ell nn + cn, \quad \text{and} \quad \lim_{n\to\infty} : \text{Prob}\left[\frac{q(n) - \frac{1}{2}n\ell nn}{n} < c\right] = e^{-e^{-2c}}.$$

Many additional issues remain unexplored, especially challenging ones being associated with the spatial as well as functional evolution of urban systems. Explicitly spatial considerations are rather new, however, in the literature. The empirical evidence on the evolution of the US urban system has not been fully explored[24] and some of the analytical material from Arthur,[7](Chaps. 3, 6, and 10), may be brought to bear on this issue.

[24]See Dobkins and Ioannides[31] for a recent inquiry.

5.2 STATISTICAL INFERENCE

Specifically, one may define a simple metric for graphs, or equivalently, as it turns out for this use, for adjacency matrices corresponding to graphs. Let $\delta_M(g_1, g_2)$ be the symmetric difference metric on graphs, defined by:

$$\delta_M(g_1, g_2) = \frac{1}{2} tr[(G_1 - G_2)^2], \tag{16}$$

which counts the number of edge discrepancies between graphs g_1 and g_2, whose adjacency matrices are G_1 and G_2, respectively. Following Banks and Carley,[8] one may define a probability measure $H(g^*, \sigma)$, over the set $\mathcal{G}_M$ of all graphs on n distinct vertices, where $g^* \in \mathcal{G}_M$ is the central graph and σ a dispersion parameter:

$$P\{g|g^*, \sigma\} = \zeta(\sigma) e^{-\delta_M(g, g^*)}, \ \forall g \in \mathcal{G}_M, \tag{17}$$

where $\zeta(\sigma)$ is a normalizing constant. It is a fortunate fact that the normalizing constant does not depend on g^*, under the above assumptions: $g^* = (1 + e^{-\sigma})^{-\binom{n}{2}}$ (Banks and Carley[8]). We are thus free to specify the central graph and thus implicitly test random graph theory. Also, Banks and Carley suggest extensions to the case of digraphs, which could be particularly useful in testing Kauffman's theory of evolution in economic webs[74] in terms of long-run changes in input-output relationships.

5.3 PRICING

It is worth mentioning in passing that the dual of who trades with whom is whose prices are inputs to whom. Consequently, much of what we have been concerned with applies to pricing models. Dynamics of pricing decisions have been examined by a number of researchers, but because of their preoccupation with representative-individual models, which has been eloquently criticized by Kirman,[80] they have not been considered as being central to macroeconomic theorizing. Notable is the work by Blanchard,[13,14] who explores the dynamics of pricing decisions in two alternative settings. In Blanchard,[13] final output is produced in a number of discrete stages, each carried out under constant returns to scale by competitive firms. Just as Blanchard allows for the number of stages to change, "leaving the [aggregate] technology unchanged," one can conceive of settings where the number of intermediate goods changes in the Dixit-Stiglitz style (cf., Blanchard[14]). One may impose patterns of intermediate goods purchases, which may generate interesting dynamics whose steady states may be not unlike the ones explored by Weitzman.[104]

5.4 MATCHING

Agent matching has been proved useful in a number of areas in economics as a simple way to express a possible dependence between frequency of trades and the number of agents. This literature has typically focussed on actual trades and, thus, avoided making assumptions about the full pattern of interaction. Matching models have been adopted in macroeconomic applications where the interest was in understanding the impact of frictions.[94,95] Recently, work in evolutionary game theory has revived interest in matching.

The typical setting in Pissarides[95] involves a homogeneous model, where matching is construed as an operation in bipartite graphs, where each part groups workers and firms. Let L denote the total number of workers, and uL and vL denote unemployed workers and the number of vacant jobs, respectively. If only unemployed workers and vacant jobs engage in matching, then the number of job matchings per unit time is given by the matching function: $x = X(uL, vL)/L$. Pissarides invokes empirical evidence to argue that the matching function is increasing in both of its arguments, concave, and homogeneous of degree 1.[25] Homogeneity implies that the rate of job contacts is given as a function of the unemployment and vacancy rates only, $x = x(u, v)$. A number of informational assumptions underlie this construction, i.e., unemployed workers and job vacancies are distinct. Workers' and firms' effectiveness in search may be modified by search intensity and advertizing, respectively.

Moving to stochastic matching, however, poses some technical problems. For example, Gilboa and Matsui[55] show that if the populations to be matched and the set of encounters are countably infinite, and purely random matching is imposed (which ensures that everyone is matched), then the distribution of encounters over time equals their common distribution as random variables with probability one: the law of large numbers holds. That is, the population as a whole is not affected by aggregate uncertainty. Boylan[20] addresses the need for detail in the specification of random matching, which is usually approximated with a deterministic model. Boylan assumes a countably infinite population $\{1, 2, \ldots, n, \ldots\}$ where each individual is to be matched with exactly one other individual. There are m different types of individuals, $S = \{s_1, \ldots, s_m\}$. Matching is such that each individual is matched exactly once and "if John is matched with Paul then Paul is matched with John." Boylan articulates a number of difficulties that random matching schemes face. For example, imposing the condition of equally likely matches leads to a contradiction; the properties of random matching do depend on the assignment of types. He gives conditions under which there exists a random matching scheme which is equivalently handled by a deterministic system. However, for a law of large numbers to hold the random matching scheme must depend on the assignment of types in the population. Boylan[21] considers a similar set of questions except that the number of matchings per individual in this note is random and converges to a Poisson

[25] For empirical evidence with US data, see Blanchard and Diamond.[15,16]

process as the population grows. He shows that the deterministic process which approximates the random matching process must have a globally stable point.

The literature is concerned primarily with bilateral matching. In contrast, several problems addressed by the literature on economies with interacting agents involve in effect *multilateral* matching. The anisotropic model developed by Kovalenko[83] may be used to model multilateral matching. Finally, some assignment and network models may actually be considered relevant to the inquiry pursued by this paper, but we shall refrain from discussing them here.

6. EQUILIBRIUM INTERACTIONS IN ANISOTROPIC SETTINGS: A SKETCH

This section demonstrates in a preliminary fashion the possibility of equilibrium models in *anisotropic,* that is, nonisotropic, settings. As we see shortly, *controlled* random fields[26] may be used here in a manner that is reminiscent of models of equilibrium search, where market structure is endogenous. In the first subsection below we take up nonadditive effects of interaction which serve two functions: one, to motivate our decision to emphasize a circular pattern of interconnections; two, to demonstrate properties of circular trade patterns which are of independent interest. In the second subsection, we offer a preliminary model of equilibrium interactions in a circular setting.

6.1 NONADDITIVE EFFECTS OF INTERACTION

Just as international trade among distinct national economies has complex effects, so does trade among interacting agents, when the patterns of interconnection are complex. A number of papers by Jovanovic[70,71,72] establish that one may design game structures which may either attenuate or reinforce the effects of independent shocks acting on each of a large number of agents. We interpret such structures as patterns of interaction. We discuss circular interaction in some detail, because it is of particular interest in this paper and is utilized in our example of equilibrium in section 6.2 below. We then take up Jovanovic's general results.

[26] In contrast, in Evstigneev's work on controlled random fields, the structure is not endogenous.

6.1.1 CIRCULAR SETTINGS. We consider $i = 1, \ldots, n$ distinct sites, each occupied by a continuum of consumers and firms, each of unit measure. Firm i uses input x_i to produce output $\theta_i x_i^\gamma$, $0 < \gamma < 1$, where θ_i and γ are parameters. Firms are owned by the consumers who consume the entirety of their profit incomes on firms' output. In a centralized setting, where owned output can be used for both consumption and investment and output may be costlessly transported across all sites, output price everywhere is equal to 1. The profit-maximizing level of output in site i is given by: $\ell n x_i = \ell n \gamma/(1-\gamma) + (1/(1-\gamma)\ell n \theta_i$. If the set of parameters $\{\theta_i\}_{i=1}^{i=n}$ are i.i.d. random variables, then aggregate output per capita obeys the law of large numbers and is thus equal to: $\ell n\gamma/(1-\gamma) + (1/(1-\gamma))E\{\ell n\theta_i\}$. While we will discuss below variations of this model, unless the pattern of trade is more complex or the shocks contain an aggregate component, the centralized setting of the Walrasian trading model precludes reinforcement of individual shocks.

Jovanovic[70] proceeds with the following restrictions on preferences. Consumers must consume goods produced in the site of their residence. Firms on site i must purchase goods produced on site $i-1$. Sites are located on a circle. For symmetry, if $i = 1$, $i - 1 = n$. Let prices be $(p_1, p_2, \ldots, p_n)$, with $p_1 = 1$ as the numeraire. Profit maximization requires that $p_{i-1}/p_i = \gamma\theta_i x_i^{\gamma-1}$. The model is closed by imposing equilibrium in the output market. Individuals spend their incomes on consumption. It turns out that firm ownership patterns do not matter and that equilibrium output in sites i and $i-1$ must satisfy $x_i = \gamma\theta_i x_{i-1}^\gamma$. It follows that consumption in site $i-1$ is equal to $(1-\gamma)\theta_{i-1}x_{i-1}^\gamma$. The model closes with the "initial" condition $x_{n+1} = x_1$, with which the solution follows:

$$\ell n x_i = \frac{\gamma \ell n \gamma}{1-\gamma} + \sum_{j=1}^{n} \frac{\gamma^{n-j}}{1-\gamma^n} \ell n \theta_{i+j}\,, \tag{18}$$

where for $i + j > n$, $\theta_{i+j} \equiv \theta_{i+j-n}$. By the change of variable $\epsilon_i \equiv (1-\gamma^n)^{-1}\ell n\theta_i$, Eq. (18) becomes $\ell n x_i = \gamma\ell n\gamma/(1-\gamma) + \sum_{j=1}^n \gamma^{n-j}\epsilon_{i+j}$. We now see a pattern of dependence [27] across outputs in all sites: the x_i's contain a common component. Some further manipulation makes that dependence even clearer. We retain the ϵ's as given parameters and vary the θ_i's as $\gamma \to 1$. This yields the result that $\ell n x_i = 1 + \sum_{j=1}^n \epsilon_j$, so that the x_i's are perfectly correlated. Furthermore, as the limit is taken the uncertainty at the individual level becomes negligible, as the variance of θ_i tends to 0. In contrast, the centralized setting when pursued in the same manner yields: $\ell n\tilde{x}_i = 1 + n\epsilon_i$, i.e., outputs are i.i.d. across sites.

[27] Jovanovic[70] does point out the fact that basic patterns of circular dependence have been studied in the context of *circular* serial correlation. Actually, the dynamics of a system exhibiting circular serial correlation of order T involve the eigenvalues and eigenvectors of a matrix $\mathbf{B}$, whose only nonzero elements are $b_{i,i+1} = 1$, $i = 1, \ldots, T-1, b_{T,1} = 1$. The eigenvalues of $\mathbf{B}$, are the Tth roots of 1 and thus naturally involve the trigonometric functions and so do the corresponding eigenvectors (Anderson,[3] pp. 278–284). Periodic dynamics associated with this system are thus inherent in it and particularly interesting. In fact, the circular interaction model that we are investigating is, as Jovanovic notes, a natural way to motivate the circular serial correlation model.

Jovanovic does emphasize that the reinforcement of local shocks is achieved by a complete absence of a coincidence of wants between any two neighbors. He concludes by expressing that "there is hope, therefore, of analyzing the role of money in such structures" (Jovanovic,[70] p. 21), but seems to be unaware of the fact that a similar set of questions have been originally addressed by Cass and Yaari[25] in a deterministic setting.

6.1.2 GENERAL SETTINGS. Jovanovic[72] (p. 399), proves a general theorem which if suitably interpreted allows us to assess the likelihood of similar results for other general patterns of interaction. Consider that $\{\theta_i\}_{i=1}^{i=n}$ are independent random variables, with θ_i being the "micro" shock to agent i' reaction function (more generally, correspondence), which may be written as $\phi(x_{\{\mathcal{I}-i\}}, \theta_i)$. A *Nash equilibrium,* is a solution to the system of structural relations $x_i \in \phi(x_{\{\mathcal{I}-i\}}, \theta_i)$. Under fairly mild conditions on the θ_i' s, any reduced-form $h(\theta_1, \ldots, \theta_n)$, which relates the endogenous x's to the exogenous θ_i's, can be generated by a family of games, with agents acting with common knowledge of all shocks. Thus, restricting attention to stochastically independent agents imposes, in general, no restriction on the distribution of observed outcomes. From the perspective of the present paper, restrictions that are associated with specific topological structures of interactions are of particular interest.

6.2 EQUILIBRIUM IN CIRCULAR SETTINGS

Let us consider now that an original random matching results in n agents ending up in a circle. We invoke a setting very much like that of section 3.6, where agent ι has as neighbors agents $\iota - 1$ and $\iota + 1$. An agent may be active or inactive, a state that is determined as the outcome of a comparison of expected benefits and costs, in a way which will be precisely specified below. That is, the interactive discrete choice model of section 3.2 is utilized here as a model of market participation.

In the context of the present discussion, the solution in section 6.1 applies only if all agents are active. The expression for x_ι is symmetric with respect to location on the circle. If we now allow for agents to be active or inactive, we quickly recognize that we must account for all possible outcomes. However, even if only one agent on the circle is inactive, the equilibrium outcome for all agents is $x_\iota = 0$, in which case consumption is also equal to 0 and $u(0) = 0$.

We can go further by restricting attention to the case of three agents, without loss of generality, as it turns out. Such a setting contains key features of the model. For that case, we have already derived in section 3.6 the equilibrium probabilities. A typical agent, say agent 1, in this setting may be found active or inactive. The probabilities of these events may be computed. Agent 1 is active with probability given by

$$\text{Prob}\{1\} = \text{Prob}[1,1,1] + 2\text{Prob}[1,1,0] + \text{Prob}[1,0,0]. \tag{19}$$

We may take $u(1)$ to be expected utility conditional on an agent's being active, which we may compute from Eq. (18) as a function of parameters, denoted by $\bar{u}$. By multiplying with Prob$[1,1,1]/$Prob$\{1\}$ we obtain expected utility conditional on an agent being active. It is now clear that a simultaneity is involved in computing the equilibrium state probabilities associated with model Eq. (13). The equilibrium state probabilities depend upon $u(1)$, expected utility conditional on an agent being active, and the latter, in turn, depends upon equilibrium state probabilities. Let the equilibrium values of the probabilities be denoted by asterisks. They satisfy:

$$\mathbf{P}^*_{ci} = \mathbf{P}_{ci}\left(\bar{u}\frac{\text{Prob}^*[1,1,1]}{\text{Prob}^*\{1\}}, J, h\right). \tag{20}$$

We may now use these results as ingredients for a dynamic model, very much along the lines of Brock and Durlauf.[24] That is, we use the above model to predict the motion of an entire economy. The economy consists of agents in three sites, who are assumed to act with static expectations and whose states are determined according to the interactive discrete choice model. The first step in the study of these dynamics is to consider that utility values $u(\omega_\iota)$ are exogenous. This will allow us to separate the effects upon the dynamics of the interactive discrete choice model from those of the endogeneity of utility associated with the state of being active. We may then proceed with the case of endogenous utilities, by assigning to those terms the values of expected utilities implied by the model. The model exhibits a fundamental simultaneity. The Gibbs state is defined in terms of the interactive choice probabilities, and the interactive choice probabilities depend upon the Gibbs state through the calculation of expected utilities. The Gibbs state for the model is defined as fixed point. As an extension of this basic framework, one could also introduce a mean field effect along with neighbor interactions.[28]

6.3 A RESEARCH AGENDA

The fact that Nash equilibrium places few restrictions on our ability to recover reaction functions (and thus, according to Jovanovic's Lemma 1,[70] p. 398, agents' preferences as well) allows us to speculate that it should be possible to obtain fairly general results for equilibrium interactions. The example of section 6.2 demonstrates that it is possible to go beyond mean field theory and characterize equilibrium interactions in anisotropic settings as long as one restricts oneself to a given topology of interactions. The more challenging problem is, of course, to let preferences, costs of interaction and costs of trading determine the topology of interactions at equilibrium.

[28]This is particularly interesting from a macroeconomic point of view in the sense that one could potentially distinguish between true "macro" shocks and aggregate shocks which result from aggregation of individual shocks.

Random graph theory allows us to make predictions about topologies which are most likely to prevail under different assumptions about the dependence upon the number of agents n of $p(n)$, the probability of an edge, or of $q(n)$, the number of edges. It remains to link these probabilities, which we see as determining the original topology of potential interaction, with the fundamentals of the problem, such as preference and cost parameters. Even before that is accomplished we speculate that a number of interesting results may be obtained; those results would rest on multiplicity and stability properties of equilibrium. Under the assumption that preferences allow a wide variety of potential interactions, the original random matching provides the boundary conditions for the interactive choice model of section 3.2. If the unique equilibrium configuration is globally stable, then the boundary conditions do not matter. If, on the other hand, there exist multiple equilibria, it matters whether or not a particular initial potential topology of interaction lies in the domain of attraction of a stable or of an unstable equilibrium. Furthermore, the original matching may not perfectly random; it might reflect specific, nonhomogeneous, features of a particular setting.

We speculate that such an approach may lead to investigations in a number of directions which we are currently pursuing. We outline a few of them here. One direction could be a model for an old problem: the appearance of money as an emergent property of an economic system. When links per trader are few, then cycles are rather unlikely and trade is restricted to bilateral or autarkic. With more links per trader, cycles become likely, and circular sequences of bilateral trades can develop. We think of such cycles as different subeconomies, each of which is equivalent to a centralized setting through the use of IOU-type money in the style of Wicksell. With even more links per trader, a hamiltonian cycle is likely, which would be associated with the emergence of a single currency. A second direction could be a model of the spatial differentiation of the economy, which may make it conducive to spontaneous growth in certain circumstances and not in others.[77] Such an approach would provide a new dimension to the growing endogenous growth literature. A third direction could be a new look at the fundamentals of the process of technological change, perhaps along the lines of Kauffman's ideas on the evolution of economic webs. A fourth direction could be to link formal theories of trade infrastructures and of relational constraints in general equilibrium.[56,60] The circle model on its own gives rise to particularly interesting dynamics. The results obtained for the deterministic case by the mathematical biology literature and for the stochastic case by the interacting particle systems literature are particularly promising. Finally, we also note the potential afforded by modeling trade based on evolving networks of interconnected traders, which we discussed in section 3 above.

7. SUMMARY AND CONCLUSION

Our review of the evolution of trading structures has attributed a central role to the literature of economies with interacting agents. It identifies two main strands in that literature, namely works that presume a given topology of interactions among agents and those that let random mechanisms determine that topology. Both are recent and very fast-growing bodies of knowledge that shares methodological tools with other social sciences.

The paper addresses two key weaknesses of these two fundamental approaches to the study of trading structures. The specified topology of interaction approach has concerned itself primarily with isotropic settings. However, some anisotropic settings are particularly interesting for economists. For example, circular patterns of interaction highlight the role of money and credit; tree-type of interactions may be seen as depicting Walrasian interactions. The random topology of interaction approach has concerned itself primarily with the sizes of trading groups, and thus has not exploited the entire range of topological properties of trading structures which may emerge.

The paper proposes an integration of those approaches which is intended to exploit their natural complementarities. In the simplest possible version, our synthesis involves individual decisions and expectations, randomness, and nature combining to fix an initial "primordial" topology of interaction. The dynamics of interaction move the economy from then on. The evolution of trading structures depends critically upon multiplicity and stability properties of equilibrium configurations of the interaction model.

The paper has also pointed to links with spatial economics and as well as with processes of growth and technological change. We hope that those suggestions will become fruitful avenues for specific applications of further research on the evolution of trading structures.

We think that ultimately it should be possible to address the problem of emergent market structures by means of mathematical tools involving controlled random fields.[29] Our analytical intuition is that just as models of controlled stochastic processes (such as search models) lead to equilibrium descriptions of the economy, it should be possible to apply the same intuition to models that generalize the notion of dependence into spatial *cum* temporal contexts. Individuals may sample over space, in a timeless context, or over space and time. The modeling problem which must be solved would require conditioning each individual's state on her neighbor's state, rather than on links between neighbors being operative or not. This intuition is very similar to the one underlying the use by search theory of sequential statistical decision theory. Just as equilibrium in search models takes the form of invariant distributions for the variables of interest, equilibrium in models involving random

[29]For mathematical theories that lend themselves to such applications, see Krengel and Sucheston,[84] Mandelbaum and Vanderbei,[88] and Evstigneev.[47,48]

fields involves probability distributions associated with the active or inactive state of subsets of relevant populations.

ACKNOWLEDGMENTS

I thank Buzz Brock, Steven Durlauf, Alan Kirman, Chuck Manski, and Douglas North for very useful comments. I also thank Hans Haller for numerous fruitful interactions during the last several years over some of the issues discussed here and Anna Hardman for numerous suggestions and comments on the paper.

REFERENCES

1. Allen, B. "A Stochastic Interactive Model for the Diffusion of Information." *J. Math. Sociol.* **8** (1982): 265–281.
2. Allen, B. "Some Stochastic Processes of Interdependent Demand and Technological Diffusion of an Innovation Exhibiting Externalities Among Adopters." *Intl. Econ. Rev.* **23** (1982): 595–608.
3. Anderson, T. W. *The Statistical Analysis of Time Series.* New York: Wiley, 1971.
4. Aoki, M. "New Macroeconomic Modelling Approaches: Hierarchical Dynamics and Mean Field Approximation." *J. Econ. Dynamics & Control* **18** (1994): 865–877.
5. Aoki, M. "Group Dynamics When Agents Have a Finite Number of Alternatives: Dynamics of a Macrovariable with Mean Field Approximation." Working Paper # 13, UCLA Center for Computable Economics, June, 1994.
6. Aoki, M. *New Approaches to Macroeconomic Modeling.* Oxford: Oxford University Press, 1996.
7. Arthur, W. B. *Increasing Returns and Path Dependence in the Economy.* Ann Arbor: University of Michigan Press, 1994.
8. Banks, D., and K. Carley. "Metric Inference for Social Networks." *J. Classification* **11** (1992): 121–149.
9. Bell, A. M. "Dynamically Interdependent Preferences in a General Equilibrium Environment." Mimeo, University of Wisconsin, Madison, WI, November, 1994.
10. Bernheim, B. D., and K. Bagwell. "Is Everything Neutral?" *J. Pol. Econ.* **96** (1985): 308–338.
11. Bairoch, P. *City and Economic Development.* Chicago: University of Chicago Press, 1988.
12. Bertsekas, D. P. "Dynamic Behavior of Shortest Path Routing Algorithms for Communication Networks." *IEEE Trans. Aut. Control* **AC–27** (1982): 60–74.
13. Blanchard, O. J. "Price Synchronization and Price Level Inertia." In *Inflation, Debt, and Indexation,* edited by R. Dornbusch and M. Simonsen, 3–24. Cambridge, MA: MIT Press, 1983.
14. Blanchard, O. J. "The Wage Price Spiral." *Quart. J. Econ.* **101** (1986): 543–565.
15. Blanchard, O. J., and P. A. Diamond. "The Beveridge Curve." *Brookings Papers on Economic Activity* **1** (1989): 1–75.
16. Blanchard, O. J., and P. A. Diamond. "The Aggregate Matching Function." In *Growth/Productivity/Unemployment,* edited by P. A. Diamond, 159–201. Cambridge, MA: MIT Press, 1990.
17. Blume, L. E. "The Statistical Mechanics of Strategic Interaction." *Games & Econ. Behav.* **5** (1993): 387–424.

18. Bollobás, B. *Graph Theory.* New York: Springer-Verlag, 1979.
19. Bollobás, B. *Random Graphs.* New York: Academic Press, 1985.
20. Boylan, R. T. "Laws of Large Numbers for Dynamical Systems with Randomly Matched Individuals." *J. Econ. Theory* **57(2)** (1992): 473–504.
21. Boylan, R. T. "Continuous Approximation of Dynamical Systems with Randomly Matched Individuals." Mimeo, Washington University, December, 1993.
22. Brock, W. A. "Understanding Macroeconomic Time Series Using Complex Systems Theory." *Struct. Change & Econ. Dyn.* **2(1)** (1991): 119–141.
23. Brock, W. A. "Pathways to Randomness in the Economy: Emergent Nonlinearity and Chaos in Economics and Finance." *Estudios Economicos* **8(1)** (1993): 3–55. Reprinted as SSRI Reprint No. 410, University of Wisconsin, Madison, WI.
24. Brock, W. A., and S. N. Durlauf. "Discrete Choice with Social Interactions." Department of Economics, University of Wisconsin, Madison, WI, June, preliminary, 1995.
25. Cass, D., and M. Yaari. "A Re-examination of the Pure Consumption-Loans Model." *J. Pol. Econ.* **74** (1966): 353–367.
26. Cohen, J. E. "Threshold Phenomena in Random Structures." *Discrete Applied Mathematics* **19** (1988): 113–128.
27. Cosslett, S. R. "Extreme-Value Stochastic Processes: A Model of Random Utility Maximization for a Continuous Choice Set." Mimeo, Department of Economics, Ohio State University, 1987.
28. Cox, D. R., and D. V. Hinkley. *Theoretical Statistics* London: Chapman and Hall, 1974.
29. de Haan, L. "A Spectral Representation for Max-Stable Processes." *The Annals of Probability* **12(4)** (1984): 1194–1204.
30. Dagsvik, John K. "Discrete and Continuous Choice, Max-Stable Processes, and Independence from Irrelevant Attributes." *Econometrica* **62(6)** (1994): 1179.
31. Dobkins, L. H., and Y. M. Ioannides. "Evolution of the U.S. City Size Distribution." Working Paper, Department of Economics, VPI&SU, July, 1995.
32. Dobrushin, P. L. "The Description of a Random Field by Means of Conditional Probabilities and Conditions on Its Regularity." *Functional Analysis and Its Applications* **2** (1968): 302–312.
33. Dobrushin, P. L. "The Problem of Uniqueness of a Gibbsian Random Field and the Problem of Phase Transitions." *Functional Analysis and Its Applications* **2** (1968): 302–312.
34. Durlauf, S. N. "Locally Interacting Systems, Coordination Failure, and the Behavior of Aggregate Economic Activity." Mimeo, Stanford University, July, 1989.
35. Durlauf, S. N. "Output Persistence, Economic Structure, and the Choice of Stabilization Policy." *Brookings Papers on Economic Activity* **2** (1989): 69–116.

36. Durlauf, S. N., "Path Dependence in Aggregate Output." SITE Technical Report No. 8, Stanford University, May, 1991.
37. Durlauf, S. N. "Multiple Equilibria and Persistence in Aggregate Fluctuations." *Amer. Econ. Rev.* (1991): 70–74.
38. Durlauf, S. N. "Nonergodic Economic Growth," *Rev. Econ. Studies* **60** (1993): 349–366.
39. Durlauf, S. N. "Bilateral Interactions and Aggregate Fluctuations," Proceedings of the First World Congress of Nonlinear Analysts, 1994.
40. Durlauf, S. N. "A Theory of Persistent Income Inequality." *J. Econ. Growth* **1** (1994): 75–93.
41. Durlauf, S. N. "Neighborhood Feedbacks, Endogenous Stratification, Income Inequality." Mimeo, University of Wisconsin, April, 1994.
42. Ellis, R. *Entropy, Large Deviations, and Statistical Mechanics.* New York: Springer-Verlag, 1985.
43. Ellison, G. "Learning, Local Interaction, and Coordination," *Econometrica* **61(5)** (1993): 1047–1071.
44. Erdös, P., and A. Renyi. "On Random Graphs I." *Publicationes Mathematicae Debrecen* **6** (1959): 290–297.
45. Erdös, P., and A. Renyi. "On the Evolution of Random Graphs." *Publications of the Mathematical Institute of the Hungarian Academy of Sciences* **5** (1960): 17–61.
46. Erdös, P., and J. H. Spencer. *Probabilistic Methods in Combinatorics.* New York: Academic Press, 1974.
47. Evstigneev, I. V. "Controlled Random Fields on an Oriented Graph." *Theory of Probability and Its Applications* **33** (1988): 433–435.
48. Evstigneev, I. V. "Stochastic Extremal Problems and the Strong Markov Property of Random Fields." *Russian Mathematical Surveys* **43** (1988): 1–49.
49. Evstigneev, I. V. "Controlled Random Fields on Graphs and Stochastic Models of Economic Equilibrium." In *New Trends in Probability and Statistics,* edited by V. V. Sazonov and T. Shervashidze, 391–412. Utrecht: VSP, 1991.
50. Evstigneev, I. V., and P. E. Greenwood. "Markov Fields Over Countable Partially Ordered Sets: Extrema and Splitting." Discussion Paper No. A-371, Sonderforschungsbereich 303, University of Bonn, June, 1993.
51. Evstigneev, I. V., and M. Taksar. "Stochastic Equilibria on Graphs, I." Discussion Paper No. A-391, Sonderforschungsbereich 303, University of Bonn, December, 1992.
52. Evstigneev, I. V., and M. Taksar. "Stochastic Equilibria on Graphs, II." Discussion Paper No. A-391, Sonderforschungsbereich 303, University of Bonn, January, 1993.
53. Georgii, H.-O. *Gibbs Measures and Phase Transitions.* New York: Walter de Gryuter, 1988.

54. Föllmer, H. "Random Economies with Many Interacting Agents." *J. Math. Econ.* **1(1)** (1974): 51–62.
55. Gilboa, I., and A. Matsui. "A Model of Random Matching," Mimeo, Northwestern University, May, 1990.
56. Gilles, R. P., H. H. Haller, and P. H. M. Ruys. "Modelling of Economies with Relational Constraints on Coalition Formation." In *Imperfections and Behavior in Economic Organizations,* edited by Robert P. Gilles and Peter H. M. Ruys, Chap. 5, 89–136. Boston: Kluwer Academic, 1994.
57. Gilles, R. P., and P. H. M. Ruys. *Imperfections and Behavior in Economic Organizations.* Boston: Kluwer Academic, 1994.
58. Griffeath, D. "Introduction to Random Fields." In *Denumerable Markov Chains,* edited by J. G. Kemeny et al. New York: Springer-Verlag, 1976.
59. Haller, H. "Large Random Graphs in Pseudo-Metric Spaces." *Math. Soc. Sci.* **20** (1990): 147–164.
60. Haller, H. "Topologies as Trade Infrastructures." In *Imperfections and Behavior in Economic Organizations,* edited by R. P. Gilles and P. Ryus, Chap. 6, 137–155. Boston: Kluwer Academic, 1994.
61. Haller, H. H., and Y. M. Ioannides. "Trade Frictions and Communication in Markets." Mimeo, Department of Economics, VPI&SU, 1991.
62. Helsley, R. W., and W. C. Strange. "Agglomeration Economies and Urban Capital Markets." Mimeo, University of British Columbia, 1988.
63. Helsley, R. W., and W. C. Strange. "Matching and Agglomeration Economies in a System of Cities." Mimeo, University of British Columbia, June 1989.
64. Henderson, J. V. *Economic Theory and the Cities.* New York: Academic Press, 1985.
65. Henderson, J. V. *Urban Development: Theory, Fact, and Illusion.* Oxford: Oxford University Press, 1988.
66. Huberman, B. A., and T. Hogg. "The Behavior of Computational Ecologies." In *The Ecology of Computation*, edited by B. A. Huberman, 77–115. Amsterdam: North Holland, 1988.
67. Ioannides, Y. M. "Random Matching, the Intermediation Externality and Optimal Risk Sharing." In *Scientific Annals,* edited by T. Georgacopoulos, 55–71. Athens School of Economics and Business, 1986.
68. Ioannides, Y. M. "Trading Uncertainty and Market Form." *Intl. Econ. Rev.* **31(3)** (1990): 619–638.
69. Ioannides, Y. M. "Residential Neighborhood Effects." Working Paper, Tufts University, July 1996.
70. Jovanovic, B. "Micro Uncertainty and Aggregate Fluctuations." Mimeo, Department of Economics, New York University, January, 1984.
71. Jovanovic, B. "Aggregate Randomness in Large Noncooperative Games." C. V. Starr Center for Applied Economics, R. R. (1985): 85–14.
72. Jovanovic, B. "Micro Shocks and Aggregate Risk." *Quart. J. Econ.* **102** (1987): 395–409.

73. Kalai, E., A. Postlewaite, and J. Roberts. "Barriers to Trade and Disadvantageous Middlemen: Nonmonotonicity of the Core." *J. Econ. Theor.* **19** (1978): 200–209.
74. Kauffman, S. A. "The Evolution of Economic Webs." In *The Economy as an Evolving Complex System,* edited by P. W. Anderson, K. J. Arrow, and D. Pines, Santa Fe Institute Studies in the Sciences of Complexity, Proc. Vol. V. Reading, MA: Addison-Wesley, 1988.
75. Kauffman, S. A. *The Origins of Order: Self-Organization and Selection in Evolution.* New York and Oxford: Oxford University Press, 1993.
76. Kauffman, S. A. *At Home in the Universe.* New York and Oxford: Oxford University Press, 1995.
77. Kelly, M. "Why not China? Smithian Growth and Needham's Paradox." Mimeo, Department of Economics, Cornell University, 1994.
78. Kindermann, R., and J. L. Snell. *Markov Random Fields and Their Applications.* American Mathematical Society, Contemporary Mathematics, Vol. 1, Providence, RI, 1980.
79. Kirman, A. P. "Communication in Markets: A Suggested Approach." *Econ. Lett.* **12** (1983): 1–5.
80. Kirman, A. P. "Whom or What Does the Representative Individual Represent?" *J. Econ. Perspectives* **6(2)** (1992): 117–136.
81. Kirman, A. P. "Economies with Interacting Agents." Mimeo, Presented at the World Congress of the Econometric Society, Tokyo, August, 1995.
82. Kirman, A. P., C. Oddou, and S. Weber. "Stochastic Communication and Coalition Formation." *Econometrica* **54** (1986): 129–138.
83. Kovalenko, I. N. "The Structure of a Random Directed Graph. " *Theory of Probability and Mathematical Statistics* **6** (1975): 83–92.
84. Krengel, U., and L. Sucheston. "Stopping Rules and Tactics for Processes Indexed by a Directed Set." *J. Mult. Anal.* (1981): 199–229.
85. Krugman, P. "On the Number and Location of Cities." *European Econ. Rev.* **37** (1993): 293–308.
86. Krugman, P. "Complex Landscapes in Economic Geography." *Amer. Econ. Rev.: Papers and Proceedings* **84** (1994): 412–416.
87. Lerman, S. R., and C. R. Kern. "Hedonic Theory, Bid Rents, and Willingness-to-Pay: Some Extensions of Ellickson's Results." *J. Urban Econ.* **13** (1983): 358–363.
88. Mandelbaum, A., and R. J. Vanderbei. "Optimal Stopping and Supermartingales Over Partially Ordered Sets." *Zeitschnft für Wahrscheinlichkeitstheorie* **57** (1981): 253–264.
89. Manski, C. F. "Identification of Endogenous Social Effects: The Reflection Problem." *Rev. Econ. Studies* **60** (1993): 531–542.
90. Manski, C. F., and D. McFadden, Eds. *Structural Analysis of Discrete Data with Econometric Applications.* Cambridge, MA: MIT Press, 1981.

91. McFadden, D. "Modelling the Choice of Residential Location." In *Spatial Interaction Theory and Residential Location,* edited by A. Karlquist, Chap. 3, 75–96. Amsterdam: North-Holland, 1978.
92. McFadden, D. "Econometric Models of Probabilistic Choice." In *Structural Analysis of Discrete Data with Econometric Applications,* edited by C. F. Manski and D. McFadden. Cambridge, MA: MIT Press, 1981.
93. Mills, E. S. "An Aggregate Model of Resource Allocation in a Metropolitan Area." *Amer. Econ. Rev.* **57** (1967): 197–210.
94. Pissarides, C. A. "Job Matching with State Employment Agencies and Random Search." *Econ. J.* **89** (1979): 818–833.
95. Pissarides, C. A. *Equilibrium Unemployment Theory.* Oxford: Blackwell, 1990.
96. Palmer, E. M. *Graphical Evolution.* New York: Wiley Interscience, 1985.
97. Preston, C. J. *Gibbs States on Countable Sets,* Cambridge, MA: Cambridge University Press, 1974.
98. Pudney, S. *Modelling Individual Choice: The Econometrics of Corners, Kinks and Holes.* Oxford: Basil Blackwell, 1989.
99. Rust, J. "Estimation of Dynamic Structural Models, Problems and Prospects: Discrete Decision Processes." In *Advances in Econometrics, Sixth World Congress,* edited by C. A. Sims, Vol. II, Chap. 4. Cambridge, MA: Cambridge University Press, 1994.
100. Schelling, T. C. *Micromotives and Macrobehavior.* New York: Norton, 1978.
101. Spitzer, F. "Markov Random Fields and Gibbs Ensembles." *Am. Math. Monthly* **78** (1971): 142–154.
102. Spitzer, F. *Random Fields and Interacting Particle Systems.* Mathematical Association of America, 1971; seminar notes, 1971 MAA Summer Seminar, Williams College, Williamstown, MA.
103. Verbrugge, R. "Implications of Local Pecuniary Interactions for Aggregate Fluctuations." Mimeo, Stanford University, January, 1995.
104. Weitzman, M. L. "Increasing Returns and the Foundation of Unemployment Theory." *The Econ. J.* **92** (1982): 787–804.
105. Weitzman, M. L. "Monopolistic Competition and Endogenous Specialization." *Rev. Econ. Studies* **61** (1994): 45–56.
106. Whittle, P. *Systems in Stochastic Equilibrium.* New York: Wiley, 1986.
107. Wicksell, K. *Lectures in Political Economy, Vol. 1.* London: George Routledge & Sons, 1934. Reprinted by Augustus Kelley, Fairfield, NJ, 1977.

David Lane* and Robert Maxfield**
*Department of Political Economy, University of Modena, ITALY; E-mail: lane@c220.unimo.it
**Department of Engineering and Economic Systems, Stanford University, Stanford, CA 95305; E-mail: maxfield@leland.stanford.edu

Foresight, Complexity, and Strategy

This chapter originally appeared in *Long Range Planning* **29(2)** (1996): 215–231, as "Strategy Under Complexity..." Reprinted by permission of Elsevier Science Ltd.

What is a strategy? The answer to this question ought to depend on the foresight horizon: how far ahead, and how much, the strategist thinks he can foresee. When the very structure of the firm's world is undergoing cascades of rapid change, and interpretations about the identity of agents and artifacts are characterized by ambiguity, we say that the foresight horizon is complex. We argue that strategy in the face of complex foresight horizons should consist of an on-going set of practices that interpret and construct the relationships that comprise the world in which the firm acts. We illustrate the ideas advanced in this chapter with a story about the entry of ROLM into the PBX market in 1975.

What is a strategy? Once upon a time, everybody knew the answer to this question. A strategy specified a *precommitment* to a particular course of action. Moreover, choosing a strategy *meant optimizing among a set of specified alternative courses of action*, on the basis of an evaluation of the value and the probability of their possible consequences. Business schools trained a generation of MBAs in optimization techniques, and strategic planning departments and consulting firms honed these tools to fit a myriad of particular circumstances. For a while, strategy as optimizing precommitment was a growth industry. And it still flourishes in some environments, both theoretical[1] and practical.[2] But it is fair to say that this notion of strategy is falling into increasing disfavor in the business world, to the extent that the CEOs of some important and successful firms will not permit the word "strategy" to be uttered in their presence.

It is not hard to find the reason why. Optimizing precommitment makes sense when a firm knows enough about its world to specify alternative courses of action and to foresee the consequences that will likely follow from each of them. But, of course, foresight horizons are not always so clear. All too often too many agents[3] doing too many things on too many different time scales make for a combinatorial explosion of possible consequences—and when that happens, foresight horizons become complicated. If you cannot even see a priori all the important consequences of a contemplated course of action, never mind carrying out expected gain calculations, you must constantly and actively monitor your world—and react quickly and effectively to unexpected opportunities and difficulties as you encounter them. As a result, an essential ingredient of strategy in the face of complicated foresight horizons becomes the *organization of processes of exploration and adaptation*.[4]

But suppose the very structure of the firm's world—that is, the set of agents that inhabit it and the set of artifacts[5] around which action is organized—undergoes cascades of rapid change. Then, the metaphor of exploration breaks down. The world in which you must act does not sit passively out there waiting to yield up its secrets. Instead, your world is under active construction, you are part of the

[1] Within university economic departments, strategy as optimized precommitment provides the nearly unchallenged foundation for microeconomic theory.

[2] Senior partners in a leading consulting firm told us that 80% of their firm's strategic consultancies still culminate in a formal optimization exercise.

[3] In this chapter, we use a few terms that may at first seem like unnecessary jargon, but for which we do not find ready, unambiguous substitutes in ordinary language. These include agent, actor, artifact, space, generative relationships, attributions, and identity. When we introduce these terms, we will present definitions—generally in footnotes—that we think will at least orient the reader to what we have in mind; the full meaning we intend these terms to bear will, we hope, emerge as we use them throughout the article. By "agent," we mean any collection of people that jointly engages in economic activity. Individuals are agents (we generally refer to them as "actors") and so are firms—as are the marketing group assigned to a particular product in a firm, a team of researchers working together to develop a new product, or an industry-level trade association.

[4] The recent burst of literature on the "learning organization" testifies to the growing importance of this conception of strategy.[5,6,10,11,12]

[5] By "artifact," we mean any object or service around which economic activity is organized—in particular, objects and services designed, produced, and exchanged by economic agents. And by "object," we mean not only cars, movies, and telephones, but software systems, architectural blueprints, and financial instruments as well.

construction crew—and there isn't any blueprint. Not only are the identities of the agents and artifacts that currently make up your world undergoing rapid change, but so also are the interpretations that all the agents make about who and what these things are. What can strategy mean, when you lack even a temporally stable language in which to frame foresight? This chapter offers a partial answer to this question.

In the first section, we distinguish between three kinds of foresight horizons: the clear, the complicated, and the complex. Our problem is to formulate a notion of strategy that makes sense for complex foresight horizons. Any such notion must be premised on an understanding of how changes come about in these situations. So in section 2, we tell a story that will help us to identify some of the mechanisms underlying these changes: in particular, we highlight the role of what we call attributions[6] and generative relationships.[7] In section 3, we extract four lessons from our story, which we apply in section 4 to arrive at a partial formulation of what strategy means in the face of a complex foresight horizon. Our discussion focuses on two intertwined kinds of strategic processes. The first is cognitive: a firm "populates its world" by positing which agents live there and constructing explicit interpretations of what these agents do. The second is structural: the firm fosters generative relationships between agents within and across its boundaries—relationships that produce new attributions and, ultimately, sources of value that cannot be foreseen in advance.

1. FORESIGHT HORIZONS

Picture an eighteenth century general perched on a hill overlooking the plain on which his army will engage its adversary the next day. The day is clear, and he can see all the features of the landscape on which the battle will be fought—the river and the streams that feed it, the few gentle hills, the fields and orchards. He can also see the cavalry and infantry battalions positioned where he and his opponent have placed them, and he can even count the enemy guns mounted on the distant

[6] An "attribution" is an interpretation of meaning by an agent to itself, to another agent or to an artifact. An agent has meaning in so far as its actions can be interpreted coherently in terms of an identity: that is, the functions toward which the agent's actions are directed and the "character" that informs the ways in which these actions are carried out. An artifact's meaning concerns the way in which agents use it. See section 4.2 for more discussion of these ideas; examples are provided there and in section 2.

[7] By "generative relationship," we mean a relationship that can induce changes in the way the participants see their world and act in it, and even give rise to new entities, like agents, artifacts, even institutions. Moreover, these changes and constructions cannot be foreseen from a knowledge of the characteristics of the participating agents alone, without knowledge of the structure and history of the interactions that constitute the relationship between them. See Lane et al.[9] for an extended introduction to this concept.

hillsides. The battle tomorrow will consist of movements of these men across this landscape, movements determined in part by the orders he and his staff and their opposite numbers issue at the beginning of the day, and in part by the thousands of little contingencies that arise when men, beasts, bullets, and shells come together. While he cannot with certainty predict the outcome of all these contingencies, nor of the battle that together they will comprise, he can be reasonably sure that one of a relatively small number of scenarios he can presently envision will actually come to pass.

Now think about a U.S. cavalry column marching through an uncharted section of Montana in the early 1870s. The commanding officer cannot know the location of the nearest river or whether there will be an impassable canyon on the other side of the hills looming over his line of march. Nor does he know where the Indian tribes who inhabit this country have established their camps or whether they are disposed to fight should he come into contact with them. He knows the general direction he wants to take his men, but it would not pay him to envision detailed forecasts of what the next days might hold, because there are too many possibilities for unexpected things to happen. Instead, he relies on his scouts to keep him informed about what lies just beyond his own horizon, and he stays alert and ready for action. He is confident that he will recognize whatever situation he encounters, when he encounters it.

Finally, imagine the situation of the Bosnian diplomat in early September 1995 trying to bring an end to the bloodshed in his country. It is very difficult to decide who are his friends and who his foes. First he fights against the Croats, then with them. His army struggles against an army composed of Bosnian Serbs, but his cousin and other Muslim dissidents fight alongside them. What can he expect from the UN Security Forces, from the NATO bombers, from Western politicians, from Belgrade and Zagreb, from Moscow? Who matters, and what do they want? On whom can he rely, for what? He doesn't know—and when he thinks he does, the next day it changes.[8]

Uncertainty figures in all three of these stories, but it does so in ways that differ in two important respects.

TIME HORIZON OF RELEVANT UNCERTAINTY. In the first story, the general's uncertainty has a clear terminal date: tomorrow, when the battle will have been fought and either won or lost. After that, other problems will surely arise and occupy his interest, but for now everything focuses on the battle, the matter at hand. The cavalry commander is concerned with getting his troops to their assigned destination, a matter perhaps of several days or weeks. For the Bosnian diplomat, there is no obvious end in view.

[8]Since we wrote this paragraph, many things have changed. But it only reinforces the point that we are trying to make to emphasize to our reader that, whenever you are reading this chapter, we are practically certain that the Bosnian diplomat is in nearly the same situation of uncertainty, but with a different cast of relevant entities.

KNOWLEDGE OF RELEVANT POSSIBLE CONSEQUENCES. The general knows what he is uncertain about: not only just which side will win the battle, but also the kind of events that will turn out to be decisive in determining the outcome—which troops will move to what locations, engaging which adversaries, under what conditions. The cavalry commander too could probably also frame propositions about almost anything likely to be relevant to the completion of his mission, but it would amount to a very long list, most items of which would not turn out to matter anyway. In contrast, the Bosnian diplomat would be at a loss to name all the actors and events that could affect the outcome of the drama of which he is a part. In fact, no one could name them, because in the working out of the drama, new actors keep getting drawn in, and they create new kinds of entities—like the Rapid Deployment Force or the abortive Moscow Peace Meeting—that simply could not be predicted in advance.

The general, the cavalry commander and the Bosnian diplomat face three different kinds of foresight horizon. The general's horizon is *clear*. He can see all the way to the end of it, and the contours of all the features from here to there stand out in bold relief. This of course does not mean that he knows what is actually going to happen; but he can be reasonably sure of what *might* happen and about how his actions will affect the likelihood of the various possible outcomes. In contrast, the cavalry commander's horizon is more *complicated*. He knows the kinds of things that might happen before he arrives at his destination, but because of the sheer number of possible geographical, meteorological, and social combinations it is difficult to imagine them all at the outset of his mission. Nonetheless, he thinks he knows how to find out about the eventualities that are likely to matter in time to respond efficaciously to them.

The Bosnian diplomat's horizon is certainly complicated, but there is more to it than that. Unlike the cavalry commander, his problem is not just to negotiate his way through a fixed landscape composed of familiar, if presently unknown, features. The social landscape through which he moves constantly deforms in response to the actions he and others take, and new features, not previously envisioned or even envisionable, emerge. To understand where he is going—and where he has been—he has to generate a new language even to describe these features, and he must also reinterpret many things that he had been regarding as fixed, but whose character has been changed by their relation to the features that have just emerged. Since his destination is always temporally beyond his current foresight horizon, the connection between what he does and where he is going is always tenuous and, hence, ambiguous. Inhabiting as he does a world of emergence, perpetual novelty, and ambiguity, the Bosnian diplomat's foresight horizon is *complex*.[9]

[9]The new sciences of complexity are concerned with describing and explaining phenomena associated with emergence and self-organization. For an engaging introduction to some of these ideas, see Waldrop,[13] while Kauffman[7] provides a synthesis of recent developments from the biological perspective. Anderson et al.,[1] Arthur,[2] and Lane[8] relate aspects of complexity to economics.

The emerging world of "multimedia"[10] well illustrates the key features of complex foresight horizons as they present themselves in business. This world, like any other business domain, is characterized by a space[11] of agents and artifacts, which are structured by various relationships—between agents, between artifacts, and between agents and artifacts. In some domains, like the textile or oil industries, these spaces are relatively stable. In contrast, multimedia's is changing very rapidly. New firms come into being almost daily. Boundaries between existing firms are reshuffled through mergers and acquisitions—and interpenetrated through strategic alliances and other kinds of joint enterprises. What is emerging is a complex cross-cutting network of collaborative and competitive relationships, and it is sometimes unclear which is which. Moreover, it is hard to classify the artifacts these agents are producing into such previously separable categories as telecommunications, computers, electronics, media, entertainment, software, or banking. Even the very identities of key multimedia artifacts are up for grabs: is the "set top box" at which the consumer will explore the information superhighway a television, a personal computer, or some entirely new device?

But more is changing in multimedia than the structure of its agent/artifact space: how agents perceive their own and each other's functionality—that is, who does what, to (and with) whom (and what)—is changing as well. Telephone companies are no longer sure about what they ought to be providing to their customers: is it primarily connectivity, or is it content? Conversely, cable companies wonder if they should be providing voice communication service. In the face of challenges from software vendors and credit card companies, banks have to rethink just what roles they'll be playing in five years—and how they'll be doing it.

Structurally, then, multimedia is emergent; cognitively, it is ambiguous. Emergent structure and cognitive ambiguity generate complex foresight horizons for firms caught up in—or plunging into—the multimedia vortex.

2. THE ROLM PBX STORY

After enough time passes, complex foresight horizons can be made to look clear, at least in retrospect. In this section, we tell an old story that illustrates some essential features of the process whereby agent/artifact spaces change their structure.

[10] Just what this world ought to be called is still under negotiation by the agents that are playing a part in its construction. Some refer to it as "the information superhighway," while others prefer a more prosaic description as the merging of voice, data, graphics, and video. From our point of view, this linguistic confusion is symptomatic of the fact that the attributions of the artifacts around which this world is being organized are still "under construction."

[11] We use "space" to denote a structured set. For agent and artifact space, the set consists of all the agents who are involved in some industry activity and the artifacts are the things these agents design, make or exchange, while the structure is provided by the various kinds of relationships that exist between these agents and artifacts.

The story describes what happened when ROLM,[12] a small California computer company, decided to enter the PBX market in 1973. The 1968 Carterphone decision had sent a shock through the telecommunications world whose effects were to be felt for years to come. In particular, foresight horizons in the PBX business became complex, and ROLM was able to take advantage of the situation. Within five years of its first PBX sale, ROLM challenged giants ATT and Northern Telecom for leadership in the then $1 billion PBX market.

ROLM had a strategy when it launched its PBX product, but events quickly invalidated many of the assumptions on which this strategy was based. In particular, the primary line of long-term product development envisioned in the strategy was shelved within a year of the first shipment. Certainly ROLM managers made some right moves as they scrambled to keep up with the consequences of what they had initiated. But the lessons of our story are not about ROLM managers as strategists; in fact, it won't even be clear what strategy has to do with what happens, a point to which we will return in section 3. Instead, our story is about how change happens in the presence of complex foresight horizons—and, in particular, about how and where the understanding of new possibilities for artifacts and the social arrangements to which they give rise come into being.

2.1 THE US PBX BUSINESS IN 1973: AGENTS AND ARTIFACTS

A PBX is a private telecommunication system that manages an organization's incoming, outgoing, and interoffice telephone calls. It connects outside lines that are provided by the local telephone company, which in 1973 would have been an ATT operating company or a so-called independent like General Telephone. Before 1968, all PBXs were provided by the local telephone company. ATT companies used PBX equipment manufactured by ATT subsidiary Western Electric; General Telephone had its own manufacturing subsidiary, Automatic Electric; and the smaller independents bought the PBXs they installed from a handful of producers, including Automatic Electric, Northern Telecom, and Stromberg Carlson. In large part because of the supply monopoly enjoyed by the local telephone companies, PBX technology had not kept pace with innovations in semiconductors and computers. Most PBXs still used electromechanical switching technology and had only a few "convenience" features like call-forwarding.

In 1968, the Federal Communication Commission's historic Carterphone decision broke the monopoly on PBXs. Now businesses and other organizations were free to lease or buy their internal telecommunications systems from any purveyors, not just the local telephone company. The Carterphone decision gave a jolt to the space of agents organized around PBXs. Many new companies entered the

[12]Our version of the story relies on various ROLM marketing and strategy documents currently in the possession of one of us (RM), a cofounder of ROLM who led its PBX division, and material collected in a Stanford Graduate School of Business case prepared by Professor Adrian Ryans.

PBX business. In particular, a whole new kind of business was created: the *interconnect distributors*, who configured, installed, and maintained business communications systems. By 1973, there were more than 300 interconnect distributors, competing with the telephone companies for shares in the nearly $500-million PBX market. These distributors came in many different sizes and shapes, from "mom-and-pop" local suppliers to $25-million-a-year-revenue national subsidiaries of large conglomerates, several of which—like ITT and Stromberg Carlson—also manufactured PBXs. Most local or regional distributors sold mainly to small companies with fewer than 30 telephone extensions, while the largest national distributors targeted big customers with up to several thousand extensions. Other distributors specialized in particular niche markets, like hotels or hospitals.

Unlike the flurry of change in distribution channels, PBX products did not change much after Carterphone. Several large established foreign PBX manufacturers entered the US market and quickly captured a small market share. But the equipment they sold was quite similar to what was already in the American marketplace, and competition was primarily centered on price. By 1973, some firms—most notably ITT in America and IBM in Europe—were producing PBXs with superior technology, like analog electronic switching, and enhanced functionality, like limited customization capacity. But the really big anticipated technological innovations were widely regarded as still being well over the horizon. In particular, it was thought that neither digital switching nor computer-based control would be feasible in terms of cost and reliability for at least seven or eight years. Digital switching would provide the means to integrate voice and data into a single system, while computer control would make possible such additional functions as least-cost routing of long-distance calls, automatic dialing, and call-detail recording—and would permit an organization to move telephone numbers from office to office just by typing commands at a keyboard, rather than by ripping out and rewiring cable.

2.2 HOW TELECOMMUNICATIONS MANAGERS AND PBXS DEFINED EACH OTHER

It is worth pausing here to ask why PBX design changed so slowly after Carterphone. "Technological limitations" is an easy answer, but it is wrong, as ROLM's success with both digital switching and computer control was soon to show. Certainly there is a lag between the decision to go with a new design and a product launch, but this cannot be the explanation either, since ROLM shipped its first system within 18 months of its decision to make a PBX—and ROLM had nothing like the human, dollar, or infrastructural resources of an ATT, an ITT, an IBM or even a Stromberg Carlson.

We think the right explanation is *social*, not technological: PBX buyers had no incentive to ask for a system that could do anything other than what existing PBX models already did. To see why this was so, we need to look at the role of the telecommunications manager (TM), the officer who was responsible for administering the PBX system in larger firms. In particular, the TM initiated PBX purchase orders, subject to the approval of his higher management. Before 1968, the TM had had few choices to make about the system his company used and little opportunity to maximize the system's cost-effectiveness. His main duty was to monitor his company's needs for outside lines and internal telephones. When a change was needed, he called the local telephone company, gave them the information, and accepted whatever recommendations the telephone company had to offer. Even though telecommunications costs were substantial, often around 3% to 5% of a company's revenues, there were almost no opportunities for effective telecommunications management. As a result, TMs tended to be ex-telephone company employees, with little incentive to be creative—or training or experience in *how* to be creative managers, even if the will were there. TMs stood in stark contrast to information systems managers, who played a major role in most companies in devising strategies to enhance the productivity and effectiveness of all parts of the company through the use of computers.

Carterphone ended the enforced dependence of the TM on the local telephone company, but it could not change *patterns of interaction* that had formed over decades between the TM and his suppliers, or between him and his higher management. Nor could it change *attributions* that the TM, his higher management, and his suppliers all shared: in particular, that a PBX is just a switchboard connected to telephone sets at one end and an outside line at the other; and that managing a PBX means making sure that there is a serviceable telephone on each desk where one is needed. After Carterphone, when vendors of PBX systems with more than 100 extensions competed for sales, they pitched their products to TMs, who passed purchase recommendations up the line to higher management. The pitch was made to and from actors who shared the attributions we have just described. Within the framework of these attributions, existing PBX systems did exactly what PBX systems ought to do.

2.3 ROLM ENTERS THE PBX BUSINESS

In 1973, ROLM was a four-year-old Silicon Valley company manufacturing militarized minicomputers, with annual revenues around $4 million and fewer than 100 employees. The firm's top managers had decided to diversify because they felt that, for a variety of reasons, they could not grow beyond about $10 million a year in the military computer market, and they were seeking larger growth opportunities. ROLM had developed competencies in designing and manufacturing computers, but it was prohibited from entering the general purpose commercial computer market by licensing restrictions on a base technology it had obtained from Data General. So top ROLM managers were seeking to find a market opportunity in which they could build a product using embedded computer technology. The PBX market attracted them, because it was large enough so that even a small market share would represent significant growth for ROLM. For ROLM management, the attraction of making a PBX was precisely the opportunity of introducing computer control and digital switching. The electrical engineers who ran ROLM were convinced that a PBX based on these technologies was feasible now.

In June 1973, ROLM hired two engineers from Hewlett-Packard (HP) to begin designing their PBX, and in August it recruited an HP marketing manager to do a market analysis. The analysis was very encouraging: computer control would not only permit a variety of desirable "convenience" features, but would result in such substantial cost savings in long-distance tolls and in moves and changes that the cost of the PBX would typically be recovered in less than two years. Since a single design could not possibly span the whole PBX market, ROLM management decided to target the 100–800 extension range, because the computer at the heart of its system would be too expensive for small installations, while systems with thousands of extensions would mean long selling cycles, few buying decisions per year, and credibility problems for a tiny company like ROLM taking on the ATT giant. In November, the market analysis and the engineering group's technical feasibility study convinced the ROLM board of directors to go ahead with the PBX project.

2.4 ROLM'S MARKETING STRATEGY: BUILDING RELATIONSHIPS WITH TELECOMMUNICATIONS MANAGERS

ROLM management now had to decide how to market their system. ROLM was simply too small a company to establish its own sales/installation/service network from the start. ROLM managers were unwilling to become dependent on a single national distributor, all of which were losing money and had poor service reputations anyway. Trying to sell to the independent telephone companies did not seem like a good first step: first, these companies had glacially slow decision-making procedures; and second, even if ROLM were able to win contracts with all of them, it would still be excluded from the major market areas under ATT control. The only remaining alternative was to contract with local and regional interconnect distributors, which is what ROLM decided to do.

But this decision posed a difficult marketing problem. Most of the local and regional distributors were only selling small systems, and so they lacked experience and contacts with the kind of companies that would need a 100–800 extension PBX. ROLM adopted a two-pronged approach to this problem. First, it sought to educate its local distributors to the special problems involved in selling to larger firms with multiple levels of decision-making. When ROLM contracted with an interconnect distributor, it required the distributor's sales people to attend a ROLM training program, staffed by instructors recruited from IBM and Xerox. Second, ROLM sought to establish direct relationships with the TMs of large firms such as General Motors, Allied Stores, and IBM. It did this by creating a group of "national accounts" representatives, ROLM employees whose mandate was to provide "liaison and support" to major customers. But the representatives' first task was to start talking to their targeted TMs, to let them know what the ROLM technology would be able to do for them, to find out what their problems were, and to figure out how ROLM could help solve them.

When the ROLM representatives first contacted the TMs, they faced an uphill task: they were asking these managers to give up a safe and easy relationship with the local telephone monopoly in favor of a new product embodying fancy new technology, manufactured by a tiny California computer company, and installed and maintained by a small, young, local interconnect distributor. Needless to say, the initial progress was slow. Many TMs were too risk-averse to look seriously at the ROLM alternative. But some were convinced, and after their initial ROLM PBX installations verified the representatives' claims of dramatic cost savings and productivity enhancements, the TMs became eager converts and pushed to purchase more systems for other company sites.

2.5 THE RELATIONSHIPS BECOME GENERATIVE: TRANSFORMATIONS

ROLM targeted the TMs because they needed to sell systems to them. But the relationships between their account representatives and the TMs yielded far greater dividends than ROLM managers had ever envisioned. Here, we recount some of the innovations—in products, attributions, and even identities—to which these relationships gave rise.

NEW PRODUCTS. The pioneering TMs who had taken a chance on ROLM found that the ROLM PBX did indeed produce substantial cost savings, which the system computer even helped to calculate. When the TMs reported these savings to their higher management, they became corporate heroes, as had their information system counterparts in the early days of computerization. As a result, these TMs were rewarded with increased budgets and responsibilities and were encouraged to continue to drive down telecommunications costs as rapidly as possible. Some responded to the new challenges and began actively to seek new opportunities for savings and productivity gains. Their search was channeled by their understanding of the technological possibilities of the digital-switched, computer-controlled PBX, derived from their ongoing discussions with their ROLM account representatives.

One TM had been interested in purchasing an Automatic Call Distribution (ACD) System, like those used by airlines to handle reservations. Such systems route a large number of incoming calls among many specially trained employees, and permit supervisory monitoring and the collection and analysis of traffic patterns. Unfortunately, existing ACD systems were very large and very expensive. So the TM posed a question to his ROLM account representative: could the ROLM PBX be programmed to do basic ACD functions? The account representative took this question back to ROLM engineers, who said yes, the idea was technically feasible—and ROLM marketers found, through a series of conversations with other national account representatives and their customers, that almost every company had the possibility, scattered around its organization, for small clusters of agents to use ACD functions (order processing, customer service, account information, and so on). Consequently, ACD features were incorporated into the third release of ROLM software, and it became a smash hit.

Another TM worked for a large retail chain with many stores in an urban area. She wanted to be able to provide 24 hour phone answering but could not justify the expense of having an operator in each store. So she asked whether it would be possible to have all the operators at a central location, with incoming calls automatically routed there by the PBX in each store, and then rerouted, when appropriate, back to a particular store. This idea led to what came to be called Centralized Attendant Service, which turned out to be useful to customers in a variety of industries.

AN EMERGENT PRODUCT ATTRIBUTION. Over the next several years, a whole set of product enhancements came out of the interactions between customer TMs and ROLM account representatives. These enhancements fulfilled new functionalities that were invented in response to new understandings about what a PBX could be, combined with an intimate knowledge of the business environment in which particular PBX systems were being used. In time, the very idea that users had about what a PBX was changed—from a telephone system, to an intelligent interface between a company and outsiders, to a tool that could be adapted to solve a wide range of "line-of-business" voice-based applications, providing productivity improvements in many aspects of customers' operations.

It is interesting to note that ROLM's idea of what a PBX could be changed too. In the original business plan for the ROLM PBX, much was made of the potential that digital switching offered for integrating a company's voice and data systems. While ROLM engineers were eager to realize this enticing vision, their efforts in this direction were side-tracked for several years, as the customer-generated voice applications were given top priority.

A NEW IDENTITY. As attributions about what a PBX was changed, so too did attitudes about the role and status of TMs. From a relatively low-level "custodian" of a company's telephonic system, the TM began to be perceived as a major factor in the ability of a company to enhance its productivity and to respond to customers. In many companies, the position became equal in importance to the information systems manager, often with a vice-presidency attached. For those TMs who personally experienced the growth in their own status and responsibilities as a result of their involvement with ROLM products, the relationship with ROLM intensified—and often these managers became more loyal to ROLM than to their own employers.

RELATIONSHIPS GENERATE RELATIONSHIPS. The "new" TMs not only bought new ROLM products for their own companies, but helped to extend ROLM's market penetration in other ways as well. For example, many were active in the International Communication Association (ICA), an organization for the TMs of large companies. Each year, the ICA held a convention, at which the members discussed topics of professional interest and product vendors were invited to display their wares. At these conventions, ROLM backers in the association gave formal presentations describing productivity-enhancing applications of the ROLM system—and afterward, over drinks in the bar, talked about the personal rewards and recognition they had won for championing ROLM inside their companies. As a result, other TMs actively sought out ROLM and its distributors.

The relationships ROLM had established between the targeted TMs and its account representatives had other positive effects for ROLM as well. First, ROLM had targeted the large, progressive companies for their initial efforts. When they succeeded with these companies, they greatly enhanced their credibility with smaller, less sophisticated customers.

More importantly, ROLM's ties to the TMs enabled its strategy for expanding distribution coverage into secondary and tertiary markets to succeed. ROLM first concentrated on the largest metropolitan markets, where it relied on local and regional interconnect companies for sales, installation and service. In smaller markets, ROLM intended to use independent telephone companies as its distributors of choice, since these companies had advantages in finance, experience, and contacts over local interconnect companies. The idea was to first win over the big independents—GTE, United Telephone, Continental Telephone. Then, the 1500 or so smaller telephone companies would be easier to convince, as they tended to look to their bigger brethren for leadership. Unfortunately, GTE was highly bureaucratic and owned its own manufacturer, Automatic Electric. To get GTE to carry the ROLM PBX, its engineering department had to certify product quality, and the approval process could take years. To accelerate the process, ROLM got a few of its "national accounts" customers with sites in several GTE territories to call their GTE marketing representatives and advise them that they would not be allowed to bid on their PBX installation projects unless GTE could propose ROLM equipment. As the number of such jobs mounted, GTE Marketing besieged Engineering with requests that the ROLM evaluation process be expedited. The tactic worked: in record time, GTE announced it was adding the ROLM PBX to its list of approved products. Shortly after, many other independents jumped on the bandwagon, significantly increasing ROLM's market coverage.

2.6 THE US PBX MARKET IN 1980

All of these transformations greatly changed the PBX market. While in the past, around 7% of the installed base of existing PBXs turned over every year, the technological innovations that ROLM introduced (and other companies adopted as quickly as they could) led to a huge bubble of demand. Within seven years of ROLM's entry into the PBX market, virtually the entire installed base was replaced. In 1980, PBX sales totaled nearly $1 billion.

ROLM's PBX business vastly exceeded management's most optimistic predictions. The original business plan's best-case forecast was $12 million in revenues in the third year; the actual figure was $50 million, and two years later revenues exceeded $200 million. Less than five years after installing its first PBX, ROLM had attained a 23% share of the US PBX market, second only to ATT. In complete contrast to the situation in 1973, the PBX market throughout the '80s had a quite stable structure: three firms—ROLM, ATT, and Northern Telecom—virtually shared market leadership, with very little left over for any other competitors.

3. LESSONS FROM THE ROLM STORY: HOW CHANGES IN AGENT/ARTIFACT SPACE HAPPEN

When foresight horizons are complex, the structure of agent/artifact space undergoes cascades of rapid change. Benefiting from these changes requires understanding something about how they happen. In this section, we highlight four lessons from the ROLM story that bear on this question, and we point out four implications of these lessons for strategy in the face of complex foresight horizons.

LESSON 1

Structural change in agent/artifact space is mediated by new attributions about the identity of agents and the meaning of artifacts.

DISCUSSION. Attributions about what an artifact "is" and what agents "do" matter. The meaning that agents give to themselves, their products, their competitors, their customers, and all the relevant others in their world, determine their space of possible actions—and, to a large extent, how they act. In particular, the meaning that agents construct for themselves constitute their *identity*: what they do, how they do it, with whom and to whom.

Of course, every action is not necessarily preceded by an explicit interpretation of the agents and artifacts that inhabit the context in which the action takes place. Far from it. Particularly in relatively stable worlds, recurrent patterns of agent interaction are triggered by the sheer familiarity of the situations in which the agents find themselves. In such situations, the role of attributions in guiding action may recede into the background, and participating actors may never consciously interpret the agents and artifacts with whom they interact.

When foresight horizons become complex, though, every situation contains some elements of novelty. If agents register this novelty, they engage in conscious deliberation to make sense of the situation before they commit themselves to act in it.[13] Of course, this deliberation does not necessarily lead to new attributions: often, instead, a situation's apparent novelty can be interpreted away by dredging up old attributions that are misidentified as "facts" or common knowledge, thus locking agents into ways of thinking and acting that blind them to emerging threats to their market position or to new opportunities for value creation. In the ROLM story, this is what happened to ITT, Stromberg Carlson, and other companies that were in the PBX business before 1973—and were out of the business by 1980.

Alternatively, agents may indeed generate new attributions when they encounter new entities or relations in agent/artifact space. These new attributions may lead to actions that, in turn, give rise to further changes in the structure of agent/artifact space. Here are some examples of this phenomenon from the ROLM story:

- ROLM managers in 1973 interpreted a PBX differently than did ATT and other manufacturers. ROLM's attribution came out of its understanding of computer control and digital switching: they saw a PBX as a flexibly configurable communication system with both voice and data potential. This attribution informed their initial product offering and their early development projects. The successful launch of the first ROLM product had a major impact on the future configuration of the PBX agent/artifact space. As a result of some of these changes, ROLM management changed its attribution about PBXs—and with it, the direction of their development efforts.
- The TMs who were early adopters of a ROLM PBX saw ROLM as an ally in innovative system-building, an attribution that no TM extended to ATT. Of course, this attribution led to the joint conception of a series of new PBX features and products.
- Changes in the way in which many companies' top management interpreted the role of TMs led to new value-creation opportunities for ROLM and its competitors in the mid-'80s.

By emphasizing the role of attributions, we do not mean to imply that causal relations between attributions and action go only one way. Clearly, attributional shifts depend upon particular action histories: for example, it would have been impossible for TMs to come to see their jobs as innovating productivity-enhancing communications systems had they not purchased and installed systems capable of generating and documenting substantial cost-savings, and subsequently worked with ROLM account representatives and applications engineers to develop modifications of these systems. We emphasize the role of attributions in causing action, rather than the other way around, because we believe that agents can learn how to "control" the process whereby they form new attributions more effectively than they can "control" directly the effects of action on attributions or on structural change—and strategy, as we shall argue in section 4, is about control.

[13]See Lane et al.[9] for an account of agent cognition that underlies what we are saying here.

LESSON 2

Generative relationships are the locus of attributional shifts.

DISCUSSION. This lesson is well illustrated by the attributional shifts that transformed the PBX from a switchboard connecting telephone sets with outside lines to a suite of productivity-enhancing line-of-business voice applications. Neither the TMs nor the various ROLM actors involved in this transformation—account representatives, development engineers, top management—could have achieved it by themselves. The transformation arose in the context of the relationships among all of these agents.

Consider, in particular, the generative relationship between TMs and ROLM account representatives. The account representatives worked to establish discursive relationships with TMs, in the context of which the TMs could come to understand the technology on which the ROLM system was based and the new functions and opportunities for cost-savings this technology offered. This understanding took time to develop. During the conversations that led up to it, the account representatives also learned a lot from the TMs, about the complex patterns of voice communication inside large companies and the ways in which TMs interacted with their managers and the suppliers and users of the systems they administered. The shared understandings that emerged from these conversations underlay the PBX feature innovations that emerged from the TM/account representative relationships.

It is important to realize why the ROLM account representative/TM relationships became generative, and those between, say, ATT salesmen and the TMs were not. In terms of their attributions about the role of PBX systems and their administration, the ATT salesmen and TMs saw everything eye-to-eye, at least before the changes unleashed by the ROLM PBX after 1975. This attributional homogeneity meant that the ATT salesmen and the TMs had nothing to learn from each other—and certainly no basis on which to establish a relationship around conversations that could challenge how each viewed what a PBX was and what it meant to administer a PBX system. The problem was *not* that ROLM listened to its customers and ATT did not: it was that the customers had nothing to say to ATT that could change how either ATT or the customers thought or acted. In fact, the interactions between ATT and the TMs were channeled into recurrent patterns in which their underlying shared attributions played no explicit role and so could never be up for negotiation.

As this example illustrates, not every relationship has the *potential* for generativeness. At a minimum, there must be some essential *heterogeneity* or distance between the participants—in attributions, competence, even access to artifacts or other agents; and then there must also be some shared *directedness*, to make the participants want to bridge the distance between them. The TMs and ROLM account representatives had very different attitudes toward and knowledge about digital technology and the problems of voice communications in large companies, but they were both directed toward a notion of a "cost-saving PBX system." In their conversations, they began to construct a common understanding according to which the ROLM PBX was, or could be made to be, such a system. As their relationship developed, they changed the whole conception of what "cost-saving" might mean, culminating finally in the idea of "productivity-enhancing."

As we discuss in more detail in section 4, monitoring and maintaining the generativeness of relationships constitute essential strategic competencies, especially in the presence of complex foresight horizons. Monitoring must be an ongoing activity, since there is no guarantee that a relationship that at one time is generative will stay so forever. In fact, there is always a danger that the very success of the relationship in bridging the distance between its participants can destroy its potential to continue to generate anything new. Just which relationships can be generative at any given time depends, in subtle ways, on the current structure of agent/artifact space, on agent attributions and competencies, and on the opportunities agents have to communicate with other agents, about what.

The firm is not the right unit in which to think about generative relationships, making the task of monitoring relationships for generativeness even more difficult. There can be generative relationships between component agents—groups, departments or divisions—of the same firm, as well as generative relationships that cross firm boundaries. Moreover, different relationships between components of the same agents may have very different generative potentials: for example, the drug safety departments of some pharmaceutical firms may have generative relationships with units of a national regulatory agency, while the firms' legal department or marketing division or top management may have nongenerative, essentially adversarial relations with the same unit or even the agency as a whole.

LESSON 3

Structural change in agent/artifact space proceeds through a "bootstrap" dynamic: new generative relationships induce attributional shifts that lead to actions which in turn generate possibilities for new generative relationships.

DISCUSSION. This is the key feature of the dynamics of structural change under complex foresight horizons: they are characterized by "constructive positive feedbacks." That is, new configurations in agent/artifact space breed further configurations, mediated by the generation of new attributions: generative relationships among agents give rise to new relations between artifacts (including new artifacts!), which provide new opportunities for generative relationships between agents. The whole ROLM story can be read as an example of this phenomenon.

These constructive positive feedbacks have an obvious cognitive counterpart: as the structure of agent/artifact space undergoes ripples of change, new agents and artifacts come into being and old ones acquire new functionalities, so identities change—and, hence, old interpretations of identity bear an increasingly strained relationship with observable actions, the facts of the world. Different agents respond differently: some respond to the resulting ambiguity by generating new attributions to make sense of experienced novelty, and so attributional heterogeneity increases—increasing further the possibility that participants in other relationships will achieve sufficient attributional diversity to become generative.

In general, sooner or later "hot" zones in agent/artifact space cool off, as other, negative feedback processes come to dominate the constructive positive feedbacks we are emphasizing here. Since our intention here is to focus on strategy in complex foresight horizons, we will not discuss these negative feedback processes. From the strategic point of view, the important thing is to learn to recognize what kind of foresight horizon is appropriate for the current situation and to employ strategic practices adapted to the requirements posed by that horizon.

LESSON 4

The "window of predictability" for the attributional shifts and structural changes that characterize complex foresight horizons are very short—and virtually nonexistent outside the particular generative relationship from which they emerge.

DISCUSSION. None can foresee what can emerge from the kind of constructive dynamics we have just described. Before a particular generative relationship has even come into being, there is no perspective from which one could imagine the new attributions it will create, much less the new possibilities for artifacts and agent relationships that these attributions will permit. Thus, during periods of rapid structural change, when the rate of creation of new generative relationships is high, windows of predictability for the phenomena to which they give rise must be very short.

Even after new structures arise in agent/artifact space, it is only the structures themselves—new agents, new artifacts, new functional relationships between agents and artifacts—that are publicly observable, not the attributional shifts that made their construction possible. Frequently, these shifts have a cumulative aspect—new attributions lead to actions that in turn induce changes in the attributions. When this aspect is most pronounced, there is a strong cognitive barrier preventing other agents from breaking into the innovative cycle, which may be exclusively confined to a particular set of generative relationships. In fact, this is what happened to ATT. They were able to imitate new features already built into ROLM products, but not to "catch up" attributionally in a way that would allow them to conceive of features ROLM was about to innovate—or to form their own generative relationships that would give rise to an alternative framing of a future direction for business communication products.

Our four lessons describe the dynamics of structural change in agent/artifact space. These dynamics have important implications for the meaning of strategy under complex foresight horizons.

IMPLICATION 1

The first requirement for successful strategizing in the face of complex foresight horizons is to recognize them for what they are. Failing to detect changes in the structure of agent/artifact space or interpreting the new structures through the lens of old attributions, are sure paths to failure.

DISCUSSION. From our after-the-fact perspective, it is clear that firms in the PBX business in 1973 faced a complex foresight horizon. Although the pace of technological change in PBXs had not yet picked up from its pre-Carterphone, monopoly-induced crawl, there was little reason to think that this situation would last much longer. In particular, the mushrooming growth of interconnect distributors since 1968 made it ever more likely that technology-intensive companies like ROLM could successfully enter the PBX business, without any previous experience in the telecommunications industry. After that happened, who could know what PBXs would become?

But existing PBX manufacturers do not seem to have interpreted their situation in this way. They acted as though they expected the rate of PBX technological change to continue to be what it had been in the past—perhaps accelerated a bit if IBM should decide to enter the American market. Their main concern seemed to be with pricing their products competitively, with some companies placing a secondary emphasis on establishing superior reputations for service. They envisioned their competition coming from exactly the same companies and on exactly the same terms as the current ones. In short, to them, the foresight horizon seemed clear, and they could afford the luxury of "optimizing" pricing and R&D-allocation strategies. As a result, most of them were out of the PBX business by 1980, and the ones that were left had to change course dramatically to survive.

How about ROLM? What they saw in 1973 was a technologically backward industry, populated by a set of firms that, for a variety of reasons, was not likely to react quickly to the technological opportunities that the ROLM managers themselves believed were already realizable. They also came to appreciate, with somewhat more difficulty, that the post-1968 changes in PBX distribution provided them with an opportunity to market their PBX system. That is, from the outside, they perceived the PBX agent space as structurally unstable. This perception motivated their decision to enter the PBX market, despite their small size, lack of communications experience, and inability to field their own distribution and service system.

IMPLICATION 2

Recognizing the existence of structural instability is not enough: it is also necessary to realize that the complex path through which some semblance of stability will eventually be attained is not predictable. It is not good strategizing to formulate and stick to a strategic plan that is premised on a particular scenario about how a complex situation will play itself out.

DISCUSSION. As the fourth lesson would lead one to expect, ROLM failed utterly to predict the effects of their entry into the PBX business. They expected to carve out a small, high-tech market niche, enough to justify their decision to enter but very far from what they actually achieved. Moreover, the trajectory of PBX development after their first product launch turned out to be quite different from what the 1973 business plan foresaw. Fortunately for ROLM, their success in the PBX market did not depend on their ability to foresee far into the future. ROLM managers abandoned the schedule for software revisions that they had set in the 1973 business plan and to which their development engineers were highly committed in favor of the "line-of-business voice applications" that arose from the TM—account representative generative relationships. ROLM managers did not allow strategic plans to channel these relationships; instead, they let the relationships channel their strategic plans.

The final two implications follow from the critical role that attributions and generative relationships play in the "bootstrap" dynamics of rapid structural change characteristic of complex foresight horizons.

IMPLICATION 3

Agents must engage in ongoing interrogation of their attributions about themselves, other agents, and the artifacts around which their activity is oriented. They must develop practices that offset the easy, but potentially very costly, tendency to treat interpretations as facts.

IMPLICATION 4

Agents must monitor their relationships to assess their potential for generativeness, and they must commit resources to enhance the generative potential of key relationships. Fostering generative relationships is especially important when foresight horizons are complex.

The implications we have just described underlie the notion of strategy under complexity that we outline in the next section.

4. STRATEGY AND COMPLEXITY

Our aim in this section is to sketch a new interpretation of what strategy means when foresight horizons are complex. Since the word "strategy" already carries a lot of interpretational baggage, we start in section 4.1 by abstracting out a core concept that we think runs through almost all the meanings that strategy has taken on: the notion of strategy as control. The realities of complex foresight horizons alter the everyday meaning of control and thus lead to a new conception of strategy as the means to achieve control. According to this conception, strategy is a process consisting of a set of practices, in which agents inside the firm[14] structure and interpret the relationships, inside and outside the firm, through which they both act and gain knowledge about their world. In sections 4.2 and 4.3, we focus on two kinds of strategic practices: *populating the world*, in which agents explicitly construct attributions about the identities of the agents and artifacts with whom they share their world; and *fostering generative relationships*, in which agents monitor their current and future relationship possibilities for generative potential and allocate human and material resources to those relationships whose generative potential appears high.

4.1 STRATEGY AS CONTROL

Most managers situate strategy within a suite of concepts that includes at least the following: vision, mission, goals, strategy, and tactics. In the language we have been using in this chapter, vision and mission together determine a firm's *directedness* in agent/artifact space. That is, they pick out a particular kind of artifact that the firm commits itself to create, identify the kind of agent to whom the firm intends to sell these artifacts, and establish the directions in which the firm would like the current structure of agent/artifact space to change. Goals specify what constitute desired outcomes for the firm: resources, returns or even particular reconfigurations of agent/artifact space. Tactics determine how the actions in which the firm intends to engage will actually be executed by its various component agents.

Strategy lies between directedness and execution. It lays down "lines of action" that the firm intends to initiate and that are supposed to bring about desired outcomes. Since outcomes depend on the interactions with and between many other agents (inside and outside the firm's boundaries), strategy really represents an attempt to *control a process of interactions*, with the firm's own intended "lines of

[14] Actually, any agent, not just a firm, can strategize: for example, a component of a firm—a division or even a research or marketing group—can have a strategy. And even when people talk about a "firm's" vision, mission, or strategy, they really mean the vision, mission, or strategy of some top managers, who are trying to make them the vision, mission, or strategy of the rest of the agents that comprise the firm. Nonetheless, for expository convenience, we refer to the strategizing agent as the "firm."

action" as control parameters. From this point of view, *the essence of strategy is control.*

How to achieve control, and how much control is achievable, depends upon the foresight horizon. When the foresight horizon is clear, it may be possible to anticipate all the consequences of any possible course of action, including the responses of all other relevant agents, and to chart out a best course that takes account of all possible contingencies. If so, strategy as control becomes the classical conception of strategy as optimizing precommitment. If foresight horizons are a little more complicated, "adequate" can substitute for "best," without surrendering the idea of control as top-down and predetermining. But as foresight horizons become even more complicated, the strategist can no longer foresee enough to map out courses of action that guarantee desired outcomes. Strategy must include provisions for actively monitoring the world to discover unexpected consequences, as well as mechanisms for adjusting projected action plans in response to what turns up. At this point, control is no longer just top-down: some control must be "delegated" to those who participate directly in monitoring, for their judgments of what constitute unexpected consequences trigger the adjustment mechanisms and, thus, affect the direction of future actions. In addition, the adjustment mechanisms themselves usually extend permissions to a variety of component agents to initiate new, experimental lines of action.

The dynamics of structural change associated with complex foresight horizons have a much more radical impact on the meaning of control. Constructive positive feedbacks make a complete nonsense of top-down control: marginal agents, like ROLM in 1973, can have huge effects; "powerful" agents like ATT may lose control over what happens to them; subordinates like the account representatives, even acting "under orders," may effect changes through their interactions with other agents that their superordinates never envisioned.

In such situations, control is not so much delegated as it is *distributed* throughout agent space. Then, the everyday way of talking about strategy can be very misleading. For example, people usually talk about strategy as something that is "set" by strategists. When control is distributed, it is more appropriate to think of it as something that *emerges* from agent interactions. As an example, think of ROLM's PBX strategy. The 1973 decision of the ROLM board of directors to make a computer-controlled, digitally switched private telephone system is a typical example of top-down, precommitting strategy setting. Within the next eight years, a strategy shift had taken place. ROLM top managers now saw the company's business as developing productivity-enhancing, line-of-business voice-data applications for the office. How did this change come about? Though top management *recognized* and *embraced* the change, they did not initiate it. Rather, the change emerged from a process of attributional shifts occurring in a network of discursive relationships that had been established to execute, not to make, strategy. This network cut across departmental and authority lines inside ROLM and across ROLM's boundaries, including, as it did, the national account TMs.

In contexts like this, the relation between strategy and control is very different from the classical conception. It is just not meaningful to interpret strategy as a *plan* to *assert* control. Rather, strategy must be seen as a *process* to *understand* control: where it resides, and how it has been exercised within each of its loci. From this point of view, strategy can be thought of as a set of practices, which are partly exploratory, partly interpretative, and partly operational. These practices can generate insight into the structure of the agent/artifact space that the agent inhabits and into the way in which control is distributed through that space. With the insights gleaned from these practices, the agent can orient itself in agent/artifact space—and figure out how it might reorient itself there, in the future.

Two kinds of strategic practices are particularly important when foresight horizons are complex. Through the first, agents seek to construct a representation of the structure of their world that can serve them as a kind of road map on which to locate the effects of their actions. Through the second, agents try to secure positions from which distributed control processes can work to their benefit. We now look a little more closely at these two fundamental classes of strategic practices.

4.2 POPULATING THE WORLD

When foresight horizons are complex, agents cannot take knowledge of their worlds for granted. They need *information*, of course—hence, the strategic need for exploration and experimentation. But information takes on meaning only through *interpretation*, and interpretation starts with an ontology: who and what are the people and things that constitute the agent's world, and how do they relate to one another? When the structure of an agent's world is changing rapidly, unexamined assumptions are likely to be out-of-date, and actions based on them ineffective. Hence, the strategic need for practices that help agents "populate" their world: that is, to identify, criticize, and reconstruct their attributions about who and what are there.

These practices have to happen in the context of discursive relationships, and so they will clearly consist, at least in part, of structured conversations. We will have something to say below about who should participate in these conversations, but first we want to be a bit more precise about their substance.[15] Populating the world goes on at a micro- and a macrolevel:

- At the microlevel, populating the world consists of constructing a list of agents and artifacts that matter and attributing an identity to each of them.

[15]We will not say anything about the shape of the conversations. Clearly, this is a practical question of the first importance. Our purpose here, though, is just to say what strategy means, not how to do it.

AGENT IDENTITY. An agent's identity has two components: its *functionality* (what it does) and its *character* (how it does it). For example, take the attributions that ROLM's founders and top managers had about their own identity. Initially, the four engineers who founded ROLM wanted primarily to build a company that could keep expanding, by finding market niches in which they could apply their engineering skills to develop a new approach to an existing problem. They saw ROLM's functionality, then, in terms of their own technological competencies. Later, after the PBX business had taken off, they reinterpreted their identity in an artifact- rather than competence-centered way, as a company that developed line-of-business voice-data applications.

In terms of character, ROLM saw itself as a fast-moving, technology-driven company. The details of this self-attribution partly derived by emulation from the ROLM founders' interpretation of their model, Hewlett-Packard; and partly derived by contrast to their interpretations of ATT's monopolistic, "bullying" character and IBM's plodding, top-down bureaucratic way of doing business.[16]

Attributions of functionality are oriented in time. The agent is not only what it does now, but what it will be doing into the foreseeable future. In the early days, ROLM interpreted what it did primarily in terms of its orientation to growth opportunities.

ARTIFACT IDENTITY. For most agents, an artifact's identity is determined by its use—who uses it, for what purposes, together with (or in place of) which other artifacts. A PBX could be regarded as a set of switches between outside lines and inside telephone sets; or as an interface between outside callers and information and services that can be supplied internally; or as a system consisting of productivity-enhancing voice- and voice-data applications. For agents who design and make artifacts, of course, a purely functional attribution is not enough: the artifact must also be related to the other artifacts that physically comprise it.

- At the macrolevel, populating the world means describing the structure of agent/artifact space.

Agents do things by interacting with other agents. The recurring patterns of agent interaction define the structure of agent space. These patterns cut across organizational boundaries. For example, two of ROLM's great sources of strength in the PBX business involved recurring interactions between people in its marketing and engineering departments—and between its national account representatives and the TMs. To understand the structure of agent/artifact space, then, "firms" cannot be taken as the primary agent unit. Instead, the focus must be on relationships defined by recurring patterns of interaction, which happen inside firms, across their boundaries, and sometimes even beyond them, in places like university departments and government agencies.

[16] Of course, these phrases are only caricatures of the much richer attributions that actually informed ROLM's actions with respect to these other companies.

Developing a vocabulary to describe the structure of agent/artifact space is no easy task. This is partly because these relationships between agents have two features that are not easy to describe together in ordinary language: they are *recursive*, in that many relationships are comprised of other relationships (as a firm is comprised of divisions and departments); but they are not *hierarchic*, in that an agent may participate in a whole set of cross-cutting relationships, with no strict inclusion-ordering among them—like a TM who works for a particular company, participates in professional activities through the International Communication Association, and collaborates with his ROLM national accounts representative on a new product idea. It would be particularly valuable to describe recursive, nonhierarchic networks of relationships in terms of a taxonomy that would help distinguish between structures that are likely to remain stable and those that are teetering on the edge of cascading change.

WHO SHOULD POPULATE THE WORLD? Every agent facing a complex foresight horizon needs to populate its world in order to understand what opportunities it has for action, what constraints affect what it can do, and what effects its actions may have. Inside the firm, the agents and relationships that matter differ from agent to agent, so populating the world is not an activity that should be reserved just for the CEO and his staff. Rather, it ought to be top management's strategic responsibility to make sure that interpretative conversations go on at all relevant levels of the company—and that sufficient cross-talk between these conversations happens so that attributions of the identities of the same agents and artifacts made by people or groups with different experiences and perspectives can serve as the basis for mutual criticism and inspiration to generate new and more useful attributions.

4.3 FOSTERING GENERATIVE RELATIONSHIPS

Generative relationships may be the key to success and even survival in complex foresight horizons, but fostering them poses two problems. First, how can agents decide which relationships have generative potential? And second, once they've determined which relationships seem promising, how can they foster them?

MONITORING RELATIONSHIPS FOR GENERATIVE POTENTIAL. If the benefits that accrue from a generative relationship are unforeseeable, on what basis can an agent decide to foster it? Part of the answer is that many relationships that *turn out* to be generative are formed for quite other reasons, reasons that justify in expectational terms the social and financial investments required to develop them—for example, the relationship between ROLM and the targeted TMs. Once a relationship begins to generate unforeseen returns, it makes sense to expend effort and funds to enlarge its scope and increase its intensity.

But relying solely on fortuitous unintended effects of existing relationships certainly ought not to be the whole story. While it may not be possible to foresee just what positive effects a particular coupling might yield, it may nonetheless be possible to determine the generative potential of a relationship. Here are some essential preconditions for generativeness:

- *Aligned directedness* The participants in the relationship need to orient their activities in a common direction in agent/artifact space. That may mean that the same kind of artifacts might be the focus of each of their activities, as was the case with the ROLM account representatives and the TMs, although in general, as this example illustrates, the participants need not have the same relationship to the focal artifact. Alternatively, the alignment may be brought about by a mutual orientation toward a particular *agent*, as is the case when a consultant works with a firm's top managers to restructure the firm to enhance its market position.
- *Heterogeneity* As we pointed out in section 3, generativeness requires that the participating agents differ from one another in key respects. They may have different competencies, attributions or access to particular agents or artifacts. We have already described some of the ways in which attributional heterogeneity can generate new attributions. Combining different competencies can generate new kinds of competence that then reside in the relationship itself. Developing ties that bridge "structural holes" in agent space can lead to brokerage opportunities (see Burt[4]), but even more can bring the bridged agents into contact with previously untapped sources of alternative attributions and competencies. An example might help illustrate this idea. The Santa Fe Institute is a research institution that brings together many scientists engaged in the study of complexity. About 25 businesses belong to SFI's Business Network. Their affiliation with the Institute brings Business Network members into contact with ideas and people that would otherwise be inaccessible to them, and out of their periodic meetings a number of joint projects between Business Network members and SFI scientists have been spawned.
- *Mutual directedness* Agents need more than common interests and different perspectives to form a generative relationship. They also must seek each other out and develop a recurring pattern of interactions out of which a relationship can emerge. Their willingness to do this depends on the attributions each has of the other's identity. It helps, but isn't necessary, for the participants to start by trusting one another. Frequently, rather than a precondition, trust is an emergent property of generative relationships: it grows as participants come to realize the unforeseen benefits that the relationship is generating.
- *Permissions* Discursive relationships are based on permissions for the participants to talk to one another about particular themes in particular illocutionary modes (requests, orders, declarations, etc.). These permissions are granted explicitly or implicitly by superordinate agents and social institutions. Unless potential participants in a relationship have appropriately matched permissions, or can arrogate these permissions to themselves, the generative potential of the relationship is blocked.

- *Action opportunities* For all our emphasis on the importance of discourse, relationships built only around talk do not usually last long or affect deeply the identities of participating agents. Engaging in joint action focuses talk on the issues and entities of greatest interest—those around which the action is organized. And action itself reveals the identities of those engaged in it. In addition, new competencies emerge out of joint action, and these competencies can change agents' functionality and, hence, identity—even leading to the formation of a new agent arising from the relationship itself.

Agents can review new and prospective relationships to gauge the extent to which these preconditions for generativeness are met. If the relationship seems promising in other respects, but one or more of the preconditions are absent, it may be possible to devise ways to improve the situation, as we describe below. For ongoing relationships, agents must monitor not only these preconditions, but also whether the relationship is actually generating new and interesting things—attributional shifts, new relationships with other agents, joint actions with the potential to change structural features of agent/artifact space. This kind of ongoing monitoring would thus be coordinated with the agent's "populating the world" practices: as the agent discovers changes in its own attributions over time, it also tries to uncover the locus of these shifts in the various relationships in which it participates—and then figures out how to foster the relationships that turn out to generate changes.

FOSTERING GENERATIVENESS. What is a truism in personal relationships applies just as well to business relationships: it takes work to build and maintain them. If an agent sees the generative potential in a possible relationship, it must make sure that the future partners in the relationship see this too.

Sometimes, the problem is simply to find incentives that make a relationship possible and attractive to potential partners. For example, Silicon Valley venture capital firm Kleiner, Perkins, Caulfield & Byers (KPCB) knows that the numerous current and retired successful Valley entrepreneurs represent a huge storehouse of competence in building companies and connections that can benefit any start-up. How can they develop relationships between these entrepreneurs and the companies in which they decide to invest? One way is to give these entrepreneurs a stake in the companies. So whenever KPCB and some other venture capital firms create a new fund, aimed primarily at institutional investors, they set up a parallel Founders Fund, in which a select group of entrepreneurs are invited to invest in exactly the same portfolio of companies, but at a lower dollar commitment than the institutional investors are required to put up. In this way, KPCB "aligns the directedness" of the start-ups and the Founders Fund investors.

In other situations, it may be necessary to enhance the "mutual directedness" of the participants in a relationship. This can sometimes be done in quite simple ways, for example by explicit expressions of gratitude and appreciation for the contributions of coparticipants toward generative interactions. From 1976–1980 ROLM's

PBX manufacturing department was under constant scaling-up pressures to meet rapidly increasing orders. The vice-president of manufacturing relied on close co-operation with the engineering, marketing, and human relations departments to carry out the scaling-up without undue shipment slippage. He made sure that the people in his department acknowledged every piece of assistance they received from these other departments and suppressed their complaints and criticisms when they couldn't get what they wanted as soon as they needed it. He arranged regular ice cream socials between units in different departments in recognition of effective co-operation, complete with appreciative plaques and public speeches. The result was the formation of informal "problem-solving" networks connecting people in these four departments and the generation of unusual and creative ways to integrate their activities. Of course, building compatibility and trust between relationship partners may require more substantial unilateral "gifts" than ice cream socials; calibrating what and how much to do is an important component of the strategic practices that go into fostering generative relationships.

Sometimes, it is necessary to bring about the creation of the very agents with whom a generative relationship is desired. For example, ROLM sponsored the formation of *user groups*, representatives of companies that had purchased ROLM systems. ROLM hosted user group meetings at company facilities, but the groups themselves decided on the agendas for the meetings, invited speakers (sometimes from ROLM), and so on. The user groups accomplished two things for ROLM. First, they provided the opportunity for different users to enter into relationships with one another, centering on ROLM's artifacts. If some of these relationships became generative, they might lead to new product ideas or new functionalities for existing ROLM products, either of which would accrue to ROLM's benefit. In addition, the user groups themselves constituted agents with whom ROLM could enter into relationships, through the meetings and the institutional structures created at the meetings (officers, board, committees, and so on). The relationships that developed between ROLM and the user groups extended the range of what could be achieved in the account representative/TM relationships, to include a much greater variety of customers, more action possibilities, and more intensive focus.

We have described only a few examples of what an agent can do to foster generative relationships. They all stand in stark contrast, in terms both of resources required and probability of enhancing generativeness, to the tendency so obvious these days in "multimedia" to think that acquisitions and mergers are the way to go. We think that acquisition is, in general, much more likely to repress generativeness than to enhance it. To realize generative potential, relationship participants must have the right permissions, time, and space to talk; they must do work together; and that work must facilitate their coming to shared understandings about each others' competencies and attributions of identity; and their relationship must be embedded in a network of other relationships that can amplify whatever possibilities emerge from their joint activities. Fostering generative relationships means managing people, projects, and events so that these conditions are realized.

4.4 CONCLUSION: STRATEGY UNDER COMPLEXITY

When agent/artifact space changes structure rapidly, foresight horizons become complex. To succeed, even survive, in the face of rapid structural change, it is essential to make sense out of what is happening and to act on the basis of that understanding. Since what is happening results from the interactions between many agents, all responding to novel situations with very different perceptions of what is going on, much of it is just unpredictable.[17] Making sense means that interpretation is essential; unpredictability requires ongoing *re*interpretation. Hence, our conclusion that the first and most important strategic requirement in complex foresight horizons is the institution of interpretive practices, which we have called *populating the world*, throughout the firm, wherever there are agents that initiate and carry out interactions with other agents—that is, at every locus of distributed control.

But of course making sense isn't enough. Agents act—and they act by *interacting* with other agents. In complex foresight horizons, opportunities arise unexpectedly, and they do so in the context of generative relationships. In this context, the most important actions that agents can take are those that enhance the generative potential of the relationships into which they enter. As a result, agents must monitor relationships for generativeness, and they must learn to take actions that foster the relationships with most generative potential. Then, when new opportunities emerge from these relationships, agents must learn to set aside prior expectations and plans and follow where the relationships lead. We call the set of strategic practices through which agents accomplish these things *fostering generative relationships*, and they constitute the second cornerstone of our conception of strategy under complexity.

Even in the most complex of foresight horizons, there will be regions of relative stability in a firm's agent/artifact space, with respect to which it still makes sense to engage in optimizing (or at least satisfying) precommitment and to organize processes of exploration and adaptation. But it is hardly possible to know where those regions are, or to which relationships precommitted actions may be entrusted, without the understandings attained through the practices we have emphasized here.

ACKNOWLEDGMENTS

We would like to thank the Santa Fe Institute and CNR (Italy) for support for our research. We benefited from discussions around the ideas presented here from many people, including especially Brian Arthur, Win Farrell, Dick Foster, John

[17]See Arthur[3] for an interesting discussion of the relation between attributional heterogeneity and what he calls indeterminacy.

Hagel, Franco Malerba, Gigi Orsenigo, John Padgett, Jim Pelkey, Howard Sherman, and Somu Subramaniam. We also appreciate helpful comments on a previous draft of this chapter from Arthur, Eric Beinhocker, Roger Burkhart, Farrell, Henry Lichstein, Kathie Maxfield, Ken Oshman, Padgett, Pelkey, Roger Purves, Sherman, Bill Sick, and Jim Webber.

REFERENCES

1. Anderson, P., K. Arrow, and D. Pines, eds. *The Economy as an Evolving Complex System.* Santa Fe Institute Studies in the Sciences of Complexity, Proc. Vol. V. Redwood City, CA: Addison-Wesley, 1988.
2. Arthur, W. B. *Increasing Returns and Path Dependence in the Economy.* Ann Arbor, MI: University of Michigan Press, 1994.
3. Arthur, W. B. "Complexity in Economic and Financial Markets." *Complexity* **1(1)** (1995): 20–25.
4. Burt, R. *Structural Holes: The Social Structure of Competition.* Cambridge, MA: Harvard University Press, 1992.
5. Garvin, D. "Building a Learning Organization." *Harvard Bus. Rev.* **July-Aug** (1993): 78–91.
6. Hayes, R., S. Wheelwright, and K. Clark. *Dynamic Manufacturing: Creating the Learning Organization.* New York, NY: Free Press, 1988.
7. Kauffman, S. *The Origins of Order: Self-Organization and Selection in Evolution.* New York, NY: Oxford University Press, 1993.
8. Lane, D. "Artificial Worlds and Economics." *J. Evol. Econ.* **3** (1993): 89–107, 177–197.
9. Lane, D., F. Malerba, R. Maxfield, and L. Orsenigo. "Choice and Action." *J. Evol. Econ.* **6** (1996): 43–76.
10. Marquardt, M., and A. Reynolds. *The Global Learning Organization.* Burr Ridge, IL: Irwin, 1994.
11. Redding, J., and R. Catalanello. *Strategic Readiness: The Making of the Learning Organization.* San Francisco, CA: Jossey-Bass, 1994.
12. Senge, P. *The Fifth Discipline: The Art and Practice of the Learning Organization.* New York, NY: Doubleday/Currency, 1990.
13. Waldrop, M. *Complexity: The Emerging Science at the Edge of Order and Chaos.* New York: Simon and Schuster, 1992.

John F. Padgett
Department of Political Science, University of Chicago, Chicago, IL 60637
E-mail: padg@cicero.spc.uchicago.edu

The Emergence of Simple Ecologies of Skill: A Hypercycle Approach to Economic Organization

INTRODUCTION

Buss[7] and Fontana and Buss[11] have cogently argued that biology's modern synthesis of genetics and Darwinian evolution achieved its impressive advances at the cost of eliding a crucial middle step—the existence of organism (or, more generally, of organization). The distribution of genetic alleles is shaped by selection pressures on the phenotypical "carriers" of those alleles. But the existence of phenotype itself is never explained. As Fontana and Buss themselves put it: "A theory based on the dynamics of alleles, individuals, and populations must necessarily assume the existence of these entities. Present theory tacitly assumes the prior existence of the entities whose features it is meant to explain." (Fontana and Buss,[11] p. 2.)

The same existence problem plagues economics. The neoclassical theory of competitive markets presumes that competition among firms (phenotypes) generates selection pressure for optimal policies (genotypes). Yet firms themselves are only posited, not explained, in this foundational theory. Advances in transaction-cost economics[8,15] and in principal-agent economics[10,13] purport to fill this gap, but only through equating the firm with a set of bilateral contracts. Such a transposition of "the firm" down to a series of dyadic negotiations overlooks the institutionalized autonomy of all stable organizations. In organisms, social or biological, rules of action and patterns of interaction persist and reproduce even in the face of constant turnover in component parts, be these cells, molecules, principals, or agents. In the constant flow of people through organizations, the collectivity typically is not renegotiated anew. Rather, within constraints, component parts are transformed and molded into the ongoing flow of action.

Viewing the firm as a dynamic organism highlights the organizational existence question not usually asked in economics: which sets of component parts and of collective actions are, in principle, mutually reproducible? This difficult question will be operationalized and investigated in this chapter within a simple, spatially grounded, hypercycle context. Hypercycle theory[9,12] is an extremely prominent approach to the topic of the molecular origins of life on earth.[1] The theory argues that life emerged through the chemical stabilization of an autocatalytic reaction loop of nucleic and amino acid transformations, each step of which reinforces the next step's reproduction and growth: $1 \rightarrow 2 \rightarrow 3 \rightarrow 4 \rightarrow 1 \rightarrow$, etc. The scientific problematic is to learn which chemical conditions (and perturbations) permit various types of hypercycles to emerge and to grow "spontaneously."

The metaphorical leap from RNA to firms is obviously enormous. But the benefit of this particular operationalization is that it will offer an extraordinarily minimalist set of assumptions within which to investigate the emergence of collective order. Not only are no assumptions about hyper-rationality required, no assumptions about consciousness of any sort are required. "Collective order" here is simply the stabilization of chains of action-reaction sequences, which fold back on each other to keep themselves alive. In the human application here, such chains wend themselves through organizationally interacting individuals, shaping and reshaping such individuals in the process. The problematic is to learn which "organizations"—that is, which joint sets of interaction patterns, hypercyclic action chains, and classes of individuals—can exist, in the sense of reproduce themselves through time. It will soon become apparent that such mutual maintenance and reproduction is no mean feat; thus, there are dynamic constraints on what firms are possible within life and, hence, within economic theory.

THE ORGANIZATION OF SKILL

Perhaps the most natural way to interpret this "hypercycle theory of the firm" is through the concept of skill. Instead of RNA and the like, each element in an "economic hypercycle" is an action-capacity or skill, which simply passively resides

[1]It is also a predecessor to the more general models in Fontana and Buss,[11] cited above.

within the individual until evoked by some compatible[2] action by another agent in that individual's immediate environment. "Work" in an organization is an orchestrated sequence of actions and reactions, the sequence of which produces some collective result (intended or not). The full set of such sequences is the human "technology" of the firm.

The hypercycle representation of "technology" emphasizes form, not content. "Elements in a sequence" are just mathematical entities that can stand for any action or rule; it is the patterning of how such actions are linked that is the focus of attention. The main constraint imposed by the literal hypercycle format adopted here is that only linear chains[3] of action sequences can be investigated. The dynamics of more complicated branching technologies is postponed for future research.

One fundamental difference between human and chemical hypercycles, in degree if not in kind, is the fact that potentially many action-capacities or skills can reside in individuals. Hence, individuals potentially can partake in many different organizational tasks. This greater versatility implies that the compositional mix of skills within individuals is variable, and can change as a function of interactional experience with others. "Learning" in this skill-centered[4] model means that never-used skills are forgotten, and that often-used skills will "reproduce," in the sense of become more frequently usable in the future. Relative rates of skill reproduction, in turn, are functions of how frequently others in your immediate environment call upon you to exercise your skills. And since this, in turn, is a function of how technologies physically become distributed across individuals, one important issue to be studied below, besides just aggregate organizational stability, is how different organizational ecologies induce different degrees of complexity in the skill mixes of the individuals that comprise them.

As determinants of organizational existence and of individual skill composition, I will focus, in the simulations below, on two central ways in which hypercycle "games" can be varied: (a) length of hypercycle—that is, two-element vs. three-element vs....vs. nine-element hypercycles; and (b) mode of skill reproduction—that is, "source-only reproduction," where only the skill of the initiator of the interaction is reproduced; "target-only reproduction," where only the skill of the recipient of the interaction is reproduced; and "joint reproduction," where both parties' skills are reproduced through compatible interaction. The first of these independent variables measures (within a linear chain setup) the complexity of the

[2] "Compatible" just means linked sequences of "if...then" rules—namely, one person's "then" is another person's "if."

[3] Hence, the "simple" in the title of this chapter.

[4] "Skill-centered" is meant to be opposed to "person-centered." Individuals certainly learn in this model, but the analytic approach is to adopt the "skill's eye" perspective on this learning—namely, to focus on the reproduction and propagation of skills through the medium of persons, rather than necessarily to assume any cognitive processing.

technology trying to be seeded into an organization. The second independent variable measures the distribution of reward in micro-interaction: source-only reproduction is "selfish" learning, in which the initiator of the action reaps the learning rewards; target-only reproduction is "altruistic or teacher" learning, in which the recipient of the action reaps the reward; and joint reproduction benefits both learners simultaneously.

A third structural dimension of potentially great interest is variation in the social network pattern of interaction among individuals. In order to better construct a neutral baseline, which maintains the focus on hypercycle length and reproduction mode, I suppress this important topic in this chapter. In particular, in the models below individuals interact with each other only in two-dimensional "geographical" space with their four nearest neighbors. This cellular-automaton spatial framework, while traditional in Santa Fe Institute research, is hardly realistic of human social networks where cliques and centrality abound. But this simplification will highlight the effects of space per se which, in contrast to fully randomized mixing, appear to be profound. How more realistic social network topologies enable or disable various hypercycle technologies is a topic that can be investigated in the future, once a simple spatial baseline has been established.

THE MODEL

The operationalization of the above ideas is as follows:

1. Assume a fixed number of individuals arrayed on a spatial grid. Each individual can interact only with its four nearest neighbors. In the simulations below, this grid will be relatively small—10 by 10, with boundaries.[5]
2. Assume a number of "action capacity" elements, or "skills," that are distributed originally at random over individuals/cells in the spatial grid. These elements are indexed by integers: $\{1,2\}, \{1,2,3\}, \ldots, \{1,2,..,t\}$, the technology set size of which is varied experimentally. Most importantly, there can be a variable number of skills resident in any one spatial site;[6] these resident elements represent the varying skill capacities of different individuals.

[5]Small size was convenient for my GAUSS program on a 486 machine. Given the tight clustering behavior of this model, small size is not as serious a constraint as it might be in other models. Spatial size clearly can be varied in future research, for those who are interested.

[6]This variable number of elements per site differentiates this model from the typical cellular automaton setup, which imposes a constraint of only one element per site. See, for comparison, the extremely high-quality work of Boerlijst and Hogeweg,[2,3,4,5] who also have studied spatial hypercycles.

3. Assume that technology element-sets are governed by "rule compatibilities," such that action capacities only become "actions" in the presence of compatible other action capacities. In particular, assume a hypercycle technology structure of compatibilities: elements 1 and 2 activate each other, elements 2 and 3 activate each other, up to elements t and 1 activate each other. Otherwise (for example, elements 1 and 3), interactions are inert.
4. Within the above framework, an iteration of the model proceeds as follows: (a) choose at random an action-capacity or skill "looking for" action, located in some individual; (b) initiate interaction (randomly) with one of the four possible interaction partners; (c) if a compatible element exists in the partner's repertoire, the interaction is "successful" and joint action occurs; (d) if not, the interaction is "inert" and nothing occurs.
5. If successful interaction occurs, reproduction of skill (or equivalently, learning within the individual) follows. This is the same as the notion of catalysis in chemistry. The reinforcement regime that governs this reproduction/learning is manipulated experimentally among (a) source-only reproduction, (b) target-only reproduction, and (c) joint reproduction, as defined above.
6. The total volume of elements in the collectivity is maintained at a fixed level (in the "standard" simulations below, this will be set at 200[7]). This means killing off a random element somewhere in the system whenever a new element is created through reproduction.
7. After setting up the model in steps 1–3, steps 4–6 are then repeated, for a large number of iterations—usually 15,000 or 20,000 interations are sufficient for equilibrium behavior to emerge.

The fixed volume constraint in step 6 is not realistic for human systems, but it allows me to investigate compositional or "structural" effects, independent of the dynamics of aggregate growth or decline.[8]

The random-element/random-interaction-partner assumption in step 4 is intended to give the model a bottom-up crescive or "emergence" flavor—as opposed, say, to imposing through formal organization a prescribed sequence of elements and partners. Social interaction here is "simple" in that it resembles self-organizing play more than it does formally-dictated work.[9] Formal organizational constraints could

[7]This particular total volume imposes an average number of action capacities per person of two. In some of the runs, I increase this total number to 400 (an average of four skills per person).

[8]This fixed volume constraint imposes competitive selection pressure on the system. It is consistent with the usual way chemical hypercycles have been operationalized, although other death regimes (for example, fixed lifetimes) certainly can be envisioned. Consistency of operationalization here insures greater comparability of findings with past studies.

[9]Of course, most close-up ethnographies reveal that much successful and engaged work in formal organizations, such as team or "organic" decision making, resembles "play" much more than standard formal organization charts imply. See the classic works of Barnard,[1] Burns and Stalker,[6] and March[14] for the distinction between formal and informal organization.

easily be added to the model, but the core existence question would remain: which sets of technologies and social networks are reproducible? Adding formal restrictions probably makes self-organization more complicated, so again it seems analytically useful to develop the baseline case first. Only then will we have a good sense of the limits to "natural" self-organization, which the right institutional constraints might (under some circumstances) be helpful in loosening.

The organizational-existence and the individual-construction conclusions that follow from this model will be presented below, first for simple and then for more complicated hypercycles.

THE SIMPLEST HYPERCYCLE: TWO ELEMENTS ONLY

Many of the findings derivable from this model can be illustrated with the simplest possible hypercycle, that of two elements only. Perhaps the most important of these conclusions is this: embedding hypercyclic interaction sequences in physical space induces a sharp asymmetry between "selfish" source-only reproduction and "altruistic" target-only reproduction. Specifically, in a spatial context, target-only reproduction/learning creates more complicated and expansive social ecologies (in equilibrium) than does source-only reproduction/learning. In sharp contrast, this asymmetry between "selfish" and "altruistic" does not exist in fully-mixed, non-spatial interaction, such as occurs in a fluid.

For the simple two-element hypercycle special case, the equilibrium outcome of the above 10×10 nearest-neighbor model is virtually always a spatially contiguous clustering of individual cells, which arrange themselves into a checkerboard pattern of 1s and 2s. Within this cluster, all individuals are fully specialized into one or the other skill exclusively; outside this cluster, all individual cells are empty or "dead." But the spatial extension of this "live" (that is, self-reproducing) equilibrium cluster varies widely as a function of mode of reproduction. For target-only reproduction, the average spatial extension of the equilibrium checkerboard cluster was 8.7 contiguous "live" individuals (an average over 10 runs of 15,000 iterations). For joint reproduction, the average spatial extension was 6.8 contiguous live individuals (again, averaged over 10 runs of 15,000 iterations). But for source-only reproduction, the average spatial extension was only 1.8 live individuals[10] (same number of runs).

The two-element hypercycle is so simple that these simulation results can be understood analytically. In the Appendix, I derive growth and decline transition

[10]This source-only average was derived from (a) five dyadic pairs of 1s and 2s, which reproduced each other, (b) one self-reproducing cluster of size four, and (c) four isolated singleton cells of either entirely 1s or entirely 2s. Isolated singletons are not "live" self-reproducing clusters, since they have no one left with whom to interact and reproduce.

rates for a single interacting dyad, under each of the spatial and nonspatial reproduction schemes mentioned above. These results, plus analogous results for a single interacting triad, are summarized in Table 1 and graphed in Figure 1.

TABLE 1 (a) Two-Element Solo Dyad: $1 \leftrightarrow 2$ ($a|b$, with $a+b=n$).

(1) Prob $(a \to a+1)$ (2) Prob $(a \to a-1)$ (3) $E(\Delta a) = (1) - (2)$	Spatial	Nonspatial
Source-only Reprod.	$(1/4)(a/n)(1-a/n)$ $(1/4)(a/n)(1-a/n)$ 0	$(a/n)(1-a/n)^2$ $(a/n)^2(1-a/n)$ $(a/n)(1-a/n)(1-2a/n)$
Target-only Reprod.	$(1/4)(1-a/n)^2$ $(1/4)(a/n)^2$ $(1/4)(1-2a/n)$	$(a/n)(1-a/n)^2$ $(a/n)^2(1-a/n)$ $(a/n)(1-a/n)(1-2a/n)$
Joint Reprod.	$(1/4)(1-a/n)^2$ $(1/4)(a/n)^2$ $(1/4)(1-2a/n)$	$2(a/n)(1-a/n)^3$ $2(a/n)^3(1-a/n)$ $2(a/n)(1-a/n)(1-2a/n)$

TABLE 1 (b) Two-Element Solo Dyad: $1 \leftrightarrow 2 \leftrightarrow 1$ ($a|b|c$, with $a+b+c=n$).

(1) Prob $(a \to a+1)$ (2) Prob $(a \to a-1)$ (3) $E(\Delta a) = (1) - (2)$	Spatial	Nonspatial
Source-only Reprod.	$(1/4)(a/n)(1-a/n)$ $(1/4)(a/n)(1-a/n)$ $+(1/4)(a/n)(b/n)$ $-(1/4)(a/n)(b/n)$	$(a/n)(b/n)(1-a/n)$ $(a/n)(b/n)(1-b/n)+c/n)$ $(a/n)(b/n)(2b/n-1)$
Target-only Reprod.	$(1/4)(b/n)(1-a/n)$ $(1/4)(a/n)$ $(1/4)(1-c/n)-(3/4)(a/n)$ $+(1/4)(a/n)(1-b/n)$	$(a/n)(b/n)(1-a/n)$ $(a/n)(b/n)(1-b/n)+c/n)$ $(a/n)(b/n)(2b/n-1)$
Joint Reprod. (1) Prob $(a \to a+1)$ (2) Prob $(a \to a-1)$ + 2 Prob $(a \to a-2)$ (3) $E(\Delta a) = (1) - (2)$	$(1/4)(1-c/n)(1-a/n)^2$ $(1/2)(a/n)(1-a/n)$ $+(1/4)(a/n)^2(1-c/n)$ $(1/4)(1-c/n)-(a/n)$ $+(1/2)(a/n)(1-b/n)$	$2(a/n)(b/n)(1-a/n)^2$ $2(a/n)(b/n)(1-b/n)+c/n)^2$ $2(a/n)(b/n)(2b/n-1)$

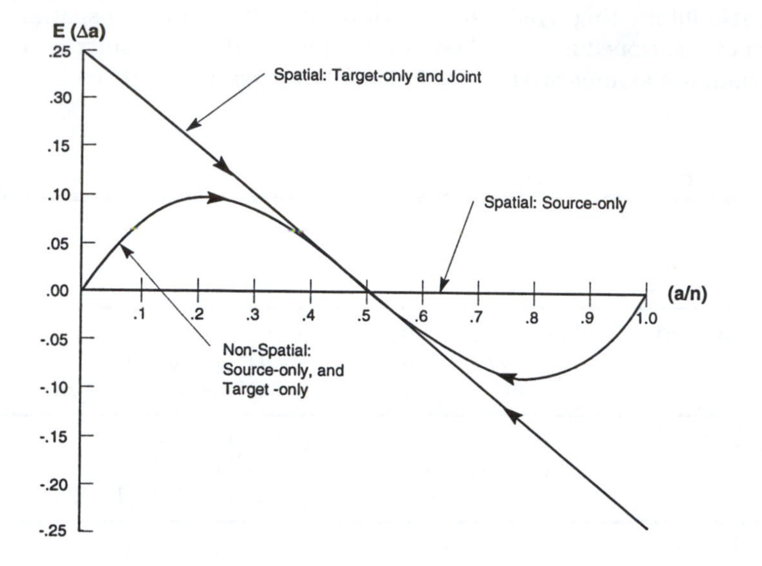

FIGURE 1 (a) Graphing Table 1a results: $1 \leftrightarrow 2$ ($a|b$, with $a + b = n$).

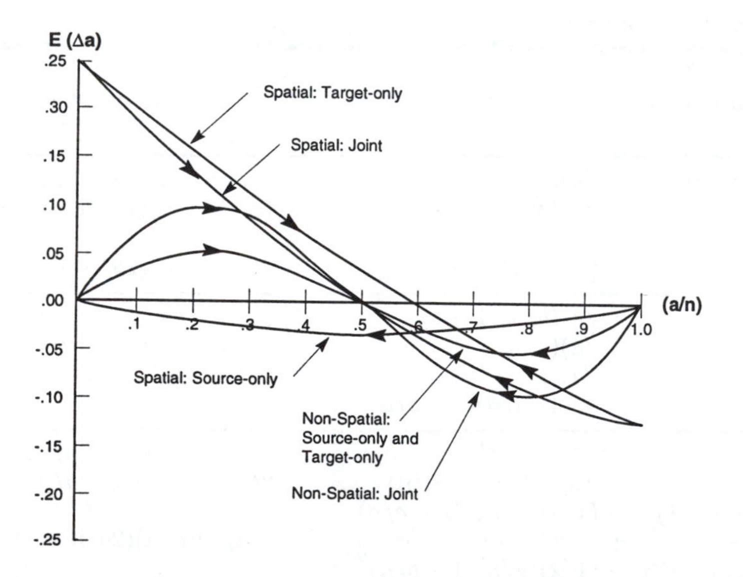

FIGURE 1 (b) Graphing Table 1b results: $1 \leftrightarrow 2 \leftrightarrow 1$ ($a|b|c$, with $a + b + c = n$). N.B. Graphed under constraint that $a = c$. Because of this constraint (a/n) cannot exceed .5.

The developmental dynamic for "selfish" source-only reproduction is depressingly simple in a spatial context. Catalytic source-only social ecologies of three or more interacting individuals inexorably and quickly collapse into isolated dyads of two specialized individuals, who then stably "feed" on each other. The force driving this, and all other collapses, is the selection pressure induced by fixing the total population volume: different pairs of interacting individuals are sharply competing against one another in the relative growth rates of their learned skills. The victorious, but isolated, final dyad is stable in the limited sense that there is no systematic tendency for one or the other of the two remaining individuals to devour the other, no matter what the relative sizes of the two individuals. But the element proportions of this final dyad perform a random walk across the (a/n) line until eventually, after a very long time (typically more than 15,000 iterations), they hit one or other of the absorbing barriers of $(a/n) = 1$ or $(a/n) = 0$. At this point the system as a whole "dies," in the sense that skills no longer reproduce—there being no one left to interact with.

Thus, over the very long run, no spatially embodied social ecology is possible under a purely "selfish" source-only reproduction/learning regime. This conclusion holds true for any level of hypercycle complexity (two-element, three-element, etc.).

The developmental dynamic for target-only reproduction is dramatically different in a spatial setting. Selection pressures still operate to drive very dispersed random distributions of skills down into tight clusters of contiguous individuals, who reproduce their skills through interaction. But stable social ecologies much larger than just two individuals are possible. The analytic results in Table 1, graphed in Figure 1, and derived in the Appendix, make it clear why this is true. Both target-only and joint-reproduction schemes exhibit very strong equilibrating or "leveling" tendencies, which cause depleted individuals to grow in their skill sets and bloated individuals to shrink. In particular, under target-only and joint reproduction, the two individuals in isolated dyads converge to equal sizes in their skill sets, rather than to the random walk, characteristic of source-only reproduction. Three individuals in isolated triads converge to a stable [.25, .50, .25] relative size distribution for joint reproduction, and to a stable [.29, .42, .29] relative size distribution for target-only reproduction.[11]

Table 1 and Figure 1 also present analogous developmental results for the nonspatial, fully-mixed "fluid" context explored by the original hypercycle theorists. For simple hypercycles, like the two-element case here, final size distributions in this nonspatial setting are similar to the results of the target-only and joint-reproduction regimes in the spatial setting. But, of course, in the nonspatial framework, there are no "individuals," because there are no physical sites within which skills reside. The

[11]The slightly greater sizes of the minority individuals produced by target-only reproduction, as compared to the qualitatively similar joint reproduction, account for the slightly greater spatial extension of the equilibrium cluster in the simulation results. Both target-only and joint reproduction allow for the emergence of social organization, but the target-only version permits more individuals stably to participate in this ecology.

primary equilibrium difference between the nonspatial and the target-only spatial setting will emerge at higher levels of hypercycle complexity: nonspatial hypercycles have a built-in stability threshold at four elements, that does not carry over to the target-only spatial framework. In other words, nonspatial hypercycles are stable at four-elements or below, but not at five-elements and above (Hofbauer and Sigmund,[12] p. 96). In contrast, stable target-only spatial hypercycles can (but not necessarily will) emerge at any level of complexity.

As can be seen in Figure 1, for simple hypercycles, the main dynamic difference between target-only spatial and nonspatial hypercycles is one of degree. The quantitative pressure toward stable equilibrium, and thus the speed of convergence, is greater in the target-only spatial context.

Besides these highly suggestive equilibrium and developmental path results, the most fundamental finding in Table 1, and in the simulations, is the asymmetry result described above—namely, that mode of reproduction/learning matters greatly in the spatial context, whereas it does not matter in the nonspatial framework. In particular, in nonspatial interaction "selfish" source-only and "altruistic" target-only reproduction behave no differently in the organizational skill ecologies they produce, whereas these learning modes produce completely different organizational ecologies in spatial grounded interaction.[12] In this chapter, I use the suggestive language selfish and altruistic because these labels exactly describe who in the model benefits, in terms of learning, from the interaction—the one who initiates the interaction or the one who responds. But, of course, in as resolutely reductionist and behaviorist a modeling framework as this one, concepts of motivation and intentionality play no causal role. The explanation of asymmetry is more basic than psychology.

The derivations in the Appendix make it clear why asymmetry exists. Successful reproduction of a skill in a hypercycle model is the product of two stages: (a) activation of action capacity/skill into an attempted action, and then (b) finding an interaction partner with a compatible other skill or action capacity with whom successfully to act. Inserting a hypercycle technology into a spatial context imposes a constraint on possible interaction partners,[13] such that differences in individuals' activation rates, a function of past learning, are preserved rather than offset in their effect on reproduction. For example, in the simplest spatial case of a two-element dyad, the probability (before deaths occur) of the successful reproduction of element 1 is $(a/n) * (1/4)$ under the source-only regime, but is $(b/n) * (1/4)$ under

[12] In the nonspatial framework, joint reproduction does differ from source-only and from target-only reproduction, but only in a tautological way. The Table 1 formulas pertaining to nonspatial joint reproduction have a "2" in them. But by definition, the joint-reproduction scheme involves two reproductions and two deaths each successful iteration, whereas both source-only and target-only involve only one reproduction and one death each successful iteration.

[13] Including a constraint against acting with oneself.

the target-only regime. In the nonspatial case, in contrast, the analogous probabilities of element 1 reproduction is $a/n) * (b/n)$ under the source-only regime, and is $(b/n) * (a/n)$ under the target-only regime.

The spatially determined importance of activation rates for reproduction implies that active individuals, under target-only learning, reproduce others, but that, under source-only learning active individuals reproduce only themselves. Under source-only reproduction (before competitive death rates are factored in), the more self is activated, the more self reproduces, the more self is activated—a cycle that, under competition, eventually drives the other out of existence, to the detriment of self. But under target-only reproduction, the more self is activated, the more (compatible) other is reproduced, the more other is activated, the more self is reproduced, with balancing tendencies even in the face of competition. This same equilibrating dynamic exists under joint reproduction.[14] In contrast, nonspatial fluidity destroys the localized concept of "individual" and, hence, any distinction between "self" and "other."

The specifics of these formulas vary as spatial arrangements change, but the general conclusion remains valid: constraints on interaction partners (spatial or network) permit past learning to be stored in current behavior (here activation rates). This storage induces reproductive variance across individuals, because localized histories vary. In a nonspatial context, in contrast, only homogeneous global history matters, because there is no local.

These findings are especially interesting when applied to humans, because they are driven entirely by the dynamic of self-reproducing skills operating through physical sites. Individuals here, in fact, are no more than passive receptacles, brought into reproductive life by the hypercycle technology of interacting skills that operates through them. In spite of a complete absence of consciousness or intent, an asymmetry between selfishness and altruism emerges as a consequence of kinetics, once spatial constraint imposes a meaning for local.

THE THREE ELEMENT HYPERCYCLE

The main causal factor that changes as one moves from two-element to three-element hypercycles is the topological fit between element/skill reproduction and individual/site interaction. The very neat checkerboard pattern of specialized individuals, so evident in two-element hypercycle technologies, is based on the compatibility of two-element reproduction with four-nearest-neighbor interaction. With three-element hypercycles, however, it is impossible for a three-step closed loop (or

[14] Differences between target-only and joint reproduction arise more dramatically at higher levels of hypercycle complexity.

indeed for any odd-step closed loop) to find siting in a square grid.[15] This means that any hypercycle with an odd number of elements must find physical instantiation through a linked pair of hypercycles, which together comprise an even number of steps—for example, $1 \rightarrow 2 \rightarrow 3 \rightarrow 1 \rightarrow 2 \rightarrow 3 \rightarrow 1$.

This complication does not imply that such a solution is difficult to find. Like in the two-element case, source-only learning within a three-element hypercycle never produces a long-term reproductive ecology of any kind. But under target-only reproduction/learning, eight out of ten of the three-element simulations discovered and reproduced this six-step paired solution, which then established a stable core for a larger equilibrium cluster of interacting individuals. Because of topological mismatch, however, these "live" equilibrium clusters were more complex in their internal structure than the simple two-element checkerboard pattern.

These three-element ecologies are more complex in two ways: (a) individuals within the core[16] frequently are not specialized, but contain more than one skill; and (b) a periphery usually exists of specialized "parasitic" individuals, who survive solely because of their attachment to the core hypercycle, but who do not destroy the hypercycle. The eight successful simulation equilibria are presented in Figure 2 in order to illustrate these two features. This more articulated core-periphery internal structure is purchased at the cost of a somewhat decreased spatial extension. The successful three-element target-only ecologies contained on average 7.5 interconnected individuals (2.6 of whom were multiskilled), as compared with the 8.7 specialized individuals that on average were contained in the always-successful two-element target-only clusters.

Three-element joint-reproduction equilibrium clusters appear similar at first glance to the target-only clusters in Figure 2. But, in fact, in only one of the ten simulations was the six-step solution found. And in two out of ten joint-reproduction cases, not even all three of the skill elements had been preserved. These results highlight the fact that joint reproduction, while a combination of source-only and target-only reproduction, is not truly a hypercycle: it is not catalytic. In the three-element hypercycle, $1 \rightarrow 2$ alone can sustain itself (without element 3), since for any adjacent pairs of numbers under joint reproduction, $1 \rightarrow 2$ effectively is $1 \leftrightarrow 2$. For this reason, the main qualitative difference between joint reproduction and target-only reproduction, in general, is the fact that target-only reproduction forces all skill elements to be maintained in the population, whereas joint reproduction does not preserve all skills except by chance. If all skills are required for some collective product, then joint reproduction, while reproductively "alive," is effectively useless. Altruistic catalysis is more effective than "fairness" in reward in maintaining complex technologies.

[15] Of course, in other spatial topologies, the details of this statement would vary. For example, in a "triangular" interactional pattern, in which every individual cell interacted with its six nearest neighbors, three-element hypercycles would form very neat "checkerboard" equilibria, while two-element hypercycles could not possibly link up into contiguous clusters of specialized individuals.

[16] "Core" is defined as the set of individuals whose skill elements participate in the hypercycle.

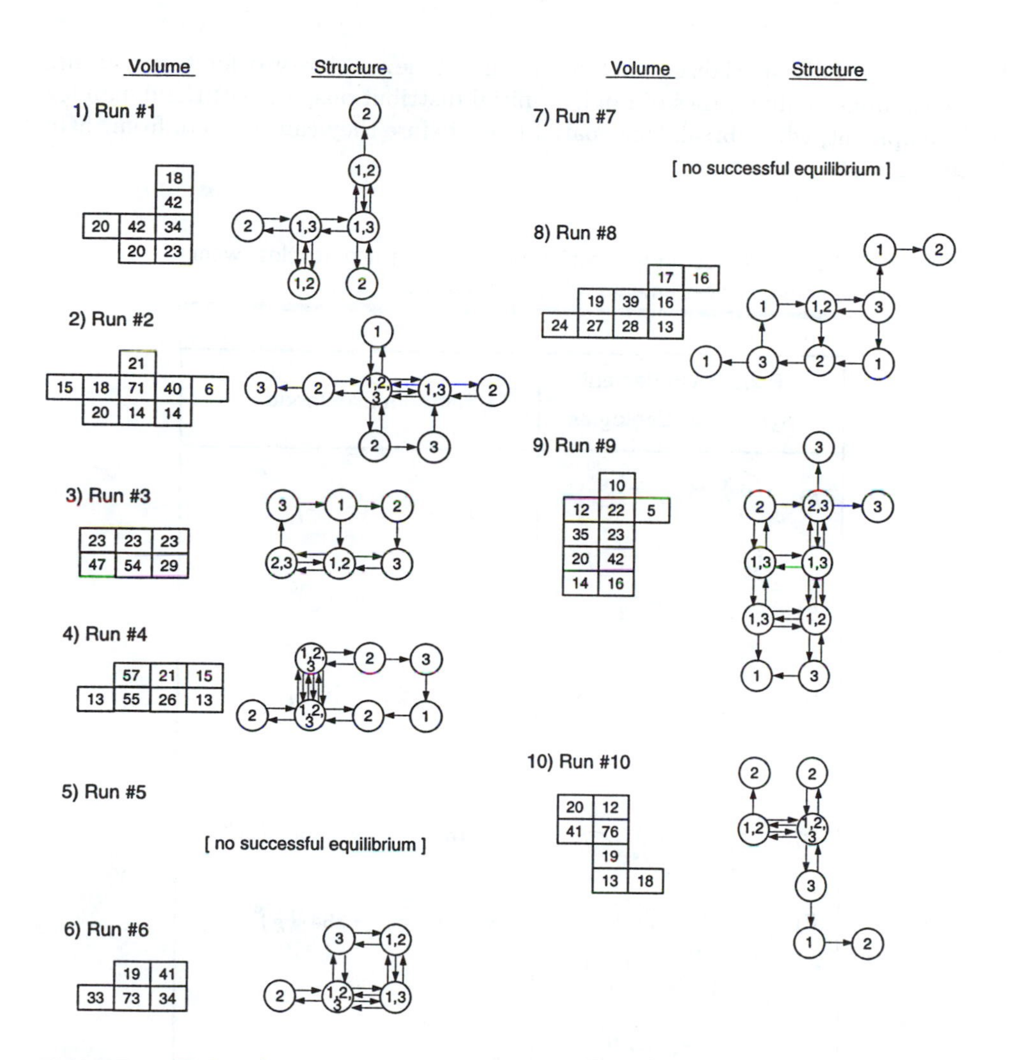

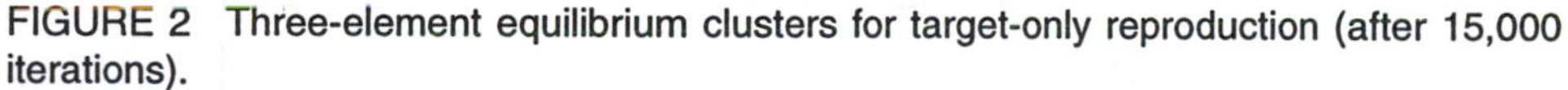
FIGURE 2 Three-element equilibrium clusters for target-only reproduction (after 15,000 iterations).

Why are complicated, multiskilled individuals created by three-element hypercycles? As already stated, a precondition here is topological mismatch, which forces the simple three-element hypercycle into a more circuitous six-step spatial loop—thereby giving that loop greater opportunity to fold back over itself as it wends its way through individual sites. (This is the "skill's eye view" of how multiple skills are laid down in individuals.) This structural possibility cannot be the complete explanation, however, since a single spatial loop of six fully specialized individuals clearly is an equilibrium. We also need to know what drives the folding. In practice, however, the fully specialized outcome is almost never observed for two reasons: (a) the simple combinatorics of random initial distributions, and (b) the dynamics of development, which break long spatial loops before they can separate from their local neighbors.

TABLE 2 Random probabilities of producing hypercycles, within the setup: three-elements within 2×3 array.

Possible 3-Element Hypercycle Ecologies	Random Likelihood*
A)	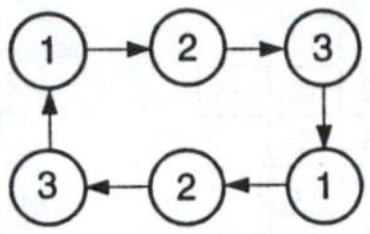 $2 \cdot 6 \cdot (\frac{1}{6})^6 = 12 \cdot (\frac{1}{6})^6$
B)	 $7 \cdot 8 \cdot (\frac{1}{6})^6 = 56 \cdot (\frac{1}{6})^6$
C)	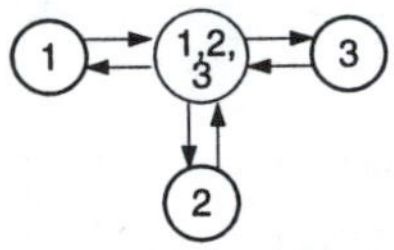 $2 \cdot 8 \cdot 6 \cdot (\frac{1}{6})^6 = 96 \cdot (\frac{1}{6})^6$
D)	 $3 \cdot 16 \cdot 10 \cdot (\frac{1}{6})^6 = 480 \cdot (\frac{1}{6})^6$
E)	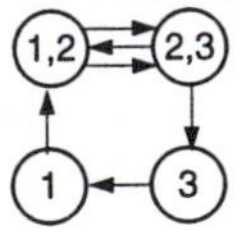 $6 \cdot 8 \cdot 8 \cdot (\frac{1}{6})^6 = 384 \cdot (\frac{1}{6})^6$
F)	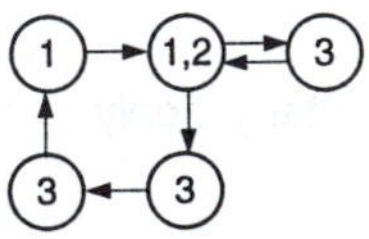
G)	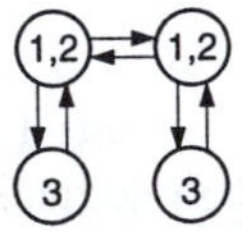 $3 \cdot 8 \cdot 12 \cdot (\frac{1}{6})^6 = 288 \cdot (\frac{1}{6})^6$

* Random likelihood calculated with all possible element and spatial permutations of the type of hypercycle indicated, within a 2x3 grid.

The combinatorial aspect is described in Table 2. This table reports the probabilities of the simple element set $\{1, 2, 3, 1, 2, 3\}$ being laid down into various reproducible patterns within a simple six-cell spatial grid, just through the operation of random processes alone.[17] Overall in such a setup, there is a 2.9% likelihood of these six elements arranging themselves into some type of reproducible spatial pattern through chance alone. But 99.1% of these reproducible patterns involve multiskilled individuals, in which the hypercycle has folded back spatially on itself. Indeed, 47% of the reproducible patterns involve at least one fully skilled individual, who possesses all three skills.

The list of "basic ecologies" in Table 2 is a logical itemization of the possible hypercycle cores inside more expansive three-element equilibrium clusters. As can be seen by comparing Figure 2 and Table 2, the successful simulation outcomes of larger population sizes (200) on a 10×10 grid differ from the basic ecologies in Table 2 only by adding parasitic "hangers on" and/or by amalgamating possibly more than one core.

While I can't quantify the magnitude of the effect, it is also clear from observing detailed developmental histories that even those specialized loops that occasionally do form randomly are easily disrupted in their development by the fact that constituent individuals are also embedded in other ecologies that are evolving simultaneously, in competition with the fully specialized loop. Such interference, of course, occurs between any overlapping ecologies, but the fully specialized hypercycle is maximally vulnerable to disruption because it is maximally extended spatially.

Thus, not only is there low likelihood to begin with for complete specialization to exist in three-element hypercycles, but also there is an "interference" dynamic mechanism that suppresses any complete specialization that does emerge. Combinatorics and interference are the two forces that translate the structural potential of topological mismatch into actual multiskilled individuals.

HIGHER-ELEMENT HYPERCYCLES

Almost all of the conclusions derived from the simplest hypercycles carry over into more complex spatially embedded technologies. In particular, (a) there is always an asymmetry between source-only and target-only reproduction; (b) joint reproduction, unlike target-only reproduction, does not preserve the full hypercycle; (c) odd-numbered hypercycles require paired coupling in order to reproduce; and (d) "live" ecologies, if found, typically involve multiskilled individuals in their hypercyclic core with specialized parasitic hangers-on. The main thing that changes as one moves to higher complexity in hypercycles is the likelihood of finding stable

[17]This was the mechanism of the initial distribution in the simulation model.

reproductive equilibria at all: it declines sharply. Since this likelihood never literally goes to zero, decline can be offset to some extent by increased density, but the greater difficulty of successfully embedding complex technologies in a spatial interaction system cannot be denied.[18]

This greater difficulty of finding and growing more complex hypercyclic solutions out of a random initial configuration can be readily documented. For target-only simulations of "standard" size (volume=200, an average of two elements per individual), the number of runs out of ten that converged on a "live" ecology were the following: for four-element hypercycles, ten; for five-element hypercycles, zero; for six-element hypercycles, five; for seven-element hypercycles, zero; for eight-element hypercycles, two; and for nine-element hypercycles, zero. This likelihood decline is monotonic once one recalls that odd-numbered hypercycles need to double in length in order to site. Joint reproduction almost always found some "live" ecology, but at the cost of implementing only a small fraction of skill elements. And source-only reproduction, as we already know, never generated stable ecologies at higher levels of technological complexity.[19]

The magnitude of this target-only decline can be moderated substantially by increasing the density of skill/action capacities per person. For example, in the case of an eight-element target-only hypercycle, doubling the volume to 400 elements (an average of four skills per individual) increased the chances of finding a reproductive ecology from two out of ten to four out of ten. In other words, not too surprisingly, increasing the average number of skills per individual increases the chances for stably reproductive social ecologies to emerge, which then better sustain these individual multiple capabilities. While straightforward, this density effect provides a kinetic foundation for more complex organizations to develop—once density somehow is increased exogenously. (Alternatively, as I speculate in footnote 18, more efficient interaction configurations may have the same effect.)

Perhaps the most important finding in the simulations of higher-element spatial hypercycles, however, is the fact that stable reproductive ecologies were found at all. As mentioned above, Hofbauer and Sigmund[12] (p. 96) have shown that nonspatial hypercycles are not dynamically stable at complexity levels of five elements and above. This sharp discontinuity does not exist in the spatial setting, even though quantitative likelihoods of stable reproduction do indeed decline as technology becomes more complex.

In the context of the current simulations, this Hofbauer and Sigmund finding was reproduced in the following sense. Nonspatial hypercycle models were created

[18]This difficulty might be one evolutionary reason for the emergence of more articulated social networks out of a simple spatial interaction framework.

[19]Even in those target-only cases that did not converge on a "live" social ecology, there was a quantitative difference between them and the analogous source-only failures. Source-only learning converged very quickly to death. Incomplete target-only learning, in contrast, lingered on well past 20,000 iteractions, reproducing subsets of the hypercycle. Such incomplete reproduction is doomed to long-term failure, but "long term" in target-only learning could be quite long indeed. This suggests (but I have not proved) a random-walk behavior.

out of the above framework simply by substituting fully mixed, random interaction for nearest-neighbor, spatial interaction. The results were: All twenty-five runs of the three-element and four-element (source-only and target-only[20]) nonspatial simulations converged on stably reproducing ecologies.[21] In sharp contrast, all twenty runs of the five-element and above (again, source-only and target-only) nonspatial simulations "crashed," in the sense that all but one element vanished in frequency. Joint reproduction behaved the same way in the nonspatial version as it did in the spatial version: namely, regardless of complexity level, the entire hypercycle splinters into just two neighboring elements who keep each other alive, even as all other elements vanish.

Locking autocatalytic skill sequences into physical space—that is, into embodied individuals in interaction—thus seems to be crucial for complicated skill sequences to emerge. The reason for this is not hard to understand: spatial clustering endogenously generates a localized density that keeps interactions reproductively focused.[22] My spatial hypercycle models do not explain where physical space itself comes from, since grids were imposed exogenously. But "individuals" were indeed derived here, in the minimalist sense of showing how autocatalytic skills sharply cluster spatially, through competitive reproduction, onto interacting physical sites.[23]

CONCLUSION

Physical space is more than just a passive receptacle for social and chemical technologies to be located within. Dynamic barriers of technological complexity can be transcended once global is transformed into the concatenation of locals. This is because reproductive variation and behavioral "memory" are created thereby. Social technologies of action/reaction rule sequences, once enabled, produce individuals as they wend themselves back and forth across sites. This social production of individuals, while spontaneous, is not accidental. The seeding of "work" into

[20] As explained above, in the nonspatial setting there is no difference between source-only and target-only reproduction.

[21] The three-element simulations converged on a simple equilibrium, which fluctuated stochastically around (1/3, 1/3, 1/3). The four-element simulations converged on a limit cycle, which persisted as long as I allowed my computer to run (20,000 iterations).

[22] Social networks could generate even higher levels of density.

[23] A later step in this derivation of "individuals" might be to demand some limited consistency among the possible skills that can coexist on site. This constraint no doubt will inhibit the emergence of "live" social ecology, but it may be a precondition for enabling movement of skill-set individuals across space. In such an extension, however, I would surely search for the minimal consistency constraints possible; otherwise, the flexibility of learning—so necessary to dynamically stable social ecology—would be eliminated.

social interaction among distinct locational units (phenotypes) permits both more complex skill sets and more multifaceted individuals to become alive.

One corollary of siting technology into physical location, discovered in this chapter, is the asymmetry between selfish and altruistic. Only the latter, true catalysis, enables "organization"—that is, the reproducible interweaving of technology, social interaction, and skilled individuals—to exist over the long run. This asymmetry has nothing to do with the usual distinction between private and collective interests. Catalysis is important because whole sets of diverse individuals can then coexist, even in the face of sharp competition.[24] Altruism, by extension, is more crucial to the life of dynamic firms than most social scientists realize, not because of selfless sacrifice for the collective good, but because of the reproductive logic of chemistry.

[24] Lest I be too one-sided in my emphasis, it should equally be noted that target-only catalysis without the discipline of competition produces nothing organized (perhaps only chaos).

APPENDIX

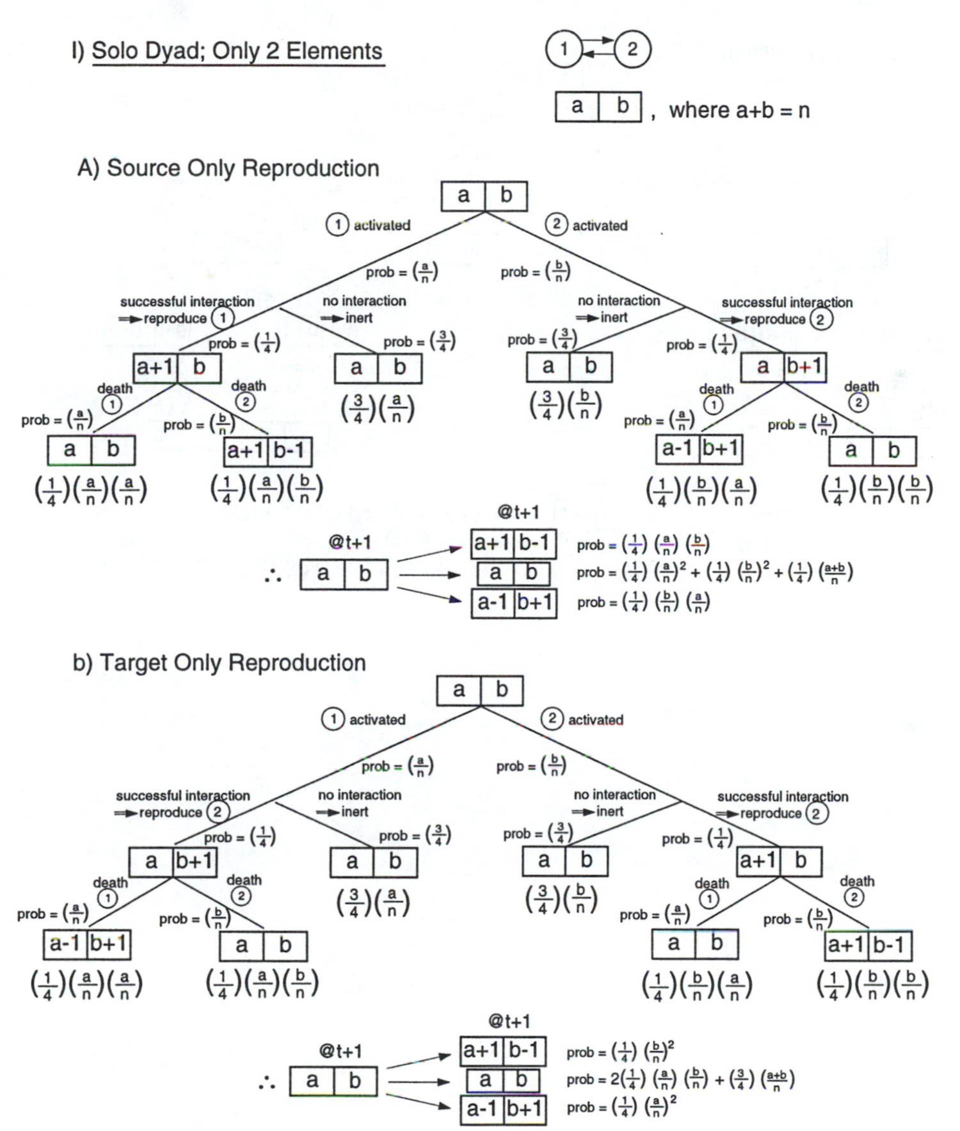
I) Solo Dyad; Only 2 Elements
1 2
a b , where a+b = n
A) Source Only Reproduction
a b
1 activated
2 activated
prob = (a/n)
prob = (b/n)
successful interaction
⇒ reproduce 1
no interaction
⇒ inert
no interaction
⇒ inert
successful interaction
⇒ reproduce 2
prob = (1/4)
prob = (3/4)
prob = (3/4)
prob = (1/4)
a+1 b
a b
a b
a b+1
death 1
death 2
(3/4)(a/n)
(3/4)(b/n)
death 1
death 2
prob = (a/n)
prob = (b/n)
prob = (a/n)
prob = (b/n)
a b
a+1 b-1
a-1 b+1
a b
(1/4)(a/n)(a/n)
(1/4)(a/n)(b/n)
(1/4)(b/n)(a/n)
(1/4)(b/n)(b/n)
@t+1
@t+1
∴ a b
a+1 b-1
a b
a-1 b+1
prob = (1/4) (a/n) (b/n)
prob = (1/4) (a/n)² + (1/4) (b/n)² + (1/4) (a+b/n)
prob = (1/4) (b/n) (a/n)
b) Target Only Reproduction
a b
1 activated
2 activated
prob = (a/n)
prob = (b/n)
successful interaction
⇒ reproduce 2
no interaction
⇒ inert
no interaction
⇒ inert
successful interaction
⇒ reproduce 2
prob = (1/4)
prob = (3/4)
prob = (3/4)
prob = (1/4)
a b+1
a b
a b
a+1 b
death 1
death 2
(3/4)(a/n)
(3/4)(b/n)
death 1
death 2
prob = (a/n)
prob = (b/n)
prob = (a/n)
prob = (b/n)
a-1 b+1
a b
a b
a+1 b-1
(1/4)(a/n)(a/n)
(1/4)(a/n)(b/n)
(1/4)(b/n)(a/n)
(1/4)(b/n)(b/n)
@t+1
@t+1
∴ a b
a+1 b-1
a b
a-1 b+1
prob = (1/4) (b/n)²
prob = 2(1/4) (a/n) (b/n) + (3/4) (a+b/n)
prob = (1/4) (a/n)²

I) Solo Dyad; Only 2 Elements

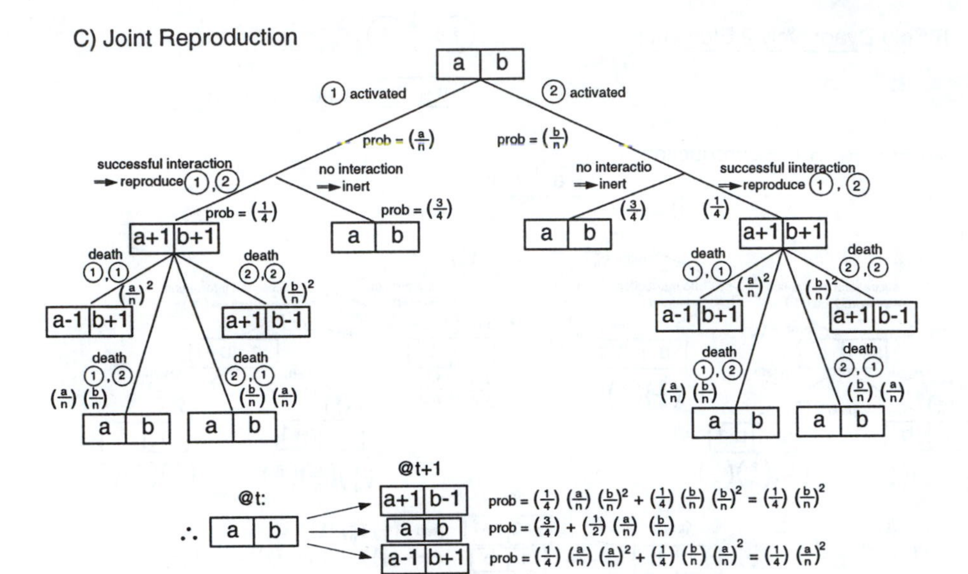

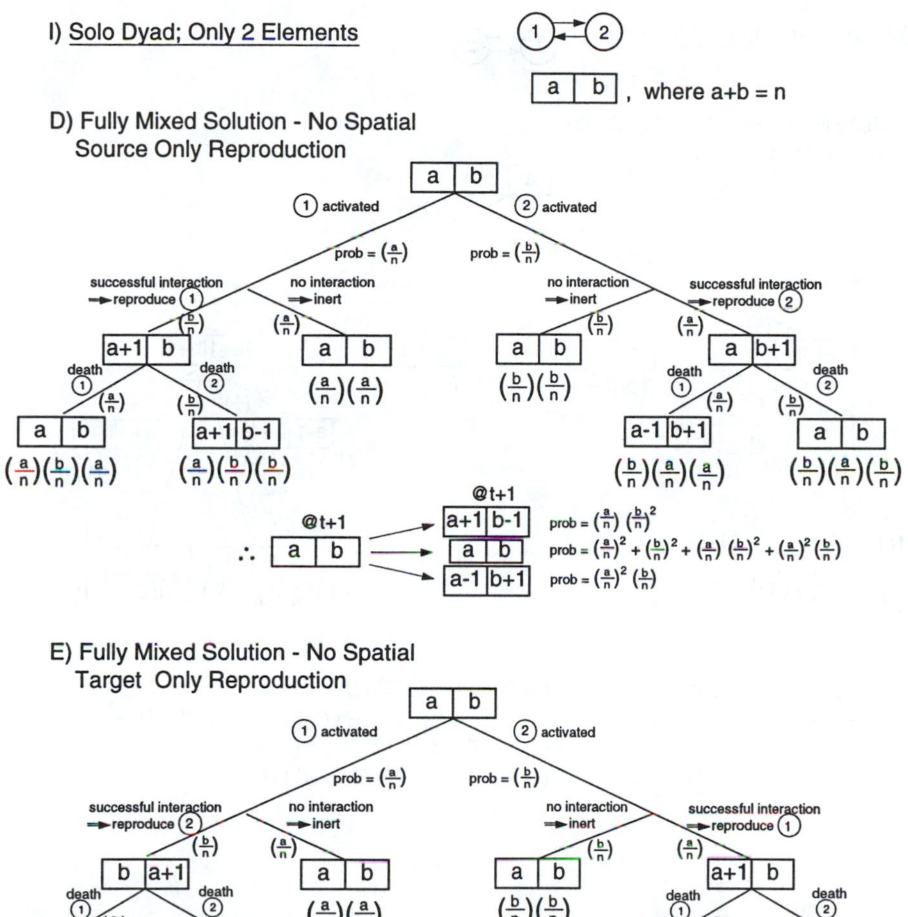
I) Solo Dyad; Only 2 Elements
1
2
a b , where a+b = n
D) Fully Mixed Solution - No Spatial
Source Only Reproduction
a b
1 activated
2 activated
prob = (a/n)
prob = (b/n)
successful interaction
reproduce 1
no interaction
inert
no interaction
inert
successful interaction
reproduce 2
a+1 b
a b
a b
a b+1
death 1
death 2
a b
a+1 b-1
a-1 b+1
a b
@t+1
a+1 b-1
a b
a-1 b+1
prob = (a/n) (b/n)^2
prob = (a/n)^2 + (b/n)^2 + (a/n) (b/n)^2 + (a/n)^2 (b/n)
prob = (a/n)^2 (b/n)

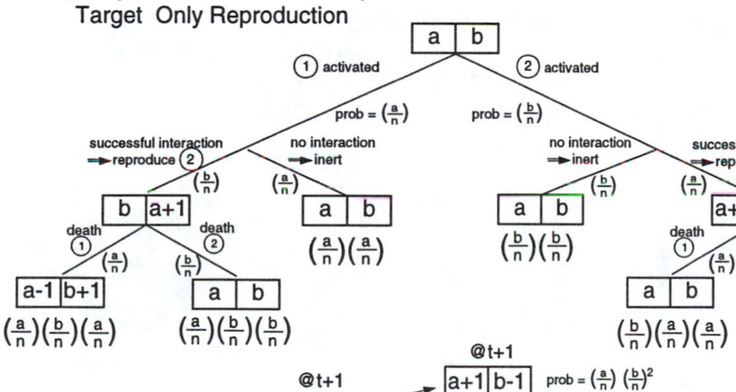
E) Fully Mixed Solution - No Spatial
Target Only Reproduction
a b
1 activated
2 activated
prob = (a/n)
prob = (b/n)
successful interaction
reproduce 2
no interaction
inert
no interaction
inert
successful interaction
reproduce 1
b a+1
a b
a b
a+1 b
death 1
death 2
a-1 b+1
a b
a b
a+1 b-1
@t+1
a+1 b-1
a b
a-1 b+1
prob = (a/n) (b/n)^2
prob = (a/n)^2 + (b/n)^2 + (a/n) (b/n)^2 + (a/n)^2 (b/n)
prob = (a/n)^2 (b/n)

I) Solo Dyad; Only 2 Elements

F) Fully Mixed solution - No Spatial
Joint Reproduction

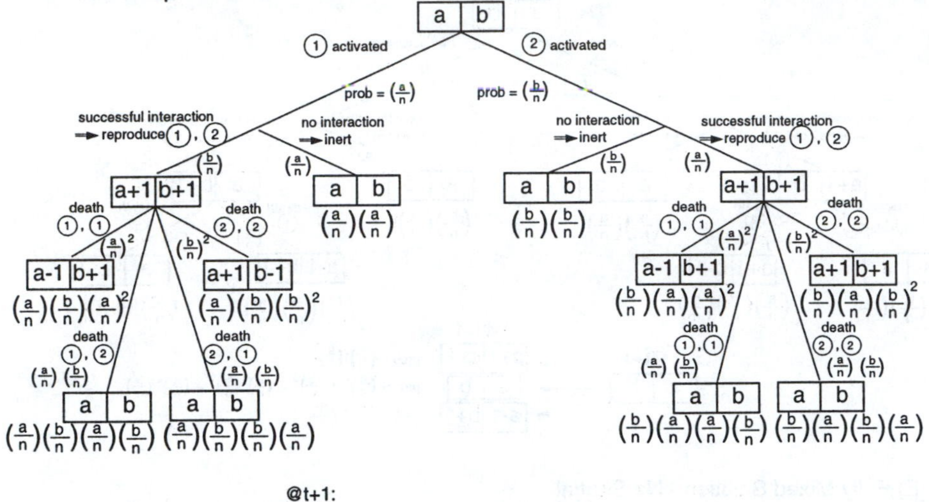

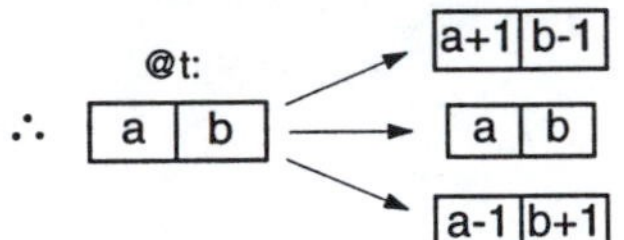

prob = $\left(\frac{a}{n}\right)\left(\frac{b}{n}\right)\left(\frac{b}{n}\right)^2 + \left(\frac{b}{n}\right)\left(\frac{a}{n}\right)\left(\frac{b}{n}\right)^2 = 2\left(\frac{a}{n}\right)\left(\frac{b}{n}\right)^3$

prob = $\left(\frac{a}{n}\right)^2 + \left(\frac{b}{n}\right)^2 + \left(\frac{a}{n}\right)\left(\frac{b}{n}\right)\left[2\left(\frac{a}{n}\right)\left(\frac{b}{n}\right)\right] + \left(\frac{b}{n}\right)\left(\frac{a}{n}\right)\left[2\left(\frac{a}{n}\right)\left(\frac{b}{n}\right)\right]$
$= \left(\frac{a}{n}\right)^2 + \left(\frac{b}{n}\right)^2 + 4\left(\frac{a}{n}\right)^2\left(\frac{b}{n}\right)^2$

prob = $\left(\frac{a}{n}\right)\left(\frac{b}{n}\right)\left(\frac{a}{n}\right)^2 + \left(\frac{b}{n}\right)\left(\frac{a}{n}\right)\left(\frac{a}{n}\right)^2 = 2\left(\frac{b}{n}\right)\left(\frac{a}{n}\right)^3$

REFERENCES

1. Barnard, C. *The Functions of the Executive.* Cambridge, MA: Harvard University Press, 1938.
2. Boerlijst, M.C., and P. Hogeweg. "Spiral Wave Structure in Pre-Biotic Evolution: Hypercycles Stable Against Parasites." *Physica D* **48** (1991): 17–28.
3. Boerlijst, M. C., and P. Hogeweg. "Self-Structuring and Selection: Spiral Waves as a Substrate for Prebiotic Evolution." In *Artificial Life II*, edited by C. G. Langton, C. Taylor, J. D. Farmer, and S. Rasmussen. Santa Fe Institute Studies in the Sciences of Complexity, Proc. Vol. X. Reading, MA: Addison-Wesley, 1992.
4. Boerlijst, M. C., M. E. Lamers, and P. Hogeweg. "Evolutionary Consequences of Spiral Waves in a Host-Parasitoid System." *Proc. Roy. Soc. London, B* **253** (1993): 15–18.
5. Boerlijst, M. C., and P. Hogeweg. "Attractors and Spatial Patterns in Hypercycles with Negative Interactions." *J. Theoret. Biol.* **176** (1995): 199–210.
6. Burns, T., and G. M. Stlker. *The Management of Innovation.* London: Tavistock Publications, 1961.
7. Buss, L. W. *The Evolution of Individuality.* Princeton, NJ: Princeton University Press, 1987.
8. Coase, R. H. "The Nature of the Firm." *Econometrica* **4** (1937) 386–405.
9. Eigen, M., and P. Schuster. *The Hypercycle: A Principle of Natural Self-Organization.* Berlin-Heidelberg: Springer, 1979.
10. Fama, E. F. "Agency Problems and the Theory of the Firm." *J. Pol. Econ.* **88** (1980): 288–307.
11. Fontana, W., and L. W. Buss. "The Arrival of the Fittest: Toward a Theory of Biological Organization." *Bull. Math. Biol.* **56** (1994): 1–64.
12. Hofbauer, J., and K. Sigmund. *The Theory of Evolution and Dynamical Systems.* Cambridge, MA: Cambridge University Press, 1988.
13. Jensen, M. C., and W. Meckling. "Theory of the Firm: Managerial Behavior, Agency Costs, and Ownership Structure." *J. Fin. Econ.* **3** (1976): 305–360.
14. March, J. G. "The Technology of Foolishness," In *Ambiguity and Choice in Organizations*, edited by J. G. March and J. Olsen. Bergen, Norway: Universitetsforlaget, 1976.
15. Williamson, O. E. *Markets and Hierarchies.* New York, NY: Free Press, 1975.

Douglass C. North
Department of Economics, Washington University, St. Louis, MO 63130-4899

Some Fundamental Puzzles in Economic History/Development

SECTION I

In this chapter I would like to confront a number of fundamental puzzles in economic history/development—puzzles that go to the heart of the nature of economic change. They can be broadly classified under two general headings: (1) how to account for the uneven and erratic pattern of both historical change and contemporary development, and (2) how to model this process of change and development. Can we use the tools at hand—i.e., the rationality assumption and growth theory we employ in economics? Let me elaborate on the nature of the puzzles and the connection between them.

The first development of agriculture took place almost four million years after human beings became separate from other primates—that is, only ten thousand years ago. In the ensuing ten thousand years the rate of change appears to have been very slow for at least the first half of that period. In the second, it has been episodic for most of the time; periods of what appear to have been economic growth in particular geographic regions have been interrupted by stagnation and decline and sometimes those geographic areas have failed to recover. In general, growth has been much more exceptional than stagnation or decline. Modern economic growth appears to have begun perhaps four hundred years ago but been confined to a small part of the earth for most of that time. Widespread growth is a recent phenomenon mostly dating since World War II. Even today large parts of the world are not experiencing growth (e.g., the republics of the former Soviet Union and sub-Saharan Africa), or growth continues to be episodic (e.g., Latin America).

The rational choice paradigm assumes that people know what is in their self-interest and act accordingly, or at the very least that competition will weed out those who make incorrect choices and reward those who make correct choices. But it is impossible to reconcile this argument with the historical and contemporary record.

Growth theory as it has evolved from neoclassical theory is equally unhelpful in explaining this historical and contemporary record. Convergence of the growth path of economies, such as Baumol maintained, has only tended to occur amongst the developed countries. Persistent divergence, as argued among some of the new growth theory variants of the Solow model, can't explain the rise of the Asian tigers or China. In fact, to put it bluntly, the growth theory stemming from neoclassical economics, old or new, suggests not only ignorance of the empirical evidence, historical or contemporary, but a failure to recognize that incentives matter; surely a remarkable position for economists whose theory is built on incentives. It is the incentive structure imbedded in the institutional/organizational structure of economies that has to be a key to unraveling the puzzle. But that entails a still deeper puzzle. Why don't economies that have institutional frameworks that are inhospitable to economic growth simply adopt the frameworks of the successful economies? They do, or at least they try to: the rush to create market economies is a ubiquitous characteristic of third world and transition economies. But look at the results. They vary enormously, from China and the Czech republic, which *so far* are successful; to the republics of the former Soviet Union, which *so far* show few signs of success; to sub-Saharan Africa, which remains a basket case.

To make sense out of the historical and contemporary evidence, we must rethink the whole process of economic growth. Current theory stems from the development of national income and growth accounting literature and explores the superficial aspects of economic growth—technology or human or physical capital—rather than the structure of incentives and disincentives that make up the institutional framework of an economy and polity. If we enquire why these incentives vary, we are driven to the various belief systems that determine the institutional framework.

We are far from being able to understand the whole process, but I do intend to suggest some explanation. Specifically, I shall assert and support three arguments that are at wide variance with the received wisdom. The first is that contrary to both the economic history literature and the economic growth literature—old and new—the primary source of economic growth is the institutional/organizational structure of a political economy and until we focus on that subject we shall not advance knowledge on economic growth. Second, that economic growth is dependent on stable political/economic institutions that provide low costs of transacting in *impersonal* political and economic markets. Third, that it is the belief systems of societies and the way they evolve that is the underlying determinant of institutions and their evolution.

SECTION II

Curiously enough institutions are front and center in all current explanations of growth or lack of it in third world or transition economies; but the explanations lack both analytical content and an understanding of the nature of institutions or the way they evolve. The implication is that they can be created at will or frequently that they are a dependent variable to getting the prices right. But recent experience provides convincing evidence that they neither can be taken for granted nor do they automatically evolve from getting the prices right. And, historical experience makes it clear that efficient economic institutions are the exception. To proceed, we must understand what institutions are and how they evolve.

Institutions provide the structure that humans impose on human interaction in order to reduce uncertainty. There is nothing about that structure that implies that institutions are efficient in the sense of inducing economic growth. Sometimes they are created to facilitate exchange, encourage technological change, and induce human capital formation and, in consequence, reduce transaction and/or transformation costs; at other times they are created to support monopoly, prevent technological change, thwart human capital development, and generally raise the costs of transacting and/or transformation. In fact, the second of these has been the far more common pattern throughout history. To provide an understanding of why, I shall state five propositions that, I believe, underlie institutional change; but first let me dispel any confusion between institutions and organizations.

Institutions are the rules of the game—both formal rules and informal constraints (conventions, norms of behavior, and self-imposed codes of conduct)—and their enforcement characteristics. Together they define the way the game is played.

Organizations are the players. They are made up of groups of individuals held together by some common objectives. Economic organizations are firms, trade unions, cooperatives, etc.; political organizations are political parties, legislatures, regulatory bodies; educational organizations are universities, schools, vocational training centers. The immediate objective of organizations may be profit maximizing (for firms) or improving reelection prospects (for political parties); but the ultimate objective is survival because all organizations live in a world of scarcity and, hence, competition.

Now to the five propositions:

PROPOSITION 1: The continuous interaction between institutions and organizations in the economic setting of scarcity and, hence, competition is the key to institutional change.

PROPOSITION 2: Competition forces organizations continually to invest in new skills and knowledge to survive. The kind of skills and knowledge individuals and their organizations acquire will shape evolving perceptions about opportunities and, hence, choices that will incrementally alter institutions.

PROPOSITION 3: The institutional framework provides the incentive structure that dictates the kinds of skills and knowledge perceived to have the maximum payoff.

PROPOSITION 4: Perceptions are derived from the mental constructs of the players.

PROPOSITION 5: The economies of scope, complementarities, and network externalities of an institutional matrix make institutional change overwhelmingly incremental and path dependent.

Let me expand on propositions 2 through 5.

PROPOSITION 2. New or altered opportunities may be perceived to be a result of exogenous changes in the external environment that alter relative prices to organizations, or a consequence of endogenous competition among the organizations of the polity and the economy. In either case, the ubiquity of competition in the overall economic setting of scarcity induces entrepreneurs and the members of their organizations to invest in skills and knowledge. Whether through learning by doing on the job or the acquisition of formal knowledge, improving the efficiency of the organization relative to that of rivals is the key to survival.

While idle curiosity surely is an innate source of acquiring knowledge among human beings, the rate of accumulating knowledge is clearly tied to the payoffs.

Secure monopolies, be they organizations in the polity or in the economy, simply do not have to improve to survive. But firms, political parties, or even institutions of higher learning, faced with rival organizations, must strive to improve their efficiency. When competition is muted (for whatever reasons), organizations will have less incentive to invest in new knowledge and, in consequence, will not induce rapid institutional change. Stable institutional structures will be the result. Vigorous organizational competition will accelerate the process of institutional change.

PROPOSITION 3. There is no *implication* in proposition 2 of evolutionary progress or economic growth—only of change. The institutional matrix defines the opportunity set, be it one that makes income redistribution the highest payoff in an economy or one that provides the highest payoffs to productive activity. While every economy provides a mixed set of incentives for both types of activity, the relative weights (as between redistributive and productive incentives) are crucial factors in the performance of economies. The organizations that come into existence will reflect the payoff structure. More than that, the direction of their investment in skills and knowledge will equally reflect the underlying incentive structure. If the highest rate of return in an economy comes from piracy we can expect that the organizations will invest in skills and knowledge that will make them better pirates. Similarly, if there are high returns to productive activities we will expect organizations to devote resources to investing in skill and knowledge that will increase productivity.

The immediate investment of economic organizations in vocational and on-the-job training obviously will depend on the perceived benefits, but an even more fundamental influence on the future of the economy is the extent to which societies will invest in formal education, schooling, the dissemination of knowledge, and both applied and pure research, which will, in turn, mirror the perceptions of the entrepreneurs of political and economic organizations.

PROPOSITION 4. The key to the choices that individuals make is their perceptions, which are a function of the way the mind interprets the information it receives. The mental constructs individuals form to explain and interpret the world around them are partly a result of their cultural heritage, partly a result of the local everyday problems they confront and must solve, and partly a result of nonlocal learning.[1] The mix among these sources in interpreting one's environment obviously varies as between, for example, a Papuan tribesman on the one hand and an American economist on the other (although there is no implication that the latter's perceptions are independent of his or her cultural heritage).

The implication of the foregoing paragraph is that individuals from different backgrounds will interpret the same evidence differently; they may, in consequence, make different choices. If the information feedback of the consequences of choices were complete then individuals with the same utility function would gradually correct their perceptions and, over time, converge to a common equilibrium; but as Frank Hahn has succinctly put it, "There is a continuum of theories that agents can hold and act upon without ever encountering events which lead them to change their theories" (Hahn,[8] p. 324). The result is that multiple equilibria are possible due to different choices by agents with identical tastes.

[1]This argument is elaborated in a cognitive science approach to the development of learning and belief systems in Denzau and North.[3]

PROPOSITION 5. The viability, profitability, and indeed, survival of the organizations of a society typically depend on the existing institutional matrix. That institutional structure has brought them into existence, and their complex web of interdependent contracts and other relationships has been constructed on it. Two implications follow. Institutional change is typically incremental and path dependent.

Why can't economies reverse their direction overnight? This would surely be a puzzle in a world that operates as neoclassical theory would have us believe. That is, in a neoclassical world abrupt, radical change should immediately result from a radical change in relative prices or performance. Now, it is true that on occasion accumulated pressures do produce an abrupt change in institutions akin to the punctuated equilibrium models in evolutionary theory. But it is simply a fact that the overwhelming majority of change is incremental and gradual. It is incremental because large-scale change would harm large numbers of existing organizations and, therefore, is stoutly opposed by them. Revolutionary change will only occur in the case of "gridlock" among competing organizations, which thwarts their ability to capture gains from trade. Revolutions are extraordinary, and even when they occur turn out, over time, to be far less revolutionary than their initial rhetoric would suggest.

Path dependence could mean nothing more than that yesterday's choices are the initial starting point for today's. But path dependence appears to be a much more fundamental determinant of long-run change than that.[2] The difficulty of fundamentally altering paths is evident and suggests that the learning process by which we arrive at today's institutions constrains future choices. The institutional structure builds in a set of constraints with respect to downstream changes that biases choices.

SECTION III

Understanding the nature of path dependence is the key to understanding the success or failure of economies in altering their competitive positions, and there is still a great deal about path dependence that we do not understand. But it is a fact of history, one of the most enduring and significant lessons to be derived from the past. It is not that economies in the past have not been aware of their inferior or declining competitive positions. They have. But perceptions about the reasons are something else again. Spain's long decline in the seventeenth century from the most powerful nation in the western world since the Roman empire to a second rate power was a consequence of recurrent war and fiscal crises. The policies that were considered feasible in the context of the institutional constraints and perceptions of

[2] The concept of path dependence has been pioneered by Arthur[1] and David.[2]

the actors were price controls, tax increases, and repeated confiscations. As for the perceptions of the actors, here is Jan De Vries's description of the effort to reverse the decline:

> But this was not a society unaware of what was happening. A whole school of economic reformers...wrote mountains of tracts pleading for new measures. ...Indeed in 1623 a *Junta de Reformacion* recommended to the new king, Philip IV, a series of measures including taxes to encourage early marriage (and, hence, population growth), limitations on the number of servants, the establishment of a bank, prohibitions on the import of luxuries, the closing of brothels, and the prohibition of the teaching of Latin in small towns (to reduce the flight from agriculture of peasants who had acquired a smattering of education). But no willpower could be found to follow through on these recommendations. It is said that the only accomplishment of the reform movement was the abolition of the ruff collar, a fashion that had imposed ruinous laundry bills on the aristocracy (De Vries,[5] p. 28).

As the foregoing quotation makes clear, there is no guarantee that economic decline will induce perceptions on the part of the actors that will lead to reversal of that decline and to improved performance. For a modern illustration we have only to look today at the evolving perceptions in the Russian republic as evidenced by the last election.

What constrains the choices of the players is a belief system reflecting the past—the cultural heritage of a society—and its gradual alteration reflecting the current experiences as filtered (and, therefore, interpreted) by that belief system. That is, in essence, why path dependence exists.

Historical success stories of institutional adaptation can only be understood in these terms. They reflected conditions where the belief system underlying an institutional matrix filtered the information players got from current experiences and interpreted that information in ways that induced choices leading to the modification, alteration, or adoption of institutions that resolved existing problems or led to improvements in competitive performance.

Take the successful development of the Netherlands and England in early modern Europe.[3] It was the lack of large-scale political and economic order that created the essential environment hospitable to political/economic development. In that competitive, decentralized environment many alternatives were pursued as each society confronted its own unique external environment. Some worked, as in the cases of the Netherlands and England; some failed, as in the cases of Spain and Portugal; and some, such as France, fell in between these two extremes. But the key to the story is the variety of options pursued and the likelihood (as compared to a single unified policy) that some would turn out to produce political/economic development. Even the relative failures in Western Europe played an essential role

[3]This section on the development of early modern Europe is elaborated at length in North.[13]

in European development and were more successful than other parts of the world because of competitive pressures.

The last point deserves special emphasis. It was the dynamic consequences of the competition amongst fragmented political bodies that resulted in an especially creative environment. Europe was politically fragmented, but it was integrated in having both a common belief structure derived from Christendom, and information and transportation connections that resulted in scientific, technological, and artistic developments in one part spreading rapidly throughout Europe. To treat the Netherlands and England as success stories in isolation from the stimulus received from the rest of Europe (and to a lesser degree Islam and China) is to miss a vital part of the explanation. Italian city states, Portugal, and Germanic states all fell behind the Netherlands and England; but banking, artistic development, improvements in navigation, and printing were just a few of the obvious contributions that the former states made to European advancement.

The Netherlands and England pursued different paths to political/economic success, but in each case, the experiences were conducive to the evolution of a belief structure that induced political and economic institutions that lowered transaction costs. In both polities, competition among the evolving nation states was a deep underlying source of change and, equally, a constraint on the options available to rulers within states. It was competition that forced the Crown to trade rights and privileges for revenue including, most fundamentally, the granting to representative bodies—variously Parliament, States General, Cortes—control over tax rates and/or certain privileges in return for revenue. But it was the evolving bargaining strength of rulers vis-à-vis constituents that was the decisive feature of their subsequent development. Three considerations were at stake: (1) the size of the potential gains the constituents could realize by the state taking over protection of property; (2) the closeness of substitutes for the existing ruler—that is the ability of rivals (both within and outside the political unit) to the existing ruler to take over and provide the same, or more, services; and (3) the structure of the economy, which determined the benefits and costs to the ruler of various sources of revenue.

Let me briefly describe the background conditions of the two polities that led up to the contrasting external environments that shaped the belief systems.

To understand the success of the Netherlands, one must cast a backward glance at the evolution of the prosperous towns of the Low Countries such as Bruges, Ghent, and Liege; their internal conflicts; and their relationship to Burgundian and Habsburg rule. The prosperity of the towns early on, whether based on the wool cloth trade or metals trade, created an urban-centered, market-oriented area unique at a time of overwhelmingly rural societies. Their internal conflicts reflected ongoing tensions between patrician and crafts and persistent conflicts over ongoing efforts to create local monopolies which, when successful, led to a drying up of the very sources of productivity which had been the mainspring of their growth. Burgundian (and later Habsburg) rule discouraged restrictive practices such as those that developed in the cloth towns of Bruges and Ghent and encouraged the growth of new centers of industry that sprang up in response to the favorable incentives embodied in the rules and property rights. In 1463 Philip the Good created a representative body, the States General, which enacted laws and had the authority to vote taxes for the ruler. The Burgundians and Habsburgs were rewarded by a level of prosperity that generated tax revenues that made the Low Countries the jewel in the Habsburg Empire.

England evolved along a route different from that of continental polities. Being an island made it less vulnerable to conquest and eliminated the need for a standing army (and undoubtedly contributed to the different initial belief structure that Macfarlane describes in *The Origins of English Individualism*). The Norman conquest, the exception to British invulnerability to external conquest, produced a more centralized feudal structure than that on the continent. The political institutions, in consequence, differed in several important respects from those of the continent. There was a single parliament for the entire country; no regional estates as in France, Spain, and the Netherlands. There were also no divisions into towns, clergy, and nobility. But the more centralized feudal structure did not ensure that the crown could not overstep the traditional liberties of the barons, as the Magna Carta attests.

We can now turn to examining the evolving bargaining strength (and the three underlying determinants) of ruler versus constituent that shaped the belief structure and the path of each polity. Take the Netherlands. The productive town economies stood to gain substantially by the political order and protection of property rights provided by the Burgundians and then by Charles V. The structure of the economy built on export trades provided the means for easy-to-collect taxes on trade, but not at a level to adversely affect the comparative advantage of those export trades. The liberty to come and go, buy and sell as they saw fit led to the evolution of efficient economic markets. But when Philip II altered the "contractual agreement" the Seven Provinces became convinced that they could prosper only with independence. The resistance was initiated by the States General which in 1581 issued the Act of Abjuration of allegiance to Philip II and claimed sovereignty for the Provinces themselves. The powers of the newly independent country resided with each province (which voted as a unit) and a unanimity rule meant that the States General could only act with the unanimous approval of the Seven Provinces. Cumbersome as that process was, this political structure survived. The polity not only evolved the elements of political representation and democratic decision rules, but equally supported religious toleration. The belief structure that had evolved to shape the independent polity was more pragmatic than intellectual, a consequence of the incremental evolution of the bargaining strength of constituents and rulers.

As with the Netherlands, it was England's external trade that provided an increasing share of Crown revenue with taxes on wine, general merchandise, and wool cloth; but it was the wool export trade that was the backbone of augmented crown revenue. Eileen Power's classic story of the wool trade[15] describes the exchange between the three groups involved in that trade: the wool growers as represented in Parliament, the merchants of the staple, and the Crown. The merchants achieved a monopoly of the export trade and a depot in Calais; Parliament received the right to set the tax; and the Crown received the revenue. Stubbs[17] summarized the exchange as follows: "The admission of the right of Parliament to legislate, to inquire into abuses, and to share in the guidance of national policy, was practically purchased by the money granted to Edward I and Edward III. . . ."

With the Tudors the English Crown was at the zenith of its power, but it never achieved the unilateral control over taxing power that the crowns of France and Spain achieved. The confiscation of monastery lands and possessions by Henry VIII alienated many peers and much of the clergy and, as a consequence, "Henry had need of the House of Commons and he cultivated it with sedulous care."[6] The Stuarts inherited what the Tudors had sown and the evolving controversy between the Crown and Parliament is a well-known tale. Two aspects of this controversy are noteworthy for this analysis. One was the evolving perception of the common law as the supreme law of the land—a position notably championed by Sir Edward Coke—and the other was the connection made between monopoly and a denial of liberty as embodied in the Crown grants of monopoly privileges.

SECTION IV

England and the Netherlands represent historical success stories in which evolving belief systems shaped by external events induced institutional evolution to provide the beginning of modern economic growth. Key institutional/organizational changes were those that permitted the growth of impersonal exchange—both economic and political. By permitted, I mean that, using game theory terminology, they altered the payoff between defection and cooperation to favor the latter in both political and economic markets. That is, personal exchange provides settings in which it typically pays to cooperate. Impersonal exchange is just the antithesis and necessitates the development of institutions to alter payoffs in favor of cooperative activity.

In the case of economic markets, recent historical research has provided analytical accounts of the evolution of the institutions that undergirded long-distance trade in the Middle Ages;[7] that led to the development of merchant codes of conduct that became the foundation of commercial law;[9] that converted uncertainty to risk, leading to the development of marine insurance;[4] and that provided the foundation of an impersonal capital market with the development of the bill of exchange and the growth of early banking organization. By the end of the sixteenth century these and other institutional/organizational innovations had created the first modern economy in Amsterdam and the Seven Provinces that formed the Netherlands.[4]

But the creation of efficient economic markets is only half, and the less puzzling half, of the story. It is much more difficult to account for and explain the growth of

[4] See North[12] and Greif[7] for a summary of the game theoretic and other analytical literature.

"efficient" political markets that are a necessary precondition to the development of efficient economic institutions. It is the polity that specifies and enforces the economic rules of the game, and our knowledge of the essential conditions for the creation of such political institutions has not progressed much from the insightful observations of James Madison in the *Federalist Papers*.

However, the historical evidence cited above provides essential clues to the evolutionary process in early modern Europe. It was the growing need for revenue in the face of the rising costs of warfare that forced monarchs to create "representative bodies"—parliament, estates general, cortes—and cede them certain rights in return for revenue. But, at this point, the stories diverge between the Netherlands and England on the one hand, and much of the rest of Europe on the other. In the former, the three conditions cited above (p. 8) led to the growth of representative government; in the latter those conditions led to the persistence or revival of absolutist regimes with centralized decision making over the economy.

The contrasting subsequent history of the New World bears striking testimony to the significance of path dependence. In the case of North America, the English colonies were formed in the century when the struggle between parliament and the Crown was coming to a head. Religious and political diversity in the mother country was paralleled in the colonies. The general development in the direction of local political control and the growth of assemblies was unambiguous. Similarly, the colonies carried over free and common socage tenure of land (fee simple ownership rights) and secure property rights in other factor and product markets.

The Spanish Indies conquest came at the precise time that the influence of the Castilian Cortes (parliament) was declining and the monarchy of Castile, which was the seat of power in Spain, was firmly establishing centralized bureaucratic control over Spain and the Spanish Indies. The conquerors imposed a uniform religion and bureaucratic administration on already existing agricultural societies. Wealth maximizing behavior by organizations and their entrepreneurs (political and economic) entailed getting control of, or influence over, the bureaucratic machinery.

One cannot make sense out of the contrasting subsequent history of the Americas north and south of the Rio Grande river without taking into account this historical background.[5]

SECTION V

Let me draw some specific implications for the study of modern economic growth from the previous two sections.

[5]The downstream consequences in the New World are discussed in more detail in North,[11] chap. 12. Robert Putnam tells a similar story of the contrasting development of different regions in Italy from the twelfth century onward in Putnam.[16]

We can briefly deal with economies that have failed to develop. These are cases in which the belief system, reflecting the historical experiences of that economy, fails to overcome the critical hurdle—the creation of institutions, economic and political, that will permit impersonal exchange. Markets remain local and/or poorly developed. As these economies beoame exposed to the larger world economy, the institutional matrix typically spawns the development of organizations whose profitable opportunities predominantly favor redistributive activities.[6] Political organizations in such cases will appear similar to those of the classic Virginia school of political economy, in which the state is little more than a theft machine. We do not have to look very far to see instances of such polities in Africa or, with some modification, in Latin America.

The obverse, the success stories, have been briefly described in the outlines above of the Netherlands and Britain—instances where external experiences reinforced evolving belief systems in the direction of productive activities by the economic organizations. Political organizations in these cases have evolved strong informal constraints—norms of behavior—that have constrained the behavior of the political actors. This last point deserves special emphasis. The gradual development of informal norms of behavior that have become deeply imbedded in the society provides the stable underpinning to the adaptive efficiency characterizing the western economies with a long history of growth. We do not know how to create such conditions in transition or third world economies.

I wish to briefly explore two groups of cases that are of particular interest to the subject matter of this chapter: the cases of the new industrial countries (NICs) and the cases of high income countries that loose their competitive edge. In both groups, it is the polity and political organizations that play a decisive role.

The NICs are important because they appear to offer a model of development to be emulated. But in order to understand these Asian cases it is necessary to reject an empty argument that reflects the ahistorical understanding of economists. The argument concerns the positive role of government in establishing the rules of the game. Did the NICs succeed in spite of government "meddling" with the economy, or was government involvement crucial to development? The argument is empty because there is an implicit assumption that the rules of a "laissez faire" economy are a natural result that occurred without the active participation of government and will fall into place by themselves. Not so. In the western world, the evolution of the legal framework underlying efficient market economies was a long incremental process stretching from the medieval law merchant to the gradual integration of those rules into common and Roman law. If the legal framework doesn't already

[6]The second economic revolution, the wedding of science to technology, which began in the last half of the nineteenth century, is the source of modern economic growth and entails enormous specialization, division of labor, urban societies, and global markets. The organizational restructuring—economic, political, and social—necessary to capture the gains from trade with this technology has necessitated fundamental changes in societies. The institutional adjustments are difficult enough for developed economies and much more difficult for transition and third world economies. See North[10] chap. 13 for further discussion.

exist, or only partially exists, it must be created. Moreover, there is not just one set of rules that will provide the proper incentives for productive activity, and finally the rules themselves must continually adapt to new technologies and opportunities. Rules that will provide low transaction and transformation costs in one context will be obsolete in another. Hayek's notion of collective learning is a powerful insight into the critical role of our cultural heritage embodying past "successes." It is the basis of path dependence. But collective learning embodying the wisdom derived from past experiences does not always equip us to solve future problems and, indeed, may prevent us from reaching solutions.

The interesting issue about the NICs is not whether they made mistakes or not—they did. Rather, it is what set of conditions will give the political organizations of a society the degrees of freedom from economic interest group pressures to pursue radically different policies? Further, how can we account for the behavior of the government officials whose objective function appears (largely) to have been maximizing the economy's growth rather than their own wealth? The usual answer to the first question is that[18] the existence of an external threat (North Korea in the case of South Korea and mainland China in the case of Taiwan) was the critical factor. The usual answer to the second question is culturally derived social norms. Neither answer is completely satisfactory. A more promising approach is some of the new literature on market-preserving federalism and the incentive system that underlies this system of decentralization.

Turning to the second group of cases: high income countries that lose their competitive edge, Mancur Olson's *The Rise and Decline of Nations*,[14] is useful, though it is incomplete since it fails to focus on the interplay between economic and political organizations. In fact, Olson fails to model the polity at all. But his emphasis on the growth of redistributive coalitions as an economy evolves is a significant contribution. It needs to be put in the context of political economy not only to see how such coalitions will affect productivity, but also to explore the prospects of altering institutions to increase competitiveness.

Successful economic growth will reflect, as described in the brief historical sketches of the Netherlands and England, productive economic organizations shaped by an institutional matrix that provides incentives to increase productivity. Political organizations will have evolved to further the goals of the economic organizations, and a symbiotic relationship between the economic and political organizations will evolve. Whether this combination will produce Olson's redistributive coalitions that strangle productivity growth is much more complicated than Olson's explanations, which focus on the stability of the society or the lack of changing boundaries or the comprehensiveness of the interest group.

A useful starting point is the path-dependent pattern that emerges from this successful evolution of political/economic organization. Path dependence in such instances reflects a degree of mutual interdependence that has both formal ties and informal interdependence. The question is, I believe, the following: with changes in external competititive conditions or technology or the evolution of internal distributive coalitions which adversely affect the productivity of the economy, how flexible is the political organizational structure necessary to change the institutional framework to improve the competitive position of the economy? To the extent that the successful evolution has bound the key political organizations to the economic organizations that must bear the brunt of the changing institutional rules, the alteration is more difficult and will only occur when the costs of competitive failure reach such proportions that they weaken the influence of the economic organizations.

To the extent that political organizations have greater degrees of freedom from the affected economic organizations, institutional reform is easier. A multiplicity of economic organizations with diverse interests, for example, can lead to a polity with some degrees of freedom of action.

SECTION VI

Let me conclude with two unresolved problems that have been on the surface or just beneath the surface throughout; problems that are central to further progress in economic history/development. (1) Economic change is a process and it is that process which we must understand. The static nature of economic theory ill fits us to understand that process. We need to construct a theoretical framework that models economic change. (2) The belief systems of the players and the nature of human learning shape the evolving institutional policies that will be pursued. The rationality assumption of neoclassical economics assumes that the players know what is in their self-interest and act accordingly. Ten millenia of human economic history says that is a wildly erroneous assumption.

ACKNOWLEDGMENTS

I am indebted to Elisabeth Case for editing this essay.

REFERENCES

1. Arthur, W. B. "Competing Technologies, Increasing Returns, and Lock-In by Historical Events." *Econ. J.* **March** (1989): 116–131.
2. David, P. "Clio and the Economics of QWERTY." *Amer. Econ. Rev.* **75(5)** (1985): 332–337.
3. Dugan, A., and D. North. "Shared Mental Models: Ideologies and Institutions." *Kyklos* **1** (1994).
4. deRoover, 1945. Cited in *Structure and Change in Economic History*, by D. North. New York: W. W. Norton, 1981.
5. De Vries. *The Economic of Europe in an Age of Crises 1600–1750.* Cambridge, UK: Cambridge University Press, 1976.
6. Elton, G., 1953. Cited in *Structure and Change in Economic History*, by D. North. New York: W. W. Norton, 1981.
7. Greif, A. "Contact Enforcement and Economic Institutions in Early Trade: The Maghribi Traders Coalition." *Amer. Econ. Rev.* **83(6)** (1993): 525–548.
8. Hahn. *Scottish J. Pol. Econ.* (1987): 324
9. Milgrom, P., D. North, and B. Weingast. "The Role of Institutions in the Revival of Trade." *Econ. & Pol.* **2** (1990): 1–24.
10. North, D., ed. *Structure and Change in Economic History.* New York: W. W. Norton, 1981.
11. North, D. *Institutions, Institutional Change and Economic Performance.* Cambridge, UK, Cambridge University Press, 1990.
12. North, D. "Institutions." *J. Econ. Perspective* **Winter** (1991): 97–112.
13. North, D. "The Paradox of the West." In *The Origin of Modern Freedom in the West*, edited by R. Davis. (1995).
14. Olson, M. *The Rise and Decline of Nations.* New Haven, CT: Yale University Press, 1982.
15. Power, E., 1941. Cited in *Structure and Change in Economic History*, by D. North. New York: W. W. Norton, 1981.
16. Putnam, R. *Making Democracy Work.* Princeton, NJ: Princeton University Press, 1993.
17. Stubbs. *The Constitutional History of England*, Vol. 2, 599. Oxford: Clarendon, 1896.
18. Weingast, B. "The Economic Role Of Political Institutions: Market Preserving Federalism and Economic Growth." *J. Law, Econ. & Org.* (1995).

Paul Krugman
Department of Economics, Stanford University, Stanford, CA 94305

How the Economy Organizes Itself in Space: A Survey of the New Economic Geography

1. INTRODUCTION

In the crumpled terrain of Northern California, each valley has a different microclimate, suited for a different crop. Thus, Napa is renowned for its wine, Castroville for its artichokes, Gilroy for its garlic. Santa Clara County happens to be ideal for apricot trees, and its produce was once famous across the United States.

But no more. The orchards have been crowded out by office parks, shopping malls, and subdivisions, and the area once known as Paradise Valley is now known as Silicon Valley.

It is hard to argue that there is anything about the climate and soil of Santa Clara County that makes it a uniquely favorable location for the microelectronics and software industries. Rather, Silicon Valley is there because it is there: that is, the local concentration of high technology firms provides markets, suppliers, a pool of specialized skills, and technological spillovers that in turn sustain that concentration. And like most such agglomerations, Silicon Valley owes its existence to small historical accidents that, occurring at the right time, set in motion a cumulative process of self-reinforcing growth.

As the example of Silicon Valley makes clear, the economic analysis of location—the geography of the economy—is the subfield of economics to which the typical buzz words of complexity apply most obviously and dramatically. The spatial economy is, self-evidently, a self-organizing system characterized by path dependence; it is a domain in which the interaction of individual decisions produces unexpected emergent behavior at the aggregate level; its dynamic landscapes are typically rugged, and the evolution of the spatial economy typically involves "punctuated equilibria," in which gradual change in the driving variables leads to occasional discontinuous change in the resulting behavior. And in economic geography, as in many physical sciences, there are puzzling empirical regularities—like the startlingly simple law that describes city sizes—which seem to imply higher-level principles at work.

There is a long intellectual tradition in economic geography (or rather several separate but convergent traditions). The last few years have, however, seen a considerable acceleration of work, especially in the development of theoretical models of the emergence of spatial structure. This chapter is a brief, but I hope suggestive survey of this work, intended to convey a flavor of the exciting ideas being developed without going into technical detail.

Consisting of five parts, the chapter begins with two very simple motivating models that illustrate why economic geography so naturally leads to the typical "complexity" themes. Each of the next three parts then describes a particular major line of research. A final part discusses implications and directions for future work.

2. TWO ILLUSTRATIVE MODELS

Probably the best way to get a sense of the style and implications of the new economic geography is to begin by looking at a couple of illustrative models. For reasons that will be discussed later, research in the field has moved on from these very simple models to more elaborate stories that dot more of the *i*'s and cross more of the *t*'s. But the simplest models already convey much of the field's essential flavor.

THE EMERGENCE OF SPATIAL STRUCTURE: SCHELLING'S SEGREGATION MODEL

Many of the themes that are now associated with complexity theory—self-organization, multiple-equilibria, path dependence, and so on—can be found, very clearly expressed, in work done a generation ago by Harvard's Thomas Schelling. His ideas on these subjects were summed up in his 1978 book *Micromotives and Macrobehavior.*[34] Among the models he described there was a deceptively simple account (originally published in Schelling[33]) of how a city might become segregated.

In Schelling's segregation model, we imagine that there are two types of individuals; one may think of them as black and white, but they can be any two types of agents that have some potential trouble getting along—butcher shops and vegetarian restaurants, auto parts stores and boutiques. These agents occupy some though not all of the points on a grid; Schelling did "simulations" by hand using a chessboard.

Each individual cares about the types of his neighbors, defined as the occupants of the eight abutting squares of the chessboard. Specifically, he requires that some minimum fraction of these neighbors be the same type as himself. If they are not, he moves to an unoccupied square (if there is one) that does satisfy this requirement.

It is obvious that such a rule will lead to segregation if each agent demands that a majority of his neighbors be the same type as himself. What Schelling showed, however, was that substantially milder preferences, preferences that on the surface are compatible with an integrated structure, typically lead to a cumulative process of segregation.

Suppose, for example, that the "city" is a torus (so that there are no edge effects) and that people are satisfied as long as at least 34% of their neighbors are the same type as themselves.[1] Then a checkerboard layout, in which each individual has four neighbors of each type, is clearly an equilibrium. But such a structure tends to be unstable. In Figure 1, a 24 × 24 checkerboard equilibrium is shaken up: 35% of the individuals are removed (among other things to open up some space for movement), and 10% of the empty squares are then filled by agents of random type. The initial structure still looks quite integrated, and most people find their neighborhoods acceptable.

But now let the city evolve, with dissatisfied individuals moving in random order to random acceptable empty locations. It turns out that when one person moves, this move often makes some other agent unhappy, either because his departure tips the balance in his old neighborhood too far against his own type, or because his arrival in a new location tips the balance there too far against agents of the other type. Thus, there is a sort of chain reaction in which each move provokes other moves. The result ends up looking like Figure 2. Even though agents have only mild preferences against being too much in the minority, the city ends up highly segregated.

It is worth noting two additional aspects of this model. First is the way that short-range interactions produce large-scale structure. Individuals care only about the type of their immediate neighbors, yet the end result is a city divided into large segregated neighborhoods. The reason, of course, is those chain reactions: my move may only directly affect my immediate neighbors, but it may induce them to move, which then affects their other neighbors, and so on. The idea that local interactions produce global structure has been regarded as a startling insight when applied to

[1]This means that if they have 1 or 2 neighbors, 1 must be the same type; if they have 3–5 neighbors, 2 must be the same; and if they have 6–8, 3 must be the same.

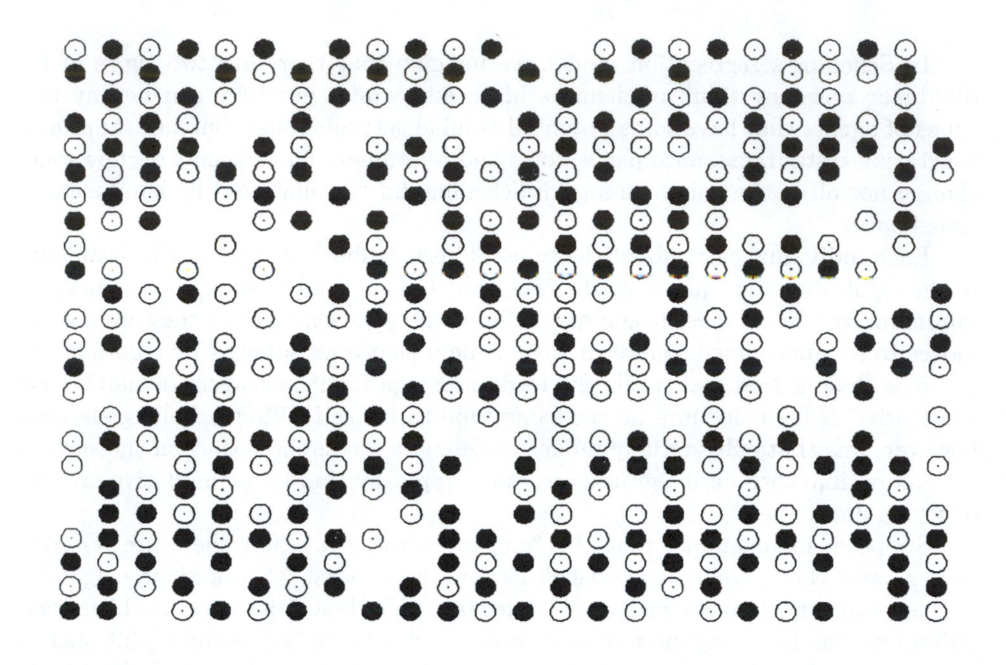

FIGURE 1 Schelling's model: An initial spatial pattern.

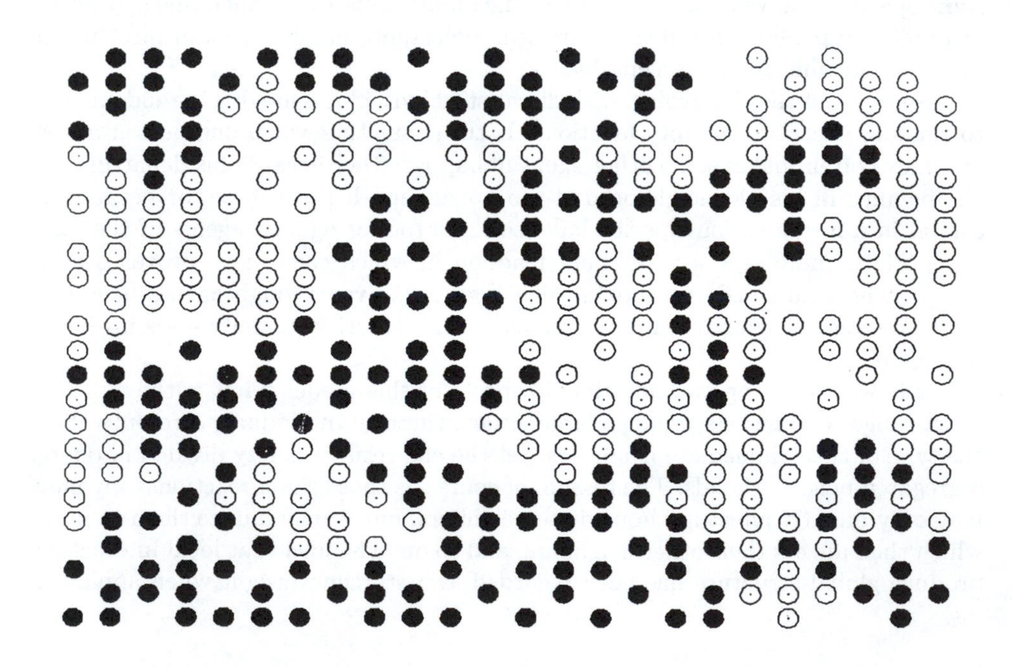

FIGURE 2 The end result: Spontaneous appearance of order.

physical systems (Prigogine and Stengers[32]); here we see it derived, in a simple and unpretentious way, in a social context.

Also note that while large-scale structure always emerges in this model, the details of that structure are very dependent on the initial conditions. The particular "geography" of the equilibrium in Figure 2 is completely dependent on subtle details both of the initial conditions shown in Figure 1 and on the sequence of moves that followed.

CRITICAL MASS AND DISCONTINUOUS CHANGE: A BASE-MULTIPLIER MODEL

Regional economists frequently make use of what is known as the "base-multiplier" model. One thinks of a region as having two kinds of economic activities. First are the activities that constitute its "export base"—activities that produce goods and services which are sold to customers elsewhere in the world (including other regions in the same country). Second are the "nonbase" activities that provide goods and services to local residents and businesses. The size of the export base is assumed to depend either on outside factors, like world demand, or on slowly changing regional advantages. The size of the nonbase activities, by contrast, depends on the size of the local economy. Thus, the Los Angeles area's base income derives from such products as sitcoms and military hardware; its nonbase activities include personal trainers, chiropractors, auto repair shops. Typically, regional scientists assume that some fixed fraction of the income generated in the region is spent on locally produced goods and services.

The base-multiplier approach in its simplest version leads to a proportional relationship between the export income of a region and its total income. Let α be the fraction of income spent locally, X the income earned in the export sector, and Y the region's total income. Then the income earned in the export sector generates local spending of αX; a fraction α of this derived income is in turn spent locally, giving rise to a second round of local spending $\alpha^2 X$; this gives rise to a third round, and so on. So we have

$$Y = X + aX + a^2X + \ldots = \frac{X}{1-a}. \tag{1}$$

But now introduce economies of scale into the story. Suppose that the range of goods and services produced locally depends on the size of the local market, because it is not worth producing some things locally unless the market is large enough. Then α will be an increasing function of Y—and that, as regional scientists such as Pred[31] have pointed out, makes the story much more interesting.

Consider the following specific example: if Y is less than \$1 billion, none of the region's income is spent locally. If it is at least \$1 billion but less than \$2 billion, 10% is spent locally. If it is at least \$2 billion but less than \$3 billion, 20% is spent locally; and so on up to 80%, which we assume is the maximum. What is the relationship between export and total income?

Well, suppose that export income is $1 billion. Then we might well guess that 10% of income will end up being spent locally, implying a total income of $1/(1-0.1)$ = $1.11 billion. If export income is $2 billion, we might guess at an α of 0.2, implying total income of $2/(1-0.2)$ = $2.5 billion. In each case our guess seems to be confirmed.

But suppose that export income is $3 billion. We might guess at $\alpha = 0.3$; but this would imply total income of $3/(1-0.3)$ = $4.28 billion, which means that our guess was too low. So we try $\alpha = 0.4$, but this implies total income of $5 billion—our guess is still too low. And so on up the line. It turns out that when export income rises from $2 billion to $3 billion, total income rises from $2.5 to $15 billion! In other words, when the export base reaches a certain critical mass, it sets in motion a cumulative process of growth, in which an expanding local economy attracts a wider range of import-substituting activities, which causes the local economy to expand still further, and so on.[2]

We can sharpen our understanding of this example by making a a continuous function of Y, and by formalizing the dynamics. Let the share of income spent locally in period t be proportional to the size of the regional economy in the previous period, up to 80%:

$$a_t = \min[0.8, 0.1Y_{t-1}]\,. \tag{2}$$

We can then define an equilibrium level of Y, given X as a level of Y, such that $Y_{t+1} = Y_t$. The relationship between the export base and equilibrium income is shown in Figure 3, with solid lines representing stable and broken lines unstable equilibria. There are three branches to this relationship. As long as $X < 2.5$, there is a stable equilibrium with

$$Y = \frac{1 - \sqrt{1 - .4X}}{0.2}\,. \tag{3}$$

As long as $X > 1.6$, there is another stable equilibrium with

$$Y = 5X\,. \tag{4}$$

And for $1.6 < X < 2.5$, there is an unstable equilibrium with

$$Y = \frac{1 + \sqrt{1 - .4X}}{0.2}\,. \tag{5}$$

[2] Joel Garreau, whose book *Edge City*[7] is an entertaining account of the growth of new urban subcenters, tells us that "Five million square feet is a point of spontaneous combustion. It turns out to be exactly enough to support the building of a luxury hotel. It causes secondary explosions; businesses begin to flock to the location to serve the businesses already there."

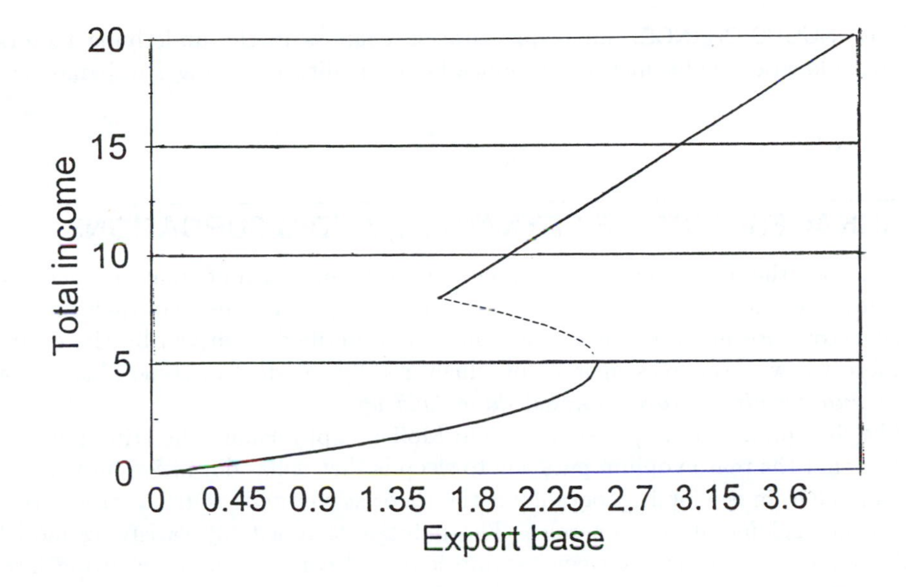

FIGURE 3 The base-multiplier model: Increasing returns create critical-mass behavior.

The critical mass aspect of the model is immediately apparent. In fact, there are two critical masses, depending on whether the regional economy is growing or shrinking. Suppose that we imagine a gradual increase in the size of the export base, starting at a low level. Then the local economy will expand gradually along the lower branch in Figure 3; but when X reaches \$2.5 billion, there will be an explosion of growth as the economy moves up to the upper branch. On the other hand, suppose that we imagine a declining economy which starts very large; then local income will decline slowly until X falls below 1.6, at which point there will be an *implosion* as falling income leads to a self-reinforcing contraction.

LESSONS FROM SIMPLE MODELS

Even these simple models suggest many of the typical features of economic geography. A far from exclusive list would surely include the following three points:

SELF-ORGANIZATION: Atomistic interactions among individual agents can spontaneously generate large-scale order;

PATH DEPENDENCE: Small differences in initial conditions can have large effects on long-run outcomes;

DISCONTINUOUS CHANGE: Small quantitative changes in the underlying factors that drive location will sometimes produce large, qualitative changes in behavior.

3. LINKAGES, COMPLEX DYNAMICS, AND BIFURCATIONS

In some ways the models described in section 2 tell you much of what you need to know about economic geography. Nonetheless, in important respects they remain unsatisfactory. Recent work in the field may be described as driven by the desire to adhere to two principles, neither of which is fully satisfied in these illustrative models: *no dormitive properties*, and *things add up*.

The first phrase comes, of course, from Moliere's physician, who triumphantly explains that the reason opium puts you to sleep is that it has dormitive properties. The equivalent in economic geography is the theorist who tells you that cities grow because of agglomeration economies. The point is that a truly satisfying model should put more distance between assumptions and conclusions, by showing how agglomeration economies emerge from the behavior of individual agents—to expand on Schelling's title, it should show how complex macrobehavior emerges from simple micromotives.

The phrase about things adding up is more often expressed by economists as a concern for "general equilibrium": an economic model should be a consistent account of how all the pieces fit together. The base-multiplier model does not quite meet this criterion. How would it work if it was meant to model, not an individual region, but the whole world? Then X would clearly be zero—all of the world's sales are to itself—and α would be 1. So world income would be $0/0$. Clearly, we have a problem.

This section describes a newer type of model that satisfies these two rules, and that also gives rise to some deeper insights.

LINKAGES AND AGGLOMERATION

In a 1991 paper,[17] I offered a general equilibrium model of economic geography, i.e., one in which things added up, designed to produce emergent spatial structure without any assumed agglomeration economies. In this model, there are two inputs into production: "farmers," assumed to be fixed in place in two or more locations, and "workers," who move over time to locations that offer them higher real wages. Farmers produce a homogeneous agricultural good with constant returns to scale. Workers produce many differentiated manufactured goods, with increasing returns in the production of each good. These manufactured goods are costly to transport. Special assumptions about technology and the structure of demand ensure, however, that despite these transport costs each manufactured good will be produced in only one location by a single, monopolistically competitive firm.

Even though no agglomeration economies are assumed, they nonetheless emerge from the logic of the model, in the form of "linkages." Because of transport costs, firms are willing to pay higher wages in locations that have good access to large markets—but markets will be large precisely where many firms chose to locate. And workers tend to prefer (and to be willing to move to) locations that offer good access to a wide range of manufactured goods—but the supply of manufactures will be large precisely where many firms, and hence, workers are located. These two forces, corresponding to the backward and forward linkages of development economics (see Hirschman[13]) create a circular logic that could lead an economy with no intrinsic differences among locations nonetheless to evolve a spatial structure in which manufacturing is concentrated in only a few of these locations.

While the model thus implies the emergence of agglomeration economies, however, it also—deliberately—implies an offsetting force, a "centrifugal force" opposing the "centripetal" force of linkages. The immobility of the farmers means that if manufacturing becomes too concentrated, it may become profitable for an individual firm to move away from the others so as to claim a chunk of the rural market all for itself. This tension between centripetal and centrifugal forces makes the model nonlinear in interesting ways, producing some striking dynamic stories.

COMPLEX DYNAMICS

The implications of the model appear most clearly if farmers are assumed equally divided among two or more equidistant locations—that is, if there are no inherent differences among locations. In that case, the dynamics are determined by the number of locations and by three parameters: transport costs, the share of manufactures in spending, and a parameter that measures the substitutability of different manufactured goods for one another (and which turns out to determine both the degree of monopoly power and the importance of economies of scale).

For some values of these parameters, manufacturing always concentrates in a single location; for others, no concentrations form at all. When the parameters are in an intermediate range, however, the model implies the existence of what Kauffman[15] has called a "rugged landscape": a dynamic system in which there are many basins of attraction, so that different initial conditions lead to different outcomes.

Figure 4 illustrates this proposition for the case of a three-location model. The distribution of manufacturing among the three locations at any point in time can be represented by a point on the unit simplex, with each corner representing concentration in one of the locations. The arrows indicate the direction and speed of "flow" on that abstract "landscape" for a particular set of parameters (as discussed below, the qualitative as well as quantitative nature of the dynamic landscape depend on the parameters). Clearly, in this case the economy has four basins of attraction: a central basin that "drains" to a point of equal division of manufacturing among regions, and three basins leading to concentration in each of the regions.

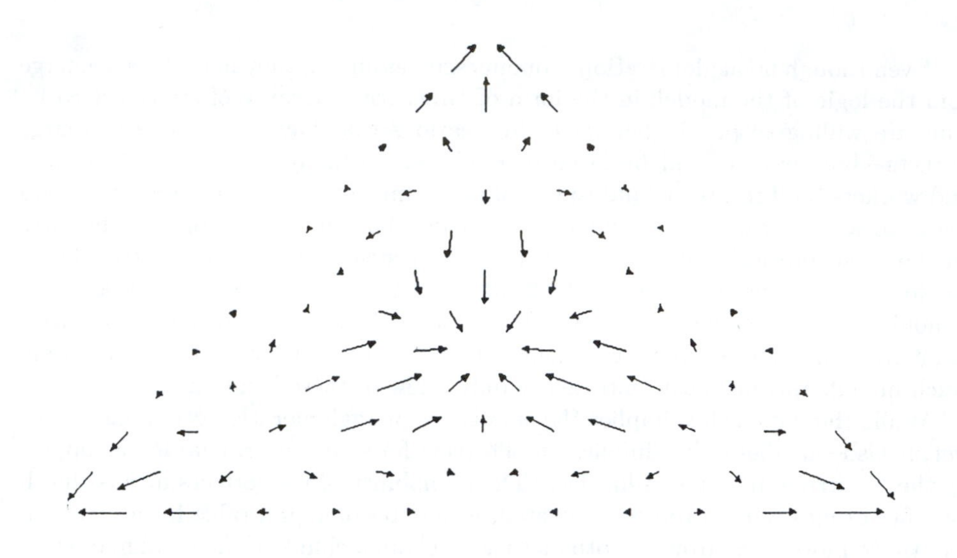

FIGURE 4 Dynamics in a three-region model (flows on the unit simplex).

The ruggedness of the dynamic landscape increases with the number of locations. Experiments with an example in which there are 12 locations in a circle, described in Krugman,[19] indicate the existence of 22 basins of attraction: 6 with two concentrations exactly opposite each other, 12 with two concentrations 5 apart, and 4 with three equidistant concentrations. The interesting properties of this "racetrack" model are described in section 4, below.

BIFURCATIONS

Quantitative changes in the parameters of the model can produce qualitative changes in the dynamic landscape, as basins of attraction appear or disappear.

A case in point is what happens as we vary transport costs. In this model low (though nonzero) transport costs are favorable to agglomeration. (This is somewhat sensitive to the specifics; in other, related models this can go the other way.) At high levels of transport costs, there is a unique equilibrium in which manufacturing is equally divided between the two regions. At low levels of transport costs, there are two stable equilibria: manufacturing must be concentrated in only one region, but it can be either one. And there is an intermediate range in which the possibilities of concentrated and unconcentrated equilibria coexist.

This story, incidentally, is strongly suggestive of historical episodes in which improvements in transportation technology seem to have led to the differentiation of economies into more and less industrialized regions, such as the locking in of the U.S. manufacturing belt around the time of the Civil War or the separation of Italy into industrial north and Mezzogiorno later in the nineteenth century.

It is typical of the new models of economic geography that they exhibit abrupt, qualitative changes in outcomes when forcing variables rise above or below some critical level. For example, Venables[37] and Krugman and Venables[26] develop a model of international trade in which falling transport costs lead to a spontaneous division of the world into a high-wage "core" and a low-wage "periphery," and then, as transport costs continue to fall, to a subsequent phase in which wages converge again. Krugman and Livas[25] develop a model inspired by the case of Mexico City, in which the tendency of a country to develop an internal core-periphery pattern depends on its trade policy: if it lowers its tariff rate, access to domestic markets and suppliers becomes less important, and when the tariff falls below a critical level the core-periphery pattern collapses.

4. THE RACETRACK MODEL

One can learn a lot by studying models of economic geography in which there are only two or three locations. Nonetheless, we would also like to know how much of the intuition from such small-scale models remains when there are many regions. But describing the dynamic behavior of arbitrary many-region models quickly becomes an uninteresting exercise in taxonomy. In order to get some insights from multiple-region models, it is necessary to impose some kind of regularity on the economy's "geometry."

One particular geometry that is useful in spite of its artificiality is what we might call the "racetrack economy": a large number of regions located symmetrically around a circle, with transportation possible only around the circumference of that circle. This setup has two useful properties. First, the economy is one-dimensional, which greatly simplifies both algebra and computations. Second, since there are no edges, and hence, no center, it is a convenient way to retain the feature that all sites are identical—which means that any spatial structure that emerges represents pure self-organization.

THE PUZZLE: UNEXPECTED REGULARITY

It is straightforward to simulate the evolution of an economy with the same economic structure as that described in section 3, but with many regions located around a circle. In Krugman,[19] I used simulation analysis to explore a racetrack economy with 12 locations. (I chose 12 because it is a fairly small number with a large number of divisors.) The basic strategy was simple: for any given set of parameters, repeatedly choose a random initial point on the 11-dimensional simplex that describes the allocation of manufacturing, and let the economy evolve to a steady state.

My original expectation was that the model would exhibit a wide range of behavior for any given parameters, and that it would be necessary to do a statistical analysis of the average relationship between the parameters and the resulting structure. And it was indeed true that for some parameter ranges there were multiple possible outcomes. For one set of parameters, for example, the model sometimes led to a steady state with two concentrations 6 apart; sometimes to a steady state with two concentrations 5 apart; and on rare occasions to three concentrations, equally spaced around the circle. As pointed out in section 3, this meant that the relevant dynamic landscape contained 22 basins of attraction.

And yet looked at another way, the results were surprisingly regular. For those parameters, the model seemed consistently to produce two or rarely three agglomerations, almost, if not exactly, evenly spaced around the circle. What was the source of this surprising regularity?

A useful clue was the fact that if one started, not from a random point on the simplex, but from a nearly uniform spatial distribution of manufacturing (i.e., a point near the center of that simplex), the results were even more regular. Indeed, when the initial conditions were sufficiently smooth, those same parameter values consistently produced the same result, time after time: two agglomerations, exactly opposite one another. The "'59 Cadillac" diagram in Figure 5 shows a frozen portrait of the full history of a typical run. (Bear in mind, when looking at Figure 5, that location 12 is next to location 1.)

TURING ANALYSIS

Considerable insight into the reasons why simulations on the racetrack model yield such regularity, and more generally into the nature of self-organization in spatial economies, can be gained by applying an analytical approach originally suggested by none other than Alan Turing.[36] Turing's concern was with morphogenesis, the process by which cells in a growing embryo become differentiated; but once one thinks about it, the parallel between this question and the question of how different locations in a city or country become differentiated is not at all an unnatural one.

Turing analysis involves focusing on the very early stages of spatial self-organization, in which the object being studied—embryo or economy—is diverging away from an almost but not quite uniform spatial distribution. Applied to the racetrack economy, it involves the following steps. First, linearize the model around a uniform spatial distribution of manufacturing. Second, represent the actual deviation of the distribution of manufacturing from uniformity as a Fourier series. That is, represent it as the sum of one sinusoidal fluctuation whose wavelength is equal to the circumference of the racetrack; a second whose wavelength is half the circumference; a third whose wavelength is a third the circumference; and so on.

The key result, which is surprisingly independent of the details of the model, is that each of these components of the Fourier series is an *eigenvector* of the linearized

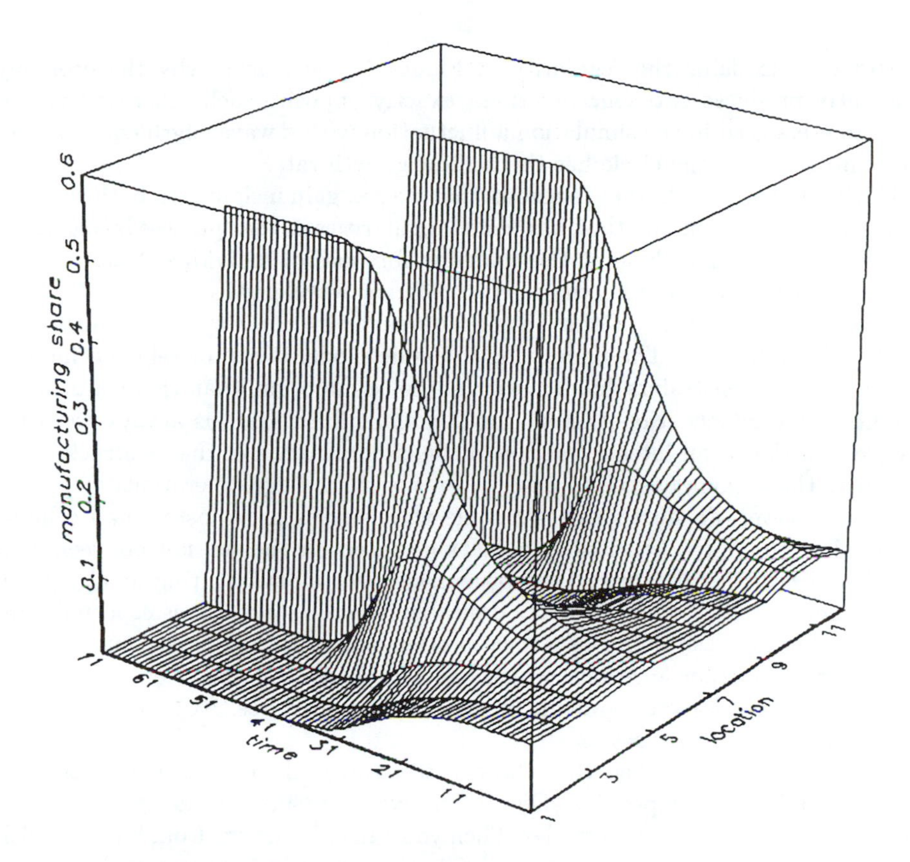

FIGURE 5 Emergence of spatial structure in a "racetrack" economy.

dynamic system, with an associated eigenvalue that depends only on its wavelength. That is, the evolution of the spatial economy near a uniform distribution of manufacturing can be, as it were, decomposed into the parallel evolution of a number of separate economies, in each of which the spatial distribution of manufacturing is a perfect sine wave, which grows (or shrinks) at a constant rate.

Now suppose that we begin with an almost flat distribution of manufacturing. If all eigenvalues are negative, this distribution will become only flatter over time: the economy will collapse toward complete uniformity. But, if some of the eigenvalues are positive, the associated components of the Fourier series will grow, creating an increasingly uneven spatial distribution—that is, the economy will experience self-organization.

But there is more. If the distribution starts off very flat, by the time it becomes noticeably uneven it will be dominated by whatever component has grown most rapidly. That is, the spatial distribution will be dominated by the fluctuation whose wavelength has the highest associated eigenvalue.

And that explains the regularity in Figure 5: the reason why the economy consistently produces two concentrations, exactly opposite each other, is that for the parameters used in the simulation a fluctuation with a wavelength equal to half the circumference of the circle has the highest growth rate.

The Turing approach, then, suggests that we can gain insight into both the process of self-organization and the unexpected regularities in that process by changing our focus: instead of looking at the changing fortunes of individual locations, we look at the growth rates of fluctuations at different wavelengths.

An interesting aspect of the Turing approach, by the way, is its surprising "quantum" implication. The underlying dynamic model is completely continuous, with its integer constraints hidden within the assumption that there are many differentiated products. Nonetheless, the dominant fluctuation is always one with a wavelength that is an integer fraction of the circumference of the racetrack—and this is why the economy tends to produce evenly spaced agglomerations.

Why, however, do fluctuations at some wavelengths grow faster than others? It is possible to offer a heuristic explanation, based on the tension between centripetal and centrifugal forces. The concentration of manufacturing at any given location increases if it is an especially desirable location, falls if it is especially undesirable. But the desirability of a location depends only on the concentration of manufacturing at other locations, which has a dual role: on one side, manufacturing concentrations elsewhere provide linkages, which are a positive spillover; on the other, they compete for markets.

The crucial point is then that the effective range of linkages is shorter than that for the adverse, competitive effect.[3] For example, suppose that goods can be shipped only a maximum of 200 miles. Then you can only benefit from linkages with another manufacturer if she is less than 200 miles away, but she can still compete for some of your customers if she is as much as 400 miles away. Thus we can think of a concentration of manufacturing as exerting positive spillovers out to some range, and negative spillovers out to some larger range.

To see why this gives rise to a "preferred wavelength" for fluctuations, imagine an economy in which the deviation of manufacturing density from complete uniformity is a perfect sine wave—as each of the "shadow" economies into which the racetrack can be decomposed must be—and ask whether a peak of this sine wave is an especially good place to locate.

Consider first a fluctuation with a very short wavelength. In this case, both above- and below-average concentrations of manufacturing in the vicinity of each peak will alternate rapidly through the ranges of both positive and negative spillovers. The net effect will, therefore, be very small: the peak will be neither especially desirable nor especially undesirable, and the fluctuation will have no strong tendency to grow.

[3] Actually, in the model there are no fixed ranges; rather, both positive and negative spillovers fade out gradually with distance. It is, however, helpful for heuristic purposes to talk as if goods can be shipped a certain distance but no further.

Alternatively, consider a very-long-wavelength fluctuation, which implies above-average concentrations throughout the ranges of both positive and negative spillovers relevant to each peak. If the positive spillovers are not too strong (a condition that has a simple algebraic representation in the model), the net effect will be negative: sufficiently long wavelength fluctuations will tend to die out over time.

Finally, imagine a fluctuation of an intermediate wavelength, chosen so as to have above-average concentrations in the range of positive spillovers, which tends to make the peaks attractive, and below-average concentrations in the range of negative spillovers, which also makes the peaks attractive. Such a fluctuation will definitely tend to grow over time.

Therefore, we can expect that the relationship between wavelength and growth rate will be an inverted U: short wavelengths will not grow, long wavelengths will actually shrink, and there will be some intermediate "preferred" wavelength that grows most rapidly.[4]

ECONOMIC "TEMPERATURE" AND CHANGES OF STATE

An interesting extension of the racetrack model comes if we introduce some "temperature"—that is, some randomness in behavior. Suppose that while firms normally move toward more profitable locations, they sometimes move in the wrong direction.

Krugman[23] shows that there is then a critical value of this temperature, at which the qualitative behavior of the economy changes. Noise tends to break down structures: because there are more firms in locations with high concentrations, random movement tends on net to be out of these locations. Indeed, in the algebra the effect of random movement appears as a "diffusion" term. Only if this randomness falls below some critical level does spatial self-organization take place.

5. URBAN HIERARCHIES

The racetrack models just described, although they provide interesting insights into the evolution of spatial structure, are in some other respects still unsatisfactory.

In the first place, the "history" that is assumed in the Turing approach, while reasonable for an embryo, is less so for an urban system. Economies do not start with an almost uniform spatial pattern, then gradually evolve cities; they start with

[4] What about the "quantum" aspect of the model? Well, it complicates matters. The algebra for a finite-sized circle is actually fairly nasty. But as Turing noted, it becomes much nicer if one takes the limit as the circle becomes arbitrarily large.

an existing set of cities, then add new ones as their population grows or as new areas are opened up.

Also, the basic racetrack models produce agglomerations that are all about the same size; the reality of the urban system is that there are cities of many sizes (indeed, there is in a very precise sense no such thing as a typical city size, as explained below).

In recent work, Mashisa Fujita and colleagues have developed an approach that offers both a somewhat more reasonable fictitious history and a way to model the formation of an urban hierarchy.

POPULATION GROWTH AND CITY FORMATION

In Fujita-type models, as in the racetrack models, the economy is one-dimensional; but instead of a closed circle, it consists of an infinitely long line. The distinction between workers and farmers is dropped; homogeneous labor can produce either manufactures or food. But agriculture requires land, so the farm population must be spread along the line.

As in the models discussed before, there are many manufactured goods produced with increasing returns to scale. In the simplest case, these goods all have the same economies of scale and transport costs; in more complex cases there are a number of different types of manufactured goods.

The initial condition for the model is a locational pattern in which all manufacturing is concentrated in a single point, with agriculture spread on either side. Provided that the population is sufficiently small, such an economic geography is a self-sustaining equilibrium. The linkages that manufacturing firms provide to each other lock the city in place, and farmers group themselves around the manufacturing center. (The equilibrium involves a land rent gradient that drops off from the city, reaching zero at the outer limit of cultivation.)

Now suppose that the population gradually increases. Initially this will simply have the effect of pushing out the agricultural frontier. Eventually, however, the most distant farmers are so far from the existing city that it becomes profitable for some manufacturing plants to move out to new locations part of the way from the center to the frontier. When some firms begin to do this, they immediately generate a cumulative process of city growth in the new locations; and these new urban locations are then locked in.

As the population continues to grow, the agricultural frontier continues to move out; and eventually the process repeats itself, with a second pair of new cities emerging, and so on.

While still obviously a very stylized story, this is clearly a better description of the process of city formation than the Turing approach. Unfortunately, it turns out that the algebra of this approach is far more complex; analytical results are very hard to derive, and even simulation analysis is fairly difficult. Thus the Turing and Fujita approaches are complements: one is analytically elegant and helps us understand the nature of self-organization, the other offers a more satisfactory account of the way self-organization happens in practice.

The real payoff to the Fujita approach comes, however, when one allows for several industries that have different characteristics.

TYPES OF CITIES

Suppose now that there are several industries, some of which have larger scale economies or lower transportation costs than others. Initially, as before, we suppose that all manufacturing is concentrated in a single center. How does the economy evolve?

Since the seminal work of Christaller,[2] geographers have argued that one gets a hierarchy of nested central places, with small cities supplying goods with high transport costs and/or low scale economies to local markets, and progressively higher-level cities supplying goods with low transport costs or large scale economies to larger areas, including the smaller cities. (The higher-level cities are bigger because they do lower-level activities too.) While the idea of such a hierarchy seems compelling, however, until recently it has never been shown how an urban hierarchy could arise from the self-interested behavior of independent firms.

Recently, however, Fujita and colleagues[5] have managed to produce simulation exercises in which the growth of population leads via a process of self-organization to the emergence of an urban hierarchy. The logic is fairly straightforward, although the algebra is quite difficult. As in the one-industry case, as population grows the agricultural frontier shifts out, and new cities emerge at some point between the existing cities and the frontier. But when these frontier cities emerge, they initially contain only industries with low scale economies and/or high transportation costs. Only when the frontier has shifted still further out, typically giving rise to several small cities in the process, does it become profitable for industries with larger scale economies or higher transport costs to start a new location. When they do, the logic of linkages usually causes them to move to an existing smaller city—not the city closest to the frontier, but a somewhat older center—which then experiences a sudden jump in population and becomes a higher-level center. This process may be repeated at many levels, producing a hierarchical urban structure.

POWER LAWS

While the Fujita approach explains the emergence of an urban hierarchy, it does not (at least as far as we currently know) explain the striking empirical regularity that governs city sizes in the United States. The distribution of sizes of metropolitan areas[5] in the U.S. is startlingly well described by the rule

$$N = kS^{-\alpha} \tag{6}$$

where N is the number of cities with populations greater than or equal to S, and α appears to be very close to one.

Figure 6 plots the log of each metropolitan area's population against the log of its rank (i.e., New York is 1, Los Angeles 2, and so on) for the 130 largest such areas in the United States in 1991. The relationship is impressively linear, and very close to a 45-degree line—the regression coefficient is −1.004. Or to put it another way, if you multiply the rank of each city by its population, you get a very nearly constant number.[6] Moreover, this relationship appears to have held, at least fairly

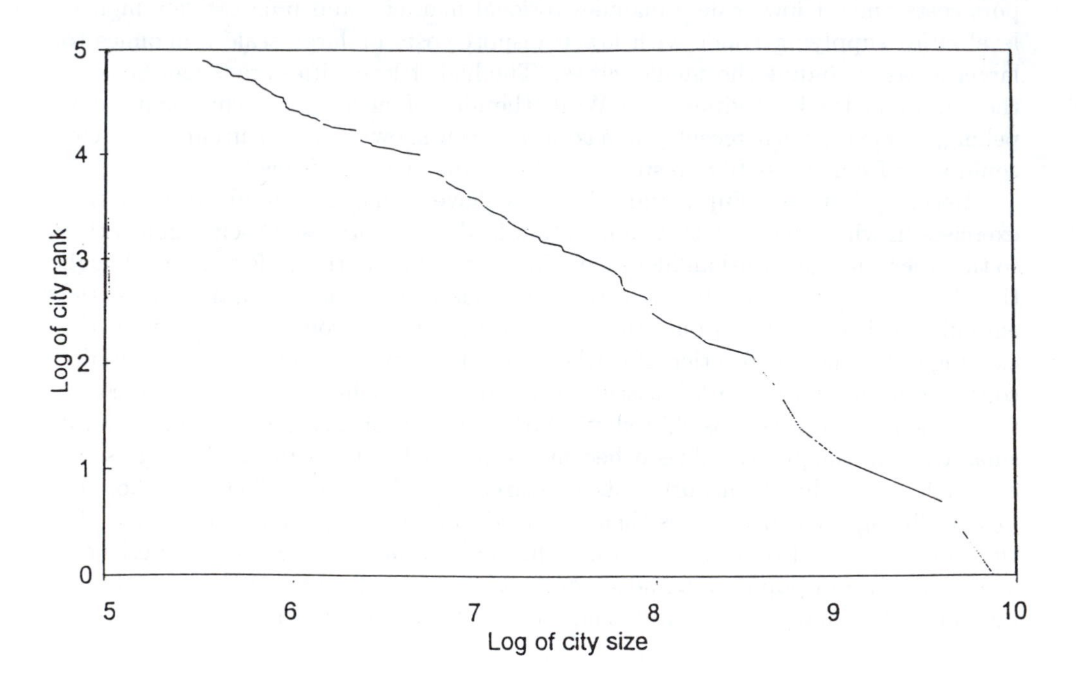

FIGURE 6 The size distribution of U.S. metropolitan areas.

[5] It is important to look at data for metropolitan areas, rather than cities as political units. San Francisco proper has fewer people than Houston.

[6] This rule may not be that obvious when one looks only at the very biggest cities. When you get a bit down the ranking, however, the exactness of the relationship is startling. The 10th largest metropolitan area in the United States is Houston, population 3.85 million; the 100th is Spokane, population 370,000.

well, for the past century. International comparisons are made difficult by data problems, but seem to suggest that the relationship (Eq. (6)) is a good description of the size distribution of cities in many countries.

Why might such a strikingly simple law describe the sizes of cities, which surely arise from a highly complex process? Herbert Simon proposed one answer 40 years ago.[35] He suggested that the regularity is the result of random city growth. Simon postulated a process in which a growing population arrives in discrete "lumps," which might correspond to new industrial agglomerations. With some probability p these lumps form the nuclei of new cities. Otherwise, they attach themselves to existing cities, with the probability that each lump attaches itself to any given city being proportional to that city's population. It is then possible to show that if this process goes on long enough, the upper tail of the city size-distribution will indeed obey the power law (Eq. (6)), with $\alpha = 1/(1-p)$.[7] If we regard the Simon process as a reasonable description of urban growth, then, the fact that α is close to one simply tells us that new city formation is a rare event.

While Simon's analysis is appealing, however—and the rank-size rule for cities demands an explanation—this stochastic growth story raises serious questions. In essence, Simon's model asserts that city size is irrelevant: the expected growth rate of a city is the same, no matter how large or small it is. Or to put it differently, a city is a collection of "lumps" that just happen to be in the same place. This is a nihilistic story, which seems to be inconsistent both with other models in economic geography and with observations, both casual and statistical, that cities of different sizes have very different economic bases. At the moment, the conflict between stochastic explanations of the rank-size rule, which seem to require constant returns to city size, and the otherwise compelling idea that cities are all about increasing returns, is unresolved.

6. FUTURE DIRECTIONS OF RESEARCH

The models described in this survey are obviously extremely stylized; they make suggestive points rather than providing a realistic picture of the actual evolution of the economy's spatial structure. Thus, there is a virtually unlimited agenda for further work. At the moment, however, three directions seem particularly compelling.

One is a seemingly trivial extension that turns out to be remarkably hard as an analytical issue: to go from the one-dimensional geometry of the "racetrack" world to two dimensions. One might think that this would be easily done with simulations, but the computational complexity of such simulations rises with astonishing speed in the two-dimensional case. (To have any reasonable chance of seeing interesting

[7] Simon's derivation of this result was quite difficult, which may have contributed to its failure to attract much attention. It turns out, however, that the result can be derived much more simply.[24]

patterns on a two-dimensional surface, one needs several hundred locations, instead of the 12 locations that worked well enough in the racetrack model. At the same time, computational time for a simulation tends to rise as roughly the square of the number of locations.) Traditional location theory (Lösch[29]) holds that central places should form a hexagonal lattice, but so far this result has not been demonstrated either analytically or in simulations. Gaspar[8] has achieved some preliminary results using a "city block" geometry, in which movement can take place only north-south or east-west; this yields a diamond-shaped rather than hexagonal lattice of central places.

A second issue concerns the role of transportation routes in influencing urban location. Obviously in reality cities are often ports, or located at the intersections of rivers; also, it is clear that manmade transportation networks are a powerful component of the self-organization of economic space. But how should this be modeled? An initial crude attempt by Krugman[21] has been followed by important new work by Fujita and Mori.[6]

Finally, although there has been substantial empirical work on metropolitan areas in recent years (e.g., Ades and Glaeser[1]), it has been at best loosely tied to the new theoretical frameworks. A closer link between theory and testing is, therefore, an important priority.

I hope that this survey has given a sense of the excitement those of us working in the new economic geography feel. This is an area in which the characteristic ideas of complexity theory lead to models that are both relevant and remarkably elegant.

APPENDIX: INTELLECTUAL TRADITIONS RELEVANT TO ECONOMIC GEOGRAPHY

An unfortunate feature of much of the more popular writing about applications of complexity and nonlinear dynamics to economics has been a tendency to claim that every insight is completely novel, and to ignore previous research that bears on the subject. In the case of economic geography, all of the typical themes of the new complexity approach have been sounded by a number of researchers over the past several generations. That is not to say that there is nothing new to be said—obviously I must believe that there is, or I would not be doing new research in this field. The point, however, is that it is important to be aware of what a number of smart and observant people have had to say about the subject in the past.

There are at least five intellectual traditions that bear strongly on the new economic geography.

CENTRAL PLACE THEORY: There is a long, largely German tradition of studying the economics of location. Much of the earliest work in this tradition concerned the problem of optimal location for a single firm given exogenously located markets and sources of materials, or the use of land surrounding a city whose existence was simply assumed. Beginning with the work of Christaller[2] and Lösch,[29] however, location theorists began to focus on the reasons why spatial structure would emerge even on a landscape on which all sites are alike.

The main limitation of central place theory is that it is descriptive rather than being a fully specified model. That is, neither Christaller and Lösch nor their followers were able to show that central place systems would actually emerge from the interacting actions of individual agents in the economy.

REGIONAL SCIENCE: During the 1950s an American version of location theory arose, led in particular by Walter Isard.[14] Regional science differed from the earlier tradition of location theory in part because of its emphasis on operational models. It also tended to place less emphasis on the geometry of equilibrium and more on the dynamics of growth.

From the point of view of those of us trying to develop a new economic geography, the most interesting theoretical contributions of regional science involved its insights into regional growth. Above all, seminal contributions by Harris[9] and Pred[31] showed, albeit in a less than rigorous fashion, how in the presence of economies of scale and transportation costs the independent location decisions of individual firms can produce emergent agglomeration economies.

URBAN SYSTEMS THEORY: There is a strong tradition within urban economics that models the determination of the sizes of cities as the outcome of a tension between the forces of agglomeration and the disadvantages of large scale—albeit usually in a framework that abstracts from space. The key works, which combine elegant theory with considerable empirical work, are those of Henderson. [11,12]

INTERNATIONAL TRADE THEORY: The role of increasing returns as a possible cause of international specialization has been known to international trade theorists for a long time. Formal models of this role go back at least to a well-known paper by Matthews.[30] Beginning in the late 1970s, theorists began to combine an increasing-returns explanation of trade with imperfectly competitive models of markets; the so-called "new trade theory" is generally considered to have begun with the simultaneous and independent papers of Krugman,[16] Dixit and Norman,[4] and Lancaster.[28] The canonical survey of this work is Helpman and Krugman.[10]

It is a fairly natural move from international trade into economic geography, especially once the ideas of increasing returns and multiple equilibria have become second nature, as they had to most international economists by the mid 1980s. In particular, the modeling techniques used to cut through the potential messiness of increasing-returns models of trade are also highly applicable to the analysis of spatial models.

GAMES OF COMPLEMENTARITY: A final tradition that feeds into the new economic geography is that of models from game theory in which the optimal strategies of players are complementary, in the sense that the actions of players can be set up in such a way that an individual's incentive to act is increasing in his expectation of another's action and vice versa. The classic discussion is that of Schelling[34]; an influential later discussion is that of Cooper and Johns.[3]

Given that there have been a number of highly suggestive discussions both of economic geography and of related subjects in the past, what is new about the current work?

The most important differences probably involve technique. In part this is a matter of advances in theory. The 1970s were marked by the development of analytically tractable models of imperfect competition, a field that had until then been viewed as impenetrable terrain for anything other than loose literary theorizing. The availability of these models played a decisive role in the development of the "new trade" theory in the late 1970s and the "new growth" theory in the mid 1980s.

It is also true that, as in so many other fields, the availability of substantial computing power on every desktop has been a factor. Simulation and, in particular, computer graphics certainly make it much easier to explore the possibilities of theory in economic geography.

Finally, the new economic geography has been informed to a greater extent than the older work by an awareness of the parallels with work outside of economics. These parallels are obvious in this chapter; they would not have been nearly as obvious in a survey carried out a decade ago.

REFERENCES

1. Ades, A., and E. Glaeser. "Trade and Circuses: Explaining Urban Giants." *Quart. J. Econ.* **110** (1995): 195–227.
2. Christaller, W. *Central Places in Southern Germany.* Jena: Fischer, 1933. English translation by C. W. Baskin. London: Prentice Hall, 1966.
3. Cooper, R., and A. John. "Coordinating Coordination Failures in Keynesian Models." *Quart. J. Econ.* **103** (1988): 441–463.
4. Dixit, A., and V. Norman. *Theory of International Trade.* Cambridge, MA: Cambridge University Press, 1980.
5. Fujita, M., P. Krugman, and T. Mori. "On the Evolution of Urban Hierarchies." Mimeo, University of Pennsylvania, 1994.
6. Fujita, M., and T. Mori. "Why are Most Great Cities Port Cities?" Mimeo, University of Pennsylvania, 1995.
7. Garreau, J. *Edge City.* New York: Anchor Books, 1992.
8. Gaspar, J. "The Emergence of Spatial Urban Patterns." Mimeo, Stanford University, Stanford, CA 1995.
9. Harris, C. "The Market as a Factor in the Localization of Industry in the United States." *Annals of the Association of American Geographers* **64** (1954): 315–348.
10. Helpman, E., and P. Krugman. *Market Structure and Foreign Trade.* Cambridge, MA: MIT Press, 1985.
11. Henderson, J. V. "The Sizes and Types of Cities." *Amer. Econ. Rev.* **64** (1974): 640–656.
12. Henderson, J. V. *Urban Development.* New York: Oxford University Press, 1988.
13. Hirschman, A. *The Strategy of Economic Development.* New Haven: Yale University Press, 1958.
14. Isard, W. *Location and Space-Economy.* Cambridge, MA: MIT Press, 1956.
15. Kauffman, S. *The Origins of Order: Self-Organization and Selection in Evolution.* New York: Oxford University Press, 1993.
16. Krugman, P. "Increasing Returns, Monopolistic Competition, and International Trade." *J. Intl. Econ.* **9** (1979): 469–479.
17. Krugman, P. "Increasing Returns and Economic Geography." *J. Pol. Econ.* **99** (1991): 483–499.
18. Krugman, P. *Geography and Trade.* Cambridge, MA: MIT Press, 1991.
19. Krugman, P. "On the Number and Location of Cities." *European Econ. Rev.* **37** (1993): 293–298.
20. Krugman, P. "Complex Landscapes in Economic Geography." *Amer. Econ. Rev.* **84** (1994): 412–416.
21. Krugman, P. "The Hub Effect: Or, Threeness in International Trade." In *Theory, Policy, and Dynamics in International Trade: Essays in Honor of*

Ronald Jones, edited by J. P. Neary. Cambridge, MA: Cambridge University Press, 1993.
22. Krugman, P. "Self-Organization in a Spatial Economy." Mimeo, Stanford University, Stanford, CA 1995.
23. Krugman, P. "A 'Slime Mold' Model of City Formation." Mimeo, Stanford University, Stanford, CA, 1995.
24. Krugman, P. *The Self-Organizing Economy.* Cambridge, MA: Blackwell, 1996.
25. Krugman, P., and R. Livas. "Trade Policy and the Third World Metropolis." *J. Devel. Econ.* **49** (1996): 137–150.
26. Krugman, P., and A. Venables. "Globalization and the Inequality of Nations." *Quart. J. Econ.* **110** (1995): 857–880.
27. Krugman, P., and A. Venables. "The Seamless World: A Spatial Model of Global Specialization and Trade." Working Paper # 5220, NBER, 1995.
28. Lancaster, K. "International Trade Under Perfect Monopolistic Competition." *J. Intl. Econ.* **10** (1980): 151–175.
29. Lösch, A. *The Economics of Location.* Jena: Fischer, 1940. English translation, New Haven, CT: Yale University Press, 1954.
30. Matthews, R. C. O. "Reciprocal Demand and Increasing Returns." *Rev. Econ. Studies* **37** (1949): 149–158.
31. Pred, A. *The Spatial Dynamics of US Urban-Industrial Growth.* Cambridge, MA: MIT Press, 1966.
32. Prigogine, I., and I. Stengers. *Order Out of Chaos.* New York: Bantam Books, 1984.
33. Schelling, T. "Dynamic Models of Segregation." *J. Math. Sociol.* **1** (1971): 143–186.
34. Schelling, T. *Micromotives and Macrobehavior.* New York: W. W. Norton, 1978.
35. Simon, H. "On a Class of Skew Distribution Functions." *Biometrika* **52** (1955): 425–440.
36. Turing, A. "The Chemical Basis of Morphogenesis." *Phil. Trans. Roy. Soc. London* **237** (1952): 37.
37. Venables, A. "Equilibrium Locations of Vertically Linked Industries." *Intl. Econ. Rev.* **37** (1996): 341–359.

Martin Shubik
Yale University, Cowles Foundation for Research in Economics, Department of Economics, P.O. Box 208281, New Haven, CT 06520-8281

Time and Money

General equilibrium is timeless, and without outside money, the price system is homogeneous of order zero. Some finite-horizon strategic-market game models are considered, with an initial issue of fiat money held as an asset. For any arbitrary finite horizon, the solution is time dependent. In the infinite horizon, time disappears with the initial issue of fiat money present as circulating capital in the fully stationary state, and the price level is determined.

1. AN APPROACH TO THE THEORY OF MONEY AND FINANCIAL INSTITUTIONS

In the past twenty-five years my basic research has been directed primarily toward the development of a theory of money and financial institutions. This chapter is devoted to an essentially nontechnical presentation of the key elements of the approach together with some of the key results. I also indicate the basic shift in paradigm away from equilibrium toward the study of process, learning, and evolution which is transforming economic analysis.

The basic approach involves a combination of game theory and experimental gaming. The economy must be modeled and considered as a "playable game," i.e., a fully defined process model whose rules can be sufficiently understood so that the game can be played.

The central feature of a theory of money and financial institutions is the minimization of the need for trust. The key role of a financial institution is information processing, aggregation and disaggregation, and evaluation. However the interpretation of data is critical. It is not what the numbers are, but what they mean. Thus, heuristic phrases such as the old banking maxim of "character, competence, and collateral" can be operationalized.

In this chapter I deal only with exchange economies with completely stationary or cyclical inputs. I believe that the basic points made here hold for growth models with some adjustments concerning the boundedness conditions on the money supply being related to the size of growth.

1.1 ON TIME AND GENERAL EQUILIBRIUM

The finite general equilibrium model of Arrow and Debreu,[1] and others, has been regarded as one of the key works of microeconomic theory, as it explains in both an elegant and abstract manner the importance of the price system in the running of a decentralized efficient economy. Unfortunately, in the masterful development of the mathematics to study equilibrium, all of the richness and context required to study process was suppressed. It is an actor poor, context poor, preinstitutional, timeless abstraction devoted to illustrating the equilibrium properties of the price system. A detailed critique of the nature of the abstraction, its uses and limitations, is discussed elsewhere (Shubik,[14] Ch. 4). In spite of its drawbacks, the approach here is deeply indebted to the general equilibrium analysis inasmuch as it presented the first coherent, consistent closed model of the economy. Because of its concern only with equilibrium, the techniques used were able to avoid the fussy details, concerning information, bankruptcy, division by zero, or unboundedness in strategy sets in the formation of prices. These details cannot be avoided when a process model is built with enough care to define the system for all points in the set of feasible outcomes. It is in these details that the institutions of the economy lie. They are part of the rules of the game and act as the carriers of process. Once they are well defined, the possibility for their modification becomes evident and the transition from an equilibrium economics to an evolutionary economics becomes clear.

1.2 THE GAMES WITHIN THE GAME

Before we limit our concern to the more or less purely economic aspects of a theory of money and financial institutions, it is desirable to have an appreciation for the broader context within which economic activity is embedded. In going from current static microeconomic analysis toward a process analysis, a key distinction must be made between "the rules of the game" or its formal structure and the motivation and behavior we consider in attempting to define a solution to the game under examination. A solution, at its most abstract and general, is nothing more than a prescription or transformation which carries some state $S(t)$ of the system being studied into a state $S(t+1)$. Nothing is necessarily implied about optimization or equilibrium in this definition of solution. An operations researcher may invent a payoff function or a microeconomist may assume the existence of a utility function, and if, as a good first-order approximation the interaction of other individuals or institutions can be ignored, they take as their solution concept some form of maximization of these functions. In much of macroeconomic analysis, instead of formulating an explicit optimization, some form of behavioral conditions are postulated and they enable the analyst to update the system and study the time series generated.

It is possible to connect optimizing behavior with an evolving environment by formalizing overlapping generations (OLG) games with government, other legal persons of indefinite lives, and real persons.[1] A player in a game of strategy is characterized by a strategy set and a payoff function. A solution of a game involves the specification of some sort of choice rule, usually based on some operation by the player on his strategy set. This operation may involve the player's attempt to maximize his payoff function. But to describe a game of strategy it is not necessary that an individual, who is a strategic player, has a utility function. Instead any player, such as government or an institution, could be equipped with a choice rule that instructs its actions in all circumstances. Even though the player may have an infinite life, the decision rule whereby it advances into the future may be based on a finite set of arguments. An appropriate way to incorporate government and corporations into an overlapping generations model is to introduce them as infinitely lived strategic players without utility functions, but with a decision or choice rule determined by the real persons. An alternative way of modeling government is as a mechanism whose move is determined as the outcome of a game played between a set of politicians and bureaucrats, where the politicians are elected every so often by the overlapping generations of real persons and the politicians appoint the bureaucrats, but for longer terms than themselves.

Figure 1 suggests the scheme for an overlapping generations economy containing three types of agent. They are: (1) the young, (2) the old, and (3) the government. The first two are the real or natural persons, who can be regarded as many in number and possibly are most conveniently represented by a continuum of agents. The third is the government and may be represented by a single large atom.

At point of time $t = 0$ there are two generations alive, those born at time $t = -1$ and $t = 0$. Government, as well as the live agents, must select its move at $t = 0$. A reasonable restriction on its choice rule is that it depends at most on all of the decisions of all of the natural persons alive at $t = 0$.

[1]See Woodford and Muller[16] for a treatment of an economic model with both finitely and infinitely lived agents.

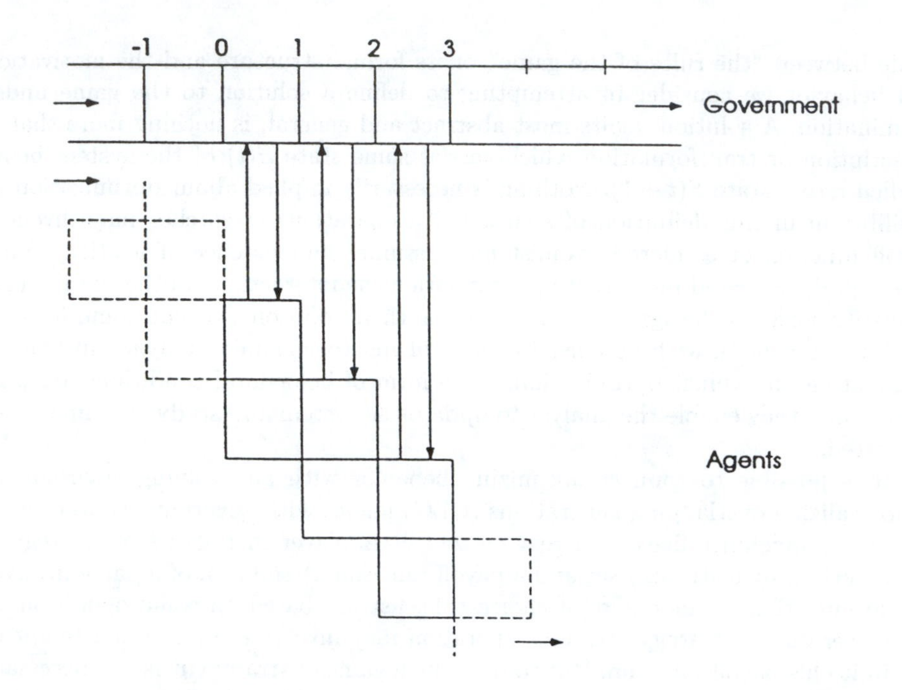

FIGURE 1 The games within the game.

At time t there are two generations alive, those born at $t-1$ and those born at t. They are concerned with maximizing their welfare.

We may consider that an individual born at t has a set of strategies S_t. Part of any strategy is an action, message or instruction that he sends to the government and part is an instruction that will depend upon the state of the system at the start of $t+1$.

Heuristically, the actions of the government at time t might depend on cultural, historical, social, political, and economic factors. If we choose to model the agents as nonsocial, nonhistorical, individualistic local maximizers we then have the most unfavorable and simplest set of conditions within which to examine the possibility that there may be structures where the needs of future generations could be served by the selfish behavior of the living.

One takes as given an institution called government whose strategy set and decision rule are given. We then consider the outcome of a game played in the context of these rules. Having considered politico-economic or economic efficiency within the context of the given institution we may then wish to ask a separate question. That is how did the institution evolve in the first place and how is its structure modified.

Government behavior can be considered in terms of both moves and strategy, but the more appropriate words are choice or action and policy or plan, designed to set the environment for the real persons. We may regard a government, at any particular point in time, as carrying out two activities. It sets taxes, subsidies, fixes the money supply, controls the central bank rate of interest, and manages the national debt, as well as taking action on many items of social welfare entailing public goods and services. The government also announces or otherwise indicates policy. Thus, for example, it indicates that it has a program or plan over the next few years to raise taxes and to lower the public debt.

Empirically, we may observe that most governments, at any point in time, appear to have some form of plan, but the plan is continually updated, modified, or otherwise changed. There is often a considerable skepticism held by the public concerning the plausibility that the stated plans of the government at time t will be consistent with its actions at time $t+1$. Yet in human affairs, even if policy changes in the future, there is a perceived need that it be announced in the present, if only to provide some indication of the gap between promise and fulfillment.

When one looks at governmental planning and policy, at least in any explicit form, it does not appear to be particularly long term. One suspects that it stretches, at most, no longer than between administrations. In a democratic society composed of individuals mainly concerned with their own welfare, it is reasonable to accept the lifetime of the youngest strategically active individual as a comfortable upper bound on the length of any announced plan.

The device of the "game within the game" allows for a reconciliation of local optimization with global evolution. The small live players behave by trying to optimize locally and taking, as a first-order approximation, their environment as given; their actions change the environment, but usually on a time scale too slow to influence their daily concerns.

1.3 QUANTITATIVE AND QUALITATIVE RESULTS

In my forthcoming book[14] the detailed description of strategic market games, together with the considerable notation and many examples and calculations, are presented. The mathematical proofs are to be found in a series of papers with colleagues and are due primarily to them.

As the purpose of this exposition is to sketch the ideas without delving into the notation, proofs or details, I concentrate on the qualitative aspects of the work rather than becoming enmeshed in the quantitative detail.

1.4 FOUR BASIC MODELING PRINCIPLES

The basic approach adopted here has been dominated by four basic methodological rules, which merit discussion. They are:

1. The insistence on the construction of well-defined process models.
2. The specification of laws of conservation.

3. The understanding of the role of symmetry.
4. The understanding of the invariant properties of the system.

A brief elaboration of these points is now given.

1. Economic dynamics are notoriously difficult. Broad intuitive insights have enabled economists such as Marx and Keynes broadly to sketch global grand dynamics, which have caught the imagination of governments and the masses and have had long-term consequences on policy. When the more cautious attempts have been made to measure phenomena, such as the propensity to consume or the Phillips curve, however, the grand dynamics appear to melt into a welter of special cases and context-dependent incidents. In contrast, the highly mathematical, and apparently precise, formulation of general equilibrium has no dynamics, as it avoids process modeling. A first step in trying to overcome the gap between the two sets of theory is to provide a process basis for the general equilibrium models. But in attempting to do so, phenomena such as thin markets, inactive markets, hyperinflation, and panics appear to be intimately connected with mathematical details such as division by zero. An insistence on well-defining the payoffs associated with every point in the state space provides an elementary check against basic error in the description of dynamics.
2. The writings on the theory of money have involved considerable vagueness concerning the definitions of money, its various properties, and near substitutes. The attitude adopted here is to avoid the type of debate concerning "what is money, really?" and instead consider that all forms of money and credit instruments have a physical existence. We may imagine them as colored chips, and devote our concern to describing how they are created, how they are destroyed, and how they move through the system. For example, we consider fiat money as blue chips, bank money as green chips, clearinghouse credits as black chips, and IOU notes as red chips. This approach avoids much of the mystery associated with processes such as bankruptcy. Red chips (IOU notes) may be destroyed, but blue chips are conserved. In many of the models investigated, in static equilibrium it is useful to assume that the quantity of fiat in the hands of the public plus the quantity of unissued fiat is a constant, or in the case of growth, that the new fiat the government is allowed to create is related to the growth.
3. All financial instruments except the original issue of fiat are created in pairs or clearly offset against a physical item such as when shares in a new corporation are issued directly against the stockholder contributions of physical resources. Double-entry bookkeeping is, in essence, an exercise in symmetry. All resources are offset. To make the books fully balance always, the accountants have developed a residual category called ownership, which is debited against any asset without an offset. The initial issue of cash is such an asset. The presence of fiat money with no offsetting debt is somewhat nonsymmetric. The restoration of full symmetry and the disappearance of time appear to be intimately related.

4. The task of formal model building, in the social sciences in general and economics in particular, is filled with difficulties involving the interpretation of the robustness and the relevance of the model. In particular, even when the formal model is built it is necessary to be specific about the solution concept to be adopted and to be in a position to know under what variations of the basic assumptions the results still hold. For example, the price system in many economic models involving an environment without exogenous uncertainty is invariant under ordinal transformations of utility functions. The pure strategy Nash equilibria of a system are invariant under transformations that preserve strategic domination, and these are even more general than the ordinal transformations of utility functions. In contrast, for an economy with exogenous uncertainty the results of the analysis are invariant only under cardinal transformations. Welfare considerations may call for both cardinality and comparability. Before policy comments can be made, the claimed invariant properties need to be specified.

1.5 THE KEY MODELS

There are 12 ($3\times2\times2$) basic models that must be examined to appreciate the basic features of an economy utilizing fiat money as a means of exchange. These basic models involve the elementary descriptions of the trading and finance structure of the economy treated to study their behavior when there is no exogenous uncertainty present and then when there are stochastic elements. All models must be also considered in a finite horizon version and then when there is an infinite horizon. The three basic models consist of: (1) an economy that uses only fiat money for transactions; all agents have a nonzero supply and there are no credit markets; (2) an economy with an initial issue of fiat, but with an outside bank, which stands ready to borrow and lend at an exogenously set rate of interest; (3) an economy with an initial issue of fiat and an inside money market for borrowing and lending where the endogenous rate of interest is formed by the competitive forces of supply of fiat and the demand for loans. Figure 2 shows the 12 basic models that are considered:

The results noted here have been obtained for players who conceivably could live forever; but with little difficulty they can be replaced by overlapping generations and the results are substantially similar for some settings. But, with OLG, one has to become specific about the birth-death processes assumed and the rules of inheritance.

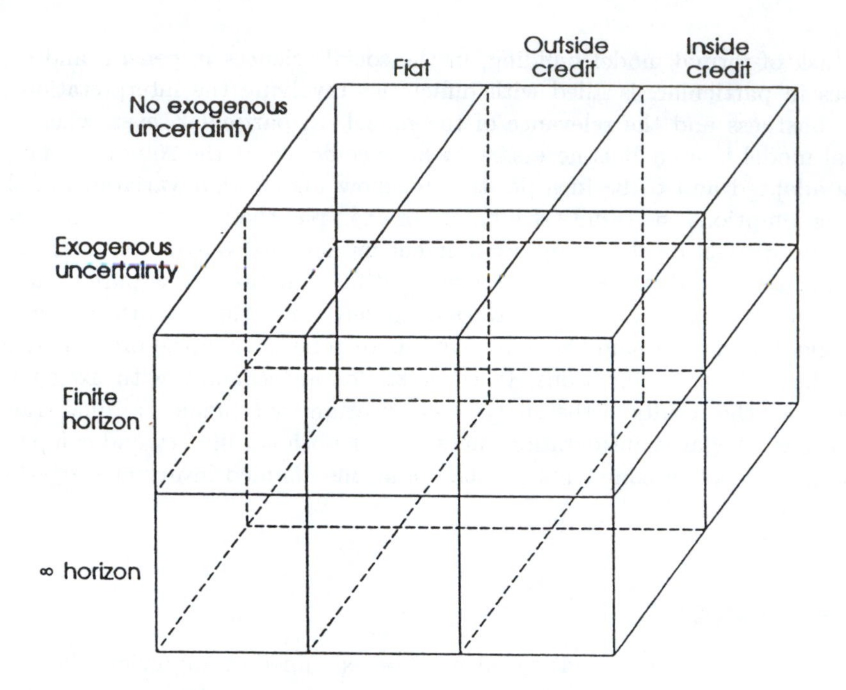

FIGURE 2 The 12 basic models.

2. THE NONSTOCHASTIC STATIONARY STATE

> In the beginning there was the big bank and fiat money was created. The economy went forward through time with an ever-growing borrowing until by the end of time all fiat money had been consumed; all motion stopped, time disappeared and perfect symmetry was restored to the economy.
>
> *On the cosmology of fiat money*

The somewhat mystical comment above is illustrated by two relatively simple examples and contains the first of the three basic points concerning the relationship between a monetary and a nonmonetary general equilibrium economy. It can be rephrased a little less poetically as follows.

The introduction of fiat money (and near monies, such as bank money) as a means of payment separates the timing of income and expenditures. It thereby provides the prerequisite degree of freedom needed to permit a trading dynamics. In a finite horizon economy, money is worthless at the end of time.

As observed by Dubey and Geanakoplos,[3] if there is an outside bank in a finite horizon economy where income is received after sales, then by the standard finite horizon backward induction argument from dynamic programming all individuals expecting income to come in at $T+1$ will try to arrange to be in debt by precisely the income they are expecting to receive in $T+1$. There are two extensions of the finite horizon model which are of importance. The first comes from the consideration of game theoretic models as contrasted with nonstrategic general equilibrium models. If individuals are permitted to borrow and there is any point in the state space of the game where an individual is unable to repay his debt, the bankruptcy and reorganization rules must be specified in order to resolve the individual's inability to repay. If at the end of the game there are resources of any sort (such as durables) left over then the rules must specify how they are to be evaluated at the end of the game.

If, for example, as is often done we utilize a utility function of the form:

$$\Phi = \sum_{t=1}^{T} \beta^{t-1} \varphi(x_t)$$

from a game theoretic strategic point of view we must add $T+2$ extra terms, one for each period indicating the bankruptcy penalty to be levied for an individual who fails to pay back a debt due at time t and an extra one at the day of settlement, period $T+1$ to indicate, what if any value is attributed to left over assets. Instead of the payoff function above, we may modify the payoff to:

$$\Phi = \sum_{t=1}^{t} \beta^{t-1} [\varphi(x_t) + \mu_1 \min[\text{Debt}_t, 0] + \beta_{\mu_1}^{T} \min[\text{Debt}_{T+1}, 0] + \mu_2 \max[\text{Asset}_{T+1}, 0] .$$

The first of the two terms involving a minimum is the period by period penalty for going bankrupt. If we were to introduce assets directly there would also be information concerning the loss of assets and garnishing of income—but, at the highest level of simplicity this term indicates that going bankrupt has unpleasant consequences. The last two terms indicate the cost of bankruptcy at the end of the game and the value of assets left over. The μ's can be regarded as simple parameters or complex expressions. For simplicity they can be regarded as simple scalars.

A term involving T is in front of this last bankruptcy term indicating that for a large T it will be extremely weak, verging on insignificant. This enables us to make a distinction between new fiat or outside money that might be introduced as a short-term loan to be paid back after a single period and the initial supply of fiat which can be regarded either as an asset, or as a loan which need only be paid back at the end of the game. But, for a game with a long duration the distinction between the two last alternatives can be made arbitrarily small. A bankruptcy penalty discounted far into the future is hardly a penalty. An immediate corollary that follows from our addition of a bankruptcy penalty to the payoff function is that a *debt which never has to be paid back is an asset.*

The term showing the value attached to the assets left over at the end is sometimes referred to in dynamic programming as the salvage value term. An alternative economic interpretation is that it portrays expectations.

We now are confronted with a paradox. If we consider the infinite horizon we do not see the extra terminal conditions involving bankruptcy and the worth of left over assets. The approach which can cover this difficulty is to study the limiting behavior of the finite horizon problem and ask whether the limiting behavior appears to approach the behavior at the limit. Fortunately this appears to be the case.

2.1 THE FINITE HORIZON

We now consider an extremely simple model for purposes of illustration. Suppose that we consider a simple T period economy where each individual tries to maximize his utility function as noted above. Each individual begins with one unit[2] of fiat money and has an ownership claim each period to the income from the sale of one unit of the only consumer perishable, all of which is put up for sale each period.

A strategy of an individual α is a vector of the form

$$b^{\alpha} = (b_1^{\alpha}, b_2^{\alpha}, \ldots) .$$

Price is formed by the amount of money chasing the amount of goods. This is:

$$P_t = \int \frac{b_t^{\alpha}}{1} .$$

Each individual obtains, as income at the start of period $t+1$, the amount p_t which equals his ownership claim from the amount of the good he has sold. For the simple example posed above, there is an elementary simple solution. All bid all of their money all of the time. This amounts to:

$$b_t^{\alpha} = 1 \;\; \forall \;\; \alpha, t$$
$$p_t = p = 1 \;\; \forall \;\; t$$
$$\text{Each obtains} : \sum_{t=0}^{T} \beta^t \varphi(1) .$$

In this set up there is an "end of the world" pathology as all traders obtain a perfectly worthless income at the start of period $T+1$ from the proceeds from the forced sale of the resource at the last period T.

[2] If I were being precise in the notation, I would distinguish between a density and a finite measure. The agents are assumed to be nonatomic, hence, their resources are in terms of densities. The central bank is atomic. As the examples are extremely simple the meaning should be clear from context.

If we had started all individuals with different amounts of fiat, the strategy would be more complex, although for a linear utility function all would have spent everything during the first period and the stationarity would settle in by the second period.[15]

We now introduce an outside bank with reserves of $B = 1$. The economy is otherwise the same as before except that the individuals can deposit or borrow from the outside bank. The outside bank sets the rate of interest exogenously. We select $1 + \rho = 1/\beta$, show the solution for this rate, and show that it has an extra property when we go to the infinite horizon that may be deemed to be desirable. The property is that in a fully stationary state for the physical economy, prices are also stationary, i.e., there is no inflation or deflation.

Suppose that at equilibrium the price is p each period. We assume that at time t an individual will borrow or lend $(+/-)$ an amount Δ_t thus we have a set of $T+1$ equations:

$$
\begin{aligned}
1 + \Delta_1 &= p \\
1 - \rho\Delta_1 + \Delta_2 &= p \\
&\vdots \\
1 - \rho\sum_{J=1}^{t-1}\Delta_j + \Delta_t &= p \\
&\vdots \\
(1+\rho)\Delta_T &= p\,.
\end{aligned}
$$

The solution to these for Δ and p yield

$$\Delta_t = \frac{(1+\rho)^{t-1}}{(1+\rho)^T - 1} \text{ and } p = 1 - \frac{(1+\rho)^T}{(1+\rho)^T - 1}\,.$$

We observe that as $T \to \infty$, $\Delta_T \to \beta$ and $p \to 1$, the system approaches a full stationary state and the need for bank reserves diminishes. For any finite first sequence of periods the amount borrowed can be made arbitrarily close to zero and the two solutions without and with the outside bank approach each other.

2.2 THE DUAL SOLUTIONS: FIAT WITH OR WITHOUT THE OUTSIDE BANK

In the examples given above for a simple trading economy, the amount of the good traded each period is the same. The introduction of the outside bank seems to be an economically irrelevant artifact. Suppose instead the economy were cyclical. For illustration we select a periodicity of 3. Instead of having $\mathcal{Q} = 1$ be the amount of the good for sale each period, we may consider an economy where the amount of the good for sale is:

$$\mathcal{Q}_{3t+1} = \mathcal{Q}_1;\, \mathcal{Q}_{3t+2} = \mathcal{Q}_2;\, \mathcal{Q}_{3t+3} = \mathcal{Q}_3;\ \text{for } t = 1, 2, \ldots\,.$$

If we now consider an example analogous to the economy in section 2.1 without a bank, but where all have an initial issue of fiat, prices can be adjusted from period to period only by hoarding. For example suppose the amounts of good were 1, 2, 3, repeated cyclically and the utility functions were:

$$\sum_{t=1}^{3T} \beta^{t-1} x_t^{\alpha} .$$

Spot prices in general equilibrium should be the same each period, but as there is a fixed amount of money in the system, this can only be achieved by hoarding. In general for a k-cycle with arbitrary equilibrium points we must consider 2^k Lagrangian multipliers to determine when the individual should hoard or spend all.[11,12,13] In this simple example, if β is small enough the present will always dominate the future, thus all will always spend all and the spot prices will be $1, 1/2, 1/3$. At the other extreme, if $\beta = 1$ the spot prices would be 1/3, 1/3, 1/3 and in the second and third periods 1/3, then 2/3 of the money supply would be hoarded. In general, prices will be $\max[1/3\beta^2, 1], \max[1/3\beta, 2/3], 1/3$. If, instead of an economy with fiat and no borrowing and lending, we introduce an outside bank and an exogenous rate of interest, the solution changes. In particular, there can be no hoarding as a nonatomic individual can always improve by depositing in the bank rather than hoarding.[3]

Suppose we set $\beta = 4/5$, then in an economy without loans the spot prices will be 25/48, 20/48, and 16/48, and the hoarding will be 23/48 in the first period, 8/48 in the second, and 0 in the third. Each individual will consume (1, 2, 3, 1, 2, 3,...) which is the general equilibrium solution. The total amount of money in the system is 1, hence, $M + B = 1 + 0 = 1$, where M is the amount held by the individuals and B is the amount held by the central bank. Here, as there is no central bank, the amount it holds is 0.

We now consider the same economy with the addition of a central bank that is prepared to borrow and lend at a given fixed rate of interest. If $1 + \rho = 1/\beta$ this gives $\rho = 25\%$ or 1/4. We may select for comparison an economy with $M + B = 1$ and the same spot price in the third period. If we do this then all the spot prices will be 1/3, 1/3, 1/3,... and the initial money distribution will be $M = .693989$ and $B = .306011$. In the first period the traders will deposit $x = .284153(52/183)$, in the second period they deposit $y = .021858(4/183)$, and in the third period they borrow $z = .306011$. We note that $x + y = z$. The cash needs each period are 1/3, 2/3, and 1. Figure 3(a) shows the hoarding and Figure 3(b) shows the depositing

[3]Surprisingly in the sell-all model (which is probably a reasonable approximation of the economy) Pareto optimality is achieved in a multidimensional economy because the inneviency or "wedge in prices" caused by a difference between buying and selling is not encountered. In a model involving bid-offer, where individuals do not have to sell their assets, general equilibrium efficiency will be lost. it is my guess that, empirically, this loss is both realistic and not very large relative to other aspects of the economy.

and borrowing for the three periods. The vertical arrow represents time. In the third period the bank reserves are depleted to zero, which is the equivalent to the no hoarding in the third period without the bank.

Empirically, we do not see the lining up of relative prices brought about by massive hoarding or intertemporal loans achieved by A staying out of the market while B buys. Instead, we see an economy where there is borrowing and lending and a rate of interest. Although in a stationary analysis the two possibilities appear to be available as substitutes, in an actual dynamic economy the economy with borrowing and lending and an interest rate appears to require far less coordination and information than the other.

2.3 THE DISAPPEARANCE OF TIME AND MONEY

There are two basic points to be gleaned from the discussion in sections 2.1 and 2.2. They are the relationship between time and the presence of fiat money, and the duality between (a) hoarding and (b) borrowing and lending.

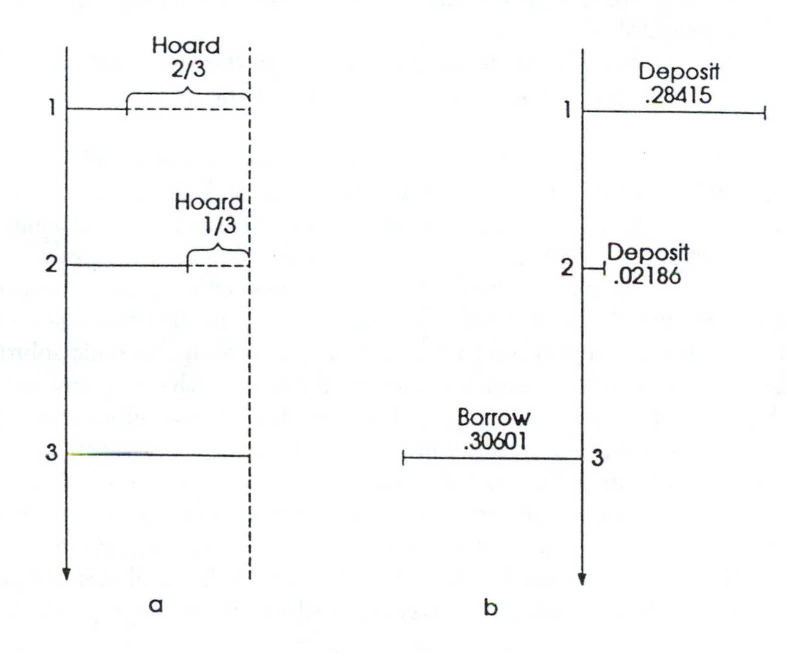

FIGURE 3 Hoarding or depositing and borrowing.

OBSERVATION 1: The presence of an initial issue of fiat, which is an asset (or the equivalent, an indefinitely long debt), creates an overall system which is slightly nonsymmetric in the sense that this money is the only financial asset without a counterclaim against it. When there is an outside bank present, the inside economy over time consumes this initial input of fiat and it flows into the bank as a net interest payment. If we limit the investigation to any finite horizon, no matter how long, by backward induction the terminal constraint is tight when there is no borrowing or lending and no salvage value for left over assets. This, in essence, fixes a price level.[4] When there is borrowing and lending, the bankruptcy conditions must be introduced to complete the definition of the model.

If the initial issue of fiat is regarded as a debt to be paid back to the government at the end of the game, then the possibility for the existence of a continuum of prices will depend on the magnitude of the bankruptcy penalty.

As long as the system is finite it remains slightly nonsymmetric, and the borrowing and lending is not perfectly stationary. When the system is infinite it becomes completely symmetric and the amount of fiat money held jointly by the economic agents and the outside bank is conserved. This amount is the circulating capital of the economy. In a world with no transactions needs whatsoever, there is no need for circulating capital.

The system becomes fully stationary or timeless in the sense that optimal strategies are based only upon a non-time-indexed state of the system.

OBSERVATION 2: In sections 2.1 and 2.2 we considered two economies with a fixed input repeated each period where the first economy had no bank and the second had an outside bank. They clearly differed for the finite horizon, but converged to the same solution for the infinite horizon. This convergence could raise questions as to the need for an outside bank. When we consider a cyclical economy with differing inputs over the cycle, even in the infinite horizon the two economies differ. They give, for the appropriate initial conditions, the same economic solution, but give different prices and are computationally distinct. In the economy without an outside bank, relative prices are adjusted by hoarding. For a sufficiently high β all money will be utilized, generically only once per cycle, for the remaining period some part of it will be in hoard.[5] The money holdings are almost always a slack variable, and price depends on cash flow inequalities as sketched in Figure 3. When a bank exists and charges a positive rate of interest, then money can never be a slack variable. Borrowing and lending from the outside bank allows the money to flow into and out of the real goods economy and the books balance at the end of

[4]There may still be some indeterminacy caused by constraints in the form of inequalities and by alternative trading and inventory possibilities.[11,12,13]

[5]We may interpret the hoarding as an adjustment of the velocity of money, or the number of times it turns over per period. In the models with an outside bank, the money not in the hands of the outside banks turns over precisely once per period. In the models without the bank, the money in the hands of the agents has a velocity of less than one.

every cycle. In the infinite horizon, the circulating capital needed to balance the books every cycle remains in transit between the bank and the agents.

The two different solutions are duals, where in one the amount of money is adjusted via hoarding and in the other it is adjusted by the appropriate level of interest, which permits a different form of overall conservation with money flowing into and out of the central bank.

2.4 AN ASIDE ON THE MONEY MARKET

Our discussion has concentrated on the contrast between an economy using fiat money for transactions with and without an outside bank. There are two basic needs for borrowing money, one is to finance transactions and the other to finance intertemporal consumption or the purchase of capital stock whose returns will come in the future. The first financing needs may call for variation in the money supply, while the second could be handled by individuals who have surplus money at period t lending directly to those who are short. It is straightforward to construct a model with a money market with the endogenous formation of the rate of interest, or a model with both a money market and an outside bank where the endogenous and exogenous interest rates interact.[5,14] The interest rate is formed by the amount of IOU notes offered in exchange for the amount of fiat put up for loan. The money market turns out to be a case intermediated between the model with no loans whatsoever, and the model with an outside bank. The endogenous rate of interest will either be positive or it will be zero, and hoarding will take place. The money market alone cannot adjust the money supply seamlessly to the variations needed for the appropriate relative prices.

OBSERVATION 3: The efficient equating of the supply and demand for money in a competitive loan market via a well-defined positive rate of interest is, in general, not possible. The belief that it is possible can only be based on the extension of trade in money substitutes which one might wish to consider as money. An inside money market trades fiat for debt, it does not create fiat.

3. INCOMPLETE MARKETS AND STOCHASTIC ELEMENTS

In section 2 we observed a duality between nonstochastic economies with and without an outside bank. When there is a stochastic element present the duality is by no means as clear. In particular, in a nonstochastic economy with loans, the role of bankruptcy is to prevent strategic default. At least around equilibrium there should be no active bankruptcy, if the default penalty is sufficiently harsh. This is no longer true when there is stochastic uncertainty present.

3.1 FIAT MONEY AS INSURANCE

The occurrence of hoarding in the model of an exchange economy utilizing money, but without loans illustrated in section 2, serves to enable the individuals to vary relative prices. A new reason to hold fiat appears when there is exogenous uncertainty present in the market. The fiat saved serves as insurance against a low income. It provides precautionary reserves against disaster. Alternatively, it provides immediate liquidity to take advantage of opportunities,[17] provides an example where in equilibrium the wealth distribution of a population of agents with the same preferences is such that a part of the monetary wealth of the economy is always in hoard.

3.2 INSTITUTIONS AND EQUILIBRIUM

In attempting to establish the existence of equilibrium in strategic market games, a persistent set of difficulties keep appearing. At first glance, they appear to be minor and seem to represent some form of casual extra institutional observation that should be omitted in early modeling by the theorist trained in abstraction and desiring high levels of generality. The institutional details can be added later after the broad outlines of a general theory have been established. In the development of an understanding of the properties of a price system, this was the approach adopted. A rigorous, frictionless, and timeless theory of price was developed in a series of models abstracting out the role of process.

In attempting to build a complete closed process model of an exchange economy utilizing fiat money, the annoying details, which could be avoided in the nonprocess analysis, appear to be of more importance than just minor frictions or quaint institutional additions to an otherwise adequate theory. They raise basic questions concerning the role of financial institutions as the carriers of process. In particular, if we model the economy as a strategic game and try to define this game for every point in the set of attainable outcomes, we must be specific as to what happens if only one side of a market is active or if neither side of a market is active. If the interest rate is formed by the amount of IOU notes offered in exchange for an amount of fiat, what happens if no fiat is offered, does the rate become infinite, or is the market active? Can individuals borrow unlimited amounts of fiat or create unbounded amounts of debt? A way to handle these difficulties is to invent agents and institutions such as the specialist stock trader whose role is to "make a market" for thinly traded stock, or a government agricultural agency who steps into the market and buys wheat when the price goes too low and reverses position with inventories when the price is too high. In some instances the agency might even burn the wheat after having "maintained an orderly market." We may invent an ideal central bank that steps in on the appropriate side of the market if the rate of interest is either too high or too low. KSS,[18,19] by using devices such as those described above, are able to establish the existence of equilibrium in strategic market game models of the economy both with secured lending and with active bankruptcy. Paradoxically, at the equilibrium the regulatory mechanism is inactive, thus it is not seen, but it would appear decisively in the dynamics. The apparatus is needed to make sure that there are not pathologies which could develop if the system were to achieve a feasible state, which might be far from equilibrium. Even with the presence of the appropriate institutional apparatus, there is no guarantee that the system started far from equilibrium would ever achieve equilibrium. All that is guaranteed is that the system cannot completely blow up.

OBSERVATION 4: At the least, the necessary conditions for the existence of an equilibrium in an economy with money and credit call for the presence of institutions and rules that preserve orderly markets and bounded variations in prices and interest rates. Mathematically, these amount to making sure that certain payoff functions are compact.

Our experience with the problems in establishing the existence of equilibrium in the competitive models of the economy with secured and unsecured lending suggests that financial institutions and laws are not a mere institutional curiosity for the institutional economist, but logical necessities for the economic system if one expects equilibrium to be feasible. Furthermore, even though at equilibrium they are apparently inactive, they are needed.

3.3 PANICS AND POWER LAWS

In section 3.2, a sketch of the reasons for the need for institutions and appropriate rules such as credit constraints and bankruptcy laws has been given. Our original concern has been with the existence of equilibrium in the financial and product markets. But it is reasonable to ask the reverse question along with several other obvious companion questions.

The basic question is what happens to the system if the regulatory rules are not sufficient to guarantee the existence of equilibrium? An immediate companion question is empirical. Are the institutions and laws of society such that they are always sufficient to guarantee the existence of equilibrium?

I conjecture that under fairly reasonable and general circumstances the type of power law suggested by Mandelbrot[8,9] which is associated with financial panics may arise in the models postulated here when the system starts in disequilibrium and the laws and regulatory agencies are not adequate.

I further suggest that there are more or less natural reasons to suspect that the institutional safeguards against the possibilities for the development of "economic earthquakes" are rarely if ever perfect. The observations here are, for the most part, consonant with the discussion given by Minsky[10] on the financial system as well as the approach of Arthur,[2] Ijiri and Simon,[4] and others in their approach to simulating economic dynamics. In essence, many of the credit conditions have a crawling peg built into them in the way that expectations are permitted to influence the granting of credit. The presence of leverage combined with the possibility for chain-reaction bankruptcy can trigger disasters of any magnitude if the credit limits are functions of inflated or deflated recent asset prices and expected prices. It is often when the economy has heated up that the credit restrictions become lax and in the downward spiral when more credit is needed to prevent the slide, the credit tightens.

The physical scientist acquainted with a power law such as that for the probability of the occurrence of an earthquake of some given magnitude knows that there is no physical way known to humans to truncate the fat tail of the distribution. The situation with the economic earthquake is somewhat different. Beyond a certain force of the economic Richter scale the stock markets close and bank holidays are declared. The economy reorganizes and the law suits stretch out for many years.

4. INSTITUTIONS AND EVOLUTION

The time for going beyond the essentially preinstitutional timeless models of general equilibrium has clearly arrived in economic theory. The ability to compute and to simulate has changed by many orders of magnitude in the last few years. These new abilities permit an exploration of dynamics that was recently impossible. But, the suggestion made here is that the natural first careful extensions beyond general equilibrium theory can be made by formulating well-defined, even though highly simplified, process models. In the act of fully defining these models, the institutions are invented as a matter of logical necessity. In the attempt to establish the conditions for the existence of equilibrium, one is also able to establish the conditions for the lack of equilibrium. Even casual empiricism is sufficient to indicate that the possibility is high that the institutions are not strong enough to avoid panics, breaks, and failures of smooth process. Thus, the conflicting view presented by conventional general equilibrium theory and its offspring, conventional finance, and the work suggested that there can be basic instabilities, are not necessarily at odds, but represent different arrays of institutional control or lack of control.

The economy is embedded in the polity and in the body of laws of the society. The disequilibriating shocks in the system trigger the changes in the rules governing the environment in which it operates.

4.1 THE NEEDED INSTITUTIONS AND INSTRUMENTS

There is a minimal set of basic institutions and instruments needed before one is in a position to construct a viable theory of money. The actors, institutions, and instruments used here are individual agents and a central bank, a goods market for price formation, a money market for endogenous interest rate formation, and an outside exogenous interest rate. Also needed are bankruptcy conditions and a terminal evaluation of assets left over at the end of any finite game. These can quite naturally be interpreted as expectations. Fiat money and short-term IOU notes are also required. It must be stressed that although there may be considerable latitude in selecting a specific institutional form, there is little longitude in covering function. For example, there are many variants of bankruptcy rules, but no matter which is chosen some method must be given for specifying what happens if an individual cannot pay back a debt.

Beyond the primary basic functions of exchange, borrowing, and repayment there are secondary functions which modify the basics. For example bank and clearinghouse money provide more flexibility than a system with fiat alone. But they are not needed in the simple models noted here; however, in any attempt at modeling an actual economy they are called for.

4.2 BEHAVIOR, STRATEGY, INFERENCE, AND EXPECTATIONS

The approach adopted here has been to begin by staying close to the equilibrium assumptions of basic general equilibrium theory in order to be able to contrast the results obtained. But because full process models are constructed, we may then consider differences as well as similarities. In particular, in models regarded as open to time we may consider solutions which are not based on backwards induction, but on extrapolating from the past forward, inferring future behavior from past performance. An open question concerns what class of expectations lead to convergence, if an equilibrium exists. The mere existence of equilibrium tells us little about the dynamics.

4.3 INSTITUTIONS: THE SELF CORRECTING SYSTEM

In section 3, I have suggested the special role for institutions and laws in serving not only as carriers of process, but in providing the appropriate bounds to guarantee the existence of equilibrium. In section 3.3, the possibility was conjectured that with imperfect institutions panics, best illustrated by power laws, could be present. But, as observed above, unlike natural power laws such as that which may govern earthquakes, individuals do have some control over financial power laws. When the game becomes too disastrous for all, they change the game. This is tantamount to postulating an elemental structure to the financial games-within-the-game. All participants take the rules of the game as given and act as context-constrained local optimizers, where the context is the set of laws and institutions of their society. When a disaster of a sufficiently large magnitude strikes, enough of the economic agents are hurt that legal, political, and social change modifies the economic environment.

I have entitled this section "The self-correcting system?" with a stress on the "?". The evidence from 1700 (when fiat money, stock markets, and modern financial institutions started to proliferate) unto this present day is that there has been a series of panics and crashes of various magnitudes, but that after each the rules have changed and (at least with Candide bifocals) it appears that the laws and institutions, by some measures may have improved.

ACKNOWLEDGMENTS

Much of the work noted here has been the result of joint collaborations over many years. My main collaborators have been Lloyd Shapley, Pradeep Dubey, John Geanakoplos, Yannis Karatzas, William Sudderth, and several others. Specific acknowledgments are given in my forthcoming volume.[14]

REFERENCES

1. Arrow, K. J., and G. Debreu. "Existence of an Equilibrium for a Competitive Economy." *Econometrica* **22** (1954): 265–290.
2. Arthur, W. B. "Inductive Reasoning, Bounded Rationality and the Bar Problem." *Amer. Econ. Rev.* (Papers and Proceedings) **84** (1994): 406–411.
3. Dubey, P., and J. Geanakoplos. "The Value of Money in a Finite Horizon Economy: A Role for Banks." Mimeo, Cowles Foundation, 1990.
4. Ijiri, Y., and H. A. Simon. *Skew Distributions and the Sizes of Business Firms.* Amsterdam: North-Holland Publishing, 1977.
5. kss, need reference.
6. Karatzas, Y., M. Shubik, and W. D. Sudderth. "Construction of Stationary Markov Equilibria in a Strategic Market Game." *JMOR* **19(4)** (1994): 975–1006.
7. Karatzas, Y., M. Shubik, and W. D. Sudderth. "A Strategic Market Game with Secured Lending." *CFDP* (1995): 1099.
8. Mandelbrot, B. "The Variation of Certain Speculative Prices." *J. Bus.* **36** (1963): 307–332.
9. Mandelbrot, B. "The Variation of Some Other Speculative Prices." *The J. Bus. of the University of Chicago* **40** (1967): 393–413.
10. Minsky, H. *Stabilizing an Unstable Economy.* New Haven, CT: Yale University Press, 1986.
11. Quint, T., and M. Shubik. "Varying the Money Supply, Part I: A General Equilibrium Model with Perishables, Storables, and Durables." *CFPP* (1995): 950511.
12. Quint, T., and M. Shubik. "Varying the Money Supply, Part IIa: Trade in Spot Markets Using Money—The Three Period Models." *CFPP* (1995): 950620.
13. Quint, T., and M. Shubik. "Varying the Money Supply, Part IIb: Trade in Spot Markets Using Money: The Infinite Horizon." *CFPP* (1997): 970221.
14. Shubik, M. *The Theory of Money and Financial Institutions.* Cambridge, MA: MIT Press, 1996.
15. Shubik, M., and W. Whitt. "Fiat Money in an Economy with One Nondurable Good and No Credit (A Noncooperative Sequential Game)." In *Topics in Differential Games*, edited by A. Blaquiere, 401–449. Amsterdam: North Holland, 1973.
16. Woodford, M., and W. J. Muller III. "Determinacy of Equilibrium in Stationary Economies with Both Finite and Infinite Lived Consumers." *JET* **46** (1988): 255–290.

John Geanakoplos
Cowles Foundation, Yale University, 30 Hillhouse Avenue, New Haven, CT 06520

Promises Promises

In the classical general equilibrium model, agents keep all their promises, every good is traded, and competition prevents any agent from earning superior returns on investments in financial markets. In this chapter I introduce the age-old problem of broken promises into the general equilibrium model, and I find that a new market dynamic emerges, based on collateral as a device to secure promises.

Given the legal system and institutions, market forces of supply and demand will establish the collateral levels that are required to sell each promise. Since physical collateral will typically be scarce, these collateral levels will be set so low that there is bound to be some default. Many kinds of promises will not be traded, because that also economizes on collateral. Scarce collateral thus creates a mechanism for endogenously determining which assets will be traded, thereby helping to resolve a long standing puzzle in general equilibrium theory. Finally, I shall show that under suitable conditions, in rational expectations equilibrium, some investors will be able to earn higher than normal returns on their investments, and the prices of goods used as collateral will be especially volatile.

The legal system, in conjunction with the market, will be under constant pressure to expand the potential sources of collateral. This will lead to market innovation.

I illustrate the theoretical points in this chapter with some of my experiences on Wall Street as director of fixed income research at the firm of Kidder Peabody.

1. INTRODUCTION

When big changes occur in one's life, one must step back and ask what one is really doing. I spent five years as director of fixed income research at Kidder Peabody & Co., departing when the firm closed in January 1995 after 130 years in business, following a scandal in its Treasury bond department and the collapse of the new-issue mortgage market. In my last two months there I, like all the other managers, went through a ritual firing of each one of my subordinates; I was then fired myself. After leaving Kidder, I became a partner along with some of the former Kidder traders in a hedge fund called Ellington Capital Management that we set up to invest in mortgages and especially mortgage derivatives. At Kidder we had dominated the collateralized mortgage obligation (CMO) market, issuing about 20% of the trillion dollars of CMOs created between 1990 and 1995. What we do now at Ellington is buy the very securities we sold while at Kidder. For a time our motto seemed to be "We made the mess, let us clean it up." What good was I doing society at Kidder Peabody? Is there any logic in switching from the sell side to the buy side, or am I just drifting with the tide of irrational market passions?

Not surprisingly, given man's natural inclination to rationalization, I shall try to argue that Kidder Peabody was part of a process that improved welfare, and that this process makes it inevitable, even if everybody is rational, that there should be periodic episodes of extraordinary investment opportunities in moving from the sell side to the buy side and back. What might be interesting is, therefore, not my conclusion, but the framework I shall try to describe in which to argue my claims, and from which I derive some other consequences of Wall Street's main activity.

I shall suggest that the main business of Wall Street is to help people make and keep promises. Over time, as more people have been included in the process, punishment and reputation have been replaced by collateral. This enabled a proliferation of promises, but has led to a scarcity of collateral. The ensuing efforts of Wall Street, in conjunction, of course, with the government and in particular the courts and the tax code, to stretch the available collateral has significant and surprising effects on the working of the economy, including the cyclical or volatile behavior of prices, which is the subject of this chapter.

Promises play a vital role in society, and this has been recognized by the great philosophers for a long time. For example, Plato's first definition of justice in the Republic is keeping promises. Socrates eventually rejects that definition because he says that it is sometimes more just not to keep one's promises.

Nietzsche begins the second essay in the *Genealogy of Morals* by asking "To breed an animal with the right to make promises—is this not the paradoxical task that nature has set itself in the case of man?" His idea is that before man could be expected to keep promises he had to develop a conscience, and the capacity to feel guilt. This capacity was bought with generations of torture and cruel punishments for transgressors. Though this history is forgotten, modern man retains the guilt. Thus, "the major moral concept guilt (*Schuld*) has its origin in the very material concept of debts (*Schulden*)."

Wall Street has emancipated the debtor relationship from guilt and reduced the amount of suffering by permitting default without punishment, but attaching collateral. One should expect that such a conversion would have important consequences for the workings of the economy and society. I hope to provide a framework for understanding some of them.

2. MORTGAGES, KIDDER PEABODY, AND ME

My subject is promises secured by collateral. The vast majority of promises, especially if they extend over a long period of time, are not guaranteed only by oaths and good intentions, but by tangible assets called collateral. Anything that has durable value can, in principle, serve as collateral, but the less sure its value, and perhaps even more importantly, the more hazardous its ownership, the less likely it will in fact be used as collateral. Short-term U.S. government bonds are an ideal collateral for short-term loans, because their value is guaranteed, and it is easy for the lender to claim the collateral if there is a default (by holding onto it from the beginning of the loan in what is called a REPO). Long-term government bonds are less appropriate for short-term loans because their value is very volatile when inflation and real interest rates are variable. Equipment, plants, and whole companies are often used as collateral for corporate bonds, though they are far from ideal, both because their market values are volatile and because much of this value can be dissipated when (or just before) there is a default. Perhaps the most plentiful and useful collateral for short and long-term loans are residential homes.

A residential mortgage is a promise to deliver a specified amount of money (perhaps over many periods) in the future, that is secured by a house. If the promise is broken, the house can be confiscated and sold, with the proceeds needed to make the promise whole turned over to the lender. The surplus, if there is any, is returned to the borrower, and the shortfall, if there is any, is typically a loss for the lender. Sometimes the defaulter faces further penalties, including liens on future income or even perhaps jail. Over time these penalties have declined in severity. Since the house cannot run away, there is little opportunity for the borrower to steal part of its value if he anticipates defaulting. Indeed the only effective way to appropriate part of the value is to refrain from making repairs over a long period of time. But defaults usually occur in the first few years of the mortgage.

The depression in the 1930s led to a huge number of defaults, especially on farm mortgages. As a result the amortizing mortgage was invented, whereby the debtor not only makes regular interest payments, but also pays back part of the principal each month. After some years, the principal has been paid down to the point where any default can easily be covered by selling off the farm. In the early years of the mortgage, the danger of default depends critically on the loan to value ratio, that is on how much is borrowed relative to the value of the farm. This ratio is set by supply and demand. By contrast, the predepression mortgages had regular payments of interest, and a large balloon payment at the end in which the debtor repaid the whole of the principal. In balloon mortgages, the probability is relatively high that the farm value will not cover the loan at the balloon payment date.

There are approximately 70 million homes in America, worth on average about $100,000. If we add to this various other kinds of multifamily and commercial real estate, we get a real estate market worth over $10 trillion, giving rise to well over $1 trillion of actively traded mortgage promises.

Until the 1970s, the banks that issued the mortgages bought and held those promises from homeowners. They, therefore, were subject to two large risks: that inflation would erode the value of their promises, and that higher real interest rates would make them regret tying up their money in a low return instrument. After the 1970s deregulation, they could rid themselves of these risks by selling the mortgages to other investors, while continuing to make money by servicing the mortgages. (The homeowner is oblivious to the fact that his mortgage promise has been sold because the same bank services his account by collecting his checks, and writing him reminders when he is late with his payments, but forwarding the money on to the new owner of the promise.)

Various government agencies called Fannie Mae, Ginnie Mae, and Freddie Mac were created to make the purchase of these securities more attractive to investors. The agencies provided insurance against homeowner default, and tough (but uniform) criteria for homeowners to get their mortgages included in the program. Most importantly, the agencies bundled together many mortgages into composite securities that investors could buy, in effect getting a small share of many different individual mortgages. As a result, the investors did not have to worry about the adverse selection problem of getting stuck with only the unreliable homeowner promises, nor did they need to spend time investigating the individuals whose promises they were buying. (Of course they did spend millions of dollars calculating average default rates and repayment rates; but that is small compared to the cost they would have incurred if they had access to the personal credit history of each homeowner.)

Wall Street took the whole operation a step further by buying big mortgage pools and then splitting them into different pieces or "tranches," which summed up to the whole. These derivative pieces are called collateralized mortgage obligations (CMOs) because they are promises secured by pools of individual promises, each of which is backed by a physical house. As the following diagram makes clear, there is a pyramiding of promises in which the CMO promises are backed by pools of promises which are backed by individual promises which are backed by physical homes. The streams of promises sometimes come together when many promises are pooled to back other promises, and sometimes split apart when one (derivative) promise is tranched into many smaller promises.

It is interesting to note that the investors buying the CMO tranches also use collateral, when they buy on margin. The investor pays a fraction of the price, borrowing the remainder, and putting up the CMO tranched itself as collateral. The fraction it is necessary to put down (the margin) is determined by the forces of supply and demand, and occasionally by government regulation.

The splitting of mortgage promises into CMO tranches increases the number of buyers in two ways. In the first place, differentiating the cash flows enables investors to purchase the flows that best suit their needs. For example, a pension plan might want cash far in the future, whereas a large winter clothing manufacturer with excess cash might be interested in finding a short-term investment with an enhanced yield. Similarly, a bank which is already committed to many long-term loans might want money when interest rates go up, whereas a credit card company that makes only short-term loans might want to invest its cash reserves in securities that pay when interest rates go down. CMO tranches accordingly differ in their temporal cash flows, and some pay more when interest rates go up (floaters), while others pay more when interest rates go down (inverse floaters). The large menu of potential investments enables the investor to better hedge his business. By spreading risk to those best able to bear it, the creation of new asset markets increases the productive capacity of the economy.

Splitting mortgage promises into pieces with cash flows contingent on exogenous variables like the interest rate increases demand in a second way, because it enables investors to bet on which way interest rates and the other exogenous variables will go. Though this may make each investor feel richer, it cannot be said to be good for the economy. It does however raise the value of a mortgage to the bank by making it easier to sell after it is split up into pieces, thereby reducing the interest rate homeowners must pay on their mortgages.

The first CMO was created by Salomon Brothers and First Boston in 1983. It consisted of four tranches, and was worth just over $20 million. Shortly thereafter, most of the Wall Street firms joined the CMO market, led by the biggest firms Lehman Brothers, Goldman Sachs, and Merrill Lynch, and joined later by Bear Stearns, Drexel (before it went bankrupt in the Michael Milken scandal), DLJ, Nomura, and others. By 1993, Kidder Peabody was tranching pools worth more than $3 billion into 100 pieces.

In the spring of 1990, when I went on leave from Yale, I decided that it would be enlightening for me, as a mathematical economist, to see first hand how the practitioners in finance were using the mathematical models that academic economists had created. In finance, more than in any other field of economics, the vocabulary

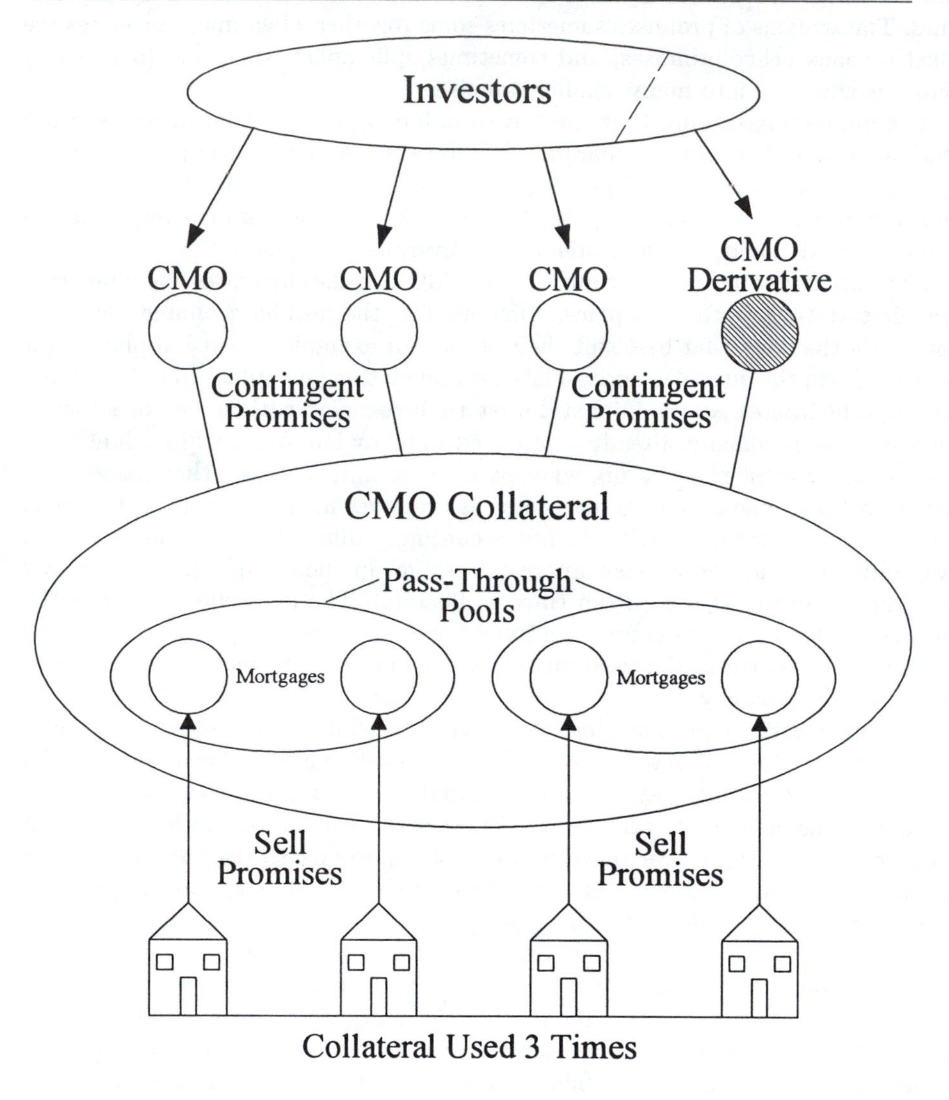

FIGURE 1 Mortgage market.

economic theorists invented to study the world has become the vocabulary of the practitioners themselves. It is never easy for a theorist to get an insider's look at the markets; I was especially afraid that if I joined a large firm for a short-term visit I would be relegated to some very specialized subdepartment. In the winter of 1989, I had spent one summer month at Kidder Peabody working on a small mathematical problem in their tiny mortgage research department. This I thought would be the perfect firm in which to spend a semester: small, yet a participant in nearly every market, and with a venerable 125 year history beginning in the last year of the Civil War. Little did I realize at the time both how lucky and how unlucky I would be. Toward the end of my stay, the firm's management decided they needed to reorient the research department in line with the growing mathematical emphasis of their competitors. They turned to me for some advice, which I did my best to give, and in October of 1990, three months after I ended my sabbatical there and returned to Yale, they offered me the job of Director (part-time) of Fixed Income Research.

During the time I spent at Kidder, the firm went from being a negligible player in the mortgage business to the greatest CMO powerhouse on the street. For each of the five years between 1990 and 1994, Kidder Peabody underwrote more CMOs than any other firm, averaging nearly 20% in a market with forty competitors. The next biggest firm averaged under 10% in that time. In those five years we created more than $200 billion of CMO tranches.

Suddenly in April 1994, the firm was rocked by a trading scandal in the government bond department, where the head trader was accused of fraudulently manipulating the accounting system into crediting him with $300 million of phantom profits. Two weeks later the Fed raised interest rates, and then did so again four more times the rest of the year. The bond market plummeted, and activity on Wall Street shrank markedly. Many firms had terrible years. The worst hit perhaps were those who had bet on interest rates continuing their downward trajectory by buying bonds such as inverse floaters. CMO tranches, like bonds in general, did terribly, but some of the more esoteric pieces called mortgage derivatives, such as the inverse floaters, did the worst of all. The newspapers were filled with stories about the dangers of investing in mortgage derivatives, and the CMO market suddenly dried up. The great Kidder money machine ground to a halt. At the same time, the trading scandal became more lurid. In January 1995, General Electric officially sold the firm to Paine Webber, and Kidder Peabody ceased to exist as an investment bank after 130 years.

The leading Kidder mortgage traders founded a new company called Ellington Capital Management, which I subsequently joined as a small partner. Instead of creating mortgage derivatives, as we did at Kidder, now we buy them. In one and a half years of operation, we have earned an extraordinarily high return on our capital. So we return to our puzzles: what good did Kidder Peabody do, and how can it be that in a world of rational investors it is possible to make money on the sell side of a market, and then on the buy side of the same market?

3. THE MODEL

We now try to formalize some of the observations we have made about the mortgage market, and we investigate the logical implications of our view of the world.

3.1 STATES OF THE WORLD

Probably the most important ingredient in our model is the idea of a state of the world, representing a complete description of what will happen to every asset and commodity and preference in the economy over its life. Furthermore, we posit a set S representing all possible states of the world.

Investors have always been aware that there are risks, and they have always sought to understand these risks. Until recently, however, much of this took the form of a qualitative analysis of important factors that might bear on their returns. For example, ten years ago an investor in the mortgage market would be told that demographics and macroeconomic business cycles affect the demand for new housing. He might have been shown graphs of previous interest rate cycles, and if he was really sophisticated, he might have been shown a history of mortgage prepayments. Nowadays, an investor will demand a computer calculated description of all his future cash flows under each of several scenarios that specify all relevant events for the next 30 years (including vectors of future interest rates and mortgage prepayments). These scenarios correspond to our states of the world.

Needless to say, there is an infinity of potential scenarios, so no matter how many the investor sees, he has at best anecdotal information about what might happen to his investment. Moreover, he cannot really make use of this information unless he also makes the additional heroic assumption that the states of the world he asked about constitute all possible states of the world. Thus, we arrive at the same point of view we take in our modeling. Our set S is in effect the union over all the investors of the states each thinks are possible. For expository simplicity, we shall suppose that there are only two time periods, today and the future.

3.2 AGENTS

Agents act on a desire to consume as much as possible. Of course, there is a variety of say L goods over which agent tastes differ, and agents must also take into account that their consumption might be different in different states of the world. The consumption space is, therefore, $R_+^L \times R_+^{SL}$, meaning that agents realize that they must choose between consumption plans that specify what they will consume of each of the L goods today, and what they will consume of each of the L consumption goods in every state next period. For each good $s\ell$ there is a market on which the good is traded at a price $p_{s\ell}$.

As we said earlier, investors trade assets primarily to bet or to hedge. We incorporate the betting motive into our model by differences in the probabilities investors assign to the states in S. Hedging is represented by supposing agents want to maximize not the expectation of consumption, but the expectation of the utility of consumption. We suppose utility is a concave function of consumption, which means that additional consumption increases utility more when there is less to start with. Thus, agents would prefer to give up some consumption in states where they will be rich in order to obtain the same amount extra in equally probable states in which they expect to be poor.

We subsume both betting and hedging by allowing agents $h \in H$ to have arbitrary concave, continuous, and monotonic utilities

$$u^h : R_+^L \times R_+^{SL} \to R$$

and arbitrary (but strictly positive) endowments

$$e^h \in R_{++}^L \times R_{++}^{SL} .$$

Utility may, but need not, take the special form

$$u^h(x^h) = v^h(x_0^h) + \sum_{s=1}^{s=S} \gamma_s^h v^h(x_s^h) ,$$

where γ_s^h represents the probability agent h assigns to state s. An agent h who does not trade will be forced to consume his endowment. He may well prefer to exchange wealth from some state s where his marginal utility of consumption $\partial u^h(e^h)/\partial x_{s\ell}$ is relatively low, either because state s is regarded to be of low probability or because consumption e_s^h is already so high there, for wealth in another state s' where his marginal utility of consumption is relatively high.

3.3 PROMISES

From the foregoing it is evident that economic welfare can be improved if agents can exchange consumption goods across states of the world. The purpose of the financial markets is to make such exchanges feasible.

Agents trade across states by making promises before the states are realized. There is evidently no point offering to give up goods in a state once everybody knows it will never occur. The receiver of a promise must provide something in return, usually a payment today. Thus, it makes sense to think of a promise as an asset sold by the promisor to the receiver in exchange for a price today. (More generally one could allow for the exchange of promises against promises, which indirectly are permitted in our model, as we shall see.) A promise $j \in J$ is denoted by a vector $A_j \in R_+^{SL}$ specifying the basket of goods that are promised for delivery in each state. Equivalently, $p_s \cdot A_{sj}$ is the money promised for delivery in state s. Notice that the promise may be contingent; there is no reason (at least a priori) that the vector A_{sj} should be the same in each state of the world. Indeed agents trade across states of the world by buying promises that are supposed to deliver in some states and selling promises that are supposed to deliver in other states.

We restrict the collection of promises to a finite set J. This is done mostly for technical reasons, so that we can work in finite-dimensional Euclidean spaces. Of course, we can take $\#J$ so large that the assets in J come close to approximating every conceivable asset on some bounded region.

3.4 DELIVERIES AND DEFAULT

An agent who makes a promise has the option of delivering less than he promised. We denote by D^h_{sj} the deliveries of money agent h actually makes in state s. The shortfall $[p_s \cdot A_{sj} - D^h_{sj}]^+$ between what is promised and what is delivered is the amount of default, measured in dollars. Note that there is no point in delivering more than is promised.

Some borrowers are more reliable than others, either because they are more honest, or because the penalties to defaulting affect them more acutely, or because they simply have more wealth. A full-blown model of default and penalties, as presented for example in Dubey-Geanakoplos-Shubik,[7] would need to take into account the adverse selection facing lenders who should be wary that anybody who wants to borrow from them is likely to be a worse than average risk. Collateral has the advantage that the lender need not bother with the reliability or even the identity of the borrower, but can concentrate entirely on the future value of the collateral. Collateral thus retains anonymity in market transactions. If there are no penalties to defaulting, then every borrower will deliver the minimum of what he owes in every state and the value of the collateral he put up to secure his promise.

A remarkable theorem of Ken Arrow[5] shows that if penalties can be taken to be infinite, so that nobody dares default, then any competitive equilibrium allocation will be efficient (i.e., optimal from the social point of view) provided there are at least as many independent promises as there are states of nature. Of course, when penalties are infinite for everybody, there is no adverse selection and anonymity is restored. The trouble is that in the real world it is impossible to make penalties too harsh; the punishment must in some sense fit the crime if it is to be politically feasible. More troubling, it is inconceivable that there will be as many assets as there are states of nature, particularly if these assets must be traded in liquid (high volume) markets. When the number of assets falls short of the number of states, then it is generally better from the social point of view to keep the penalties at intermediate levels of harshness, as shown in Dubey-Geanakoplos-Shubik.[7] If the harshness of penalties is limited, some people will default, and adverse selection becomes an issue. Collateral will again become useful together with the penalties in order to guarantee more promises. In the interest of simplicity, we are led to preserve anonymity and restrict attention to collateral in place of penalties.

3.5 COLLATERAL

The difficulty with promises is that they require some mechanism to make sure they are kept. This can take the form of penalties, administered by the courts, or collateral. As we mentioned at the outset, more and more often collateral has displaced penalties. In this chapter we shall exclusively deal with collateral, by supposing that there is no penalty, legal or reputational, to defaulting. Of course, even collateral requires the courts to make sure the collateral changes hands in case of default.

The simplest kind of collateral is pawn shop collateral—valuable goods like watches or jewelry left with third parties (warehoused) for safekeeping. Financial markets have advanced as the number of goods that could function as collateral has increased, from watches and jewelry, to stocks and bonds. A further advance occurred when lenders (instead of warehouses) held collateral, like paintings, that afforded them utility. This required a more sophisticated court system, because the lender had to be obliged to return the collateral if the promise was kept. The biggest advance, however, was in allowing the borrower himself to continue to hold the collateral. This enabled houses, and later cars, to be used as collateral, which again is only possible because of a finely tuned court system that can enforce the confiscation of collateral.

More recently the complexity of collateral has taken several more giant steps forward. Pyramiding occurs when an agent A puts up collateral for his promise to B, and then B in turn uses A's promise to him, and hence in effect the same collateral for a promise he makes to C, who in turn reuses the same collateral for a promise he makes to D. Mortgage pass through securities offer a classic example of pyramiding. Pyramiding naturally gives rise to chain reactions, as a default by Mr. A ripples through, often all the way to D.

Still more complex is tranching, which arises when the same collateral backs several promises to different lenders. Needless to say, the various lenders will be concerned about whether their debts are adequately covered. Tranching usually involves a legal trust that is assigned the duty of dividing up the collateral among the different claims according to some contractual formula. Again collateralized mortgage obligations offer a classic example of tranching.

Every one of these innovations is designed to increase or to stretch the available collateral to cover as many promises as possible. We shall see later that active default is another way of stretching the available collateral.

For the formal analysis in this chapter we shall avoid pyramiding and tranching. All collateral will be taken by assumption to be physical commodities. Collateral must be put up at the moment the promise is sold, even if the delivery is not scheduled for much later. Agents are not allowed to pledge their future endowment as collateral, because that would raise questions in the minds of lenders about whether the borrowers actually will have the endowments they pledged and, therefore, it would once again destroy the anonymity of markets.

3.6 ASSETS

To each promise j we must formally associate levels of collateral. Any good can potentially serve as collateral, and there is no reason why the single promise j cannot be backed by a collection of goods. The bundle of goods that is required to be warehoused for asset j is denoted $C_j^W \in R_+^L$, the vector of goods that the lender is allowed to hold is denoted $C_j^L \in R_+^L$, and the vector of goods the borrower is obliged to hold is denoted $C_j^B \in R_+^L$. An asset j is defined by the promise it makes *and* the collateral backing it, $(A_j, C_j^W, C_j^L, C_j^B)$. It is quite possible that there will be many assets which make the same promises $A_j = A_{j'}$, but trade at different prices because their collateral levels are different $(C_j^W, C_j^L, C_j^B) \neq (C_{j'}^W, C_{j'}^L, C_{j'}^B)$. Similarly the two assets might require exactly the same collaterals, but trade at different prices because their promises are different.

The price of asset j is denoted by π_j. A borrower sells asset j, in effect borrowing π_j, in return for which he promises to make deliveries according to A_j.

3.7 PRODUCTION

Collateral is useful only to the extent that it is still worth something when the default occurs. Durability is a special case of production, so we introduce production into our model, and allow all goods to be durable, to varying degrees.

For ease of notation we shall suppose that production is of the fixed coefficient, constant returns to scale variety. One unit of commodity ℓ becomes a vector of commodities next period. A house may become a house that is one year older, wine may become a wine that is one year older, grapes may become wine one year later and so on. In these examples, one good became a different good the next period, but there is no reason not to permit one good to become several goods if it splits.

The transformation of a commodity depends of course on how it is used. Again for simplicity, we suppose that each commodity ℓ is transformed into a vector $Y_{s\ell}^0 \in R_+^L$ in each state s if it is used for consumption (e.g., living in a house, or using a light bulb). If ℓ is warehoused, then we assume that it becomes a vector $Y_{s\ell}^W \in R_+^L$ in each state s. Likewise, if it is held as collateral by the lender it becomes a vector $Y_{s\ell}^L \in R_+^L$ in each state s, while if it is held by the borrower it becomes the vector $Y_{s\ell}^B \in R_+^L$ in each state s. The $SL \times L$ dimensional matrices Y^0, Y^W, Y^L, Y^B summarize these different durabilities.

Observe that we have allowed for differential durability depending on the use to which the commodity is put. But we have not allowed the durability to be affected by the identity of the user, or by the intensity of its use (which would amount to the same thing). In this way the anonymity of markets is maintained, and our modeling problem becomes easier. In Geanakoplos-Zame-Dubey[11] the more general case of penalties and individual differences in the treatment of collateral is permitted.

Given the collateral requirements (C_j^W, C_j^L, C_j^B) for each promise j, the security they provide in each state s is

$$p_s \cdot [Y_s^W C_j^W + Y_s^L C_j^L + Y_s^B C_j^B] .$$

No matter who holds the collateral, it is owned by the borrower but may be confiscated by the lender (actually by the courts on behalf of the lender) if the borrower does not make his promised deliveries. Since we have assumed that the borrower has nothing to lose but his collateral from walking away from his promise, it follows that the actual delivery by every agent h on asset j in state s will be:

$$D_{sj}^h = \min\{p_s \cdot A_s^j, p_s \cdot [Y_s^W C_j^W + Y_s^L C_j^L + Y_s^B C_j^B]\} .$$

4. COLLATERAL EQUILIBRIUM

We are now ready to put together the various elements of our model. An economy E is defined by a vector

$$E = ((u^h, e^h)_{h \in H}, (A^j, C_j^W, C_j^L, C_j^B)_{j \in J}, (Y^0, Y^W, Y^L, Y^B))$$

of agent utilities and endowments, asset promises and collateral levels, and the durability of goods kept by buyers, warehouses, lenders, and borrowers, respectively.

In keeping with the standard methodological approach of general equilibrium and perfect competition, we suppose that *in equilibrium* agents take the prices (p, π) of commodities and assets as given. We do not mean to suggest by this that the prices are actually set somewhere else, but only that competition among many agents guarantees that no one of them has the power to set the prices himself, so that, paradoxically, it appears to each and every one of them that the prices are set elsewhere.

These equilibrium prices cannot be known in advance, independent of the preferences and endowments of the agents, because they must be precisely chosen so that when agents believe in them, their optimizing trades will lead to a balance of supply and demand in all the commodity markets and all the asset markets.

Our price-taking hypothesis also has the implication that agents have rational expectations about future prices, for these are taken as given as well. Agents in our model have perfect conditional foresight, in that they anticipate at time 0 what the prices p_s will be, depending on which state s prevails at time 1. Since they know the collateral that has been put up, and they know the production technology, they also understand in each state how much each asset will actually pay.

It might seem, therefore, that we could simply replace each asset promise A_j with an actual delivery vector, and thereby bypass the complications of collateral. But this is not possible, since whether an asset defaults or not in state s depends on whether the promise or the collateral is worth more. Since both are vectors, this cannot be known in advance until the prices $p_s \in R_+^L$ have been determined in equilibrium.

4.1 THE BUDGET SET

Given the prices (p, π), each agent h decides what net trades of commodities $(x_0 - e_0^h)$, what asset purchases θ, and what asset sales φ he will make at time 0. Note that for every promise φ_j that he makes, he must put up the corresponding collateral $(C_j^W, C_j^L, C_j^B)\varphi_j$. The value of all his net trades at time 0 must be less than or equal to zero, that is, the agent cannot purchase anything without raising the money by selling something else (initial endowments of money are taken to be zero).

After the state of nature is realized in period 1, the agent must again decide on his net purchases of goods $(x_s - e_s^h - Y_s^0 x_0)$. Recall that the goods x_0 whose services he consumed at time zero may be durable, and still available (in the form $Y_s^0 x_0$) for consumption at time 1 in each state s. These net expenditures on goods can be financed out of sales of the collateral that the agent put up in period 0, and from the receipts from assets j that he purchased at time 0, less the deliveries the agent makes on the assets he sold at time 0. Putting all these transactions together, and noting again that the agent cannot buy anything without also selling something else of at least equal value, we derive the budget set for agent h:

$$
\begin{aligned}
&B^h(p,\pi) = \{(x,\theta,\varphi) \in R_+^L \times R_+^{SL} \times R_+^J \times R_+^J : \\
&\quad p_0(x_0 - e_0^h) + \pi(\theta - \varphi) + p_0 \sum_{j\in J}(C_j^W + C_j^L + C_j^B)\varphi_j \le 0 \text{ and for all } s \in S, \\
&\quad p_s(x_s - e_s^h - Y_s^0 x_0) \le \sum_{j\in J} \varphi_j p_s \cdot [Y_s^W C_j^W + Y_s^L C_j^L + Y_s^B C_j^B] + \\
&\quad \sum_{j\in J}(\theta_j - \varphi_j)\min\{p_s \cdot A_s^j, p_s \cdot [Y_s^W C_j^W + Y_s^L C_j^L + Y_s^B C_j^B]\}\}\,.
\end{aligned}
$$

4.2 EQUILIBRIUM

The economy $E = ((u^h, e^h)_{h\in H}, (A_j, C_j^W, C_j^L, C_j^B)_{j\in J}, (Y^0, Y^W, Y^L, Y^B))$ is in equilibrium at macro prices and individual choices $((p,\pi), (x^h, \theta^h, \varphi^h)_{h\in H})$ if supply equals demand in all the goods markets and asset markets, and if given the prices, the designated individual choices are optimal, that is if:

$$
\sum_{h\in H}\left(x_0^h - e_0^h - \sum_{j\in J}(C_j^W + C_j^L + C_j^B)\varphi_j^h\right) = 0, \tag{1}
$$

$$
\sum_{h\in H}\left(x_s^h - e_s^h - Y^0 x_0 - \sum_{j\in J}\varphi_j^h[Y_s^W C_j^W + Y_s^L C_j^L + Y_s^B C_j^B]\right) = 0 \tag{1'}
$$

$$\sum_{h \in H} (\theta^h - \varphi^h) = 0 \tag{2}$$

$$(x^h, \theta^h, \varphi^h) \in B^h(p, \pi) \tag{3}$$

$$(x, \theta, \varphi) \in B^h(p, \pi) \Rightarrow u^h\left(x_0 + \sum_{j \in J}[C_j^B \varphi_j + C_j^L \theta_j], \bar{x}\right) \tag{4}$$

$$\leq u^h\left(x_0^h + \sum_{j \in J}[C_j^B \varphi_j^h + C_j^L \theta_j^h], \bar{x}^h\right).$$

We write $x^h = (x_0^h, \bar{x}^h)$, so consumption at time 0 is $x_0^h + \sum_{j \in J}[C_j^B \varphi_j^h + C_j^L \theta^h]$.

5. BENCHMARK ECONOMIES

It is easiest to see the role collateral plays in the economy and the realism of our model by considering three alternative benchmark economies in which collateral is absent, either because there is no need to worry about default, or because the court system is so backward that there is no available provision for recovering collateral in case of default. We can simultaneously cover all three benchmark cases with one formal model that drops collateral and default, by reinterpreting in various ways the set J of assets.

Consider an economy in which the promises are exactly as before, but in which agents automatically keep all their promises (perhaps because implicitly they recognize that there would be a horrible punishment if they did not). Without default or collateral, only net asset trades matter and the budget set would simply become

$$B^h(p, \pi) = \{(x, \theta, \varphi) \in \mathbf{R}_+^L \times \mathbf{R}_+^{SL} \times \mathbf{R}_+^J \times \mathbf{R}_+^J : p_0(x_0 - e_0^h) + \pi(\theta - \varphi) \leq 0,$$
$$p_s(x_s - e_s^h - Y_s^0 x_0) \leq p_s A(\theta - \varphi) \text{ for all } s \in S\}.$$

In equilibrium demand would have to balance supply for commodities and assets, and every agent would have to be choosing optimally in his budget set. The equilibrium conditions then are:

$$\sum_{h \in H} (x_0^h - e_0^h) = 0, \sum_{h \in H} (x_s^h - e_s^h - Y_s^0 x_0) = 0, \quad \forall s \in S \tag{1}$$

$$\sum_{h \in H} (\theta^h - \varphi^h) = 0 \tag{2}$$

$$(x^h, \theta^h, \varphi^h) \in B^h(p, \pi) \tag{3}$$

$$(x, \theta, \varphi) \in B^h(p, \pi) \Rightarrow u^h(x) \leq u^h(x^h) \tag{4}$$

We describe each of our benchmark economies as special cases of this model below. In a later section, we shall consider two examples which further illuminate the difference between these models and our collateral model. Perhaps the most important difference is that without explicitly considering the possibility of default and the safeguards the social system designs to counteract it, one has no good theory to explain which asset markets are actively traded and which are not.

5.1 ARROW-DEBREU COMPLETE MARKETS EQUILIBRIUM

In our first benchmark economy we presume that every conceivable promise is available to be traded in J. It can then easily be shown, as Arrow and Debreu noted in the 1950s, that equilibrium efficiently spreads risks and that there is nothing any government agency can do to improve general welfare. A government which observes a complete set of asset markets should, therefore, concentrate on punishing people who default by setting draconian penalties so that there will be no default. There is no need for a government in such a world of complete markets to make any arrangements for compensating lenders against whom there is default.

In Arrow-Debreu equilibrium, the value of durable commodities will be at a peak. Durable goods can have tremendously high potential value because they provide services over a long period of time. If there were sufficient assets available, agents could in principle sell promises to pay their entire future endowments in order to get the money to buy the durable good today. An agent can thus buy a house for say $200,000 even though he has only a few thousand dollars of savings, because he can borrow the rest of the money by selling promises to pay people later out of his future endowment.

The value of a durable good can fluctuate over time or across states of nature because tastes for the good have changed across the states, or because people expect tastes to change (in a multiperiod model). There is no reason for prices to change because relative income changes across states of nature, since with complete markets, agents will insure themselves so that relative incomes will not change dramatically across states of nature.

In Arrow-Debreu equilibrium there is no connection between the value of promises outstanding, and the value of durable goods. An agent can, for example, sell a promise to deliver a large number of goods tomorrow for a very high price today, even though he owns no durable goods of any kind.

Arrow, Debreu, and McKenzie noted that a complete markets equilibrium always exists, given the assumptions we have already made on endowments and utilities.

5.2 ASSETLESS EQUILIBRIUM

At the other extreme from the Arrow-Debreu complete markets economy is an economy in which no provision is made for enforcing promises. In that case investors would rationally anticipate that all promises would be broken, and the economy would effectively reduce to one without any assets: $J = \phi$. In these circumstances it is easy to see that durable goods might have very low prices, since the potential buyers might not be able to give much immediately in return, and by hypothesis they are forbidden from promising anything in the future. Holding durable goods would also be very risky, because it would be impossible to hedge them except by holding other durable goods. Not only might prices change across states of nature because tastes changed or because tastes were expected to change, but more importantly, changes in the distribution of income could affect durable goods prices.

Equilibrium would always exist, but risk would not be distributed efficiently. The government in such an assetless economy should work hard at creating new markets, perhaps by subsidizing them, or even by guaranteeing delivery if an agent should default. Only at a later stage of economic development can a country expect to benefit by constructing a court system capable of recovering collateral and so on.

5.3 THE GEI INCOMPLETE MARKETS MODEL

Lastly we turn to an intermediate and more realistic setting in which there are some assets, but far fewer than the astronomical number it would take to complete the asset markets. At first glance one would expect the equilibrium behavior of this kind of economy to be intermediate between the two just described. In some respects it is. For example, GEI equilibria are typically (though not always) better from a welfare point of view than assetless economies, but less good than complete markets economies. On the other hand, the presence of some but not all asset markets creates the possibility of new phenomena that are smoothed over when markets are complete or completely absent. For example, the volatility of durable goods prices may be greater with incomplete markets than it is with either complete markets or assetless economies. And the government may be able to intervene by taxes or subsidies to improve welfare in a GEI economy, but not in an assetless or Arrow-Debreu economy. Some of the rough edges of a GEI economy may be rounded out by replacing the (implicit) infinite penalties for default with collateral.

If asset markets are missing (that is if $\#J$ is taken to be smaller than S but greater than 1 by assumption), then the equilibrium GEI allocation does not share risk in the ideal way, despite the absence of default. Somewhat more surprising, it is possible in some GEI equilibria to make everybody better off without adding any assets simply by letting people default. The prohibition against default is presumably enforced by draconian and inflexible penalties. By substituting collateral or weaker penalties, the system may be improved despite the fact that defaults appear. For example, if there were one highly improbable state of the world in which an agent could not pay, then he could never sell an asset that promised something in every state. The loss to the lenders from his default in just this one state, plus any penalty he might be required to pay for defaulting, could well be less important than the gains to both the lenders and to him from being allowed to borrow the money and pay it back in the other states.[7] The moral is: when asset markets are complete, set high penalties, but when some asset markets are missing, set low or intermediate levels of default penalties, when penalties are available, or better yet, use collateral whenever possible. We shall see this in our example.[1]

A shortage of available assets will tend to reduce the value of durable goods. If assets are missing, then the agents only have available the income from today's endowment and a few promises to spend on the durable good; the demand for the durable good is effectively curtailed, and its price will typically be between the complete markets price and the assetless economy price. More surprising, the volatility of durable goods prices will often be higher than in complete or assetless economies on account of the possibility of leverage. We discuss this in the next section.

Unless the assets in J are redundant, typically they will all be traded in the GEI equilibrium. The reason is that in equilibrium the marginal utility to each agent of buying or selling a little more of the asset must equal the price. The chances that every agent will happen to have the same marginal utility at zero trade are infinitesimal. The GEI model, therefore, cannot explain why we observe so few assets actually traded. In order to match reality, it must simply assume that many assets are missing from J. That is how the theory gets its name (general equilibrium with incomplete markets). But it begs the question: which asset markets are missing and why?

The GEI theory, like the complete markets economy, allows for large trades in the asset markets. An agent might take a huge positive position in one asset, and a huge negative position in another asset that nearly offsets all his debts. In general, the gross value of the promises requiring delivery in each state might be much larger than the value of all the outstanding goods in that state, which in turn is greater than the value of the durable goods held over from the previous period. Agents would be able to keep their promises because other agents kept their promises to them. If each agent were forced to pay out of his own income before receiving payments from others, the equilibrium would break down. Of course, some promise structures give rise to larger promises than others, even for the same underlying economic preferences and endowments. This possibility for arbitrarily large asset transactions sometimes compromises the existence of GEI equilibrium.[12]

[1]When some asset markets are missing it can also be shown[10] that an outside agency that knows everybody's utilities can almost always make every agent better off, without introducing any new assets, if it has the power to coerce agents into trading a little differently on each asset than they would left to their own devices.

6. PROPERTIES OF COLLATERAL EQUILIBRIUM

6.1 THE ORDERLY FUNCTION OF MARKETS

We return now to our model of equilibrium with default. Compared to the benchmark model of general economic equilibrium, the agents we have described must anticipate not only what the prices will be in each state of nature, and not only what the assets promise in each state of nature, but also what they will actually deliver in each state of nature. The hypothesis of agent rationality is therefore slightly more stringent in this model than in the conventional models of intertemporal perfect competition. Nevertheless, equilibrium always exists in this model, (under the assumptions made so far), yet in the standard model of general equilibrium with incomplete asset markets, equilibrium may not exist. The following theorem is taken from GZD.

THEOREM 1 (GEANAKOPLOS, ZAME, DUBEY[11]). Under the assumptions on endowments and utilities already specified, equilibrium must exist, no matter what the structure of assets and collateral.

If an asset contains no provision for collateral whatsoever, then, of course, everybody will rightly anticipate that it will deliver nothing, and its equilibrium price will be 0. Indeed the economy would function exactly the same way if it were not available at all. For assets with some nonzero collateral, agents will not be able to sell arbitrarily large quantities, because they will not be able to obtain the required collateral. This limiting factor helps to guarantee the existence of equilibrium.

6.2 ENDOGENOUS ASSETS

One of the major shortcomings of the GEI model is that it leaves unexplained which assets are traded. Generically, all the assets exogenously allowed into the model will be traded. When default can only be avoided by collateral, the situation is different and much more interesting.

The crucial idea is that without the need for collateral, the marginal utility $m_j^h(B)$ to an agent h of buying the first unit of an asset j is almost exactly the same as the marginal utility loss $m_j^h(S)$ in selling the first unit of the asset; we can call both m_j^h. Only by an incredible stroke of luck will it turn out that $m_j^h = m_j^{h'}$ for different agents h and h' and, hence, asset j will almost surely be traded in a GEI equilibrium. When collateral must be provided by the seller, the disutility of making a promise goes up, sometimes by as much as the consumption foregone by buying the collateral. If the required collateral is borrower held, and if it is something that agent h planned to hold anyway, then there is no extra utility loss from selling the first unit of asset j. But if agent h did not plan to hold the collateral for consumption, or if all that he intended to hold as consumption has already been allocated as collateral for other promises, then the loss in utility from selling even the first unit of asset j would be larger than the marginal utility from buying the first unit of asset j, $m_j^h(S) > m_j^h(B)$. It might well transpire that

$$\min_{h \in H} m_j^h(S) > \pi_j > \max_{h \in H} m_j^h(B)$$

and, hence, that asset j does not trade at all in equilibrium.

This situation can be most clearly seen when the value of the Arrow-Debreu promises in some state exceeds the salvage value of all the durable goods carried over into that state. It is then physically impossible to collateralize every socially useful promise up to the point that every delivery is guaranteed without exception. The market system, through its equilibrating mechanism, must find a way to ration the quantity of promises. This rationing is achieved by a scarcity of collateral. The resulting wedge between the buying marginal utility of each asset and the selling marginal utility of the asset, however, not only serves to limit the quantity of each promise, but more dramatically, it chokes off most promises altogether, so that the subset of assets that are actually traded is endogenous and potentially much smaller than the set of available assets.

Consider a set $\Pi = \{a \in Z_+^{SL} : a_{s\ell} \leq 10^{10}\}$ of potential promises, restricted to integer coordinates. Note that Π is a finite set, but it includes virtually every conceivable promise (since only the relative size of the promises matter). Let $\mathcal{C} = \{(c^W, c^L, c^B) \in Z_+^L \times Z_+^L \times Z_+^L : c_\ell^i \leq 10^{100}\}$ be a finite set of (virtually) all potential collateral levels. We can then take $J = \Pi \times \mathcal{C}$. In equilibrium, all of these many assets will be priced, but only a very few of them will actually be traded. The rest will not be observable in the marketplace and, therefore, the appearance will be given of many missing markets. The untraded assets will lie dormant not because their promises are irrelevant to spreading risk efficiently, but because the scarce collateral does not permit more trade.

It would be interesting to catalogue the rules by which the market implicitly chooses one promise over another, or one level of collateral over another. This issue is more fully developed in Geanakoplos-Zame-Dubey,[11] but let us note some things here. The easiest way of economizing on collateral is by allowing default in some states of nature. Moreover, if one vector of collaterals guarantees full delivery in every state of nature, there is no point in trading the same promise collateralized by greater levels of collateral. Finally, if a vector of promises is very different from the vector of its collateral values across the states of nature, the asset is not well drawn. In some states there will be too much collateral, and in others not enough. One might suspect that such an asset would also not be traded. The general principle is that the market chooses assets that are as efficient as possible, given the prices. We make this precise in the next section.

6.3 CONSTRAINED EFFICIENCY

It is to be expected that an increase in available collateral, either through an improvement in the legal system (e.g., borrower held collateral), or through the increased durability of goods, will be welfare improving. More subtly, we might wonder whether government intervention could improve the functioning of financial markets given a fixed level of available collateral. After all, the unavailability of collateral might create a wedge that prevents agents from trading the promises in J that would lead to a Pareto improving sharing of future risks. If the government transferred wealth to those agents unable to afford collateral, or subsidized some market to make it easier to get collateral, could the general welfare be improved? What if the government prohibited trade in assets with low collateral levels? The answer, surprisingly, is no, at least under some important restrictions.

CONSTRAINED EFFICIENCY THEOREM.[11] *Each collateral equilibrium is Pareto efficient among the allocations which* (1) *are feasible and* (2) *given whatever period 0 decisions are assigned, respect each agent's budget set at every state* s *at time 1 at the old equilibrium prices, and* (3) *assume agents will deliver no more on their asset promises than they have to, namely the minimum of the promise and the value of the collateral put up at time 0, given the original prices.*

In particular, no matter how the government redistributes income in period 0, and taxes and subsidizes various markets at time 0, if it allows markets to clear on their own at time 1, then we can be sure that if at time 1 market clearing relative prices are the same as they were at the old equilibrium, then the new allocation cannot Pareto-dominate the old equilibrium allocation. This will be illustrated in our examples.

6.4 VOLATILITY, THE DISTRIBUTION OF WEALTH, AND COLLATERAL

In any general economic equilibrium, the price of a good depends on the utilities of the agents and the distribution of wealth. If the agents who are fondest of the good are also relatively wealthy, the good's price will be particularly high. Any redistribution of wealth away from these agents toward agents who like the good less will tend to lower the price of the good.

To a large extent, the value of durable goods depends on the expectations, and, when markets are incomplete, on the risk aversion of potential investors, as well as on intrinsic utility for the good. These multiple determinants of value make it quite likely that there will be wide divergences in the valuations different agents put on durable goods.

For example, farms in 1929 could be thought of as an investment, available to farmers and bankers, but to farmers there is a superior intrinsic value that made it sensible for them to own them and use them at the same time. Since the farmers did not have enough money to buy a farm outright, they typically borrowed money and used the farm as collateral. Similarly, mortgage derivatives in the 1990s are worth much more to investors who have the technology and understanding to hedge them than they are to the average investor.

Needless to say, the value of many durable assets will be determined by the marginal utilities of those who like them the most. (This is literally true if one cannot sell the asset short.)

Since the 1929 stock market crash, it has been widely argued that low margin requirements can increase the volatility of stock prices. The argument is usually of the following kind: when there is bad news about the stocks, margins are called and the agents who borrowed against the stocks are forced to put them on the market, which lowers their prices still further.

The trouble with this argument is that it does not quite go far enough. In general equilibrium theory, every asset and commodity is for sale at every moment. Hence, the crucial step in which the borrowers are forced to put the collateral up for sale has, by itself, no bite. On the other hand, the argument is exactly on the right track.

We argue that, indeed, using houses or stocks or mortgage derivatives as collateral for loans (i.e., allowing them to be bought on margin) makes their prices more volatile. The reason is that those agents with the most optimistic view of the assets' future values, or simply the highest marginal utility for their services, will be enabled by buying on margin to hold a larger fraction of them than they could have afforded otherwise.

The initial price of those assets will be much higher than if they could not be used as collateral for two reasons: every agent can afford to pay more for them by promising future wealth, and second, the marginal buyer will tend to be somebody with a higher marginal utility for the asset than would otherwise be the case.

As a result of the margin purchases, the investment by the optimistic agents is greatly leveraged. When the asset rises in value, these agents do exceedingly well, and when the asset falls in price, these agents do exceedingly badly. Thus, on bad news the stock price falls for two reasons: the news itself causes everyone to value it less, and this lower valuation causes a redistribution of wealth away from the optimists and toward the pessimists who did not buy on margin. The marginal buyer of the stock is, therefore, likely to be someone less optimistic than would have been the case had the stock not been purchased on margin, and the income redistribution not been so severe. Thus, the fall in price is likely to be more severe than if the stock could not have been purchased on margin.

The properties of collateral equilibrium are illustrated in two examples.

7. EXAMPLE: BORROWING ACROSS TIME

We consider an example with two agents[2] $H = \{A, B\}$, two time periods, and two goods F (food) and H (housing) in each period. For now we shall suppose that there is only one state of nature in the last period.

We suppose that food is completely perishable, while housing is perfectly durable. Thus, the consumption-durability technology is $Y_1^0 = \binom{0\,0}{0\,1}$, meaning that 1 unit of F becomes 0 units of both goods, and 1 unit of H becomes 1 unit of H in period 1.

We suppose that agent B likes living in a house much more than agent A,

$$u^A(x_{0F}, x_{0H}, x_{1F}, x_{1H}) = x_{0F} + x_{0H} + x_{1F} + x_{1H},$$
$$u^B(x_{0F}, x_{0H}, x_{1F}, x_{1H}) = 9x_{0F} - 2x_{0F}^2 + 15x_{0H} + x_{1F} + 15x_{1H}.$$

Furthermore, we suppose that the endowments are such that agent B is very poor in the early period, but wealthy later, while agent A owns the housing stock

$$e^A = (e_{0F}^A, e_{0H}^A, e_{1F}^A, e_{1H}^A) = (20, 1, 20, 0),$$
$$e^B = (e_{0F}^B, e_{0H}^B, e_{1F}^B, e_{1H}^B) = (4, 0, 50, 0).$$

7.1 ARROW-DEBREU EQUILIBRIUM

If we had complete asset markets, so that $J = \{A_{1F}, A_{1H}\}$, with infinite default penalties and no collateral requirements, then it is easy to see that there would be a unique equilibrium (in utility payoffs):

$$p = (p_{0F}, p_{0H}, p_{1F}, p_{1H}) = (1, 30, 1, 15), \pi = (1, 15)$$
$$x^A = (x_{0F}^A, x_{0H}^A, x_{1F}^A, x_{1H}^A) = (22, 0, 48, 0),$$
$$x^B = (x_{0F}^B, x_{0H}^B, x_{1F}^B, x_{1H}^B) = (2, 1, 22, 1),$$
$$u^A = 70\ ;\ \ u^B = 62.$$

Housing has a very high price in period 0 because it is a durable good which provides a lot of utility to agent B in period 0 and in period 1. Even though agent B is relatively poor in period 0, he can afford to pay the high price of 30 for the house because he can sell a promise to deliver 28 units of food in period 1 for a price of 28 today, which he then repays out of his endowment of food in period 1. The price of housing naturally falls over time as it ages. The fact that the relative endowments of A and B change over time has no bearing on the fluctuation of housing prices.

Notice that the equilibrium is welfare efficient: the agents who most like housing own the whole housing stock.

[2] In order to maintain our hypothesis of many agents, we shall actually suppose that there are a great many agents of type A, and the *same* large number of agents of type B. Agents of the same type will take identical actions.

7.2 GEI EQUILIBRIUM

Suppose now that instead of the two Arrow-Debreu-state-contingent commodity assets there is just the first asset $A_j = \binom{1}{0}$ which promises 1 unit of food in period 1, and no housing. Suppose, in addition, that there are infinite default penalties for this asset, so that full delivery is assured even with no collateral. As Arrow[5] pointed out, since there are as many assets as states of nature, the GEI equilibrium allocation will be identical to the Arrow-Debreu allocation computed above. To describe the equilibrium, we need only add to the Arrow-Debreu equilibrium the portfolio trades $\theta^A = 28, \varphi^A = 0, \theta^B = 0, \varphi^B = 28$. The difference between Arrow-Debreu equilibrium and GEI equilibrium will not emerge until there are more states than assets.

7.3 NO COLLATERAL AND NO PENALTIES EQUILIBRIUM, I.E., ASSETLESS EQUILIBRIUM

Without the sophisticated financial arrangements involved with collateral or default penalties, there would be nothing to induce agents to keep their promises. Recognizing this, the market would set prices $\pi_j = 0$ for all the assets. Agents would, therefore, not be able to borrow any money. Thus, agents of type B, despite their great desire to live in housing, and great wealth in period 1, would not be able to purchase much housing in the initial period. Again it is easy to calculate the unique equilibrium:

$$\begin{aligned}
\pi_j &= 0 \\
p &= (p_{0F}, p_{0H}, p_{1F}, p_{1H}) = (1,\ 16,\ 1,\ 15), \\
x^A &= (x^A_{0F}, x^A_{0H}, x^A_{1F}, x^A_{1H}) = \left(20 + \frac{71}{32}, 1 - \frac{71}{32 \cdot 16}, 35 - \frac{71 \cdot 15}{32 \cdot 16}, 0\right), \\
x^B &= (x^B_{0F}, x^B_{0H}, x^B_{1F}, x^B_{1H}) = \left(\frac{57}{32}, \frac{71}{32 \cdot 16}, 35 + \frac{71 \cdot 15}{32 \cdot 16}, 1\right), \\
u^A &= 56; \;\; u^B = 50 + \frac{71 \cdot 87}{32 \cdot 16} \approx 62.6.
\end{aligned}$$

Agent A, realizing that he can sell the house for 15 in period 1, is effectively paying only $16 - 15 = 1$ to have a house in period 0, and is, therefore, indifferent to how much housing he consumes in period 0. Agents of type B, on the other hand, spend their available wealth at time 0 on housing until their marginal utility of consumption of x_{0F} rises to 30/16, which is the marginal utility of owning an extra dollar's worth of housing stock at time 0. (The marginal utility of an extra unit of housing is worth $15 + 15$, and at a cost of \$16 this means marginal utility of 30/16 per dollar.) That occurs when $9 - 4x^B_{0F} = 30/16$, that is, when $x^B_{0F} = 57/32$.

The assetless equilibrium is Pareto inferior, since the agents of type B do not get to consume the bulk of the housing stock even though it matters relatively more to them. They simply cannot use their last period wealth to command more buying power over the initial period housing stock. For the same reason, the price of housing in the first period is much lower in this equilibrium than it is in Arrow-Debreu equilibrium.

7.4 COLLATERAL EQUILIBRIUM

We now introduce the possibility of collateral in a world where there are no penalties. As in the GEI model, we assume there is only one asset j that promises one unit of food in period 1 and nothing else. We suppose that the collateral requirement for each unit of asset j is 1/15 unit of housing and the borrower holds the collateral,[3] $C_j^B = \binom{0}{1/15}$, $C_j^L = C_j^W = \binom{0}{0}$. We also suppose that borrower-held collateral has the same durability as consumption $Y_{1j}^B = Y_1^0 = \begin{pmatrix} 0 & 0 \\ 0 & 1 \end{pmatrix}$. The unique equilibrium is then:

$$\begin{aligned}
\pi_j &= 1, \\
p &= (p_{0F}, p_{0H}, p_{1F}, p_{1H}) = (1, 18, 1, 15), \\
x^A &= (x_{0F}^A, x_{0H}^A, x_{1F}^A, x_{1H}^A) = (23, 0, 35, 0) \\
\theta_j^A &= 15; \quad \varphi_j^A = 0; \\
x^B &= (x_{0F}^B, x_{0H}^B, x_{1F}^B, x_{1H}^B) = (1, 1, 35, 1), \\
\theta_j^B &= 0; \quad \varphi_j^B = 15 \; ; \\
u^A &= 58; \quad u^B = 72.
\end{aligned}$$

Agent B borrows 15 units of x_{0F}, and uses the 15 units of x_{0F} plus 3 he owns himself to buy 1 unit of x_{0H}, which he uses as collateral on the loan. He returns 15 units of x_{1F} in period 1, which is precisely the value of the collateral. Since, as borrower, agent B gets to consume the housing services while the house is being used as collateral, he gets final utility of 72. Agent A sells all his housing stock, since the implicit rental of $3 = 18 - 15$ is more than his marginal utility (expressed in terms of good F) for housing services for a single period. Agent B is content with purchasing exactly one unit of housing stock at the implicit rental of 3, since at $x_{0F}^B = 1$, his marginal utility of consumption of x_{0F} is $9 - 4(1) = 5$; and $5 = 15/3$, his marginal utility of a dollar of housing for a single period. (Recall that by giving up 3 units of x_{0F} he could effectively own the house for one period.)

Notice that the collateral equilibrium is Pareto superior to the assetless equilibrium, but not on the Pareto frontier. Although the housing stock goes to the agents of type B who value it the most, they are forced to give up too much food consumption in period 0 to get the housing stock. Their marginal utility of food

[3] It is important to note that without the invention of borrower-held collateral (e.g., if we supposed collateral had to be warehoused), there would be no improvement on the assetless economy.

consumption in period 0 relative to period 1 is higher than the agents of type A. Both could be made better off if the type A agents consumed δ less in period 0 and $\delta/(1-\delta)$ more in period 1 while the agents of type B did the reverse. This cannot be arranged in the collateral economy because agents of type B do not have the ability to borrow more money than \$15 because they do not have the collateral to put up for it. To borrow more money, they would have to get more housing to use as collateral, which they could only do if they gave up still more food consumption in time 0.

The difficulty of borrowing when collateral must be put up keeps the price of the durable housing down to 18, above the assetless price but well below the Arrow-Debreu price. Even though there is no default in the collateral equilibrium, the shadow of default dramatically changes the equilibrium. The inconvenience of putting up collateral cuts the borrowing down from 28 to 15.

It is interesting to observe that although neither the assetless equilibrium nor the collateral equilibrium is Pareto efficient, neither is dominated by the Arrow-Debreu equilibrium. The B agents who are buyers of durable goods are paradoxically helped by limitations in the asset markets because then they cannot borrow as easily, so they cannot compete with each other as fiercely over the housing stock, hence, the price is lower, hence, they end up better off. In general it is the owners of durable goods who will have the most to gain by loosening credit controls.

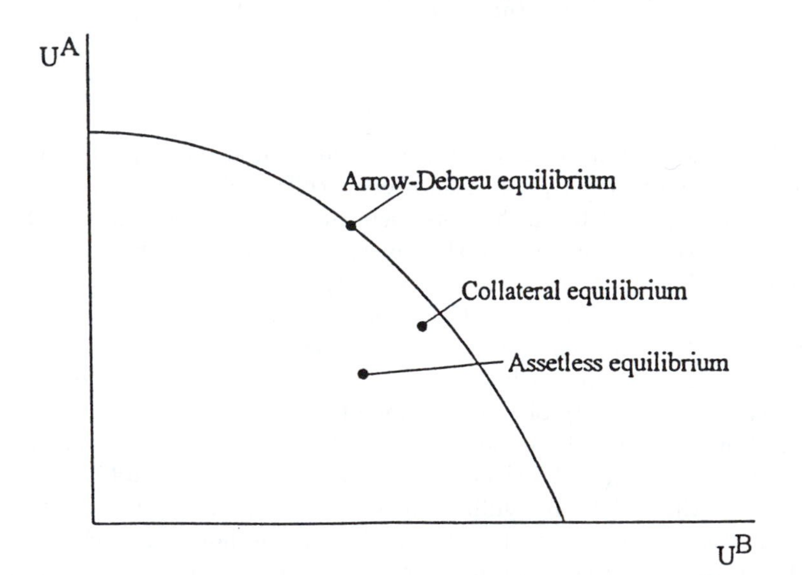

FIGURE 2

ENDOGENOUS COLLATERAL REQUIREMENTS. We can extend our example by adding more assets j that promise the same 1 unit of food in period 1, but which require $1/j$ units of housing as borrower-held collateral. Say $J = \{1, \ldots, 10^{10}\}$. (With this notation, our previous economy consisted of a single asset, namely $j = 15$.) A collateral equilibrium will now have to specify a price π_j for each of these assets j. It is easy to see that the equilibrium is essentially unique: $\pi_j = \min\{1, 15/j\}$. Each asset $j < 15$ forces agents to put up more than 1/15th of a house as collateral for each unit of food owed and, hence, will not be traded since it fetches no higher price than $j = 15$, and causes the borrower-seller additional hardship. Assets $j > 15$ enable borrowers to put up less collateral, but they fetch proportionately lower prices and, hence, amount to exactly the same assets as $j = 15$. Thus, we see that the market forces of supply and demand pick out of the multiplicity of assets a single one which agents trade. The others are all priced, but not traded.

8. EXAMPLE: BORROWING ACROSS STATES OF NATURE, WITH DEFAULT

We consider almost the same economy as before, with two agents A and B, and two goods F (food) and H (housing) in each period. But now we suppose that there are two states of nature $s = 1$ and $s = 2$ in period 1. The extra state allows for uncertainty and permits a distinction between complete markets and GEI equilibrium. The example turns on the assumption that in state 2, agents of type B have smaller endowments than in state 1. This objective change in the economy would tend to lower the price for houses in state 2, because it is agents of type B who especially like houses. The interesting phenomenon illustrated in this example is that the drop in state 2 housing prices is much greater in the GEI equilibrium and the collateral equilibrium than it is in either the complete markets equilibrium or the assetless equilibrium. The price volatility is less in the collateral equilibrium than in the GEI equilibrium, and welfare is higher.

As before, we suppose that food is completely perishable and housing is perfectly durable,

$$Y_s^0 = \begin{pmatrix} 0 & 0 \\ 0 & 1 \end{pmatrix} \quad \text{for} \quad s = 1, 2$$

meaning that 1 unit of F becomes 0 units of both goods, and 1 unit of H becomes 1 unit of H, in both states in period 1.

We assume

$$\begin{aligned} &u^A(x_{0F}, x_{0H}, x_{1F}, x_{1H}, x_{2F}, x_{2H}) \\ &\quad = x_{0F} + x_{0H} + (1-\varepsilon)(x_{1F} + x_{1H}) + \varepsilon(x_{2F} + x_{2H}), \\ &u^B(x_{0F}, x_{0H}, x_{1F}, x_{1H}, x_{2F}, x_{2H}) \\ &\quad = 9x_{0F} - 2x_{0F}^2 + 15x_{0H} + (1-\varepsilon)(x_{1F} + 15x_{1H}) + \varepsilon(x_{2F} + 15x_{2H}). \end{aligned}$$

Furthermore, we suppose that

$$e^A = (e^A_{0F}, e^A_{0H}, (e^A_{1F}, e^A_{1H}), (e^A_{2F}, e^A_{2H})) = (20, 1, (20,0), (20,0)),$$
$$e^B = (e^B_{0F}, e^B_{0H}, (e^B_{1F}, e^B_{1H}), (e^B_{2F}, e^B_{2H})) = (4, 0, (50,0), (3,0)).$$

To complete the model, we suppose as before that there is one asset A_j with $A_{sj} = \binom{1}{0}$, $\forall s \in S$ promising one unit of good F in every state $s = 1$ and 2. We suppose that the collateral requirement is $C^B_j = \binom{0}{1/15}$, as before, and that borrower-held houses are perfectly durable $Y^B_s = \binom{0\,0}{0\,1}$ for $s = 1, 2$.

It turns out that it is very easy to calculate the various equilibria for arbitrary ε.

8.1 ARROW-DEBREU EQUILIBRIUM

The unique (in utility payoffs) Arrow-Debreu equilibrium is:

$$p=((p_{0F}, p_{0H}), (p_{1F}, p_{1H}), (p_{2F}, p_{2H}))=((1, 30), (1-\varepsilon)(1,15), \varepsilon(1,15)),$$
$$x^A=((x^A_{0F}, x^A_{0H}), (x^A_{1F}, x^A_{1H}), (x^A_{2F}, x^A_{2H}))=\left((22,0), \left(20+\frac{28}{(1-\varepsilon)}, 0\right), (20,0)\right),$$
$$x^B=((x^B_{0F}, x^B_{0H}), (x^B_{1F}, x^B_{1H}), (x^B_{2F}, x^B_{2H}))=\left((2,1), \left(50-\frac{28}{1-\varepsilon}, 1\right), (3,1)\right),$$
$$u^A=70; \quad u^B = 62 - 47\varepsilon \,.$$

Notice that agent B transfers wealth from period 1 back to period 0 (i.e., he borrows), and also transfers wealth from state 1 to state 2. The great drop in agent B wealth in state 2 has a small effect on the economy (if ε is not too large). The reason is that with the complete markets, agent B is able to completely insure himself against this loss in wealth. The price of the durable good housing does not change at all. The only effect is a diminution in utility for agents of type B, reflecting the probability that their wealth might be smaller. They are forced to forego more consumption in state 1 in order to insure against this eventuality.

8.2 NO-COLLATERAL AND NO-PENALTIES EQUILIBRIUM, I.E., ASSETLESS EQUILIBRIUM

$$\pi_j = 0$$

$$p = ((p_{0F}, p_{0H}), (p_{1F}, p_{1H}), (p_{2F}, p_{2H})) = \left((1,16), (1,15), \left(1, \frac{3}{1-(71/32\cdot 16)}\right)\right)$$
$$\approx ((1,16), (1,15), (1,3.6)) \ ,$$
$$x^A = ((x^A_{0F}, x^A_{0H}), (x^A_{1F}, x^A_{1H}), (x^A_{2F}, x^A_{2H}))$$
$$= \left(\left(20 + \frac{71}{32}, 1 - \frac{71}{32\cdot 16}\right), \left(35 - \frac{15\cdot 71}{32\cdot 16}, 0\right), (23,0)\right),$$
$$x^B = ((x^B_{0F}, x^B_{0H}), (x^B_{1F}, x^B_{1H}), (x^B_{2F}, x^B_{2H}))$$
$$= \left(\left(\frac{57}{32}, \frac{71}{32\cdot 16}\right), \left(35 + \frac{15\cdot 71}{32\cdot 16}, 1\right), (0,1)\right),$$
$$u^A = 56 + \varepsilon\frac{15\cdot 71}{32\cdot 16} - 12\varepsilon \approx 56 - 9.9\varepsilon; \;\; u^B = 50 + \frac{87\cdot 71}{32\cdot 16} - \left(35 + \frac{15\cdot 71}{32\cdot 16}\right)\varepsilon$$
$$\approx 62 - 37.1\varepsilon \ .$$

In the assetless equilibrium there is a large drop from 15 to 3.6 in the price of the durable good in state 2 because the agents of type B cannot insure against their loss of wealth. They carry over part of the housing stock, and in state 2 spend all their wealth acquiring more housing.

8.3 COLLATERAL EQUILIBRIUM

We can exactly calculate the unique collateral equilibrium by noting that agents of type B will default in state 2 and then spend all of their endowment e^B_{2F} on good $2H$, giving a price $p_{2H} = 3$. The asset price is then $\pi_j = (1-\varepsilon) + \varepsilon\frac{3}{15}$

$$\pi_j = 1 - \frac{4}{5}\varepsilon,$$
$$((p_{0F}, p_{0H}), (p_{1F}, p_{1H}), (p_{2F}, p_{2H})) = ((1, 18-12\varepsilon), (1,15), (1,3)),$$
$$x^A = ((x^A_{0F}, x^A_{0H}), (x^A_{1F}, x^A_{1H}), (x^A_{2F}, x^A_{2H})) = ((23,0), (35,0), (23,0)),$$
$$\theta^A_j = 15; \;\; \varphi^A_j = 0;$$
$$x^B = ((x^B_{0F}, x^B_{0H}), (x^B_{1F}, x^B_{1H}), (x^B_{2F}, x^B_{2H})) = ((1,1), (35,1), (0,1)),$$
$$\theta^B_j = 0; \;\; \varphi^B_j = 15;$$
$$u^A = 58 - 12\varepsilon; \; u^B = 72 - 35\varepsilon.$$

To see that this is the correct equilibrium, note that agent B can sell 15 units of asset j in period 0, use the money $(1-4/5\varepsilon)15 = 15 - 12\varepsilon$ he receives in exchange, plus three units of $0F$ from his endowment to purchase one housing unit, which gives him consumption utility of 15 in period 0 and enables him to put up the whole of the necessary collateral for his loan. Thus, on the margin, 3 units of good $0F$ can be transformed into 15 utiles. But the marginal utility of consumption of good $0F$ is also $9 - 4x_{0F}^B = 5$.

EXCESS VOLATILITY. One striking aspect of this example is that the volatility of housing prices is much higher when there is scarce collateral than when there are complete markets or totally incomplete markets. When the housing stock can be purchased on margin (i.e., used as collateral), agents of type B are enabled to purchase the entire housing stock, raising its price from 16 (where it would have been without collateral) to 18. In the bad state these agents default and all their holdings of the housing stock are seized. Although they end up buying back the entire housing stock, their wealth is so depleted that they can only bid up the housing prices to 3.

When there is no collateral the agents of type B can afford to purchase only a fraction $\alpha = 71/(32)(16)$ of the housing stock at time 0. But they own that share free and clear of any debts. Thus, when bad news comes, they do not lose any more on account of past debts. They can apply their wealth to purchasing the remaining $1-\alpha$ of the housing stock, which forces the price up to approximately 3.6. Thus, when there is no collateral (and no other penalty for defaulting), the housing prices are never as high nor never as low as when the housing stock can be used as collateral.

The volatility could have been even more extreme if agents could not walk entirely away from their debts after their collateral was seized. In that case, agents of type B would lose even more wealth in state 2, and the price of housing would drop still further. Of course anticipating this further loss of wealth, agents of type B would probably borrow less, but that would mitigate and not prevent their wealth from going down, at least for robust examples. We demonstrate this in our example of GEI equilibrium in section 8.4.

ENDOGENOUS COLLATERAL AND ENDOGENOUS ASSETS. Consider again the collateral example, but with assets A_j, for $j = 1, \ldots, J$, defined as before so that each asset j promises 1 unit of good sF in each state $s = 1, 2$, and requires $1/j$ units of good $0H$ as borrower-held collateral. Any asset j with $j < 15$ is more secure than asset A_{15}, paying the same amount in state 1 and strictly more in state 2. The question is, will lenders insist on the more secure assets, or will they be content in equilibrium to loan money on terms that are more likely to lead to default?

The reader can check that the unique equilibrium of this enlarged economy is identical to the equilibrium we just calculated: none of the assets j with $j < 15$ will be traded at all, and the assets of type $j > 15$ are identical to asset $j = 15$. In equilibrium, each asset j will be priced at $\pi_j = (1-\varepsilon)\min\{1, 15/j\} + \varepsilon \min\{1, 3/j\}$, since at this price agents of type A are just indifferent to buying or not buying the asset. For $j > 15$, agents who sell asset j will default in both states and so the deliveries of the asset are proportional to $1/j$, hence their prices must be as well, and the assets are all effectively the same. Any asset j with $j < 15$ will bring only a miniscule increase in price (if ε is small), but cause a huge inconvenience to agents of any type who might want to sell it. Thus, it will not be traded in equilibrium.

Thus, we see that the free market will not choose levels of collateral which eliminate default. Borrowers, for obvious reasons, and lenders, because it increases demand for loans and for the durable goods they wish to sell, like surprisingly low collateral levels. We are left to wonder whether the collateral levels are in any sense optimal for the economy: does the free market arrange for the optimal amount of default?

In our example with the augmented list of assets, the government could mandate that loans be carried out only at a collateral level of 1/14, thereby reducing default. Since the resulting equilibrium would maintain the same price ratios of 1:15 in state 1 and 1:3 in state 2 between the two goods F and H, our constrained efficiency theorem implies that the resulting equilibrium could not Pareto dominate the original equilibrium at which trade took place exclusively at collateral levels 1/15.

Our theorem suggests that the free market does not expose this economy to unnecessary risks from default. However, it relies on the hypothesis that future relative prices are not affected by current collateral requirements. In general this will prove to be false. Indeed, if the collateral requirement were set high enough in our example, there would be no financial market transactions, and the future relative prices would change. In the example, this only hurts the economy still more but, in general, the relative price change caused by government mandated, stringent collateral requirements may be beneficial.

The reader can also check that if two new assets i (an Arrow security paying food only in state 1) and k (an Arrow security paying food only in state 2) were introduced with promises $A_{1i} = \binom{1}{0}, A_{2i} = \binom{0}{0}, A_{1k} = \binom{0}{0}, A_{2k} = \binom{1}{0}$, then no matter what the collateral requirements were for assets i and k, they would not be traded. Suppose, for example, that without loss of generality we take the collateral on asset i to be 1/15 housing units. Its price would then have to be $(1-\varepsilon)$. Then an agent could borrow $15(1-\varepsilon)$ units of food, put up $3+3\varepsilon$ units of his own endowment to attain the housing price of $18 - 12\varepsilon$, gain 15 utiles by living in the house and putting it up as collateral to back his loan, and another 15ε utiles from living in the house in state 2. This is no improvement. The ideal arrangement would be for B to sell j and buy asset i. But A would never sell asset i because he would not want to put up the collateral, since he gains little utility from living in the house. Both assets i and k waste collateral in the sense that the deliveries in some states are less than the value of the collateral there. In this example that suffices to eliminate them from trade.

Thus, our example also illustrates how scarce collateral rations the number and kind of promises that actually get made in equilibrium.

SUPERIOR RETURNS. Consider an infinitesimal agent of type B who preserved wealth equal to 50 units of food in state 2 instead of 3 units of food. Then, in state 2, he would spend all his wealth on housing, getting a fantastic return on his money, and doing much better than in state 1 even though objectively speaking to him, the housing stock was no different. Let us now try to interpret this parable.

The agents of type B are responsible for maintaining the price of housing. They may care more about housing because they get special utility from living in houses, as our model explicitly presumes (and like farmers did from farms in the 1930s), or because they are more optimistic about the future of housing in some unmodeled future, or because they are less risk averse about the future, or because they know how to hedge this kind of asset better than others, as perhaps Ellington Capital Management and other hedge funds do with respect to mortgage derivatives. Needless to say, a sudden drop in the wealth of this class of people will lower the price of housing in the short run. But at the same time it will make the return to each dollar invested in housing earn a higher rate of return. If somebody in the class has not suffered a wealth hit, then he can earn a higher total return as well. Incomplete markets permit leverage, which magnify exogenous shocks. We see a more dramatic example next.

8.4 GEI EQUILIBRIUM

Suppose, as before, that there is just one asset j with $A_{sj} = \binom{1}{0}$ for each $s \in S$, which promises 1 unit of food in period 1 in both states, and no housing. Suppose, in addition, that there are infinite default penalties for this asset, so that full delivery is required even with no collateral.

Remarkably, it turns out that the price volatility of houses is even greater than when there was collateral. The reason is that in state 2 the type B agents not only lose their housing stock, but since default is not permitted, they also must make up all of their debts from their endowment. This reduces their wealth even more, and forces type A agents to become the marginal buyers of housing in state 2, which, of course, lowers the housing price all the way to 1. We see here the full force of leverage, without any ameliorating notion of limited liability.

Of course, the agents of type A and B anticipate that all B's wealth will be confiscated in state 2 to pay the debt, and this limits the amount of money that B can borrow at time 0.

The price of the asset is clearly 1, since delivery is guaranteed. Moreover, the price of housing in period 0 can be expressed exactly as $1+(1-\varepsilon)15+\varepsilon 1 = 16-14\varepsilon$, since the marginal buyer in period 0 will be a type A agent, who realizes that he gets only 1 utile from the house in period 0 and can then sell it off at prices 15 or 1 in states 1 and 2, respectively. Given this price for the asset, agent B will realize that borrowing is a great deal for him as long as $\varepsilon < 1/30$ and he has the wealth to pay back the loan in the last period, since by borrowing 1 unit of food he gets $30/(16-14\varepsilon)$ utiles of consumption and gives up only $(1-\varepsilon)1+\varepsilon 15$ utiles in the last period to pay back the loan. Therefore, he will continue to borrow until the point at which he entirely runs out of wealth in state 2. The closed form formula for the rest of the equilibrium with $\varepsilon > 0$ is messy, so we give an approximate calculation for very small ε.

$$\pi_j = 1$$

$$((p_{0F}, p_{0H}), (p_{1F}, p_{1H}), (p_{2F}, p_{2H})) = ((1, 16 - 14\varepsilon), (1, 15), (1, 1)),$$

$$x^A = ((x^A_{0F}, x^A_{0H}), (x^A_{1F}, x^A_{1H}), (x^A_{2F}, x^A_{2H})) \approx \left(\left(22.2, \frac{10.5}{16}\right), (33.1, 0), (23, 1)\right),$$

$$\theta^A_j \approx 3.3; \quad \varphi^A_j = 0;$$

$$x^B = ((x^B_{0F}, x^B_{0H}), (x^B_{1F}, x^B_{1H}), (x^B_{2F}, x^B_{2H})) \approx \left(\left(\frac{57 - 63\varepsilon}{32 - 28\varepsilon}, \frac{5.5}{16}\right), (36.9, 1), (0, 0)\right),$$

$$\theta^B_j = 0; \quad \varphi^B_j \approx 3.3;$$

$$u^A \approx 58; u^B \approx 61.6.$$

It is interesting to note that the GEI equilibrium is also Pareto dominated by the collateral equilibrium. (Had we made agents of type A risk averse this would have been more dramatic.) The reason is that with infinite default penalties, there is much less flexibility in the credit markets, since once an agent sees that he cannot pay off in some state, no matter how unlikely, he must forego further borrowing.[4]

9. CONCLUSION

We have argued that Wall Street directs much of its activity to the stretching of collateral to cover as many promises as possible. We saw concretely, in our example, how making more collateral available made it possible to make more promises, and improve social welfare. It also typically increases the prices of houses. Perhaps with some justice, Kidder Peabody can claim to have played a role in this enterprise.

We also saw how the scarcity of collateral rations the kind of promises that are made, and fixes the levels of collateral that are required for any given promise. The kinds of promises and collateral levels that are traded in the market are not made at the whim of one seller like Kidder Peabody, but are the result of forces of supply and demand.

We saw how incomplete markets leads agents to leverage, which in some states can change the distribution of income, or magnify changes that would have occurred anyway. When agents have heterogeneous valuations of durable assets, these

[4] Indeed, it is clear that there is a discontinuity in the GEI equilibrium, since for $\varepsilon = 0$, presumably we would not require full delivery in a state that has zero probability, and we would be back in the no uncertainty situation from the last section in which we got a vastly different equilibrium.

redistributions of income can create large fluctuations in prices. Inevitably this in turn will create situations where some agents have much greater than normal opportunities to invest, at least compared to the opportunities they might have taken had events been different, or had collateral requirements been different.

There are two aspects of this investigation that call for further work. In this chapter I have presumed that there was no tranching or pyramiding of promises. Only physical commodities were allowed to collateralize promises. In Geanakoplos-Zame-Dubey[11] the model allows for both tranching and pyramiding, but further work is probably merited. This line of research touches on the Modigliani-Miller theorem, for example.

Finally, we have limited ourselves to two periods. With more time periods, we might have discovered an interesting dynamic. In states in which the physical collateral suffers a precipitous drop in price, the agents who most like it suddenly have their best investment opportunity. Assuming some of them are still around to take advantage of the opportunity, their wealth will tend to grow over time. But of course they will tend to use this increasing wealth to bid up the price of the very same physical collateral. This dynamic suggests a cyclical behavior to prices, and suggests that the opportunity to earn extraordinary returns will eventually disappear.

ACKNOWELDGMENTS

The analysis in this chapter is based on the paper Geanakoplos-Zame-Dubey.[11]

REFERENCES

1. Aiyagari, S. R., and M. Gertler. "Overreaction of Asset Prices in General Equilibrium." Mimeo, New York University, 1995.
2. Akerlof, G. "The Market for Lemons: Qualitative Uncertainty and the Market Mechanism." *Quart. J. Econ.* **84** (1970): 488–500.
3. Allen, F., and D. Gale. *Financial Innovation and Risk Sharing.* Cambridge, MA: MIT Press, 1994.
4. Arrow, K. J. "An Extension of the Basic Theorems of Classical Welfare Economics." In *Proceedings of the Berkeley Symposium on Mathematics and Statistics*, edited by J. Neyman. Berkeley, CA: University of California Press, 1951.
5. Arrow, K. J. "Generalization des Theories de l'Equilibre Economique General et du Rendement Social au Cas du Risque." *Econometrie* (1953): 81–120.
6. Baron, D. P. "Default and the Modigliani–Miller Theorem: A Synthesis." *Amer. Econ. Rev.* **66** (1976): 204–212.
7. Dubey, P., and J. Geanakoplos. "Default and Efficiency in a General Equilibrium Model with Incomplete Markets." Discussion Paper, No. 773R, Cowles Foundation, New Haven, CT, 1990.
8. Dubey, P., and M. Shubik. "Bankruptcy and Optimality in a Closed Trading Mass Economy Modeled as a Noncooperative Game." *J. Math. Econ.* **6** (1979): 115–134.
9. Gale, D. "A Walrasian Theory of Markets with Adverse Selection." *Rev. Econ. Studies* **59** (1992): 229–255.
10. Geanakoplos, J., and H. Polemarchakis. "Existence, Regularity, and Constrained Suboptimality of Competitive Allocations when the Asset Market Is Incomplete. " In *Uncertainty, Information, and Communication, Essays in Honor of Ken Arrow*, edited by W. Heller, R. Starr, and D. Starrett, 65–96. Cambridge, MA: Cambridge University Press, 1986.
11. Geanakoplos, J., W. Zame, and P. Dubey. "Default, Collateral, and Derivatives." Mimeo, Yale University, New Haven, CT, 1995.
12. Hart, O. "On the Existence of Equilibrium in Securities Models." *J. Econ. Theory* **19** (1974): 293–311.
13. Hellwig, M. F. "Bankruptcy, Limited Liability, and the Modigliani–Miller Theorem." *Amer. Econ. Rev.* **71** (1981): 155–170.
14. Hellwig, M. F. "A Model of Borrowing and Lending with Bankruptcy." *Econometrica* **45** (1981): 1879–1905.
15. Hellwig, M. F. "Some Recent Developments in the Theory of Competition in Markets with Adverse Selection." *European Econ. Rev. Papers & Proc.* **31** (1987): 319–325.
16. Kiyotaki, N., and J. Moore. "Credit Cycles." Mimeo, LSE, 1995.
17. Modigliani, F., and M. H. Miller. "The Cost of Capital, Corporation Finance, and the Theory of Investment." *Amer. Econ. Rev.* **48** (1958): 261–297.

18. Smith, V. L. "Default Risk, Scale and the Homemade Leverage Theorem." *Amer. Econ. Rev.* **62** (1972): 66–76.
19. Stiglitz, J. E. "The Irrelevance of Corporate Financial Policy." *Amer. Econ. Rev.* **64** (1974): 851–866.
20. Stiglitz, J., and A. Weiss. "Credit Rationing in Markets with Imperfect Information." *Amer. Econ. Rev.* **72** (1981): 393–410.
21. Wilson, C. "A Model of Insurance Markets with Incomplete Information." *J. Econ. Theory* **16** (1977): 167–207.
22. Zame, W. "Efficiency and the Role of Default when Security Markets Are Incomplete." *Amer. Econ. Rev.* **83** (1993): 1142–1164.

Axel Leijonhufvud
Center for Computable Economics, and Department of Economics, UCLA, 405 Hilgard Avenue, Los Angeles, CA 90095; E-mail: axel@ucla.edu

Macroeconomics and Complexity: Inflation Theory

INTRODUCTION

The complaints about today's macroeconomics are familiar to all. The economy is not modeled as an evolving system exhibiting emergent properties. Behavior is represented as precalculatedly optimal rather than adaptive. Actions are assumed to be prereconciled by rational expectations rather than brought into consistency by market interaction. Heterogeneity of goods or of agents is all but completely denied.

The infinite-horizon optimizing representative agent model is not altogether representative of today's macroeconomics. We also have models with two types of mortal agents (overlapping generations) and some models with asymmetric information or externalities capable of producing coordination failures of a sort. But generally speaking these, too, tend to be subject to the same complaints.

An overview of today's macroeconomics might only too easily deteriorate into catalogue of models to which these complaints and reservations apply in varying combinations. Rather than tire you with such a litany, I will describe a set of (somewhat atypical) empirical phenomena, explain why I do not think they are handled well by current macroeconomics, and end with a few suggestions about how they might be attacked. If these suggestions were to start a critical discussion, I would find that helpful.

THE EXTREMES OF MONETARY INSTABILITY

The ensuing material is drawn altogether from my recent book with Daniel Heymann, entitled *High Inflation.*[5] His contributions to what I know about the subject are very great but he is to be absolved from any misuse or abuse that I might commit in this particular chapter.

A word or two on the perspective from which we have approached the subject of the book. In a nonexperimental field, it is often a good research strategy to pay particular attention to observational "outliers." A study of atypical or pathological cases may reveal properties of a complex system of which we would remain unaware as long as we always concentrate on well-ordered states. In macroeconomics, this means paying particular attention to great depressions, high inflations, and (now) the transition from socialism.

The extremes of monetary instability have been my particular interest throughout my career. I find it natural to refer to such system states as "disequilibria" (or "far-from-equilibrium states") but in economics these have become fighting words in recent years, so I will avoid the usage so as to spare you any semantic or methodological digressions.

CONFRONTING "STANDARD" MONETARY THEORY WITH THE HIGH INFLATION EVIDENCE

As a start, two definitions: First, Heymann and I define an inflation as "high" when people living through it quote the inflation rate in percent per month and treat annual rates as meaningless (except for historical purposes). Inflation rates quoted in per annum terms refer to "moderate" inflations.

Second, by "standard" monetary theory I will mean the type of constructions that have dominated the monetary literature from Patinkin through Lucas and also very much dominate the two volumes of the *Handbook on Monetary Economics.*[2] This standard theory "adds" money to a "real" general equilibrium model within which relative prices and the allocation of resources are determined without any reference to money or financial institutions. Modern finance theory is also done in this manner.

Applied to the problem of inflation, models of this family basically have but one channel through which money affects real income, relative prices and resource allocation, namely the so-called "inflation tax," i.e., the rate at which real balances lose real purchasing power. Like ordinary taxes, the inflation tax will induce some distortions in the allocation of effort and of resources. But the most important implication is not the oft-mentioned "shoe-leather costs of too many trips to the bank," even though in high inflations the efforts and resources devoted to economizing on cash balances are quite substantial. It is, rather, the strangulation of financial intermediation. The monetary liabilities of the banking system as a ratio to GNP may be on the order of 1/4 or 1/5 of what is normal under monetary stability. This important channel for the intermediation of saving is almost shut down, therefore.

Another point that is often missed is the regressivity of the inflation tax. The well-to-do can avoid it more easily than the poor. The point is worth making because the view that anti-inflationary policies tend to hurt the working classes is so deeply ingrained in the English language literature.

One further comment of a general nature before turning to the high inflation phenomena I want to concentrate on. In my opinion, models that attempt to explain high inflations as the result of governments calculating and implementing the "optimal" inflation tax time profile are, of limited, if any, value. The individuals that find themselves in positions of fiscal or monetary authority in a high inflation are "reaping the whirlwind"—and almost always a whirlwind sown by others. They are not in control. Control has been lost and the intermittent attempts to stabilize are in effect desperate attempts to regain a measure of control. In the extreme, on the verge of hyperinflation, the inability to finance the public sector turns into inability to govern when not even the resources needed to maintain the proverbial "nightwatchman state" can be mustered.

We can now turn to the phenomena not predicted by standard theory.

SURPRISE NO. 1: THE SURVIVAL OF DOMESTIC MONEY

Domestic money stays in general use even at inflation rates on the order of 30+% per month. These are pretty high "tax" rates, corresponding to 2–3000% per annum—taxes that would kill many a market.

Standard general equilibrium theory, in contrast, does not require money, monetary instruments, or monetary institutions. A reason for money to "hang in there" has to be invented and the most frequently used devices by which money is pasted onto the basic structure are less than persuasive. That such exceedingly costly moneys survive deserves to be listed as a surprise.

Domestic money stays in circulation, but this should not be taken to mean that it plays all the roles that we expect money to play in a country enjoying monetary stability. It is neither the only unit in which prices are quoted, nor the medium in which all transactions are settled. Instead, a triple monetary standard will develop: the markets for the stuff of daily existence deal with domestic money, but some are dollarized and others are indexed. Multiple standards of this sort give rise to numerous inconsistencies in the overall price system as, for example, when exchange rate movements change the price of all goods quoted in dollars relative to those whose prices are fixed by a (backward-looking) index.

SURPRISE NO. 2: THE UNIT OF ACCOUNT MATTERS

High inflation renders accounts kept in the domestic monetary unit virtually meaningless. No very good or generally shared substitute emerges. This, of course, is a "surprise" because in monetary general equilibrium theory the "choice of numeraire" is arbitrary.[15] But the social and legal convention of tying accounting to the unit of government issued money is not so easily changed.

The unit of account matters, most obviously, for accounting—for measuring the "real" result of business operations. Various devices are tried. Some firms, for example, construct indices of their own input prices so as to enable them to determine whether sales proceeds would suffice to repurchase the quantities of inputs used to produce the output sold. An indicator of whether you are breaking even or not is certainly worth having, but it does not amount to a high inflation information system that would help one determine whether a firm is on its minimum cost schedule and is producing at the profit-maximizing output rate.

The inflationary economy does not gravitate toward the use of a single index as the common unit of account. Consider a contract between two firms each one of which would prefer to make future payments predictable in "real" terms as measured by its own input basket. But customers and suppliers, by definition, do not have the same input baskets. So some compromise index may have to be negotiated. But the customer in this contract is a supplier in others where different compromises have to be made, etc. So the result is a multiplication of contract indices in use. Before the 1994 *real* stabilization in Brazil, for example, some three dozen contract indices were said to be in widespread use.

Less obviously perhaps, the unit of account matters for accountability, that is, for monitoring the performance of agents in a wide variety of principal-agent relationships. This is most obvious in the public sector. When governmental budgets can no longer be drawn up for a fiscal year in a meaningful way, the executive can no longer be held fiscally accountable to parliaments, and departments or provincial and municipal governments can no longer be held accountable to the national government. Instead, everyone operates in effect under "soft budget constraints."[7]

Monetary budgeting and monetary accounts and reports are vital components of monitoring systems for all sorts of private sector principal-agent relationships as well. This includes shareholder control over corporate management.

SURPRISE NO. 3: THE DISAPPEARANCE OF INTERTEMPORAL MARKETS

American readers may recall that the very moderate inflation in the United States in the 1970s saw the disappearance of the market for 30-year bonds and the virtual demise of the 30-year fixed-rate mortgage. That moderate inflation never exceeded 15% per year.

Higher inflations will kill off markets for far shorter maturities. In Argentina in 1985, the longest nominal rate maturity was 45 days—and that was a thin market. Typical trade credit was 7 days, not the 90 days that are normal in stable circumstances.

A foreshortening of temporal perspective is "built into" the definition of high inflation from which we started. But the all but total lack of temporal depth of the high inflation economy will nonetheless qualify as a "surprise" because standard monetary and finance theory do precisely nothing to prepare us for it. Nominal contracts become risky in inflation and the more risky the higher the inflation rate and the longer the maturity. But finance theory will deny that high variance is a reason for an asset completely to disappear from the efficient portfolio. Moreover, the heterogeniety of expectations that definitely characterizes high inflations should cause more trades, not less.

It is an additional surprise that high inflations more or less kill off stock markets as well. The old belief used to be that risk-averse people would avoid nominal placements in inflationary times and switch to "real" assets, equity shares in particular. But stock markets tend to lapse into near total inactivity; no new issues are floated and the volume of trades is extremely low.[1] We conjecture that the reasons for this have to do with the accounting difficulties noted above. Corporations are in effect unable to report their earnings in a way that the market can reliably evaluate.

In brief, the kind of finance theory that prices assets in "real terms" first and introduces money later to find the "nominal scale" of those prices encounters a number of anomalies in the high inflation evidence.[2]

SURPRISE NO. 4: THE FRAGMENTATION OF SPOT MARKETS

Intertemporal markets disappear but spot markets fragment. Arbitrage conditions that, under monetary stability, tend to fix, for example, the relative prices of the same good sold in different locations are constantly violated in high inflations. Innumerable relative prices, that we normally do not think about because they are

[1] In high inflation cases, the volume of trading may in fact be roughly consistent with the life-cycle theory of saving.

[2] Recall, also, the strangulation of intermediation discussed above.

(more or less) constant, are instead constantly jiggling. So, despite the disappearance of intertemporal markets, the number of markets that the Hicksian aggregation theorem would oblige us to recognize is most probably multiplying.

The lack of sufficient spatial arbitrage to keep these relative prices within tight bounds is readily understandable once one takes a realistic look at the two sides of the market. On the supplier side, buying low (in one location) and selling high (in another) always takes some time. If you are not "quick enough," high inflation conditions will turn routine arbitrage into speculative transactions and highly risky ones at that. On the consumer side, the price dispersion at any one time is much greater than normal and that is an incentive to shop around. But the price discrepancies found through comparison shopping are not autocorrelated. The information gained is depreciating very fast. Under monetary stability, this kind of everyday search activity is an investment yielding a long stream of returns. Under high inflation, the price found by visiting a store may be changing behind the shopper's back as he leaves the premises.

This fourth surprise is but one aspect of No. 5.

SURPRISE NO. 5: EXCESS RELATIVE PRICE VARIABILITY

The disappearance of markets seems to have received no attention at all in the theoretical literature.[3] The phenomenon of excess relative price variability, in contrast, has been the subject of a fair number of theoretical papers. The trouble, we believe, is that this literature starts from a misidentification of the phenomenon that is being modeled.

Consider two alternative interpretations—two different ways of "stylizing the facts," if you so will. The principal difference between the two inheres in the roles in the process that they assign, respectively, to "fix-prices" and "flex-prices" (to use John Hicks' terms).

The friction interpretation blames fix-prices. What is reported as "the" inflation rate is the rate of change of some index. It is assumed that all flex-prices move in pace with the index since, by definition, they are not subject to friction. The variability of relative prices in excess of the adjustments that would be observed under monetary stability is attributed to the spasmodic and intermittent adjustment of fix-prices (Figure 1).

The friction interpretation is the one generally accepted in the literature. It fits a common preconception among economists, namely, that if anything goes wrong with the price system it must be because someone, somewhere interferes with market

[3]The sizable and growing "missing markets" literature deals with an entirely different set of problems. "Missing markets" are imposed as a constraint on permissible trades. The problem here is that people chose not to trade in a particular maturity, for example, so that the transactions volume shrinks until it no longer pays anyone to make a market in it.

forces and causes some price or prices to "stick." This stickiness is usually explained by invoking so-called "menu costs," i.e., costs which the seller has to incur when he changes his price.

The turbulence interpretation sees flex-prices as becoming too volatile. The official inflation rate is taken to reflect mainly the adjustment of fix-prices while the flex-prices fluctuate around it (Figure 2).

If the friction hypothesis fits economists' preconceptions, the turbulence hypothesis goes against the grain: here the preconception is that money can be made by "buying low and selling high" so as to smooth the price paths. Any excess volatility of flex-prices ought to be smothered by profitable speculation.

Both these interpretations are too clean-cut to fit the reality. We know there is some friction, because indexation would create it even if it did not exist before. But Heymann and I believe that there is more turbulence than friction in the high inflation process.

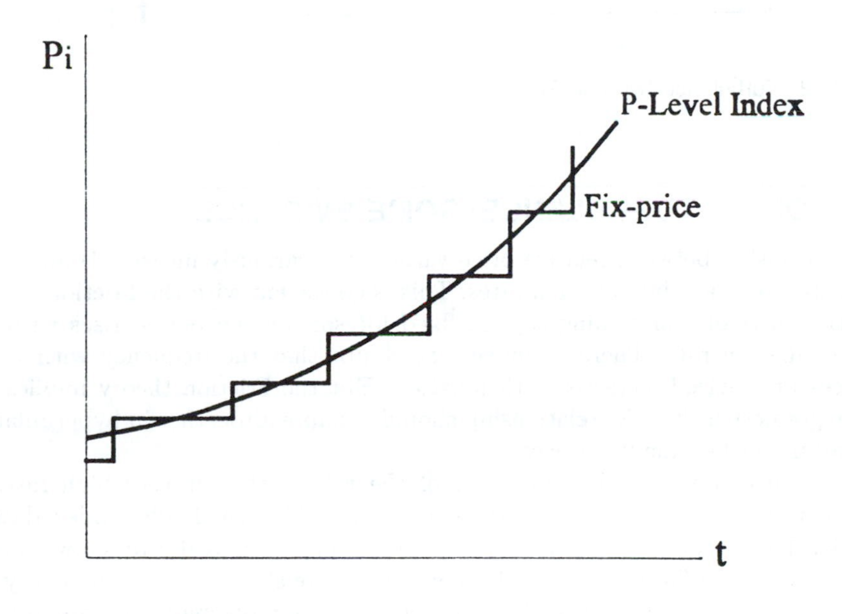

FIGURE 1 Friction hypothesis.

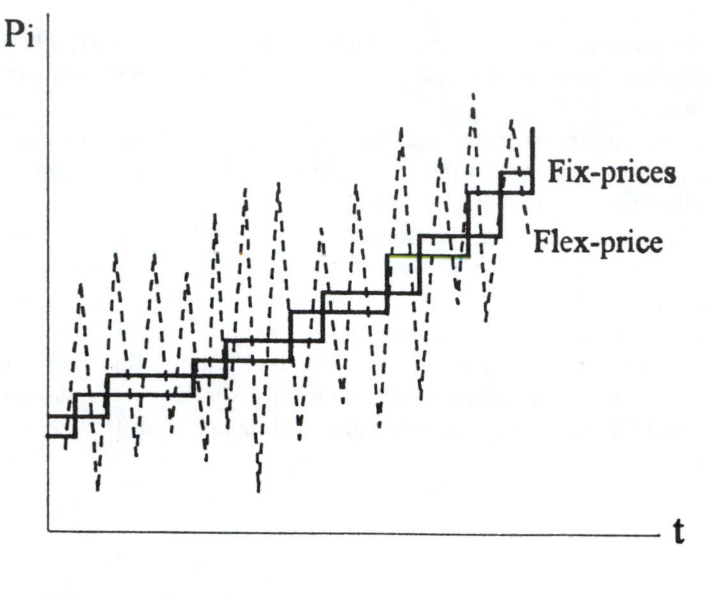

FIGURE 2 Turbulence hypothesis.

FRICTION OR TURBULENCE: SOME EVIDENCE

The relationship between relative price variability (variously measured) and inflation flattens out at high inflation rates. This is consistent with the Friction theory, since the costs of maintaining a price fixed for some given period rises with the general inflation rate. There is, indeed, no doubt that the frequency with which fix-prices are revised increases with inflation. But the Friction theory implies the further prediction that the relationship should collapse altogether in hyperinflation when all prices become flex-prices.

As it turns out, the flattening out of the relationship at very high rates of inflation is in all probability a statistical artifact. The usual price index data is monthly. This is too low a frequency of observation for us really to know what is going on at high inflation rates. In his comprehensive study of more than 40 years of Mexican data, Angel Palerm[13,14] experimented with reducing the data frequency from monthly to quarterly to annual. He found that so doing shifts the curve relating relative price variability to inflation progressively downward and makes it flatter (Figure 3). The inference is plausible, therefore, that most of the price variability at very high inflation rates escapes notice at the monthly frequency of observation.

A careful study by Dabus[1] of the (monthly) Argentine data that had been adduced as supporting the "menu cost" theory shows the relationship between relative price variability and inflation to be monotonic. There is no tendency in these data for relative price variability to decline even in the 1989 Argentine hyperinflation episodes.

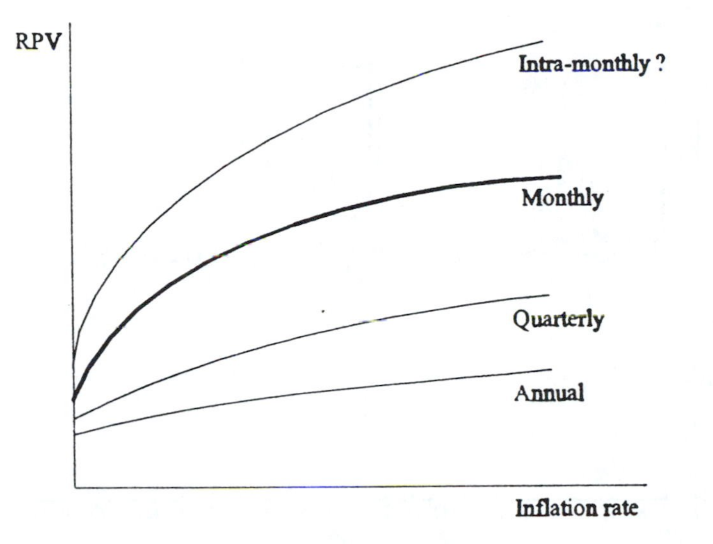

FIGURE 3 Observational frequency of relative price variability (RPV).

Some samples of higher frequency data have been collected and studied.[4] A particularly noteworthy result of Tommasi's[16] work with weekly data from a number of Buenos Aires supermarkets is his finding that nearly 1/5 (17%) of the price changes were negative. Most such price cuts are presumably made to correct for previous overshooting. Their high frequency in a period during which the inflation rate averaged above 4% per week strongly suggests turbulence, as opposed to friction, in the price formation process.

Further high frequency evidence can be drawn from the foreign exchange market. Figure 4 shows day-to-day changes of the Argentine exchange rate over the two-year period 1989–1990. The two hyperinflation episodes separated by an abortive stabilization attempt are clearly marked by the dramatic increase in the volatility of the exchange rate. The onset of hyperinflation occurs as a sharp phase transition which is an interesting problem in its own right. Here, however, the point is simply the inferences we can draw from these data for the behavior of relative prices. Recall the "triple standard": some prices are dollarized, others are indexed, yet others remain quoted in local currencies. In hyperinflation, the number of dollarized prices increases while the number of prices that are set by reference to backward-looking indices decline. The dollarized prices will move with the extreme frequency and amplitude of Figure 4, the indexed ones will not, while local currency prices will vary widely with regard to the extent that they are or are not adjusted daily (or hourly) with reference to the latest dollar quotations.

[4] In addition to the Tommasi study, see e.g., Lach and Tsiddon.[8]

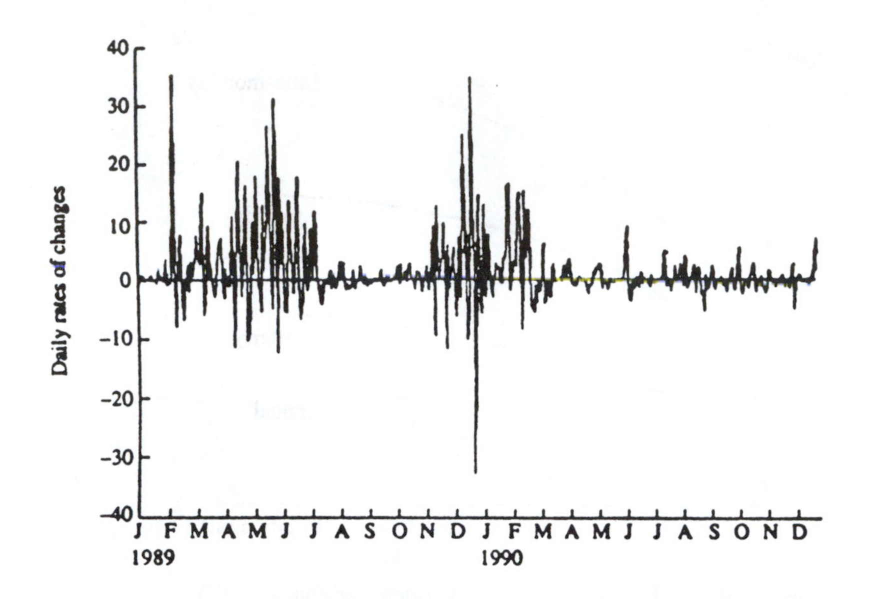

FIGURE 4 Argentine dollar exchange rate: day to day changes. Source: Heymann and Leijonhufvud.[5]

TURBULENCE AND THE DISAPPEARANCE OF MARKETS

The disappearance of intertemporal markets starts in the long end, extinguishes shorter and shorter contract maturities as inflation rises, and reaches its logical limit with the outbreak of true hyperinflation when many spot transactions are affected. In the first 1989 outbreak of hyperinflation in Argentina, some store owners put out the sign "Closed for the Lack of Prices," thus declaring their refusal to sell in a situation where the prices at which they might replenish their inventories had become totally unpredictable. This phenomenon is the counterpart in the market for groceries to the simultaneous behavior of the market for foreign exchange. The extreme volatility of the exchange rate in the hyperinflation episodes of Figure 4 reflects a market from which all the big financial institutions have basically withdrawn. With the institutions unwilling to take positions on the price of tomorrow, no smoothing of the price from day to day takes place.

In the hyperinflationary limit, therefore, the general unwillingness to take intertemporal positions and the excess variability of relative prices are seen to be two aspects of the same behavioral adaptation. But the point applies not only at this limit.

Behavior representations in economics range over a spectrum of models. At one end of this spectrum, we would put the infinite-horizon (stochastic) optimization models which in recent years have become the macroeconomist's regulation equipment. At the other end, put the simplest of adaptive models, that is, models of behavior governed by feedback and relying on next to no memory and no foresight. Neither extreme is particularly palatable: one leaves no room for learning and adaptation and makes timepaths of consumption and prices far "too smooth"; the other has no vestige of deliberate rationality and produces chaotic dynamics far too easily. Independent of theoretical fashion, the world we live in has its being somewhere away from either extreme.

The interpretation of the evidence that Heymann and I propose is that high inflations shift people's behavior away from the long-term intertemporal optimizing mode toward a short-memory/short-foresight adaptive mode. The twin results of this adaptation are the disappearance of intertemporal markets and turbulence in spot markets.[5]

The turbulence interpretation of relative price variability has this to recommend it: that it leads to a unified explanation of these two phenomena whereas the friction interpretation will require an entirely unrelated hypothesis to account for the disappearance of markets. But the unified explanation also has a problem, namely, that standard theory does not contemplate—and is perhaps unlikely to countenance—such shifts in the general mode of behavior. Thus, the proposed explanation raises a deeper and more difficult question: How are we to understand the shift away from long-term intertemporal optimization toward shorter-horizon adaptive behavior?

At this stage an attempted answer has to be both tentative and speculative.[6] What is going on at the individual level, I believe, is that agents adapt to the increased complexity of their environment by going to simpler, more flexible[7] strategies. A strategy is more flexible if it reduces or shortens time-spanning commitments and allows the agent to reallocate his resources freely in response to information that he will receive in the future but cannot anticipate (even probabilistically) in the present. When large numbers of agents respond in this way the result, on the system level, is an even more complex environment.[8]

"Complexity" is an ill-defined term, however. What does it mean? It will not do to think of it just as "increased risk." High inflation, contrasted to "monetary stability," is not simply a mean-preserving increase in variance for an otherwise unchanged structural model. To the individual agent, increased complexity means rather that the model that he or she requires in order to predict the utility-relevant (future) outcome of a present action has more variables and requires knowledge of more parameters. This may occur, for example, as the result of a change in the monetary regime.

[5] "Turbulence" here, of course, in the literal sense of nonlinear chaotic behavior.

[6] For a somewhat fuller discussion, see Heymann and Leijonhufvud[5] and also Leijonhufvud.[10]

[7] I use "flexibility" in the exact sense of Hicks,[6] pp. 37 ff.

[8] The text leaves open the question of how this "vicious circle" of deteriorating performance originates. In general, the story will start with a government bent on spending beyond its means and using the inflation tax to do so.

When the regime changes, the true specification of the novel system structure will not be known to agents. They must try to infer it. But experience with the now more complex regime is brief. The "usable memory" is short and the "useful forecasts" will be equally so.

In the standard optimization paradigm, the individual response to increased complexity is commensurably more sophisticated strategies. But this presumes that the agent understands the structure of the more complex environment as well as that of the less complex one and also that his quantitative information is as adequate to the new situation as it was to the old. When these presumptions do not hold, however, the individual response to increased complexity will be to fall back on simpler strategies, utilizing less rather than more information.[3,4]

In our context, this means (among other things) a general foreshortening of the time horizon over which commitments are presently made and an increase in the frequency with which plans are adapted in response to unfolding events. At the macro level, this change in the behavior of agents makes the system more complex, for example by making its dynamics strongly nonlinear where it may have been very "nearly linear" before.

RETHINKING FIX-PRICES AND PRICE "STICKINESS"

One of the lessons to be drawn from the above analysis concerns price flexibility or the lack of it. Macroeconomics, in particular, seems to have gotten totally hung up on the notion that (more) price flexibility is always a good thing and price stickiness always a source of trouble.[9]

But fix-price behavior is not generally an "imperfection" in the price system. Price setters take a position on tomorrow's price. This is elementary "smoothing" and will suffice to suppress the highest frequency price oscillations. In posting prices, sellers "linearize" the environment for others.

When producers and middlemen are unwilling to post a price and "stick" with it for a while, ordinary agents find themselves in an environment where "fragments" of markets "hog-cycle" out of sync with each other. Such an environment poses horrendous information problems. There is almost no correlation of prices between locations and almost no autocorrelation of prices at any one location. In such complex, nonlinear settings, ordinary people find their ability to achieve the least-cost household budget or to plan ahead in a coherent manner to be "bounded."

Price setting behavior is probably an essential aspect of how "order" emerges. When tomorrow's prices have all become predictable, and day-to-day volatility suppressed, traders will look for patterns in the next highest frequency of price

[9] Thus, "menu costs" have become a welcome new toy also for those who would revive "Keynesian" economics of the price stickiness variety.

fluctuations and take positions in the day-after-tomorrow's prices, etc. Meanwhile, emergent autocorrelation of prices at various locations will make spatial arbitrage increasingly feasible. In this way, a coherent price system spreads across space and into the future. Under extremely favorable conditions, the economy will become sufficiently "linearized" so that markets for 30-year bonds and mortgages appear.[10]

The spread of the coherent price system may be viewed as a form of coevolution. Simple, profit-motivated adaptation by individual agents makes a fitness landscape emerge that is sufficiently smooth so that relatively simple adaptive behavior does not get trapped on very local optima. This is not a guarantee of a global optimum, of course. The spread of markets—of thick markets—always stops far short of Arrow-Debreu. But a quite high degree of collective efficiency becomes feasible as long as this coevolution is allowed to run its course under favorable conditions. Monetary stability, of course, is only one of the relevant conditions.

Whether it is accurate to say that the spread of markets takes the system to the "edge of chaos," I am not qualified to say. What the high inflation evidence shows (on our interpretation) is that as the disappearance of intertemporal markets edges closer to the present, irregular high frequency price oscillations are seen to appear. These are suppressed under more normal monetary conditions. We may conjecture, therefore, that the further into the future people plan and coordinate their plans in markets, the longer the wavelength of the nonlinear movements that would "survive" in the system.

ADAPTIVE BEHAVIOR AND "BOUNDED RATIONALITY"

The main conclusion of Heymann's and my book is that inflation destroys institutional arrangements and routines on which people depend in order to come close to optimizing (in some objective instrumental sense). To argue that they *depend* on stable institutions is, of course, to say that their rationality is "bounded."

Most economists heartily dislike references to bounded rationality, less perhaps because they do not know what to make of it than because they do not know what to do with it, given the economist's inherited box of tools. I have no handy recipe for how to model boundedly rational behavior either. But I want to conclude with a note[11] on that subject.

Our rationality is surely "bounded" but not necessarily by our ability to reason or to calculate. We are much better at processing large amounts of information correctly and rapidly at a subliminal level than at a consciously reasoning level. The visual cortex, for example, appears to handle on the order of 10 million bits per second whereas our conscious mind chugs along at a pace of about 50 bits per

[10] Under the most favorable conditions of all—Queen Victoria on the throne—we get consols.

[11] Much influenced by reading Tor Noerretranders,[12] a book that deserves an English translation.

second or less.[12] But our subliminal capabilities to perform complicated tasks (such as to hit a golf ball) have to be trained and, for noninstinctual activities, conscious reasoning effort goes into the training.

Now, the proof that Arrow-Debreu plans are effectively not computable[11,18] is theoretically interesting in itself. But the effective reason that people do not precalculate and precommit their actions in detail over long periods with multiple contingencies is not necessarily that their computing capabilities are limited, but rather that it is impossible to repeat the experiment a sufficient number of times to train yourself to do it well. In the Arrow-Debreu world, in fact, "you only go around once."

REFERENCES

1. Dabús, C. "Inflación y Precios Relativos: El Caso Argentino 1960–1990." Ph.D. dissertation, Universidad Nacional del Sur, Bahia Blanca, 1993.
2. Friedman, B. M., and F. H. Hahn, eds. *Handbook of Monetary Economics*, 2 vols., Amsterdam: North-Holland, 1990.
3. Heiner, R. "The Origin of Predictable Behavior." *Amer. Econ. Rev.* **83(4)** September (1983): 560–595.
4. Heiner, R. "Uncertainty, Signal-Detection Experiments, and Modelling Behavior." In *Economics as Process: Essays in the New Institutional Economics*, edited by R. Langlois. Cambridge, MA: Cambridge University Press, 1986.
5. Heymann D., and A. Leijonhufvud. *High Inflation.* Oxford: Oxford University Press, 1995.
6. Hicks, J. *The Crisis in Keynesian Economics.* New York: Basic Books, 1974.
7. Kornai, J. "The Soft Budget Constraint." *Kyklos* **39(1)** (1986): 3–30.
8. Lach, S., and D. Tsiddon. "The Behavior of Prices and Inflation: An Empirical Analysis of Disaggregated Price Data." *J. Pol. Econ.* **100(2)** (1992): 349–389.
9. Leijonhufvud, A. "Towards a Not-Too-Rational Macroeconomics." *Southern Econ. J.* **60(1)** (1993): 1–13.
10. Leijonhufvud, A. "Adaptive Behavior, Market Processes, and the Computable Approach." *Revue Économique* **46(6)** (1995): 1497–1510.
11. Lewis, A. A. "On Effectively Computable Realizations of Choice Functions." *Math. Soc. Sci.* **10** (1985): 43–80.
12. Noerretranders, T. *Märk världen: En bok om vetenskap och intuition.* Stockholm: Bonnier Alba, 1993. (Orig. Danish edn. 1991.)
13. Palerm, A. "Price Formation and Relative Price Variability in an Inflationary Environment." Ph.D. dissertation, UCLA, 1990.
14. Palerm, A. "Market Structure and Price Flexibility." *J. Devel. Econ.* **36(1)** (1991): 37–53.
15. Patinkin, D. *Money, Interest, and Prices.* New York: Harper and Row, 1965.
16. Tommasi, M. "Inflation and Relative Prices: Evidence from Argentina." In *Optimal Pricing, Inflation, and the Cost of Price Adjustment*, edited by E. Sheshinski and Y. Weiss. Cambridge, MA: MIT Press, 1992.
17. Velupillai, K. *Computable Economics: The Fourth Arne Ryde Lectures.* Oxford: Oxford University Press, 1995.

Kristian Lindgren
Institute of Physical Resource Theory, Chalmers University of Technology and Göteborg University, S-412 96 Göteborg, Sweden; E-mail: frtkl@fy.chalmers.se

Evolutionary Dynamics in Game-Theoretic Models

A number of evolutionary models based on the iterated Prisoner's Dilemma with noise are discussed. Different aspects of the evolutionary behavior are illustrated: (i) by varying the trickiness of the game (iterated game, mistakes, misunderstandings, choice of payoff matrix), (ii) by introducing spatial dimensions, and (iii) by modifying the strategy space and the representation of strategies. One of the variations involves the finitely iterated game that has a unique Nash equilibrium of only defecting strategies, and it is illustrated that when a spatial dimension is added, evolution usually avoids this state. The finite automaton representation of strategies is also revisited, and one model shows an evolution of a very robust error-correcting strategy for the Prisoner's Dilemma game.

INTRODUCTION

The term "evolutionary dynamics" often refers to systems that exhibit a time evolution in which the character of the dynamics may change due to internal mechanisms. Such models are, of course, interesting for studying systems in which variation and selection are important components. A growing interest in these models does not only come from evolutionary biology, but also from other scientific disciplines. For example, in economic theory, technological evolution as well as evolution of behavior in markets may call for models with evolutionary mechanisms.

In this chapter, we focus on dynamical systems described by equations of motion that may change in time according to certain rules, which can be interpreted as mutation operations. In models of this type, the variables typically correspond to the number or the mass of certain components (individuals/organisms), or in more mechanistic (or spatial) models the position of certain components.

In the models that we discuss here, the equations of motion for the different variables (individuals) are usually coupled, which means that we have *coevolutionary* systems. The success or failure for a certain type of individual (species) depends on which other individuals are present. In this case, there is not a fixed fitness landscape in which the evolutionary dynamics climbs toward increasing elevation, but a position that at one time is a peak may turn into a valley. This ever changing character of the world determining the evolutionary path allows for complex dynamic phenomena.

Coevolutionary dynamics differ, in this sense, from the common use of the genetic algorithm,[18] in which a fixed goal is used in the fitness function and where there is no interaction between individuals. In the genetic algorithm, the focus is on the final result—what is the best or a good solution? In models of coevolutionary systems, one is usually interested in the transient phenomenon of evolution, which in the case of open-ended evolution never reaches an attractor. There are variations, though, where interactions between individuals have been suggested as a way to improve on the genetic algorithm, see, e.g., Hillis.[16] For a recent review article on genetic algorithms in an evolutionary context, see Mitchell and Forrest.[33]

We make a distinction between evolutionary systems and adaptive systems. The equations of motion in an evolutionary system reflect the basic mechanisms of biological evolution, i.e., inheritance, mutation, and selection. In an adaptive system, other mechanisms are allowed as well, e.g., modifications of strategies based on individual forecasts on the future state of the system. But, increasing the possibilities for individualistic rational behavior does not necessarily improve the outcome for the species to which the individual belongs in the long run. An example of this difference is illustrated and discussed for one of the evolutionary lattice models.

In the next section, we briefly discuss some evolutionary models in the literature. The main part of this chapter is devoted to a discussion of a number of evolutionary models based on the iterated Prisoner's Dilemma game as the interaction between individuals, see, e.g., Axelrod.[3,4] Different aspects will be illustrated:

(i) by varying the trickiness of the game (iterated game, mistakes, misunderstandings, choice of payoff matrix), (ii) by introducing spatial dimensions, and (iii) by modifying the strategy space and the representation of strategies.

The results in this chapter concerning the finite memory strategies have, to a large extent, been reported before; for the mean-field model, see Lindgren,[26] and for the lattice model, see Lindgren and Nordahl;[27] see also the review article by Lindgren and Nordahl.[28] The model based on the finitely iterated game as well as the models using finite automata strategies are new. A detailed analysis of these models will be reported elsewhere.

MODELS OF EVOLUTION

Evolutionary models can be characterized both by the level at which the mechanisms are working and the dimensionality of the system. Usually these characteristics are coupled to some extent, so that low-dimensional models usually reflect mechanisms on a higher system level. An example of that regime is an evolutionary model in which the variables are positions of phenotypic characters in phenotype space, and the dynamics is determined by an ordinary differential equation that uses the distance to fitness maximum.[48] At the other end of the scale, there are models based on "microscopic" components that interact and organize, e.g., in the form of catalytic networks. The work by Eigen,[11] Shuster,[45] Fontana,[15] and many others[6,13,14,21] lies in this region. Some of these models have similarities to spin-glass models in statistical mechanics, see, e.g., the models for evolution of RNA.[1,43,47] Other examples can be found in, e.g., Weisbuch,[51] and one of the most well-known models is the *NK*-model by Kauffman; for a review, see Kauffman.[22] Such spin-glass-like models could also be interpreted as models of interacting individuals or species.

A special class of models is based on instructions floating in a virtual computer memory. Such instructions may eventually organize in a structure that can be interpreted as an organism. Then, the interactions between organisms will not be fixed by the model, but will be a result of the evolution. This is the approach taken by, for example, Rasmussen and coworkers[41] and Ray.[42] Models of this type are obviously high dimensional.

There are many examples of other models, some of them mentioned in Figure 1. The models that will be described in more detail in this paper have been constructed based on mechanisms at the level of individual organisms and their interactions. There is a large number of models of this kind, many of them using various game-theoretic approaches for the interaction between individuals.[4,26,27,30,32,38,46]

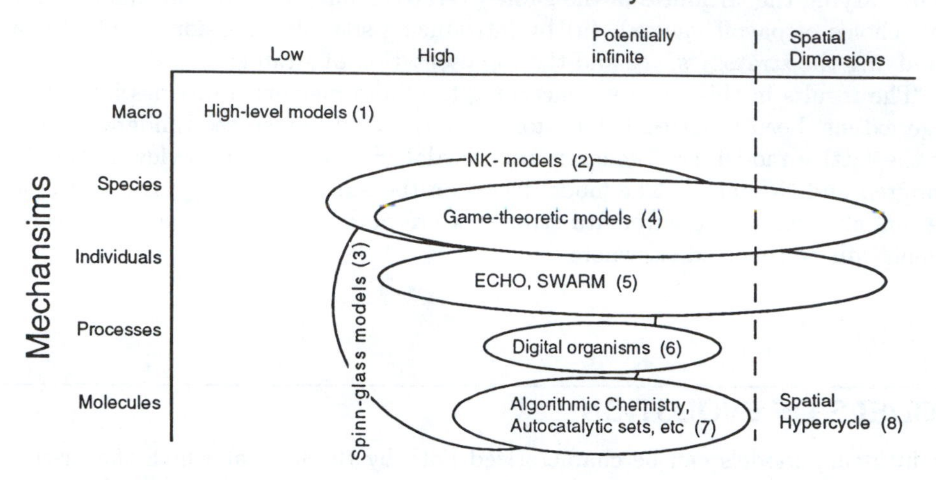

FIGURE 1 Evolutionary models characterized by level and dimensionality. The numbers in the figure refer to various models: (1) Stenseth and Maynard-Smith[48]; (2) Kauffman[22]; (3) Anderson[1] and Stein and Anderson[47]; (4) Matsuo,[30] Axelrod,[4] Miller,[32] Lindgren,[26] Nowak and Sigmund,[38] Stanley et al.,[46] and Lindgren and Nordahl[27]; (5) Holland[18,19] and Langton[24]; (6) Rasmussen et al.[41] and Ray[42]; (7) Eigen,[11] Schuster,[45] Fontana,[15] Kauffman,[21] Farmer et al.,[14] Eigen et al.,[13] and Bagley et al.[6]; (8) Boerlijst and Hogeweg[7,8]; see also Eigen and Schuster.[12]

As a basis for the models presented in the following sections, we use the three necessary mechanisms usually associated with biological evolution:

- inheritance of genetic information
- variation (mutations)
- the feedback effect of natural (or artificial) selection

Furthermore, to get a potential for evolution, the population needs to be provided with a difficult "task" to solve, i.e., complex interactions with environment (including other organisms).

GAME-THEORETIC MODELS AND DYNAMICS

In the choice of interaction model, one has to find a problem that cannot be "solved" too easily in the chosen genetic representation. A solution here means that there is a strategy that is evolutionarily stable (or a set of strategies that together form a fixed point in the evolutionary dynamics). An evolutionarily stable strategy (ESS) cannot be invaded by any other strategy that initially is present at an arbitrarily small amount.[31]

The Prisoner's Dilemma game, described below, is a suitable model for the interaction between individuals in an evolving population. There are many variations of the game that offer nontrivial problems for the individuals, and still this interaction can be solved analytically given the genetic code for the strategies.

In the Prisoner's Dilemma (PD) game, two players simultaneously choose to cooperate (C) or to defect (D), without knowing the opponent's choice. If they both cooperate they share the highest total payoff and get R points each, but there is a temptation to defect because defection scores $T > R$ against a cooperating opponent, who in that case gets the lowest score S. If both defect they share the lowest total payoff and get P points each. If $T > R > P > S$ and $2R > T + S$, there is a dilemma, since, in an isolated game, rational players choose to defect and then share the lowest total payoff (see Table 1). Cooperative behavior is possible in the iterated game when the same players meet in a series of rounds. In that case, it has been shown that tit-for-tat (TFT), a strategy that mimics the opponent's last action (and starts with C), is capable of establishing cooperation.[3]

Many dynamical systems and evolutionary models have been constructed with the PD game as a model for the interaction between individuals. Axelrod[4] applied a genetic algorithm (GA) to the iterated Prisoner's Dilemma and used a bit-string representation of finite memory strategies. In another GA study, Miller[32] let the bit strings represent finite automata playing the iterated PD game. Miller also introduced noise in the game, but did not find the evolution of the error-correcting strategies like we have seen in the models described below. Stanley et al.[46] also used finite automata to study strategies that could choose their opponents. For recent work along these lines, see Tesfatsion.[50] A third type of strategy representation, suggested by Ikegami,[20] uses a tree structure with varying depth. Strategies could also choose actions with certain probabilities. This has been analyzed by Molander,[34] and population dynamics as well as evolutionary dynamics has been studied for these strategies.[37,38]

TABLE 1 The payoff matrix of the Prisoner's Dilemma game, showing the scores given to player 1 and 2, respectively.

	player 2 cooperates	player 2 defects
player 1 cooperates	(R , R)	(S , T)
player 1 defects	(T , S)	(P , P)

In all models described in this chapter, we have used variations of the Prisoner's Dilemma that can be solved analytically. This reduces the computational power required and allows for long simulations (in terms of generations) of the models.

DYNAMICS BASED ON THE ITERATED PRISONER'S DILEMMA WITH MISTAKES

The iterated game is easily modified to a more complicated one by introducing noise. In this way the game is made more "tricky," and very simple cooperative strategies, like TFT, are not very successful. Noise could either disturb the actions so that the performed action is opposite to the intended one (trembling hand), or the information telling a player about the opponent's action is disturbed (noisy channels).[5] In any case, the complication is sufficient to give rise to interesting evolutionary behavior.

In the first interaction model, there is no noise, but a dilemma is introduced by iterating the game a fixed (known) number of rounds. In that case, the final round is an ordinary Prisoner's Dilemma, and then the next to last round will get a similar status, and so on.

In the second and third models, noise is introduced in the form of mistakes, i.e., the action performed differs from the intended one (given by the strategy), with a certain probability. On the other hand, the problem with the dilemma of the final round is removed, and the game is made infinitely iterated. In the second model, the individuals are equipped with finite memory strategies, while in the third model the more powerful representation of finite automata is used. In all cases the average score per round is calculated, for the finitely iterated game and for the infinitely iterated game, respectively, and used in the dynamics to determine the next generation of the population.

We shall demonstrate various evolutionary phenomena, for all three models by studying the behavior in two completely different "worlds," the mean-field model (all interact with all, i.e., no spatial dimensions), and the cellular automaton (CA) model (local interaction on a lattice with synchronous updating).

In the mean-field model, the dynamics is based on the simple equations,

$$x'_k = x_k + dx_k \left(\sum_i s_{ki} x_i - \sum_{ij} s_{ij} x_i x_j \right), \qquad k = 1, \ldots, N, \tag{1}$$

where x_k is the fraction of the population occupied by strategy k, and s_{ij} is the score for strategy i against j. The prime on the left-hand side denotes the next time step, i.e., we are assuming discrete time. Since the double sum $(\Sigma s_{ij} x_i x_j)$ equals the average score in the population, the total population size is conserved. The number of different strategies N may change due to mutations, which are randomly generated after each step of the population dynamics equations (1). We have used continuous variables x_k, with a cut-off value corresponding to a single individual, below which the strategy is considered extinct and removed from the population. The equations reflect a change of abundancy for a species that is proportional to its own score minus the average score in the population. Each time step is viewed as one generation and, in addition to the equations, mutations are added. This means that generations are nonoverlapping and that reproduction is asexual.

In the CA model, it is assumed that each individual interacts with its four nearest neighbors on a square lattice. The average score for each individual (or lattice site) is calculated, and this determines which individuals will be allowed to reproduce. In each neighborhood (consisting of a cell with its four nearest neighbors), the individual with the highest score reproduces in the middle cell, see Figure 2. (Ties are broken by adding a small random number to each player's score.) The offspring inherits the parent's strategy, possibly altered by mutations to be described below in detail for the different models.

Depending on the choice of payoff matrix, the evolutionary paths may look very different. In the CA model, since the dynamics is determined only by score differences, there are only two independent parameters in the payoff matrix.[27] Therefore, we have chosen to study the parameter region given by $R = 1$, $S = 0$, $1 < T < 2$, and $0 < P < 1$. In the mean-field model, the *value* of the score difference is also important, since the dynamics depends on how fast a score difference changes the composition of the population. Therefore, as a third parameter, we can use the growth constant d in Eqs. (1), while we keep R and S fixed and vary T and P as in the CA model. In the mean-field simulations we have assumed $d = 0.1$. (This results in a growth that is slow enough to view Eqs. (1) as rough approximations to a set of ordinary differential equations, even though this is not the intention.) In all models where noise is present, we have used an error probability of $e = 0.01$.

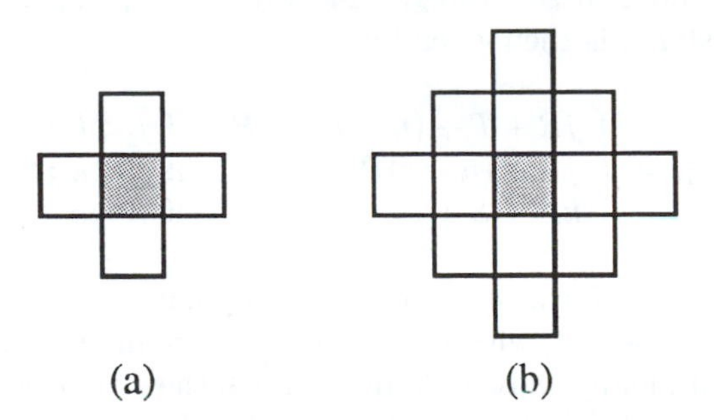

FIGURE 2 (a) In the cellular automaton (CA) model, each individual plays the iterated Prisoner's Dilemma with each of its four nearest neighbors. Simultaneously, for each neighborhood, the player with the highest average score per round reproduces in the "gray" cell. (b) Since the score of the nearest neighbors depends on the strategies present in the next nearest cells, we get a CA with interaction radius 2.

FINITE NUMBER OF ROUNDS

If the iterated game is finite and the number of rounds is known, the dilemma of the single-round PD game remains. Assume that we have two players cooperating, for example, by using the tit-for-tat strategy. When the two players enter the final round of their iterated game, they face the single-round dilemma and profit-maximizing behavior calls for the defect action. But, if we know that our opponent is using a rational strategy, the next to last round will be a dilemma of the same type, and we should defect in that round too. Then the third round from the end will be under consideration, and so on. This implies that the only Nash equilibrium is given by both players defecting all rounds. This does not mean, though, that defecting all the time is the best strategy. Which strategy is the best depends on which other strategies are present in the population. If the population is dominated by, for example, strategies that play TFT for all rounds except the last two when they defect, it is, of course, a bad strategy to start with defection from the beginning. Instead, if one would know the opponent's strategy, one should start with TFT, i.e., cooperate, but switch to always defect (ALLD) the round before the opponent is to defect.

The situation faced by a player entering a population playing the finitely iterated game is an ill-defined one, in the sense discussed by Arthur,[2] where one cannot say which strategy is the best. Like in most real situations, the success or failure of a certain behavior depends on what other individuals or strategies are present.

There are various ways to construct an evolutionary model of the PD game iterated n rounds. Here, we pick a simple example and look at a strategy set containing strategies playing TFT for k rounds and playing ALLD for the remaining $n-k$ rounds. We denote these strategies F_k, with $k = 0, \ldots, n$. The score s_{jk}, for a strategy F_j against F_k, is then given by

$$s_{jk} = \begin{cases} jR + T + (n-j-1)P, & \text{if } j < k \; ; \\ jR + (n-j)P, & \text{if } j = k \; ; \\ kR + S + (n-k-1)P, & \text{if } j > k \; . \end{cases} \tag{2}$$

The population dynamics for a fully connected population (mean-field model) can then be described by Eqs. (1), with mutations at a rate of 10^{-5}, so that a strategy F_i is replaced by a randomly chosen F_j. In Figure 3, the evolutionary path, starting with the strategy F_n, or TFT, shows that cooperative behavior is exploited and that the evolution ends in the ALLD state.[1]

[1]This result also follows from a theorem,[44] saying that an *iteratively strictly dominated strategy* is eliminated in any *aggregate monotonic* selection dynamics. A recent generalization of this result shows that this also holds for a larger class of selection dynamics termed convex-monotone dynamics.[17] In this example, F_n is strictly dominated by F_{n-1}, and F_k is iteratively strictly dominated by F_{k-1}, for $0 < k < n$.

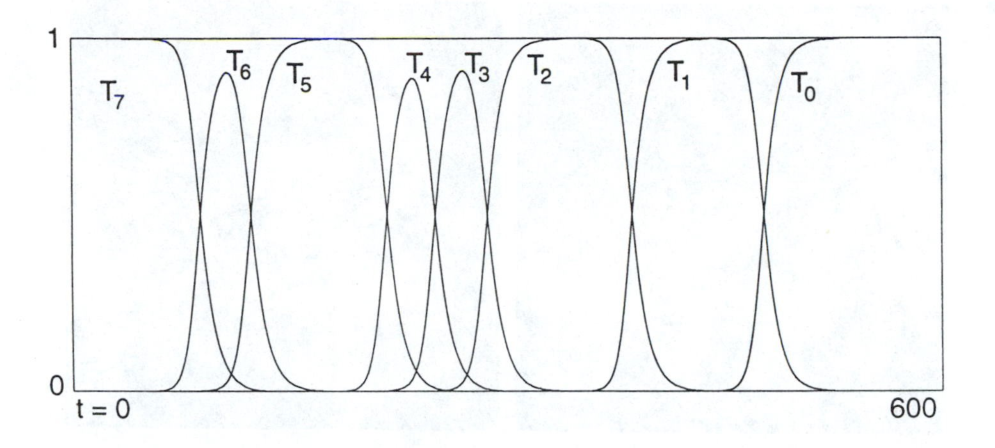

FIGURE 3 The interaction is given by the PD game iterated $n = 7$ rounds. Initially, the population (1000 individuals) consists only of the strategy F_7, or TFT, but due to the dilemma of the fixed number of rounds, the evolution leads to strategies defecting earlier in the iterated game. The final state is then a fixed point, in which all play F_0, or ALLD.

Cooperation cannot be maintained since there will always appear exploiting strategies that grows at the expense of the more cooperative ones. It is well known that the introduction of spatial dimensions may increase the possibility for cooperative behavior.[7,8,27,37,35]

In the CA model, the individuals are put on a 128 × 128 square lattice with periodic boundary conditions. Contrary to what we saw in the mean-field case, cooperation is maintained for a large part of parameter space defining the PD game. This is made possible either by the presence of spatiotemporal patterns, like waves of strategies sweeping over the system, see Figure 4b and 4c, or by formation of stable islands of more cooperative strategies in a background of slightly exploiting strategies, see Figure 4a and 4d.

FINITE MEMORY STRATEGIES

In the infinitely iterated Prisoner's Dilemma with mistakes, a simple type of strategy to consider is the deterministic finite memory strategy. This type of strategy may take into account the actions that have occurred in the game a finite number of rounds backward, and deterministically choose a certain action.

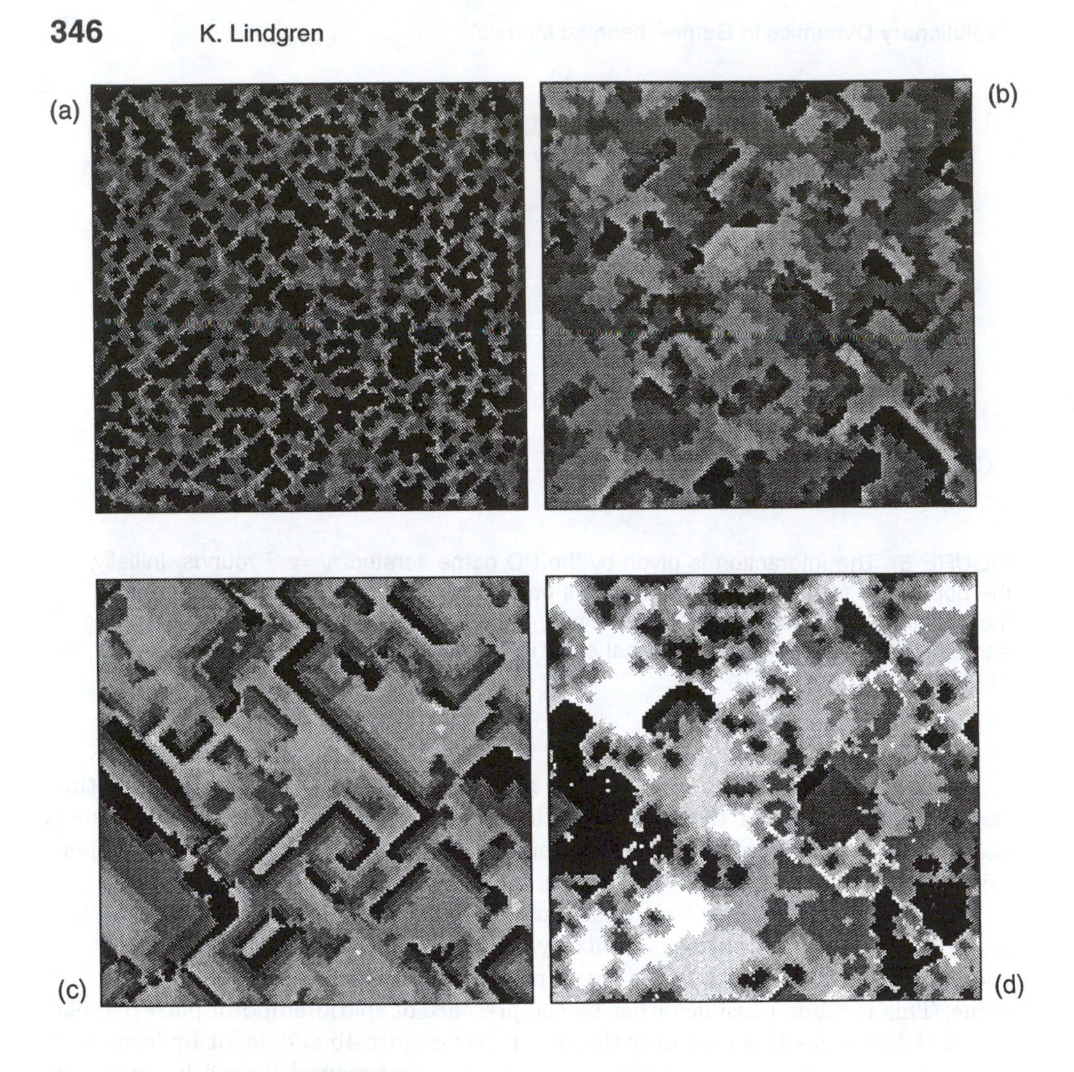

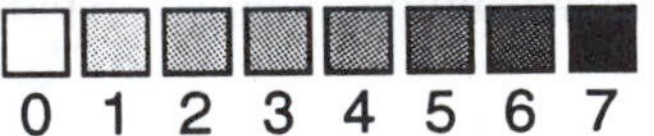

FIGURE 4 The evolution of F_k strategies playing the 7-round iterated Prisoner's Dilemma on a 128×128 lattice is shown for four different parameter values ($R = 1$, $S = 0$, T and P are varied). In all cases, we see that the Nash equilibrium of a homogenous ALLD population is avoided. In (a), $T = 1.3$, $P = 0.3$, the state is frozen and dominated by F_7 (TFT), F_6, and F_5. In (b), $T = 1.667$, $P = 0.333$, and in (c), $T = 1.3$, $P = 0.3$, and there are waves running across the system constantly changing the pattern. In (d), $T = 1.7$, $P = 0.7$, and there are frozen patterns that occasionally are modified by waves of TFT. Here, all strategies from F_0 (ALLD) to F_7 (TFT) are present, and they are represented in gray scale from white for ALLD to black for TFT.

A strategy can then be viewed as a look-up table, in which each entry corresponds to each of the possible finite-length histories that can be memorized by the strategy, see Figure 5. The symbols in the output column of the table denote the action to be performed, but there is a small chance (due to noise) that the action is altered.

The game between two finite memory strategies can be described as a stationary stochastic process, and the average score can, therefore, be calculated from a simple system of linear equations. The number of equations necessary equals 2^b, where b is the size of the history needed to determine the next pair of actions. For example, for two memory-1 strategies (as well as memory 2), both actions in the previous round are required and, therefore, the number of equations is 4.

We use three types of mutations. The *point mutation* alters the symbol in the genome. The *gene duplication* attaches a copy of the genome to itself. In this way the memory capacity increases, but it should be noted that this mutation is neutral—the strategy is not changed, but a point mutation is required to make use of the increased capacity. The *split mutation* randomly removes the first or second half of the genome.

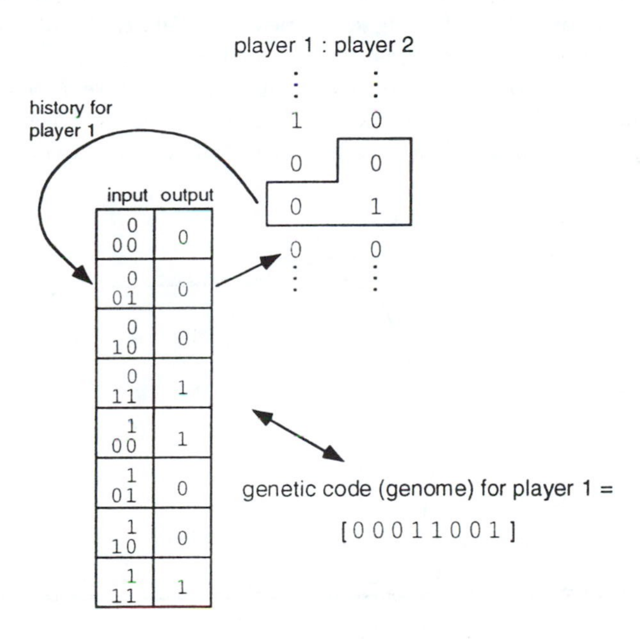

FIGURE 5 The genetic coding of finite memory strategies uses the look-up table. The length of the genome (size of the table) then carries the information on the memory capacity of the strategy, since the table doubles in size when memory increases by 1.

In the simulations, the initial population consists of equal parts of the four possible memory-1 strategies: [00] is ALLD (always defects), [01] is TFT, [10] is ATFT (i.e., anti-tit-for-tat), and [11] is ALLC (always cooperates). Due to mutations, more complex strategies may emerge, and this offers the possibility for evolution of strategies that correct for accidental mistakes.

MEAN-FIELD MODEL

In the mean-field model all interact with all, and the dynamics is given by Eqs (1). The number of individuals is 1000, the point mutation rate is $2 \cdot 10^{-5}$ and the other mutation rates are 10^{-5}.

In Figures 6 and 7, the evolutionary path is shown for two different choices of payoff matrices. The first one, $T = 1.6$ and $P = 0.3$, shows that the system finds a very stable state dominated by the strategy 1001. In the second simulation, at $T = 1.7$ and $P = 0.4$, this strategy does not succeed, but the evolution continues to strategies with longer memory. It turns out that the final state in Figure 7, consists of a large group of strategies having the same mechanism for correcting for mistakes in the iterated PD game. In both cases, the final population is cooperative and cannot be exploited by defecting strategies. For a more detailed discussion on simulations of the finite memory mean-field model, see Lindgren.[26]

Both the strategy 1001 and the final strategies in Figure 7 belong to a class of strategies that can deal with mistakes. These strategies are based on retaliation and synchronization. An accidental defection is followed by a certain number of

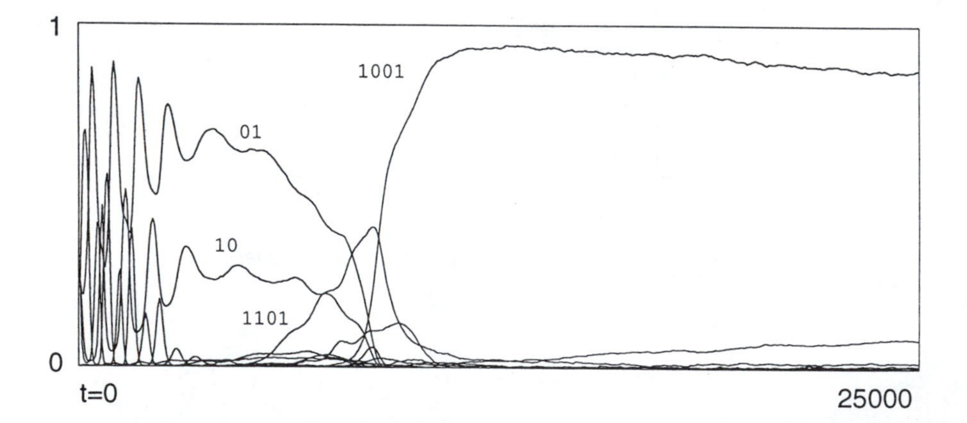

FIGURE 6 The evolution of finite memory strategies in the mean-field model is shown for 25,000 generations, with the parameter values $T = 1.6$, $P = 0.3$. There is a metastable coexistence between 01 (TFT) and 10 (ATFT) before strategies of memory 2 take over. The dominating 1001 strategy is actually evolutionarily stable for these parameter values.

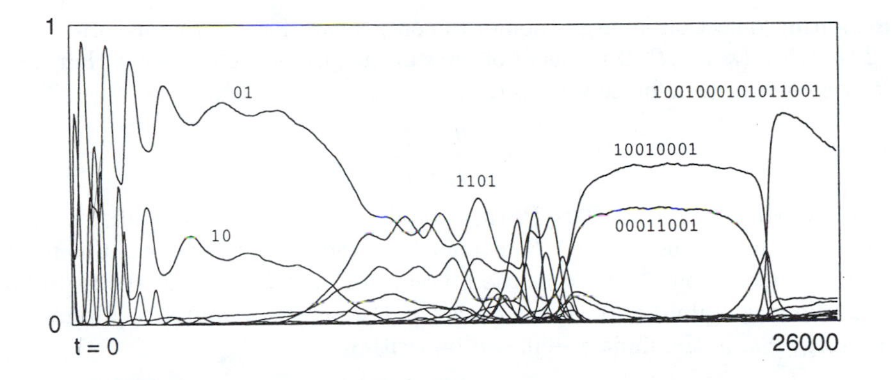

FIGURE 7 The evolution of finite memory strategies in the mean-field model is shown for 26,000 generations, with the parameter values $T = 1.7$, $P = 0.4$. Again, the coexistence between 01 (TFT) and 10 (ATFT) is present, but 1001 does not appear. For these parameter values, longer memories seem to be required to get cooperative and stable (or metastable) strategies. The simulation ends with a strategy that has an unexploitable error-correcting mechanism. The coexistence between the two memory-3 strategies 10010001 and 00011001 is an example of mutualism.[26]

rounds of mutual defection. The rounds of defection both punish possible exploiting strategies and synchronize the strategies before cooperative behavior is established again. The strategies can be described by the simple mechanism:

> After a defection by either of the players, defect until there have been n consecutive rounds of mutual defection, and then start cooperating again.

We shall denote such a strategy by S_n. The first, S_1, often referred to as Simpleton[40] (or Pavlov[38]), is a strategy that cooperates if both players chose the same action in the previous round, and this one is identical to 1001 in our notation. This strategy exhibits the following pattern in case of an accidental defection $\underline{\text{D}}$: $(\ldots, \text{CC}, \text{CC}, \underline{\text{D}}\text{C}, \text{DD}, \text{CC}, \text{CC}, \ldots)$. For the standard parameters used by Axelrod,[3] i.e., $R = 3$, $S = 0$, $T = 5$, and $P = 1$, a population of Simpletons can be invaded, since ALLD (always defects) can exploit it—a single round of mutual defections is not enough to punish an exploiting strategy. The second strategy in this class is S_2, which answers a defection (by either of the players) by defecting twice in a row, and after that returns to cooperation. This behavior leads to the pattern $(\ldots, \text{CC}, \text{CC}, \underline{\text{D}}\text{C}, \text{DD}, \text{DD}, \text{CC}, \text{CC}, \ldots)$, which is the mechanism we find in the population of strategies that dominate at the end of the simulation in Figure 7.

A population of S_n cannot be invaded by a strategy that tries to get back to cooperation faster than does S_n—the invader will only be exploited if that is tried. The only chance for a single invader (or mutant) to succeed is by exploiting the cooperative behavior. Thus, an invader will fail if the number n of defections after its exploiting defection is large enough to compensate for the temptation score T, i.e., $T + nP < (n+1)R$, if terms of order e are neglected. This means that, in the limit of $e \to 0$, n must be chosen so that

$$n > \frac{T-R}{R-P}. \tag{3}$$

In other words, since $T > R > P$, there is always an error probability e and an integer k, for which the strategy S_k cannot be invaded. Note that $e > 0$ is needed, since it has been shown[9,10,29] that the presence of mistakes are necessary to allow for evolutionary stability. With our choice of $R = 1$, the stability requirement for the strategy S_k, in the limit $e \to 0$, can be written

$$P < 1 + \frac{1}{n} - \frac{T}{n}. \tag{4}$$

These lines are drawn in Figure 8. The two examples discussed above (Figures 6 and 7) are on separate sides of the first line. Parameter choices toward the bottom left hand corner appear to make the situation easier for cooperative strategies, since the temptation to defect is relatively small. In the upper right-hand corner, however, the score difference between cooperating and defecting is smaller, and a strategy that punishes defectors may need a longer memory to be able to avoid exploitation.

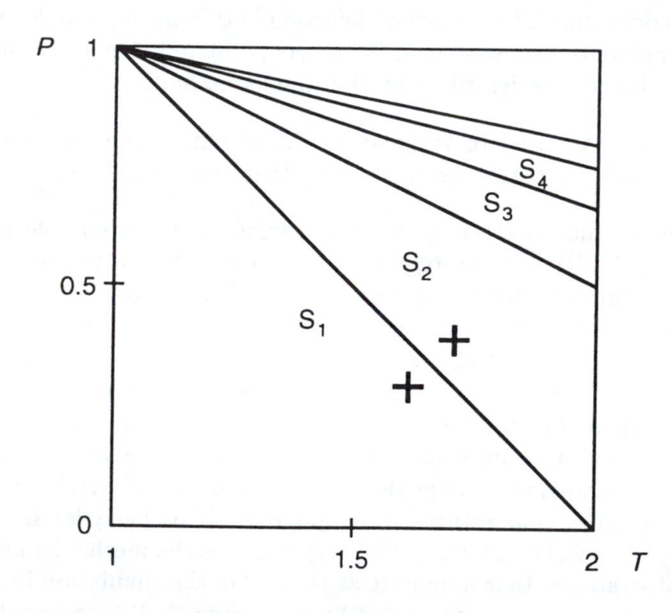

FIGURE 8 The parameter region, $1 < T < 2$ and $0 < P < 1$, is divided into subregions in which the smallest evolutionarily stable strategy of the S_n type is shown, according to the inequality requirement Eq. (4). The two parameter values corresponding to the simulations of Figures 6 and 7 are shown as crosses.

THE CA MODEL

The introduction of spatial dimensions, so that individuals only interact with those in their neighborhood, may affect the dynamics of the system in various ways. It has been demonstrated, for example, in the study by Boerlijst and Hogeweg,[7,8] that parasitic species may be rejected by spiral waves consisting of several species. Other forms of cooperative behavior are also possible, for example regions of cooperative strategies surrounded by exploiting cheaters. The possibility of spatiotemporal structures may allow for global stability where the mean-free model would be unstable. The presence of these various forms of spatiotemoral phenomena may, therefore, also alter the evolutionary path compared with the mean-field case and we may see other strategies evolve.

In Lindgren and Nordahl,[27] we performed a detailed analysis of the dynamic behavior of memory-1 strategies for the parameter region $1 < T < 2$ and $0 < P < 1$. Already at memory 1, there are several spatiotemporal phenomena that may lead to global stability while the system is locally unstable. Spatial coexistence in frozen patterns is also possible. In Figure 9, four examples of spatial patterns from simulations of memory-1 strategies for different parameter choices are shown.

In Figure 10, the strategy abundancy is shown for the first 800 generations of the CA model, with $T = 1.6$ and $P = 0.3$. (In the examples presented here, we have used a 64×64 lattice, and the mutation rates are $p_{\text{point}} = 2 \cdot 10^{-3}$ and $p_{\text{dupl}} = p_{\text{split}} = 10^{-3}$.) It is clear that the strategy 1001 (or S_1) is successful, even though we are close to the line above which it can be exploited, cf. Figure 8. Before 1001 takes over, the system is dominated by 00 and 11, i.e., ALLD and ALLC. This is contrary to what we found in the mean-field case, where 01 (TFT) and 10 (ATFT, does the opposite to TFT) dominate before memory-2 strategies appear. The reason is the possibility to form cooperating regions in the spatial model. Here we get islands of ALLC surrounded by ALLD, similar to the pattern in Figure 9(d). The ALLC individuals at the edges of the islands are certainly exploited by the ALLD individuals and, in the reproduction step, they are removed and replaced by the offspring of the most successful individual in their neighborhood. Now, an ALLC individual that is not in contact with ALLD has the highest score and, therefore, the edge individuals are replaced by an offspring of their own kind. This can be viewed as "kin selection," in which the individual at the edge sacrifices herself since that is beneficial for her own strategy type.

If we would choose a more "rational" procedure for selecting the new strategy, e.g., by taking the strategy in the neighborhood that maximizes the score in the particular cell in the next time step (generation), ALLD would invade the island and wipe out the ALLC strategy from the population. It would be interesting to investigate the difference between "rational" and "natural" selection in this type of model, and to see how the evolutionary dynamics changes.

In Figure 11, the evolutionary path is shown for $T = 1.7$ and $P = 0.4$. Here, 1001 (S_1) should not be stable, according to Eq. (4), but instead, strategies of type

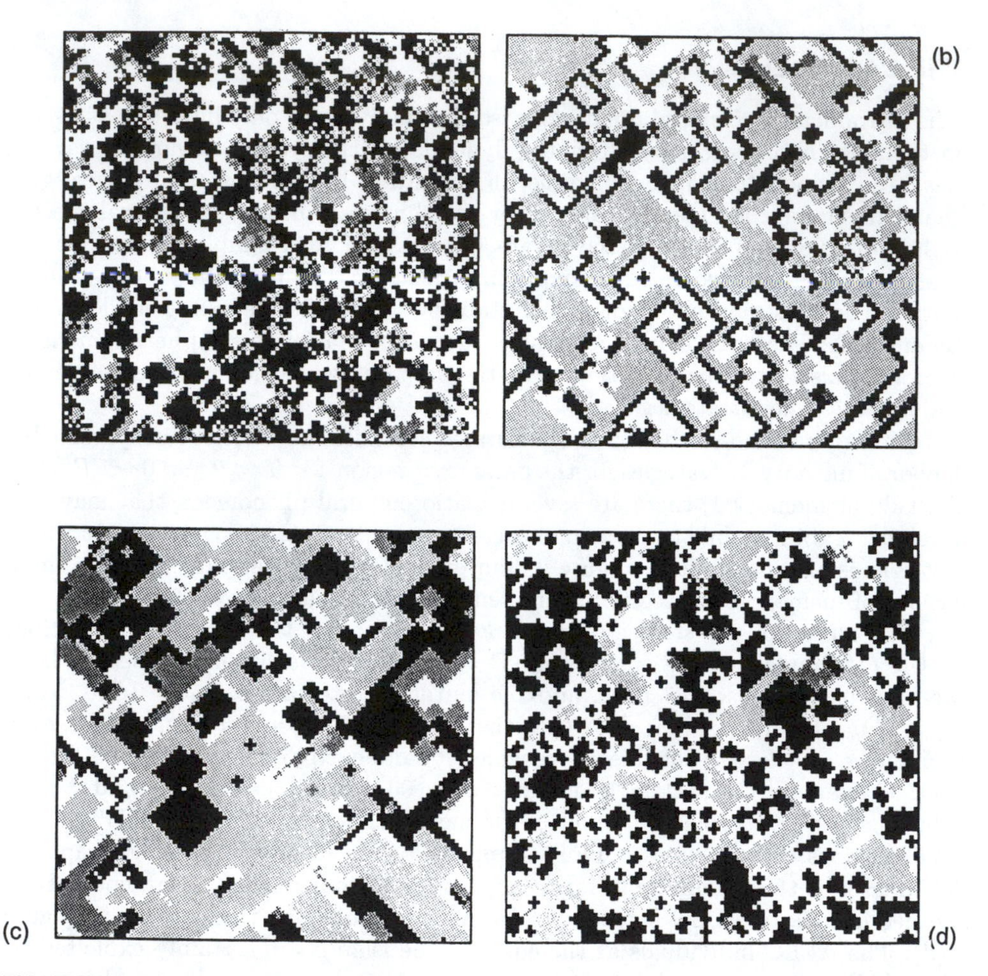

FIGURE 9 Examples of four different types of patterns from simulations of memory-1 strategies on the lattice. The payoffs are (a) $(T, P) = (1.4, 0.05)$, (b) $(T, P) = (1.9, 0.2)$, (c) $(T, P) = (1.9, 0.8)$, (d) $(T, P) = (1.4, 0.5)$. The strategies ALLD, TFT, ATFT, and ALLC are represented in gray-scale from white to black.

S_2 should be able to successfully deal with both mistakes and exploiting strategies. The simulation ends with one of the strategies using the S_2 pattern for error correction, i.e., two rounds of mutual defection after a mistake. There is a large number of strategies in this group, and longer simulations show that in the CA model there is usually a certain memory-5 strategy that takes over the whole population, since even very small score differences may lead to the total dominance of one of the strategies. In the CA model there are also examples of spatial coexistence between strategies of longer memory, that were not observed in the mean-field model. For some examples of these phenomena, see Lindgren and Nordahl.[27]

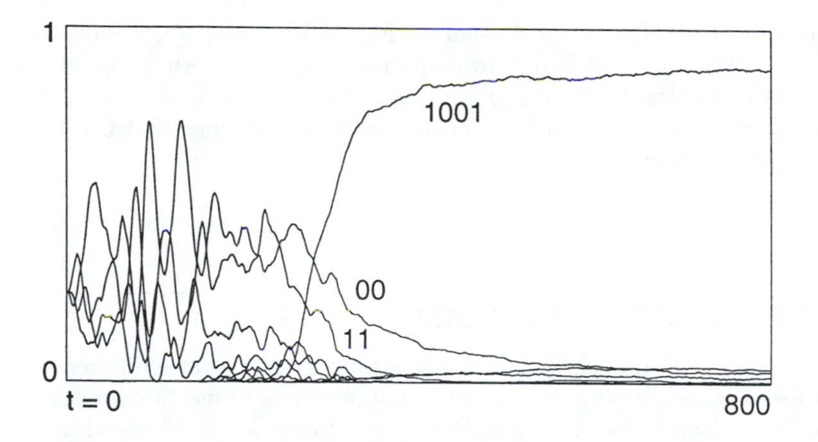

FIGURE 10 The evolution of finite memory strategies in the CA model is shown for 800 generations, with the parameter values $T = 1.6$, $P = 0.3$, cf. Figure 6. There is a metastable coexistence between 00 (ALLD) and 11 (ALLC), which is based on a spatial pattern of ALLC islands surrounded by ALLD, similar to what is seen in Figure 9(d). Still the 1001 strategy dominates at the end.

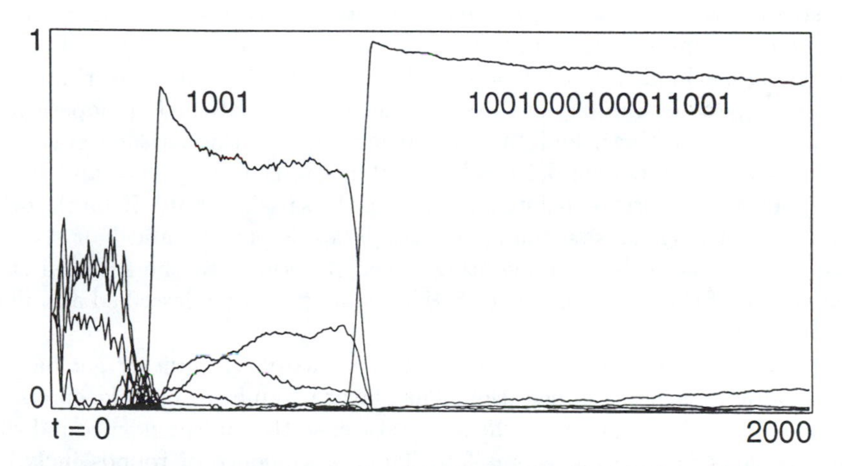

FIGURE 11 The evolution of finite memory strategies in the CA model is shown for 2000 generations, with the parameter values $T = 1.7$, $P = 0.4$, cf. Figure 7. Initially, during the first 200 generations, there is a metastable coexistence between TFT, ALLC, and ALLD in the form of spiral waves, similar to the pattern in Figure 9(b). The spatial dimension allows 1001 to dominate in a metastable state, due to the possibility to form cooperative regions. At the end, the S_2-like strategy from Figure 7 comes to dominate.

Before 1001 dominates the system, in Figure 11, there is a period with three strategies coexisting. This is due to a spatiotemporal pattern of spiral waves involving the strategies TFT, ALLD, and ALLC, similar to Figure 9(b). Again, we see an example of a locally unstable composition of strategies that can be globally stable in the CA model.

FINITE AUTOMATA STRATEGIES

As a final modification of the previous models, we shall look at a different representation of strategies for the game. A natural approach is to use finite automata, and this has been done before by Miller[32] and by Stanley et al.,[46] see also Nowak et al.[39] The novelty in the approach presented here is the combination of an evolvable genome size, the analytically solved iterated Prisoner's Dilemma with mistakes, and the finite automata representation.

The reason why this representation could be interesting to investigate is that there are reasons to believe that more successful strategies may evolve. In fact, there is an extremely simple, cooperative, uninvadeable, and error-correcting strategy in this representation.

This strategy was suggested by Sugden,[49] as a robust strategy that can maintain cooperation in the presence of mistakes. We call the strategy "Fair," since the error-correcting mechanism is built on the idea of a fair distribution of the scores. The Fair strategy can be in one of three possible states: "satisfied" (cooperating), "apologizing" (cooperating), and "angry" (defecting). In the satisfied state Fair cooperates, but if the opponent defects, Fair switches to the angry state and defects until the opponent cooperates, before returning to the satisfied state. If, on the other hand, Fair accidentally defects, the apologizing state is entered and Fair stays in this state until it succeeds in cooperating. Then it returns to the satisfied state again, regardless of the opponent's action. The strategy can be described as a finite automaton, see Figure 12.

This means that two players using the Fair strategy will quickly get back to cooperation after an accidental defection. The player making the mistake enters the apologizing state and the opponent the angry state, so that in the next round they will switch actions making the scores fair. Thus, a sequence of rounds including an accidental defection $\underline{\text{D}}$ looks like $(\ldots, \text{CC}, \underline{\text{D}}\text{C}, \text{CD}, \text{CC}, \text{CC}, \ldots)$. (Only if the apologizing player fails to cooperate, for example by another mistake, they will both stay in the apologizing and angry state, respectively, until the apologizing player succeeds in a C action.)

Before we go on discussing the evolutionary simulations based on finite automata, we shall take a look at the game-theoretic properties of the Fair strategy. The stability properties of the Fair strategies were investigated by Sugden,[49] and

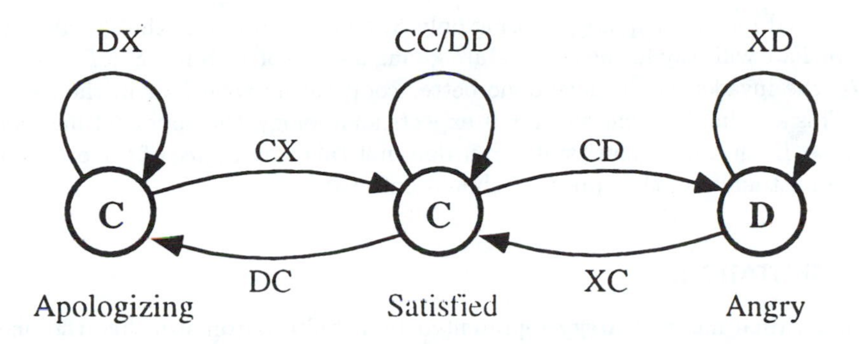

FIGURE 12 The Fair strategy can be represented as a finite automaton with three states. The symbols in the state circles denote the action that is intended, though a mistake may alter the action. After each round of the game, the performed actions are inspected and the strategy enters the state given by the appropriate transition. (On the transition arcs, the left and right symbols correspond to the action performed by the strategy itself and the opponent, respectively, where an X denotes an arbitrary action.) The anti-symmetry assures that when two Fair players meet they will either both be in the satisfied state, or they will be in the Apologizing and the Fair state, respectively, from where they simultaneously return to the satisfied state.

the analysis was extended by Boyd,[9] who labeled the strategy "contrite-tit-for-tat." Wu and Axelrod[52] showed the advantage of the Fair strategy in a population dynamics simulation, starting with a population mainly consisting of the 63 strategies submitted to the second tournament organized by Axelrod.[3]

In the following analysis, we are assuming the infinitely iterated PD with a small probability e for mistakes, i.e., the performed action is not the intended one. In the game between two Fair strategies, the probability is $1/(1+2e)$ to be in the satisfied state and $e/(1+2e)$ for both the apologizing and the angry state.[2] This leads to an average score of $R + 2e(S + T - 2R)$, if terms of order e^2 or higher are neglected. (We are assuming small e, and in the following we shall make a first-order analysis, omitting terms of order e^2 or higher.) For the standard parameter values used by Axelrod ($R = 3$, $T = 5$, $S = 0$, and $P = 1$) and $e = 0.01$, the score is 2.98. This is close to the score that two ALLC strategies get when they meet,[3] $R + e(S + T - 2R)$, or, for the same parameters, 2.99.

For a small probability for mistakes (in the limit $e \to 0$), a new player entering a population of Fair players must adopt the same strategy not to be out-competed: If Fair is in the satisfied state, a defect action (by the invader) scoring T must be

[2] A detailed analysis will be presented elsewhere.

[3] Even if ALLC is more cooperative, it cannot invade a population of Fair players for small values of e, since the score received by an invading ALLC, $R + e(2S + T - 3R)$, differs by $e(R - T) < 0$ compared to the score for Fair, if terms of order e^2 are neglected.

compensated for by an apology scoring only S, before Fair enters the satisfied state again, or Fair will stay in the angry state giving a score of P. Since $S+T < 2R$ and $P < R$, the invader would have done better cooperating with Fair in the satisfied state. This also implies that when Fair expects an apology, the opponent had better cooperate. In the apologizing state, Fair does not take any notice of the opponent's action and, therefore, the opponent should defect.[4]

REPRESENTATION

Each individual has a strategy represented by a finite automaton like the one in Figure 12. The size of the automaton is not fixed in the evolution, but nodes may be added by mutation increasing the number of internal states. Other mutations may alter the actions associated with the internal states, move transition arcs, and change start node.

The game between two FA strategies is a stochastic process on a finite automaton representing the game between the two strategies. This "game" automaton has, as its internal states, the pairs of possible internal states for the two players, and the transitions are given by the transitions of the two strategies. The average score for the players is then calculated from probability distribution over the internal states of the game automaton.

Note that all finite memory strategies can be represented by FA, but that the FA representation results in a larger evolutionary search space. As in the finite memory models, we start with a population consisting of equal numbers of the memory-1 strategies, ALLD, ALLC, TFT, and ATFT, which can be represented as FA like in Figure 13.

The mutations we allow are the following. The *node mutation* changes the action associated with a certain node. The *transition mutation* changes the destination node of a transition. The *start mutation* changes the start node of the FA. The *growth mutation* adds a node to the FA. This is done in a neutral way by adding a copy of a node that has a transition to itself. After the mutation, this transition connects the original node with a new one that is identical to the original one, see Figure 14. There is also a *delete mutation* that randomly removes a node of the FA. The mutation rate is p_{mut} per node for all mutations, except for the transition mutation that has a rate of $4p_{\text{mut}}$ per node, i.e., p_{mut} per transition. In the mean-field model we have used a rate of $p_{\text{mut}} = 10^{-5}$. In the lattice model, we have used the same mutation rate per individual for all types of mutations, $p_{\text{ind}} = 10^{-3}$.

[4]The Fair strategy can only be invaded if the constants of the payoff matrix are close to the limits, e.g., the punishment score P for mutual defection is not much less than the reward for mutual cooperation R, $R \approx P$, or if the risk for mistake is high. Fixing $S = 0$, one finds that the mistake probability must be of the order $e \approx (1 - P/R)/(4 - 3P/R)$ for ALLD to exploit the Fair strategy. The punishment score P must not be less than $0.99R$, for ALLD to invade with an e of 1%. For the standard parameter values, an error probability of $e \approx 2/9 \approx 22\%$ is required for ALLD to invade.

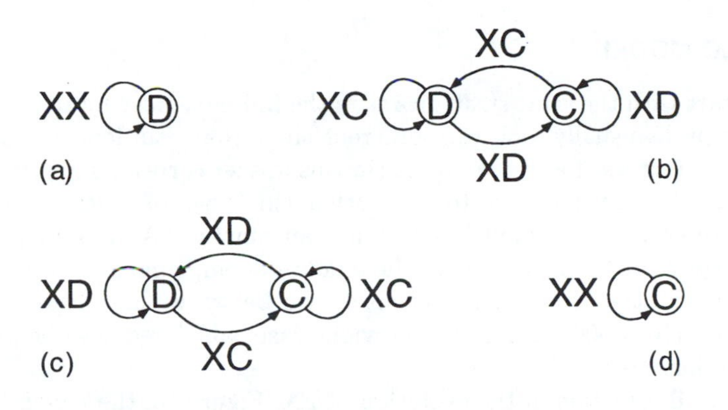

FIGURE 13 The initial state of the finite automata (FA) models consists of equal numbers of four FA that correspond to the memory-1 strategies: (a) ALLD, (b) ATFT, (c) TFT, and (d) ALLC.

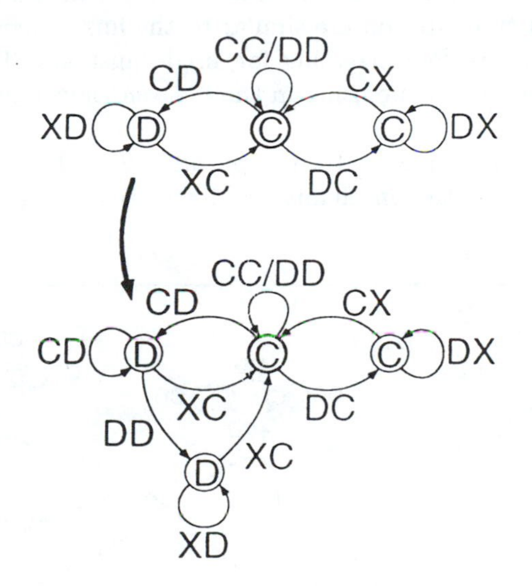

FIGURE 14 The growth mutation can only be applied if the parent FA has a node with a transition to itself. In that case, the mutation creates a copy of the original node with transitions *from* the copied node identical to the original one. The self-transition of the original node is directed to the copy, while the self-transition of the copy remains. This results in a mutation that increases the number of internal states without changing the strategy—the mutation is neutral.

MEAN-FIELD MODEL

Even if we start with the same strategies as in the finite memory representation, the evolutionary paths usually look very different since the evolutionary search space is dramatically increased and the points (in this space) corresponding to the finite memory strategies are not close to each other (in terms of mutations). It turns out that in the mean-field model with finite automata (FA) representation, the ALLD strategy more often takes over the whole system, compared with the finite memory representation. In the simulations shown below, we have chosen a smaller population size ($N = 500$) than in the previous case, which reduces the probability for ALLD to dominate.

In the first illustration of the evolution of FA, Figure 15, the payoff is $T = 1.6$ and $P = 0.3$. After the initial oscillations dominated by TFT, a strategy (not of finite memory type) that only differs from TFT by a transition mutation appears. This strategy is slightly better than TFT since it does not leave the cooperative state if both players defected last round, while in the defecting state it requires a cooperation from the opponent to get back to its cooperating state. The strategies that dominate the picture later on are similar to the finite memory strategy 1001, showing the same pattern for correcting for single mistakes. The slowly growing strategy at which the arrow is pointing in the bottom right-hand corner is the FA representation of 1001.

In the second example, Figure 16, the payoff is $T = 1.7$ and $P = 0.4$, which means that the error-correcting mechanism of 1001-like strategies can be exploited.

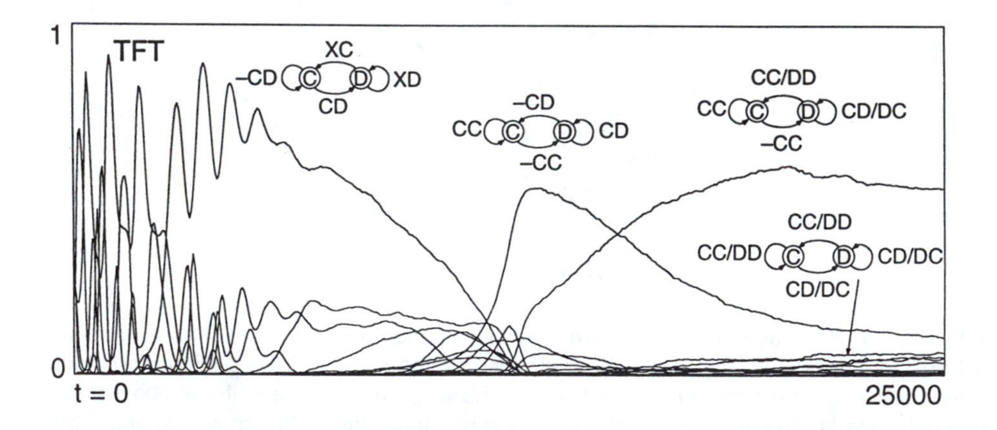

FIGURE 15 The evolution of finite automata (FA) is shown for 25000 generations in the mean-field model, with payoff parameters $(T, P) = (1.6, 0.3)$. Some of the automata that evolve are drawn on top of their abundancy curves, respectively. The FA, drawn at the bottom toward the end of the simulation, is the FA representation of the finite memory strategy 1001. Also, the dominating strategy at the end has an error-correcting mechanism similar to 1001.

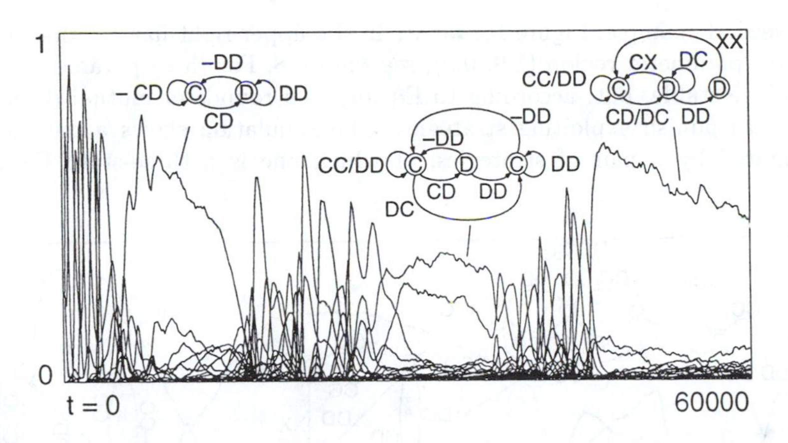

FIGURE 16 The evolution of finite automata (FA) is shown for 60000 generations in the mean-field model, with payoff parameters $(T, P) = (1.7, 0.4)$. Here, strategies corresponding to 1001 (S_1) should not be stable, cf. Figure 8. The automaton shown in the middle is related to the Fair strategy in its mechanism for dealing with accidental defections, but it can be exploited. Instead, an FA that defects twice, after an isolated defect action, dominates at the end similar to the finite memory strategy S_2.

The strategy shown in the middle of the figure, dominating for more than 10000 generations, resembles the Fair strategy, but it is too forgiving, which implies that it will be exploited. The strategy shown at the end, dominating for more than 15000 generations, has an error correction similar to S_2, and it is possible (but not certain) that this simulation would end with a mixture of strategies of this type.

LATTICE MODEL

The spatial dimension again increases the chances for cooperative strategies to succeed in the population. Many simulations end with cooperative strategies, like Fair, dominating the whole lattice for large parts of the parameter region. This is in accordance with the analysis of the Fair strategy above.

In the first example, Figure 17, we have the parameters $(T, P) = (1.7, 0.4)$ on a 64×64 lattice. After the initial oscillations we again see the S_1-like strategy that dominated at the end of the mean-field simulation of Figure 15. This behavior is not stable here, according to Eq. (4), and the simulation continues with a sequence of FA with three and four states, all having an S_2-like mechanism for correcting mistakes. There are many FAs that coincide with the pattern of two mutual defect actions after an isolated mistake, but that differ when it comes to second-order effects, i.e., when there is a mistake within the error-correcting pattern. These second-order effects may be sufficient to give a decisive advantage so that only one of them will dominate.

In the second example, Figure 18, we are in the upper right-hand corner of the payoff matrix parameter region (1.9, 0.9), see Figure 8. For these parameters, not even S_9 will be stable, but, according to Eq. (4), ten rounds of mutual defection are required to punish exploiting strategies. The simulation shows a metastable state dominated by a pair of strategies, of which one is a three-state FA that

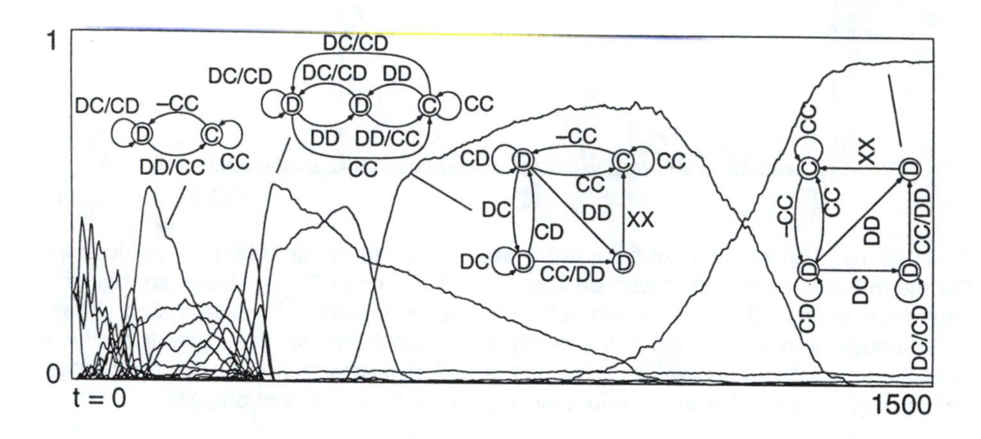

FIGURE 17 The evolution of finite automata for the cellular automaton model is shown for 1500 generations, with the parameter values $(T, P) = (1.7, 0.4)$. The automata of three and four states that evolve all show the S_2-mechanism of the finite memory strategies for error correction.

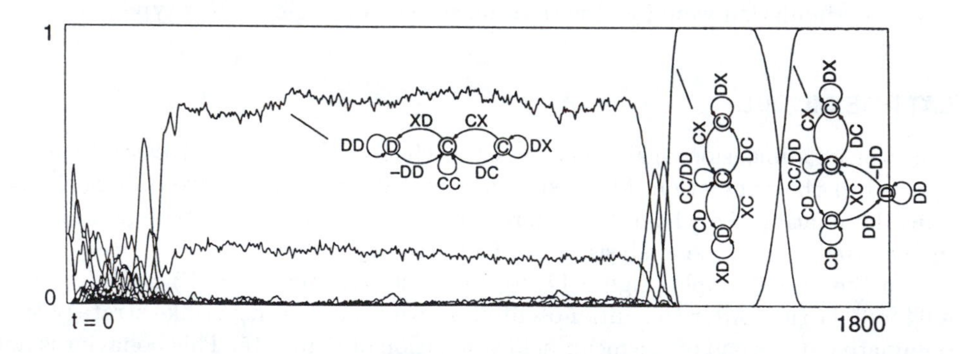

FIGURE 18 The evolution of finite automata for the cellular automaton model is shown for 1800 generations, with the parameter values $(T, P) = (1.9, 0.9)$. This is close to the upper right-hand corner of the payoff parameter region, see Figure 8. The chance that an S_n strategy would evolve here is small since the stability requirement Eq. (4) implies that n would be 10. In the simulation, we see that a predecessor to Fair dominates for a long time until Fair appears, but at the end there is a four-state variation of Fair that manages to take over.

resembles the Fair strategy. The Fair strategy is two transition mutations away, and the intermediate FA can be seen as one of the two narrow peeks before Fair appears, cf. Figure 12. However, there is a four-state variation of Fair that succeeds in out-competing the original Fair. The difference lies in a fourth-order effect, i.e., involving three mistakes in the apology pattern following one mistake. This strategy only differs from Fair by a (neutral) growth mutation and a transition mutation.

Other simulations, from various parts of the parameter region of the payoff matrix in Figure 8, show that Fair is successful and not very sensitive to the parameter choices. In most cases, the simulations end with one strategy dominating the whole lattice, a result that is not general in the CA simulations of the finite memory strategies. From the evolutionary point of view, these simulations may seem less interesting in the sense that the evolutionary transients appear to be short. On the other hand, they demonstrate the important difference between the spatial and the nonspatial models, as well as between different evolutionary search spaces, and that cooperative behavior may not be given the chance unless there is some compartmentalization possible, here in the form of spatial locality.

The examples presented here are, of course, preliminary, and a more detailed study of the evolution of finite automata strategies will be reported elsewhere.

MISUNDERSTANDING VS. MISTAKES

As a possibility for future modifications of these models, one could consider a different kind of noise. Mistakes are a relatively simple form of noise to deal with in games, since both players have the same history of actions to look at. Misunderstandings, on the other hand, leave the two players with different histories, and one cannot be sure whether the distribution of scores is fair or not even for a few rounds. This is of course critical for strategies like Fair, since a misunderstanding may lead them into situations where one player is in a satisfied state while the other one is in an angry state, which may result in sequences of alternating defect and cooperate actions, as in the TFT strategy.

The class of evolutionarily stable strategies (S_n) in the case of mistakes, discussed before, turn out to be error correcting also in the case of misunderstanding, at the same time as there is a possibility of avoiding both exploiting strategies and strategies that are more cooperative.

In the case of *misunderstanding*, two players using, e.g., the S_2 strategy, will show the same pattern as in the case of mistake, $(\ldots, \mathrm{CC}, \mathrm{CC}^*, \underline{\mathrm{D}}\mathrm{C}, \mathrm{DD}, \mathrm{DD}, \mathrm{CC}, \mathrm{CC}, \mathrm{CC}, \ldots)$, where C* denotes a C action that is perceived as a D action by the opponent. As before, the length of the "quarrel" must be large enough to prevent potential exploiters to defect on purpose, which leads to the inequality requirement of Eq. (3).

But when misunderstandings are involved, there is another possibility to invade a population of S_n strategies, which leads to an additional requirement for the choice of n. If a player ignores a perceived D action from its opponent and still cooperates the score may increase, since the action could have been C, and then the cost of quarrel could have been be avoided, e.g., $(\ldots, \mathrm{CC}, \mathrm{CC}^*, \mathrm{CC}, \ldots)$. On the other hand, if the action was truly D, then the opponent will defect again, and the return to cooperation will be delayed. For example, ignoring a defect action from S_2 leads to the pattern $(\ldots, \mathrm{C}^*\mathrm{C}, \mathrm{C}\underline{\mathrm{D}}, \mathrm{CD}, \mathrm{DD}, \mathrm{DD}, \mathrm{CC}, \mathrm{CC}, \ldots)$, where $\underline{\mathrm{D}}$ is a defect action due to S_2 misunderstanding the previous action C*. Since the risk for misunderstanding is equal for both players, the two patterns above that can result from ignoring a perceived D action occur with equal probability. The expected score for ignoring instead of answering directly is $\sigma_0 = (n+2)R/2 + (2S + nP)/2$, to compare with the score for S_n, $\sigma_n = (S+T)/2 + nP + R$, for the rounds following a defect action. Thus, for S_n to avoid invasion, the inequality $\sigma_n > \sigma_0$ must be fulfilled. This can be rewritten as a requirement on n, $n < (T-S)/(R-P)$, or $n < (T-R)/(R-P) + 1 + (P-S)/(R-P)$. Since the third term is positive, there is always an integer n satisfying both conditions

$$\frac{T-R}{R-P} < n < \frac{T-R}{R-P} + 1 + \frac{P-S}{R-P}. \tag{5}$$

For the parameters used by Axelrod ($R = 1, S = 0, T = 5, P = 1$), we get $n = 2$, and the score is 2.92. Contrary to Fair, the error correction of these strategies works for both mistake and misunderstanding, but at a cost of a lower average score.

DISCUSSION

By showing examples from various models based on the Prisoner's Dilemma as the interaction between individuals, I have illustrated some aspects of evolutionary models. The differences between the mean-field models and the lattice models are evident in all these cases. The possibility of spatiotemporal phenomena may give rise to a stable coexistence between strategies that would otherwise be outcompeted. These spatiotemporal structures may take the form of spiral waves, irregular waves, spatiotemporal chaos, frozen patchy patterns, and various geometrical configurations.[28] This is very clear in the first model (finite game) where ALLD is the only fixed point in the mean-field model, but where ALLD seldom appears in the simulations of the CA model. Space also allows for increased cooperative behavior by a mechanism that can be described as kin selection. For example, regions of ALLC can be stable in an ALLD environment, since the ALLC individuals at the border are replaced (in the next generation) by more successful ALLC individuals from the interior of the region.

The choice of representation is important, as the difference between the simulations of the finite memory and FA strategies illustrates. Still, the existence of a stable fixed point, e.g., in the form of the Fair strategy, does not necessarily mean that the evolutionary dynamics always finds it. Of course, the regions of attraction of the fixed points are important to characterize the evolutionary search space. This may, however, be a very difficult task since the space of strategies is potentially infinite. The final example also illustrates that the region of attraction may increase dramatically if the spatial dimension is added. The increased probability for the Fair strategy to succeed probably depends on the advantage for cooperative strategies in the spatial world. The Fair strategy may more easily survive in a local situation from where it may then spread.

The models presented here all have in common that they are very simple and can be easily simulated. Despite their simplicity, a number of complex phenomena in the form of evolutionary dynamics and spatiotemporal patterns can be observed. Models similar to these (in terms of simplicity) could be constructed in a more economic context to study economic phenomena like, e.g., trading and price formation. Some of the ingredients of the models presented here should be considered in that work, for example, the advantage of starting small in combination with neutral mutations. (This has been used in another context, where evolving artificial neural networks were selected for recognizing strings of a regular language.[26]) The workshop has inspired us to start constructing economics models along these lines, and we hope to report on these in future contributions.

ACKNOWLEDGMENT

This research is supported by the Swedish Natural Science Research Council. I thank Robert Axelrod for pointing out the reference to Sugden concerning the Fair strategy. I also thank the Santa Fe Institute for supporting several visits during which parts of the work presented here were done.

REFERENCES

1. Anderson, P. W. "Suggested Model for Prebiotic Evolution: The Use of Chaos." *Proc. Natl. Acad. Sci.* **80** (1983): 3386–3390.
2. Arthur, W. B. "Asset Pricing Under Endogenous Expectations in an Artificial Stock Market." In *The Economy as an Evolving Complex System II,* edited by W. B. Arthur, S. N. Durlauf, and D. Lane, 1–30. Santa Fe Institute Studies in the Sciences of Complexity, Proc. Vol. XXVII. Reading, MA: Addison-Wesley, 1997.
3. Axelrod, R. *The Evolution of Cooperation.* New York: Basic Books, 1984.
4. Axelrod, R. "The Evolution of Strategies in the Iterated Prisoner's Dilemma." In *Genetic Algorithms and Simulated Annealing,* edited by L. Davis, 32–41. Los Altos, CA: Morgan-Kaufmann, 1987.
5. Axelrod, R., and D. Dion. "The Further Evolution of Cooperation." *Science* **242** (1988): 1385–1390.
6. Bagley, R. J., J. D. Farmer, and W. Fontana. "Evolution of a Metabolism." In *Artificial Life II,* edited by C. G. Langton, C. Taylor, J. D. Farmer, and S. Rasmussen, 141–158. Santa Fe Institute Studies in the Sciences of Complexity, Proc. Vol. X. Redwood City, CA: Addison-Wesley, 1991.
7. Boerlijst, M. C., and P. Hogeweg. "Spiral Wave Structure in Prebiotic Evolution: Hypercycles Stable Against Parasites." *Physica D* **48** (1991): 17–28.
8. Boerlijst, M. C., and P. Hogeweg. "Self-Structuring and Selection: Spiral Waves as a Substrate for Prebiotic Evolution." In *Artificial Life II,* edited by C. G. Langton, C. Taylor, J. D. Farmer, and S. Rasmussen, 255–276. Santa Fe Institute Studies in the Sciences of Complexity, Proc. Vol. X. Redwood City, CA: Addison-Wesley, 1991.
9. Boyd, R. "Mistakes Allow Evolutionary Stability in the Repeated Prisoner's Dilemma Game." *J. Theor. Biol.* **136** (1989): 47–56.
10. Boyd, R., and J. P. Lorberbaum. "No Pure Strategy Is Evolutionarily Stable in the Repeated Prisoner's Dilemma Game." *Nature* **327** (1987): 58–59.
11. Eigen, M. "Self-Organization of Matter and the Evolution of Biological Macro-Molecules." *Naturwissenschaften* **58** (1971): 465–523.
12. Eigen, M., and P. Schuster. *The Hypercycle: A Principle of Natural Self-Organization.* Berlin: Springer-Verlag, 1979.
13. Eigen, M., J. McCaskill, and P. Schuster. "Molecular Quasi-Species." *J. Phys. Chem.* **92** (1988): 6881–6891.
14. Farmer, J. D., S. A. Kauffman, and N. H. Packard. "Autocatalytic Replication of Polymers." *Physica* **22D** (1986): 50–67.
15. Fontana, W. "Algorithmic Chemistry." In *Artificial Life II*, edited by C. G. Langton and C. Taylor, 313–324. Santa Fe Institute Studies in the Sciences of Complexity, Proc. Vol. X. Redwood City, CA: Addison-Wesley, 1991.
16. Hillis, D. "Coevolving Parasites Improve Simulated Evolution as an Optimization Procedure." *Physica D* **42** (1990): 228–334.

17. Hofbauer, J., and J. W. Weibull. "Evolutionary Selection Against Dominated Strategies." Paper presented at The Sixth Conference of the International Joseph A. Schumpeter Society, 1996.
18. Holland, J. H. *Adaptation in Natural and Artificial Ecosystems.* 2nd ed. Cambridge, MA: MIT Press, 1992.
19. Holland, J. H. "Echoing Emergence: Objectives, Rough Definitions, and Speculations for Echo-Class Models." Working Paper # 93-04-023, Santa Fe Institute, Santa Fe, NM, 1993.
20. Ikegami, T. "From Genetic Evolution to Emergence of Game Strategies." *Physica D* **75** (1994): 310–327.
21. Kauffman, S. A. "Autocatalytic Sets of Proteins." *J. Theor. Biol.* **119** (1986): 1–24.
22. Kauffman, S. A. *The Origins of Order: Self-Organization and Selection in Evolution.* New York: Oxford University Press, 1993.
23. Kauffman, S. A., and S. Johnsen. "Coevolution to the Edge of Chaos: Coupled Fitness Landscapes, Poised States, and Coevolutionary Avalanches." In *Artificial Life II,* edited by C. G. Langton, C. Taylor, J. D. Farmer, and S. Rasmussen, 325–369. Santa Fe Institute Studies in the Sciences of Complexity, Proc. Vol. X. Redwood City, CA: Addison-Wesley, 1991.
24. Langton, C. G. Paper presented at The Economy as an Evolving Complex Systems, Santa Fe Institute, Santa Fe, NM, 1995.
25. Lindgren, K. "Evolutionary Phenomena in Simple Dynamics." In *Artificial Life II,* edited by C. G. Langton, C. Taylor, J. D. Farmer, and S. Rasmussen, 295–311. Santa Fe Institute Studies in the Sciences of Complexity, Proc. Vol. X. Redwood City, CA: Addison-Wesley, 1991.
26. Lindgren, K., A. Nilsson, M. G. Nordahl, and I. Råde. "Regular Language Inference Using Evolving Recurrent Neural Networks." In *COGANN-92 International Workshop on Combinations of Genetic Algorithms and Neural Networks,* edited by L. D. Whitley and J. D. Schaffer, 75–86. Los Alamitos, CA: IEEE Computer Society Press, 1992.
27. Lindgren, K., and M. G. Nordahl. "Evolutionary Dynamics of Spatial Games." *Physica D* **75** (1994): 292–309.
28. Lindgren, K., and M. G. Nordahl. "Cooperation and Community Structure in Artificial Ecosystems." *Artificial Life J.* **1** (1994): 15–37.
29. Lorberbaum, J. *J. Theor. Biol.* **168** (1994): 117–130.
30. Matsuo, K. "Ecological Characteristics of Strategic Groups in 'Dilemmatic World.'" In *Proceedings of the IEEE International Conference on Systems and Cybernetics*, 1071–1075, 1985.
31. Maynard-Smith, J. *Evolution and the Theory of Games.* Cambridge, MA: Cambridge University Press, 1982.
32. Miller, J. H. "The Coevolution of Automata in the Repeated Iterated Prisoner's Dilemma." Working Paper No. 89-003, Santa Fe Institute, Santa Fe, NM, 1989.

33. Mitchell, M., and S. Forrest. "Genetic Algorithms and Artificial Life." *Artificial Life J.* **1** (1994): 267-289.
34. Molander, P. "The Optimal Level of Generosity in a Selfish, Uncertain Environment." *J. Conflict Resolution* **29** (1985): 611–618.
35. Nowak, M. A., and R. M. May. "Evolutionary Games and Spatial Chaos." *Nature* **359** (1993): 826–829.
36. Nowak, M. A., and R. M. May. "The Spatial Dilemmas of Evolution." *Intl. J. Bif. & Chaos* **3** (1993): 35–78.
37. Nowak, M. A., and K. Sigmund. "Tit-For-Tat in Heterogenous Populations." *Nature* **355** (1992): 250–253.
38. Nowak, M. A., and K. Sigmund. "A Strategy of Win-Stay, Lose-Shift that Outperforms Tit-For-Tat in the Prisoner's Dilemma Game." *Nature* **364** (1993): 56–58.
39. Nowak, M. A., K. Sigmund, and E. El-Sedy. "Automata, Repeated Games, and Noise." *J. Math. Biol.* **33** (1995): 703–722.
40. Rapoport, A., and A. M. Chammah. *Prisoner's Dilemma.* Ann Arbor: University of Michigan Press, 1965.
41. Rasmussen, S., C. Knudsen, and R. Feldberg. "Dynamics of Programmable Matter." In *Artificial Life II,* edited by C. G. Langton, C. Taylor, J. D. Farmer, and S. Rasmussen, 211–254. Santa Fe Institute Studies in the Sciences of Complexity, Proc. Vol. X. Redwood City, CA: Addison-Wesley, 1991.
42. Ray, T. S. "An Approach to the Synthesis of Life." In *Artificial Life II,* edited by C. G. Langton, C. Taylor, J. D. Farmer, and S. Rasmussen, 371-408. Santa Fe Institute Studies in the Sciences of Complexity, Proc. Vol. X. Redwood City, CA: Addison-Wesley, 1991.
43. Rokhsar, D. S., P. W. Anderson, and D. L. Stein. "Self-Organization in Prebiological Systems: Simulation of a Model for the Origin of Genetic Information." *J. Mol. Evol.* **23** (1986): 110.
44. Samuelson, L., and J. Zhang. "Evolutionary Stability in Assymmetric Games." *J. Econ. Theory* **57** (1992): 363–391.
45. Schuster, P. "The Physical Basis of Molecular Evolution." *Chemica Scripta* **26B** (1986): 27.
46. Stanley, E. A., D. Ashlock, and L. Tesfatsion. "Iterated Prisoner's Dilemma with Choice and Refusal of Partners." In *Artificial Life III,* edited by C. G. Langton, 131–175. Santa Fe Institute Studies in the Sciences of Complexity, Proc. Vol. XVII. Redwood City, CA: Addison-Wesley, 1993.
47. Stein, D. L., and P. W. Anderson. "A Model for the Origin of Biological Catalysis." *Proc. Natl. Acad. Sci.* **81** (1984): 1751–1753.
48. Stenseth, N. C., and J. Maynard-Smith. "Coevolution in Ecosystems: Red Queen Evolution or Stasis?" *Evolution* **38** (1984): 870–880.
49. Sugden, R. *The Economics of Rights, Co-operation and Welfare.* Oxford: Basil Blackwell, 1986.
50. Tesfatsion, L. "How Economists Can Get ALife." In *The Economy as an Evolving, Complex System II,* edited by W. B. Arthur, S. Durlauf and

D. Lane, 533–565. Santa Fe Institute Studies in the Sciences of Complexity, Proc. Vol. XXVII. Reading, MA: Addison-Wesley, 1997.

51. Weisbuch, G. *C. R. Acad. Sci. III* **298** (1984): 375–378.
52. Wu, J., and R. Axelrod. "How to Cope with Noise in the Iterated Prisoner's Dilemma." *J. Conflict Resolution* **39** (1995): 183–189.

Charles F. Manski
Department of Economics, University of Wisconsin-Madison, Madison, WI 53706;
E-mail: manski@ssc.wisc.edu

Identification of Anonymous Endogenous Interactions

1. INTRODUCTION

In theoretical studies of social interactions, we hypothesize a process and seek to deduce the implied outcomes. In inferential studies, we face an inverse logical problem. Given observations of outcomes and maintained assumptions, we seek to deduce the actual process generating the observations.

Econometricians have long found it useful to separate inferential problems into statistical and identification components.[9] Studies of identification seek to characterize the conclusions that could be drawn if one could use a given sampling process to obtain an unlimited number of observations. Studies of statistical inference seek to characterize the generally weaker conclusions that can be drawn from a finite number of observations. Analysis of identification logically comes first. Negative identification findings imply that statistical inference is fruitless: it makes no sense to try to use a sample of finite size to infer something that could not be learned even if a sample of infinite size were available. Positive identification findings imply that one should go on to examine the feasibility of statistical inference.

Throughout the modern development of the social sciences, analysis of the problem of empirical inference on social interaction processes has lagged far behind our willingness and ability to theorize about these processes. This asymmetry is unfortunate. Theoretical studies hypothesizing alternative realities are ultimately sterile if we do not also ask how these alternative realities are empirically distinguishable.

The classic economic example of the asymmetric state of theory and inference is the analysis of competitive markets. The theory of equilibrium in markets with price-taking consumers and price-taking (or quantity-taking) firms was reasonably well-understood a century ago, but the corresponding econometric problem of identification of demand and supply from observations of equilibrium market transactions was only dimly understood until the 1940s.[6] Even then, the econometric literature developed under the strong maintained assumption that demand and supply are linear functions of price and quantity. The present-day econometric literature offers only scattered findings about the identifiability of nonlinear demand and supply functions. See Fisher[4] (Ch. 5), Roehrig,[15] and Manski[12] (section 6.4).[1]

Several years ago, I became aware of another sharp contrast between the state of theory and inference. The story begins with an empirical regularity: It is often observed that persons belonging to the same group tend to behave similarly.

For at least fifty years, sociologists and social psychologists have sought to explain this regularity by theorizing the existence of social interactions in which the behavior of an individual varies with the distribution of behavior in a "reference group" containing the individual. Depending on the context, these interactions may be called "social norms," "peer influences," "neighborhood effects," "conformity," "imitation," "contagion," "epidemics," "bandwagons," or "herd behavior." See, for example, Hyman,[7] Merton,[13] and Granovetter.[5]

Economists have increasingly contributed to this theory, and have done much to formalize it in a coherent manner. See, for example, Schelling,[16] Pollak,[14] Jones,[8] Benabou,[1] Bernheim,[2] and Brock and Durlauf.[3] Economists have also become aware that various economic and physical processes share the same formal structure. Oligopoly models posit reaction functions, wherein the output of a given firm is a function of aggregate industry output. Endogenous growth models assume dynamic learning processes in which the output of a given firm depends on the lagged value of aggregate industry output. Biological models of contagion suppose that the probability of an uninfected person becoming infected varies with the fraction of the population already infected. In physics, mean-field theory explains magnetism by hypothesizing that the orientation of a particle in space varies with the mean orientation of surrounding particles.

I shall use the term *anonymous endogenous interactions* to describe the class of social processes in which the behavior of an individual varies with the distribution of behavior in a group containing the individual. The interactions are endogenous because the outcome of each group member varies with the outcomes of the other

[1]It is important to recognize that the work on estimation of nonlinear simultaneous equations that began in the 1970s and matured into the literature on method-of-moments estimation in the 1980s does not address the question of identification. This literature assumes identification and then focuses on the statistical aspects of inference (see Manski,[10] pp. 93–94).

group members, not with other attributes of the group. In contrast, the sociological literature on *contextual* interactions assumes that individual outcomes are influenced by the exogenous characteristics of the group members.[17] The interactions are anonymous because they may be described without naming the members of the group or otherwise specifying the internal structure of the group. What matters is the distribution of outcomes in the group, not who experiences what outcome. In contrast, the interactions studied in the literatures on social networks and on local interactions require specification of the internal structure of the group.

Despite the large body of theory on anonymous endogenous interactions, I found no work on the corresponding problem of identification from observations of outcomes. So I set out to study the problem.[11] In particular, I wanted to understand the conditions under which these interactions can be distinguished empirically from other hypotheses that have often been advanced to explain the empirical regularity that individuals belonging to the same group tend to behave similarly. Three familiar hypotheses are

1. **endogenous interactions**, wherein the propensity of an individual to behave in some way varies with the behavior of the group.
2. **contextual** (exogenous) **interactions**, wherein the propensity of an individual to behave in some way varies with the exogenous characteristics of the group.
3. **correlated effects**, wherein individuals in the same group tend to behave similarly because they have similar individual characteristics or face similar institutional environments.

To illustrate, consider the determination of achievement in school. We have endogenous interactions if, all else being equal, individual achievement tends to vary with the average achievement of the students in the youth's school, ethnic group, or other reference group. We have contextual interactions if achievement tends to vary with the socioeconomic composition of the reference group. We have correlated effects if youth in the same school tend to achieve similarly because they have similar family backgrounds or because they are taught by the same teachers.

In Manski[11] I showed that, given the maintained assumption of a linear-in-means interaction model, a researcher who observes equilibrium outcomes and the composition of reference groups cannot empirically distinguish endogenous interactions from contextual interactions or correlated effects. The researcher may or may not be able to determine whether there is some form of social interaction at work. Furthermore, a researcher who observes equilibrium outcomes but not the composition of reference groups cannot distinguish among a host of alternative interaction models, all of which hold tautologically.

I paraphrase these findings in section 2. I then extend the analysis in some respects in section 3. Section 4 gives conclusions. Here is the main notation used throughout the paper:

$t = 0, \ldots, T$	dates at which outcomes are observed
x	individual's time-invariant reference group
$w_t = (z_t, u_t)$	individual's covariates at date t
y_t	individual's outcome at date t
$P(x, w_t, y_t, t = 0, \ldots, T)$	distribution of groups, covariates, and outcomes
$P(y_t \mid x)$	distribution of outcomes in group x at date t
$P(w_t \mid x)$	distribution of covariates in group x at date t.

The reader should take note of two maintained assumptions embedded in this notation. First, outcomes and covariates are potentially time-varying but reference groups are not. Second, each individual belongs to exactly one reference group, not to multiple reference groups. In other words, the population is assumed to partition into a set of mutually exclusive and exhaustive groups.

2. A LINEAR-IN-MEANS MODEL

In this section, we assume that the outcomes at each date $t = 1, \ldots, T$ are generated by the *linear-in-means* model

$$y_t = \alpha + \beta E(y_{t-1} \mid x) + E(z_{t-1} \mid x)'\gamma + z_t'\lambda + u_t\,, \tag{1a}$$

$$E(u_t \mid x, z_t) = x'\delta\,, \tag{1b}$$

or, equivalently,

$$E(y_t \mid x, z_t) = \alpha + \beta E(y_{t-1} \mid x) + E(z_{t-1} \mid x)'\gamma + x'\delta + z_t'\lambda\,. \tag{2}$$

Here y_t and u_t are scalars, while x and z_t are vectors. The scalar parameter β measures the strength of the endogenous interaction among the members of group x, which is transmitted through the lagged group-mean outcome $E(y_{t-1} \mid x)$. The parameter vector γ measures the strength of the contextual interaction among the members of group x, which is transmitted through the lagged group-mean covariate vector $E(z_{t-1} \mid x)$. The parameter vector δ measures the direct variation of outcomes with x, and so expresses the idea of correlated effects. The parameter vector λ measures the direct variation of outcomes with the contemporaneous individual covariates z_t.

Integrating Eq. (2) with respect to $P(z_t \mid x)$, we find that the law of motion for $E(y_t \mid x)$ is

$$E(y_t \mid x) = \alpha + \beta E(y_{t-1} \mid x) + E(z_{t-1} \mid x)'\gamma + x'\delta + E(z_t \mid x)'\lambda\,. \tag{3}$$

If $\beta \neq 1$ and the group-mean covariate vector $E(z \mid x)$ is time-invariant, there is a unique steady state, or *social equilibrium*, namely

$$E(y \mid x) = \frac{\alpha}{1-\beta} + E(z \mid x)' \frac{\gamma + \lambda}{1-\beta} + \frac{x'\delta}{1-\beta}. \quad (4)$$

Now consider the inferential problem faced by a researcher who knows that outcomes satisfy Eq. (2), who observes a random sample of realizations of $(x, z_t, y_t, t = 0, \ldots, T)$ but not u_t, and who wants to infer the parameters. Given minimal regularity conditions, the observational evidence enables the researcher to consistently estimate the conditional expectations $[E(y_t \mid x, z_t), E(y_t \mid x), E(z_t \mid x)]$ at each value of (x, z_t) realized in the population (technically, on the *support* of the regressors).[2]

I want to focus on the identification component of the inferential problem and abstract from statistical concerns, so let us suppose that the researcher knows these conditional expectations. In particular, I want to focus on the situation in which $\beta \neq 1$, $E(z \mid x)$ is time-invariant, and each group x is observed in its unique equilibrium. Inspection of Eq. (4) shows that the equilibrium group-mean outcome $E(y \mid x)$ is a linear function of the vector $[1, E(z \mid x), x]$; that is, these regressors are perfectly collinear. This implies that the parameters $(\alpha, \beta, \gamma, \delta)$, are unidentified. Thus, endogenous effects cannot be distinguished from contextual effects or from correlated effects.[3]

One may or may not be able to learn if there is some form of social interaction. Inserting Eq. (4) into Eq. (2) yields the linear reduced form

$$E(y \mid x, z) = \frac{\alpha}{1-\beta} + E(z \mid x)' \left[\frac{\gamma + \lambda\beta}{1-\beta}\right] + x' \frac{\delta}{1-\beta} + z'\lambda. \quad (5)$$

The solutions to Eq. (5) contain a unique value of the composite parameter $(\gamma+\lambda\beta)/(1-\beta)$ if $E(z \mid x)$ is linearly independent of the vector $(1, x, z)$. In particular, $E(z \mid x)$ must vary nonlinearly with x. Although identification of $(\gamma+\lambda\beta)/(1-\beta)$ does not enable one to distinguish between endogenous and contextual interactions, it does enable one to learn if some social interaction is present. If the composite parameter is nonzero, then either β or γ must be nonzero.

The foregoing assumes that the researcher observes the reference groups into which the population is sorted. Suppose that these groups are unobserved. Can

[2] In particular, consistent estimation is possible if these conditional expectations vary continuously with (x, z_t), or if (x, z_t) has discrete support.

[3] This finding does not imply that the three hypotheses are indistinguishable in principle. They are distinguishable if groups are observed out-of-equilibrium or if an environmental or policy shift should change the parameters of model (2). To illustrate the implications of a parameter change, consider again the determination of achievement in school. Suppose that a technological advance in instruction increases the value of α to, say, $\alpha + k$, where $k > 0$. By Eq. (4), equilibrium mean achievement rises by $k/(1-\beta)$ rather than by k. Thus, endogenous interactions generate a "social multiplier" that is not generated by contextual interactions or correlated effects.

the researcher infer the composition of reference groups from observations of the outcomes y_t and covariates z_t?

The answer is necessarily negative if $\beta \neq 1$, z is time-invariant (not just $E(z \mid x)$), and each group x is observed in its unique equilibrium. To see this, consider any hypothesized specification of reference groups such that z is a function of x, say $z = z(x)$. Then the equality

$$E[y \mid x, z(x)] = E(y \mid x) \tag{6}$$

holds tautologically. Comparing Eqs. (6) and (2) shows that, with $E(y \mid x)$ being time-invariant in equilbrium, Eq. (6) is the special case of Eq. (2) in which $\beta = 1$ and $\alpha = \gamma = \delta = \lambda = 0$.

3. PURE ENDOGENOUS INTERACTIONS MODELS

Let Y denote the space of possible outcome values y, W the space of possible covariate values w, and Γ the space of all distributions on Y. In this section, we assume that the outcomes at each date $t = 1, \ldots, T$ are generated by a response function of the form

$$y_t = f[P(y_{t-1} \mid x), w_t], \tag{7}$$

where $f(\cdot,\cdot) : \Gamma \times W \to Y$. Thus, we assume that an individual's outcome y_t is determined in part by anonymous endogenous interactions, acting through the lagged group outcome distribution $P(y_{t-1} \mid x)$, and in part by the individual's contemporaneous covariates w_t. The members of group x may have similar covariates, so the model incorporates the idea of correlated effects. Contextual interactions, however, are assumed absent.

The time path of the group outcome distribution $P(y_t \mid x)$ is determined by the form of $f(\cdot,\cdot)$, by the initial condition $P(y_0 \mid x)$, and by the time path of the covariate distribution $P(w_t \mid x)$. Let A denote a measurable subset of Y. The law of motion for $P(y_t \mid x)$ is

$$P(y_t \epsilon A \mid x) = P\{w_t : f[P(y_{t-1} \mid x), w_t] \epsilon A \mid x\}, \quad \text{all } A \subset Y. \tag{8}$$

Given a time-invariant group covariate distribution $P(w \mid x)$, a time-invariant outcome distribution $P(y \mid x)$ is a social equilibrium if $P(y \mid x)$ solves the system of equations

$$P(y \epsilon A \mid x) = P\{w : f[P(y \mid x), w] \epsilon A \mid x\}, \quad \text{all } A \subset Y. \tag{9}$$

The set of equilibria is determined by the form of $f(\cdot,\cdot)$ and by the covariate distribution $P(w \mid x)$. The process may have a unique equilibrium in group x, multiple equilibria, or no equilibrium at all.

3.1 SOME ILLUSTRATIONS

Before discussing inference, I describe some models that illustrate alternative forms of the response function. These models express a range of ideas about the nature of interactions, yet all are quite simple.

LINEAR-IN MEAN MODEL: Let us begin with a modest variation on the model examined in section 2. We assume that a scalar outcome varies linearly with the lagged group-mean outcome and separably with the individual's covariates. Thus

$$y_t = \beta E(y_{t-1} \mid x) + g(w_t)\,. \tag{10}$$

As before, β expresses the strength of the endogenous interaction. Now $g(\cdot) : W \to R^1$ expresses the manner in which covariates influence outcomes. The law of motion for $E(y_t \mid x)$ is

$$E(y_t \mid x) = \beta E(y_{t-1} \mid x) + E[g(w_t) \mid x]\,. \tag{11}$$

Given a time-invariant covariate distribution $P(w \mid x)$, a time-invariant mean outcome $E(y \mid x)$ determines a social equilibrium if $E(y \mid x)$ solves the equation

$$E(y \mid x) = \beta E(y \mid x) + E[g(w) \mid x]\,. \tag{12}$$

Provided that $\beta \neq 1$, this equation has the unique solution

$$E(y \mid x) = \frac{E[g(w) \mid x]}{(1-\beta)}\,. \tag{13}$$

Hence, the unique equilibrium outcome distribution is

$$P(y \epsilon A \mid x) = P\left[w : \frac{\beta E[g(w) \mid x]}{(1-\beta)} + g(w) \epsilon A \mid x\right], \quad \text{all } A \subset Y\,. \tag{14}$$

DISPERSION-DEPENDENT STRENGTH OF NORM: The strength of the effect of a social norm on individual behavior may depend on the dispersion of behavior in the reference group; the smaller the dispersion, the stronger the norm. A straightforward way to express this idea is to replace the parameter β in Eq. (10) with some function of the variance of outcomes in the reference group. Thus, consider the model

$$y_t = s[V(y_{t-1}) \mid x] \cdot E(y_{t-1} \mid x) + g(w_t)\,. \tag{15}$$

Here $s(\cdot) : R^+ \to R^1$ expresses the strength of the endogenous interaction as a function of the lagged group outcome variance $V(y_{t-1} \mid x)$.

The laws of motion of the mean and variance of the group outcome distribution are

$$E(y_t \mid x) = s[V(y_{t-1}) \mid x] \cdot E(y_{t-1} \mid x) + E[g(w_t) \mid x] \tag{16}$$

and

$$V(y_t \mid x) = V[g(w_t) \mid x]\,. \tag{17}$$

Given a time-invariant covariate distribution $P(w \mid x)$, a time-invariant mean outcome $E(y \mid x)$ determines a social equilibrium if $E(y \mid x)$ solves the equation

$$E(y \mid x) = s\{V[g(w) \mid x]\} \cdot E(y \mid x) + E[g(w) \mid x]\,. \tag{18}$$

Thus, group x has a unique equilibrium if $s\{V[g(w) \mid x]\} \neq 1$.

COVARIATE-DEPENDENT STRENGTH OF NORM: The strength of a social interaction may vary across individuals. A straightforward way to express this idea is to replace the parameter β in Eq. (10) with some function of the individual covariates w_t. Thus, consider the model

$$y_t = s(w_t) \cdot E(y_{t-1} \mid x) + g(w_t)\,. \tag{19}$$

Here $s(\cdot) : W \rightarrow R^1$ expresses the manner in which the strength of the social interaction varies with the covariates w.

The law of motion of the group mean outcome $E(y_t \mid x)$ is

$$E(y_t \mid x) = E[s(w_t) \mid x] \cdot E(y_{t-1} \mid x) + E[g(w_t) \mid x]\,. \tag{20}$$

Given a time-invariant covariate distribution $P(w \mid x)$, a time-invariant mean outcome $E(y \mid x)$ determines a social equilibrium x if $E(y \mid x)$ solves the equation

$$E(y \mid x) = E[s(w) \mid x] \cdot E(y \mid x) + E[g(w) \mid x]\,. \tag{21}$$

Thus, group x has a unique equilibrium if $E[s(w) \mid x] \neq 1$.

LINEAR-IN-LOCATION MODEL: An obvious extension of the linear-in-mean model is to assume that y varies linearly with some location parameter $L(\cdot)$ of the lagged group outcome distribution, not necessarily its mean. Thus, let

$$y_t = \beta L(y_{t-1} \mid x) + g(w_t)\,. \tag{22}$$

The location parameter $L(y_{t-1} \mid x)$ could, for example, be the median or some other quantile of $P(y_{t-1} \mid x)$. This model has the same dynamics as does the linear-in-mean model, with the location operator $L(\cdot)$ replacing the expectation operator $E(\cdot)$.

BINARY RESPONSE MODELS: Models with nonlinear interactions are generally not as transparent as are the semilinear models described above. Perhaps the most familiar nonlinear models are binary response models of the form

$$\begin{aligned} y_t &= 1 \text{ if } \beta P(y_{t-1} = 1 \mid x) + g(w_t) > 0 \,. \\ &= 0 \text{ otherwise} \,. \end{aligned} \tag{23}$$

As before, β is a parameter expressing the strength of the social interaction and $g(\cdot) : W \to R^1$ expresses the manner in which covariates influence outcomes.

The law of motion for $P(y_t = 1 \mid x)$ is

$$P(y_t = 1 \mid x) = P[w_t : \beta P(y_{t-1} = 1 \mid x) + g(w_t) > 0 \mid x] \,. \tag{24}$$

Given a time-invariant covariate distribution, a time-invariant response probability $P(y = 1 \mid x)$ is a social equilibrium in group x if $P(y = 1 \mid x)$ solves the equation

$$P(y = 1 \mid x) = P[w : \beta P(y = 1 \mid x) + g(w) > 0 \mid x] \,. \tag{25}$$

Manski[11] shows that if the distribution function of $P[g(w) \mid x]$ is continuous and strictly increasing, then Eq. (25) has at least one solution. The equilibrium is unique if $\beta \leq 0$, but there may be multiple equilibria if $\beta > 0$. Brock and Durlauf[3] show that if $P[g(w) \mid x]$ has a logistic distribution and $\beta > 0$, then there is either a unique equilibrium or three equilibria, depending on the location and scale parameters of $P[g(w) \mid x]$.

AUTOREGRESSIVE MODELS: Although our primary concern is with social interactions, it is worth noting that Eq. (7) encompasses autoregressive models of the form

$$y_t = f(y_{t-1}, w_t) \,. \tag{26}$$

In this case, each person is his or her own reference group, with a distinct value of x. The distribution $P(y_{t-1} \mid x)$ is degenerate with all its mass at the person's realization of y_{t-1}.

3.2 NONPARAMETRIC IDENTIFICATION WITH COMPLETE DATA

The severity of the problem of inference on the response function $f(\cdot, \cdot)$ is determined by three factors:

1. the researcher's maintained assumptions restricting the form of $f(\cdot, \cdot)$
2. the data available to the researcher
3. the values of $[P(y_{t-1} \mid x), w_t]$ realized in the population.

There are a vast set of configurations of these factors that might usefully be explored. The researcher might assume nothing a priori about the form of the response function or might assume that $f(\cdot,\cdot)$ belongs to a tightly restricted family of functions. The data available to the researcher might be a sample of observations of some outcomes and covariates, or might be the entire distribution $P(x, w_t, y_t, t = 0, \ldots, T)$. The values of $[P(y_{t-1} \mid x), w_t]$ realized in the population, termed the *support* of $[P(y_{t-1} \mid x), w_t]$, might be a single point or the entire domain $\Gamma \times W$ of the response function.

Although all three factors combine to determine the feasibility of inference on the response function, there is a fundamental difference between the first two factors and the third. Whereas the maintained assumptions and the available data characterize the researcher's contribution to inference, nature determines the support of $[P(y_{t-1} \mid x), w_t]$. The actual interaction process reveals the value of $f(\cdot,\cdot)$ on the support but not at other points in $\Gamma \times W$. Henceforth, the support of $[P(y_{t-1} \mid x), w_t]$ will be denoted Ω.

I focus here on the role of Ω in inference when the researcher makes no assumptions restricting $f(\cdot,\cdot)$, but observes the entire distribution $P(x, w_t, y_t, t = 0, \ldots, T)$; a setting with unobserved covariates will be examined later. To simplify the exposition, I assume henceforth that the distribution $P(x, w_t, t = 1, \ldots, T)$ of reference groups and covariates has discrete support. Let X_s denote the support of $P(x)$; thus, $x \epsilon X_s$, if and only if group x contains a positive fraction of the population. For $x \epsilon X_s$, let W_{xt} denote the support of $P(w_t \mid x)$. Let $Ws \equiv (\cup W_{xt}, t = 1, \ldots, T, x \epsilon X_s)$ denote the set of covariate values realized at some date by a positive fraction of the population. Then Ω has the form

$$\Omega = \{[P(y_{t-1} \mid x), w_t], w_t \epsilon W_{xt}, t = 1, \ldots, T, x \epsilon X_s\}. \tag{27}$$

The shape of Ω determines what can be learned nonparametrically about the response function Ω, and, consequently, determines the extent to which inference must rely on assumptions restricting the form of $f(\cdot,\cdot)$. Under the maintained assumption (7), the shape of Ω is constrained by the law of motion (8). I have called this constraint on Ω the *reflection problem.*[11] The term reflection is appropriate here because, at each date $t, P(y_{t-1} \mid x)$ is the aggregated image of the most recent outcomes of the members of group x. The problem of identification of $f(\cdot,\cdot)$ is similar to the problem of interpreting the movements of a person standing before a mirror. Given data on the person's movements, one might ask what can be inferred about the way the person reacts to his most recent image in the mirror. Analogously, given data on the time path of outcomes in group x, we ask what can be inferred about the way the group members react at date t to their most recent aggregated image $P(y_{t-1} \mid x)$.

I want to focus on the shape of Ω when the population is observed in equilibrium. Suppose that in each group x, the covariate distribution $P(w \mid x)$ is time-invariant and the initial condition $P(y_0 \mid x)$ is an equilibrium. Then $P(y_t \mid x) = P(y_0 \mid x)$ at each date t and

$$\Omega = \{[P(y_0 \mid x), w], w \epsilon W_x, x \epsilon X_s\}. \tag{28}$$

With no time-series variation in covariates or interactions, inference on $f(\cdot,\cdot)$ must rely solely on cross-sectional variation.

When the population is observed in equilibrium, the shape of Ω depends on the unknown form of $f(\cdot,\cdot)$, the known distribution $P(x, w)$ of reference groups and covariates, and the known initial conditions $P(y_0 \mid x)$. Let us focus on the distribution $P(x, w)$, which expresses how the population is sorted into reference groups. A simple but interesting finding is that, if x and w are either statistically independent or perfectly dependent, then Ω is a curve on W_s; that is,

$$\Omega = [(\Theta_w, w), w \epsilon W_s] , \qquad (29)$$

where $\Theta_w \epsilon \Gamma$ is a distribution that may vary with w.

Equation (29) means that, at each covariate value $w \epsilon W_s, f(\cdot, w)$ is identified only at the single point Θ_w. Hence, we can learn nothing about how $f(\cdot, w)$ varies over Γ. See Proposition 3 of Manski[11] for a proof under the assumption that interactions are transmitted through group-mean outcomes and that each group has a unique equilibrium. A more general proof applying to models of the form (7) is given in the Appendix to the present paper.

To illustrate, consider the determination of pregnancy among teenage girls. Let the pregnancy outcome y_t of a girl in reference group x be determined by the lagged pregnancy rate $P(y_{t-1} = 1 \mid x)$ of girls in the group and by the girl's covariates w_t. Let a girl's reference group x be the neighborhood in which she lives. Let the relevant covariate w_t be the girl's religion. Let the population of teenage girls be observed in pregnancy equilibrium.

In this setting, nonparametric study of the effect of endogenous interactions on pregnancy among girls of a given religion is impossible if every neighborhood has the same religious composition (statistical independence of x and w), if neighborhoods are segregated by religion (w perfectly dependent on x), or if all members of the given religion live in the same neighborhood (x perfectly dependent on w). Each of these findings taken alone is simple, almost trivial. Yet the three findings taken together are intriguing. They imply that nonparametric inference is feasible only if neighborhoods and religions are "moderately associated" in the population.

3.3 NONPARAMETRIC IDENTIFICATION OF THE REDUCED FORM

Equation (7), which states the structural dependence of outcomes on group outcome distributions and individual covariates, implies the reduced-form

$$y_t = f_t^*(x, w_t) , \qquad (30)$$

where $f_t^*(x, w_t) \equiv f[P(y_{t-1} \mid x), w_t]$. It is important to distinguish the problem of identifying the response function $f(\cdot,\cdot)$ from that of identifying the reduced-form functions $f(\cdot,\cdot), t = 1,\ldots,T$. Whereas $f(\cdot,\cdot)$ is nonparametrically identified on the set Ω defined in Eq. (27), $f(\cdot,\cdot)$ is nonparametrically identified on the set

$$\Omega_t^* \equiv [(x, w_t), w_t \epsilon W_{xt}, x \epsilon X_s]\,. \tag{31}$$

If the support W_{xt} of $P(w_t \mid x)$ is time-invariant for each $x \epsilon X_s$, then so is Ω_t^*, with

$$\Omega^* = [(x, w), w \epsilon W_x, x \epsilon X_s]\,. \tag{32}$$

Nonparametric study of the reduced-form variation of outcomes with reference groups is a simpler task than is nonparametric study of the structural variation of outcomes with reference-group outcome distributions. Suppose, for simplicity, that W_{xt} is time-invariant for each $x \epsilon X_s$, so Eq. (32) holds. We showed earlier that Ω is a curve on W_s if x and w are either statistically independent or perfectly dependent. In contrast, Ω^* is a curve on W_s only if x is perfectly dependent on w.

Suppose that a researcher estimating the reduced form finds that, for each $w \epsilon W_s$ and $t = 1,\ldots,T$, $f_t^*(\cdot, w)$ does not vary with x. This empirical finding does not suffice for the researcher to conclude that social interactions are absent. Constancy of $f_t^*(\cdot, w)$ may reflect constancy of the outcome distributions $P(y_{t-1} \mid x)$ across x and t.

3.4 IDENTIFICATION OF RESPONSE FUNCTIONS SEPARABLE IN UNOBSERVED COVARIATES

There are many empirically relevant ways to weaken the idealized assumption that the researcher fully observes the population distribution $P(x, w_t, y_t, t = 0,\ldots,T)$ of reference groups, covariates, and outcomes. This section examines the problem of structural inference when some covariates are unobserved and the response function is separable in these unobserved covariates. Inference on separable response functions is a longstanding concern of the literature on econometric methods (see Manski,[10] sections 2.5 and 6.1).

Formally, let $w_t \equiv (z_t, u_t)$, where $W \equiv Z \times U$. Let the response function $f(\cdot,\cdot)$ be known to have the separable form

$$f[P(y_{t-1} \mid x), z_t, u_t] = h[P(y_{t-1} \mid x), z_t] + u_t\,, \tag{33}$$

where $h(\cdot,\cdot) : \Gamma \times Z \rightarrow Y$. The researcher observes the distribution $P(x, z_t, y_t; t = 0,\ldots,T)$ but does not observe u_t.

Inference on $f(\cdot,\cdot)$ is impossible in the absence of assumptions restricting the unobserved covariates. This obvious fact is often referred to as the "omitted-variables" problem. It is common in econometric studies to assume that the distribution of unobserved covariates conditional on observed covariates has zero mean, median, or other location parameter. In the present case, let

$$L(u_t \mid x, z_t) = 0\,, \tag{34}$$

where L denotes a specified location parameter. Taken together, assumptions (7), (33), and (34) are equivalent to assuming that at each date $t = 1, \ldots, T$,

$$L(y_t \mid x, z_t) = h[P(y_{t-1} \mid x), z_t] . \tag{35}$$

Comparison of Eqs. (7) and (35) shows that the problem of nonparametric identification of $h(\cdot, \cdot)$ is precisely analogous to the problem of nonparametric identification of $f(\cdot, \cdot)$ studied in section 3.2. For each $x \epsilon X_s$, let Z_{xt} denote the support of $P(z_t \mid x)$. Let Ω_h denote the support of $[P(y_{t-1} \mid x), z_t]$. That is, let

$$\Omega_h \equiv \{[P(y_{t-1} \mid x), z_t], z_t \epsilon Z_{xt}, t = 1, \ldots, T, x \epsilon X_s\} . \tag{36}$$

Observation of $P(x, z_t, y_t; t = 0, \ldots, T)$ reveals $h(\cdot, \cdot)$ on Ω_h but not elsewhere on the domain $\Gamma \times Z$.

Identification of $h(\cdot, \cdot)$ on $\Omega_h$implies identification of the values of the covariates u. Thus, assumption (35) overcomes entirely the identification problem generated by unobserved covariates. Consider a researcher who does not observe u, but knows $f(\cdot, \cdot)$ to be separable in u and knows u to have zero conditional location parameter. From the perspective of identification (although not from the perspective of statistical inference), this researcher is in the same position as one who fully observes reference groups, covariates, and outcomes.

4. CONCLUSION

This paper concludes with some questions answered. Analysis of the linear-in-means model in section 2 shows that endogenous interactions, contextual interactions, and correlated effects are generally not distinguishable given observations of the outcomes and covariates of groups in equilibrium. These hypotheses are, however, distinguishable in principle (see footnote 3). The analysis of pure endogenous interactions models in section 3 emphasizes that the law of motion of the interaction process constrains the data generated by this process, and so constrains the feasibility of nonparametric inference.

We also conclude with important issues unresolved. We have had almost nothing to say about inference when groups are observed out of equilibrium. The time path of a process that is not in equilibrium is, by definition, more variable than the time path of a process that is in equilibrium. This suggests that observation of an out-of-equilibrium process generically enhances the feasibility of inference. What is missing thus far, however, is a constructive way to characterize the inferential benefits of out-of-equilibrium observations.

We have had too little to say about the way that the composition of reference groups affects the feasibility of inference. We have reported that nonparametric inference is impossible if x and w are statistically independent or perfectly dependent, but we have not characterized the possibilities for inference in the vast array of intermediate situations where x and w are moderately associated. We also have had too little to say about the vexing, and quite common, problem of inference when the researcher does not observe the composition of reference groups.

Some readers will naturally be interested in other models of social interactions: ones with time-varying reference groups or multiple reference groups, models with local rather than anonymous interactions, and so on. To the extent that such models constitute generalizations of the ones examined here, the negative identification findings reported here extend immediately. Our positive findings, however, do not necessarily extend to more general models.

Stepping further back from the particular concerns of this paper, I would reiterate my observation in the Introduction that the theoretical study of social interactions has long dominated the attention given to inference. Recent theoretical advances in our understanding of the economy as a complex evolving system, whether achieved by algebraic or computational means, are to be applauded. It is unfortunate that these advances have not been accompanied by corresponding progress in our grasp of the problem of empirical inference on actual economic processes.

APPENDIX

Proposition 1 gives the simple abstract reasoning underlying the finding, reported in section 3.2, that nonparametric inference on endogenous interactions is infeasible if x and w are either statistically independent or perfectly dependent. Three corollaries apply this reasoning to the cases of interest. The proofs are immediate.

PROPOSITION 1: Let $v \epsilon W_s$. Let $X_v \equiv (x \epsilon X_s : v \epsilon W_x)$. Let there exist a distribution Q_v on W such that $P(w_t \mid x) = Q_v$, $t = 1, \ldots, t$ and $x \epsilon X_v$. Let $\Theta_v \epsilon \Gamma$ solve the social equilibrium equation

$$\Theta_v(y \epsilon A) = Q_v[w : f(\Theta_v, w) \epsilon A], \quad \text{all } A \subset Y. \tag{A1}$$

Let $P(y_0 \mid x) = \Theta_v$, all $x \epsilon X_v$. Then $P(y_t \mid x) = \Theta_v$, $t = 0, \ldots, t$ and $x \epsilon X_v$. Hence, $f(\cdot, v)$ is identified only at Θ_v. □

COROLLARY 1.1. (w statistically independent of x): Let there exist a distribution Q on W such that $P(w_t \mid x) = Q$, $t = 1, \ldots, t$ and $x \epsilon X_s$. Let $\Theta \epsilon \Gamma$ solve the social equilibrium equation

$$\Theta(y \epsilon A) = Q[w : f(\Theta, w) \epsilon A], \quad \text{all } A \subset Y. \tag{A2}$$

Let $P(y_0 \mid x) = \Theta$, all $x \epsilon X_s$. Then $P(y_t \mid x) = \Theta$, $t = 0, \ldots, t$ and $x \epsilon X_s$. Hence, $f(\cdot, w)$ is identified only at Θ, all $w \epsilon W_s$. □

COROLLARY 1.2. (w perfectly dependent on x): Let $v \epsilon W_s$. Let $X_v \equiv (x \epsilon X_s : v \epsilon W_x)$. Let almost all members of the groups X_v have covariate value v; that is, let $P(w_t = v \mid x) = 1$, $t = 1, \ldots, t$ and $x \epsilon X_v$. Let $\Theta_v \epsilon \Gamma$ solve the social equilibrium equation

$$\begin{aligned} \Theta_v(y \epsilon A) &= 1 \text{ if } f(\Theta_v, v) \epsilon A \\ &= 0 \text{ otherwise}, \quad \text{all } A \subset Y. \end{aligned} \tag{A3}$$

Let $P(y_0 \mid x) = \Theta_v$, $x \epsilon X_v$. Then $P(y_t \mid x) = \Theta_v$, $t = 0, \ldots t$ and $x \epsilon X_v$. Hence, $f(\cdot, v)$ is identified only at Θ_v. □

COROLLARY 1.3. (x perfectly dependent on w): Let $v \epsilon W_s$. Let almost all persons with covariate value v belong to the same reference group, denoted $x(v)$; that is, let $P[x = x(v) \mid w_t = v] = 1$, $t = 1, \ldots, T$. Let there exist a distribution Q_v on W such that $P[w_t \mid x(v)] = Q_v$, $t = 1, \ldots, t$. Let $\Theta_v \epsilon \Gamma$ solve the social equilibrium equation

$$\Theta_v(y \epsilon A) = Qv[w : f(\Theta_v, w) \epsilon A], \quad \text{all } A \subset Y. \tag{A4}$$

Let $P[y_0 \mid x(v)] = \Theta_v$. Then $P[y_t \mid x(v)] = \Theta_v$, $t = 0, \ldots, t$. Hence, $f(\cdot, v)$ is identified only at Θ_v. □

ACKNOWLEDGMENTS

This research is supported by National Science Foundation Grant SBR92-23220.

REFERENCES

1. Benabou, R. "Workings of a City: Location, Education, and Production." *Quart. J. Econ.* **108** (1993): 619–652.
2. Bernheim, B. "A Theory of Conformity." *J. Pol. Econ.* **102** (1994): 841–877.
3. Brock, W., and S. Durlauf. "Discrete Choice with Social Interactions." Department of Economics, University of Wisconsin-Madison, Madison, WI, 1995.
4. Fisher, F. *The Identification Problem in Econometrics.* New York: McGraw-Hill, 1966.
5. Granovetter, M. "Threshold Models of Collective Behavior." *Amer. J. Sociol.* **83** (1979): 1420–1443.
6. Hood, W., and T. Koopmans, eds. *Studies in Econometric Method.* New York: Wiley, 1953.
7. Hyman, H. "The Psychology of Status." *Archives of Psychology* **269** (1942).
8. Jones, S. *The Economics of Conformism.* Oxford: Basil Blackwell, 1984.
9. Koopmans, T. "Identification Problems in Economic Model Construction." *Econometrica* **17** (1949): 125–144.
10. Manski, C. *Analog Estimation Methods in Econometrics.* London: Chapman & Hall, 1988.
11. Manski, C. "Identification of Endogenous Social Effects: The Reflection Problem." *Rev. Econ. Studies* **60** (1993): 531–542.
12. Manski, C. *Identification Problems in the Social Sciences.* Cambridge, MA: Harvard University Press, 1995.
13. Merton, R. *Social Theory and Social Structure.* Glencoe: The Free Press, 1957.
14. Pollak, R. "Interdependent Preferences." *Amer. Econ. Rev.* **78** (1976): 745–763.
15. Roehrig, C. "Conditions for Identification in Nonparametric and Parametric Models." *Econometrica* **56** (1988): 433–447.
16. Schelling, T. "Dynamic Models of Segregation." *J. Math. Sociol.* **1** (1971): 143–186.
17. Sewell, W., and J. Armer. "Neighborhood Context and College Plans." *Am. Soc. Rev.* **31** (1966): 159–168.

William A. Brock
Department of Economics, University of Wisconsin, Madison, WI 53706

Asset Price Behavior In Complex Environments

This chapter first surveys empirical patterns that appear in asset returns. The main patterns are the following. (i) Returns are hard to predict out of sample. The correlation dimension is rather large and is unstable; i.e., returns are generated by a "complex" process. (ii) Detrended volume of trading activity is very persistent, i.e., the autocorrelation function is positive and decreases slowly as lag length increases. (iii) Volatility of returns is also very persistent and the autocorrelation function of volatility is rather similar to that of detrended volume. (iv) Detrended volume and volatility are highly correlated contemporaneously. (v) The cross autocorrelation function of detrended volume and volatility falls off rapidly in leads and lags with a large peak at 0. (vi) More sophisticated bootstrapping type tests have found evidence of some conditional predictability of returns out of sample provided the conditioning set is chosen carefully.

Secondly, this chapter sketches development of adaptive evolutionary theoretic financial asset pricing models that nest the usual rational expectations type of models (e.g., a class of models that are versions of the *efficient market hypothesis*). But rational expectations beliefs are costly and compete with other types of beliefs in generating net trading profits as the system moves forward in time. The chapter briefly reviews some studies of the types of structures of belief heterogeneity that is needed to generate equilibrium returns, volume of trading, and volatility of returns that is consistent with the stylized facts reviewed above. An "ensemble" approach is developed, rather in the spirit of statistical mechanics, which seems to be useful in developing analytical results for large evolutionary stochastic dynamical systems.

1. INTRODUCTION

Eight years ago, the Santa Fe Institute organized a conference called "The Economy as an Evolving Complex System" and produced a book with the same title. The purpose of that conference was to bring together "free spirits" who were willing to examine the economy using techniques from "complex systems theory."

Complex systems theory is hard to define, but I shall use the term to refer to recent methods of time series analysis, which are inspired by dynamical systems theory as well as computer-assisted analysis of complex adaptive systems consisting of many interacting components. Complex systems analysis, as used here, includes neural nets, genetic algorithms, and artificial life analysis such as Arthur et al.[7] I shall sometimes speak loosely and use "complex systems theory" to denote the research practices of complex system theorists including those at Santa Fe.

This chapter gives a brief review of the use of complex systems techniques in the study of asset price dynamics and related subjects since the SFI conference. The chapter is organized by listing the facts we wish to explain, followed by explanation of works that address those facts. For lack of space, I shall concentrate on works by myself, my co-authors, and my students. The reader should be warned that finance is a huge field that contains many researchers and that the use of "complex systems techniques" in this field has grown rapidly. Hence, there is a vast amount of work that we shall not be able to discuss in this article. I shall give frequent reference to recent surveys to give the reader a port of entry into the literature. Many of the works discussed here give extensive discussion of the work of others.

1.1 TIME SERIES ANALYSIS

One set of research practices which are associated with complex systems scientists, in which SFI affiliated researchers played a major role in developing, supporting, and popularizing, is the study of time series data using tools inspired by dynamical systems theory and related methods that are tailored to the study of "deep" nonlinearity.

A survey of this type of work up to 1991 is contained in the book by Brock, Hsieh, and LeBaron.[17] Two recent surveys with an emphasis on methods of testing for "deep" nonlinearity are contained in Abhyankar et al.[1] and LeBaron.[48] Excellent general surveys on financial time series findings, which include findings on nonlinear dynamics, are Goodhart and O'Hara,[36] and Guillaume et al.[37]

SFI-affiliated researchers also played a major role in developing statistical methods to detect nonlinearities in data. Many of these methods use versions of bootstrapping-type ideas from statistical theory. Examples of this kind of work include Brock, Lakonishok, and LeBaron,[20] LeBaron,[48,49] and Theiler et al.[65]

Let us be clear on what we are talking about here. At an abstract level, *all* serious analysis of time series data is inspired by "dynamical systems theory." We are concerned here, with methods that study, for example, nonlinear stochastic dynamical systems where the nonlinearity is so "deep" that it cannot be dealt with by minor modifications of linear techniques such as (i) change of variables after "detrending," (ii) adaptation of linear techniques to conditional variances (e.g., ARCH-type models, cf. Bollerslev, Engle, and Nelson[11]), and (iii) stochastic analogues of "Taylor"-type expansions where all remaining nonlinearity is tucked into the "remainder" terms and treated as "errors" in econometric models.

INITIAL REACTIONS TO EMPHASIS ON "DEEP" NONLINEARITIES

Before I give a discussion of research findings, let me give a few remarks on the initial reaction to the new methods that were inspired by dynamical systems theory. Eight years ago, I think it is fair to say that there was some controversy over whether application of dynamical systems theory and related nonlinear techniques would add much value to received methodology in time series econometrics in economics and finance. I believe the main reason for this opposition is the following.

Many economists believe that the strength of "deep" nonlinearities in financial data is small enough relative to the size of nonforecastible nonstationarities that reliable detection could not be had with the datasets available. After all, if such predictively useful nonlinearities were present, they would be found and exploited by expert trading houses. Such activity would cause asset prices to move to eliminate any predictive structure in such nonlinearities. Furthermore, nonlinear prediction algorithms were widely available before methods were developed which were based upon dynamical systems theory, so there was no reason to believe that these newer methods would find structure that had not been already found by earlier, perhaps cruder, methods.

I believe there is rough evidence that opposition to using "fancier" technology including methods based upon dynamical systems theory, neural nets, genetic algorithms, and the like, has fallen faster in areas where better data has become available at a more rapid rate.

For example, "high technology" groups such as Olsen and Associates (cf. Corcella[26] for a cover story on Olsen), which are quite common in high-frequency financial applications, seem less common in low-frequency financial applications as well as in aggregative macroeconomics applications, such as business cycle analysis and analysis of economic growth.

However, while academics initially debated the usefulness of the new methods, at least some practitioners appeared to rapidly adopt and use the new methods. Not only did some researchers associated with the Santa Fe Institute form their own company (The Prediction Company), many other like-minded companies and groups have emerged in recent years (see, for example, Ridley[61]).

While the debate on the value added of methods such as neural nets, genetic algorithms, and dynamical systems theoretic time series analysis has not yet died down in academia, it appears that the rate of adoption of the new methods by academics has increased rapidly in the last few years. The recent surveys by Abhyankar, Copeland, and Wong[1]; LeBaron[48]; Creedy and Martin[27]; Goodhart and O'Hara[36]; and Guillaume et al.[37] list many recent studies in academia that use "complex system techniques" as defined here.

Why did this rather silly (viewed from hindsight from the vantage point of the widespread adoption by sectors that "have to meet the bottom line") opposition (especially in academia) to the use of "high nonlinear" technology occur and what has caused the change in attitude? Manski,[54] in his well-known paper, stresses that when there is not enough data to resolve issues one way or another, divergent beliefs can be held very firmly. When large and widely held investments in established techniques are threatened, opposition can run deep.

But, in finance, the availability of high-quality high-frequency datasets such as that collected and made widely available by the Olsen Group (cf. Corcella[26] and Ridley[61] for popular articles on the Olsen Group) have made it possible to detect usuably predictable "nonlinearities" in financial data. In fields such as aggregative macroeconometrics, there is still a widely held attachment to linear methods. But aggregative macroeconometrics as a field does not have access to data of such quality and plenitude as does high-frequency finance.

Of course, much of the opposition will melt away against those techniques which prove useful in more conventional practices which have nothing to do with "deep" nonlinearities such as chaos. An example is the use of the BDS test (which originally had its roots in chaos theory) as a specification test for GARCH-type models which have nothing at all to do with complex systems theoretic techniques. See Bollerslev, Engle, and Nelson[11] for a discussion of the use of the BDS test as a specification test.

Someone once cynically said that scientific advances take place "funeral by funeral." vs. when the powerful old guard dies off, younger and more open minded scientists can fill the vacated niche space. While there may be some truth to this, I believe instead that economic science advances dataset by dataset rather than funeral by funeral because the evidence can be adduced with more precision when more and better data become available.

This is especially true in situations where the goal function of scientific activity is more precise. For example, corporations and government agencies may be subject to a stronger practical performance discipline to understand how the "system actually works" (e.g., better "prediction and control") than academics. Even in that much maligned institution, government, there is strong pressure to understand how the system really works.

In agencies such as private corporations and government departments, academic paradigm attachment may not matter as much as delivery of useful results. This may be so because there is an incentive structure operating in these environments that resembles more the profit motive in conventional competition analysis then the incentive structure in academic environments.

I also believe that the SFI meeting of 1988 and subsequent activities by SFI-affiliated researchers played a key role in popularizing new "high-technology" ideas in economics and finance. Turn now to a list of stylized facts we wish to explain.

A LIST OF STYLIZED FACTS WE WISH TO EXPLAIN

Before any theorizing is done we should ask the following questions: (i) What are the facts and regularities that we wish to explain? (ii) What are the goals of the explanation? While a standard commercial goal is prediction for the purpose of generating trading profits, we shall take the position here that we are only interested in understanding the stochastic properties of stock returns and trading volume and the forces that account for those properties. This basic scientific understanding is essential to design of intelligent regulatory policy. It also has practical commercial value for the management of risk.

Here is a list of stylized facts. Most of these concern high-frequency data, ranging from the tic-by-tic frequency to the weekly frequency. The facts are taken from the book by Brock, Hsieh, and LeBaron,[17] and articles by Guillaume et al.,[37] Goodhart and O'Hara,[36] LeBaron,[48,49] and Brock and de Lima.[15] We especially stress the article by Guillaume et al.[37] for an excellent list of high-frequency facts.

FACT 1: COMPLEXITY. Using estimated correlation dimension and lack of predictability as a measure of complexity, financial asset returns are highly complex.

That is, the estimated correlation dimension is high (and is also unstable across subperiods, in some cases) and there is little evidence of low-dimensional deterministic chaos. Evidence against low-dimensional chaos includes not only high-estimated (and unstable) correlation dimension, but also very little evidence of out of sample predictability using nonlinear prediction methods that have high power against low-dimensional deterministic chaos. Here the term "low-dimensional deterministic chaos" is used for chaos that is not only "low-dimensional" but also "usably" low dimensional in the sense that nonlinear short term prediction can be conducted out of sample.

We doubt if anyone in finance believes that asset returns are wholly a deterministic chaos. The issue is whether there might be a chaos of low enough dimension and "regular" enough lurking in the sea of nonstationarity and noise that marginally effective short term prediction might be possible. Unfortunately the evidence for even this rather limited notion of chaos is weak.

FACT 2: NONLINEARITY. There is good evidence consistent with nonlinearity, but effort must be made to avoid confusion of evidence of nonlinearity with evidence of nonstationarity. Much of the evidence of nonlinearity is also consistent with evidence of neglected nonstationarity.

Although evidence for low-dimensional deterministic chaos is weak, there is, however, evidence of stochastic nonlinearity. Let "i.i.d." denote "Independently and Identically Distributed." There is extremely strong evidence against i.i.d.-linearity and there is weaker evidence against MDS-linearity. Here a stationary stochastic process $\{X_t\}$ is said to be i.i.d.-linear (MDS-linear) if $X_t = \sum a_j \varepsilon_{t-j}, \sum a_j^2 < \infty$, where $\{\varepsilon_j\}$ is an i.i.d. stochastic process (Martingale Difference Sequence, i.e., $E\{\varepsilon_t | \varepsilon_{t-1}, \varepsilon_{t-2}, \ldots\} = 0$, all t). While the evidence may be due to nonstationarity rather than nonlinearity, many studies reviewed in, for example, Abhyankar et al.,[1] Hsieh,[42] and Guillaume et al.[37] attempt to control for nonstationarities by adjusting for known seasonalities from higher to lower frequencies such as daily seasonalities in the bid/ask spread, volatility and volume over the trading day to the January Effect.

Guillaume et al.[37] introduce a rescaling of time, which they call "theta" time, which shortens periods with little trading activity and magnifies periods with a lot of trading activity. They show that this device is very useful for controlling for nonstationarities. It is also sensible from an economic point of view and plays an important role in the Olsen Group's vision of how the markets work—a vision which emphasizes trader heterogeneity at all time scales.

Studies such as Abhyankar et al.,[1] and Hsieh[42] adjust for other types of nonstationarities by studying the pattern of rejections of nonlinearity by frequencies. At a very rough level of expository approximation one may describe their results as showing that the pattern of rejections of linearity have a "self similar" structure. That is, the pattern of rejection is similar at all frequencies. This forces some discipline on heterogeneous belief or heterogeneous characteristics theories of traders.

Evidence of nonsymmetry such as skewness allows rejections of many models of the form $X_t = \sum a_j \varepsilon_{t-j}$, for example $\{\varepsilon_t\}$ Gaussian. Gallant, Rossi, and Tauchen[35] (p. 873), adduce evidence against a Vector AutoRegressive model driven by AutoRegressive Conditionally Heteroskedastic errors, abbreviated as VAR-ARCH in stock price and volume data by showing that the symmetry imposed by such a model conflicts with "the intrinsic nature of the price-volume relationship." In general, as Gallant, Rossi, and Tauchen[35] show for asset returns and volume data, and as Potter[60] shows for macroeconomic data, nonlinear dynamic impulse response analysis coupled with a bootstrapping type of statistical significance analysis is a good way to adduce evidence against linear models.

FACT 3: PREDICTABILITY. Evidence of out of sample predictability is weak.

The studies of Diebold and Nason,[28] Meese and Rose[58] put a damper on the effort to find reliable out of sample predictability in asset returns. They showed that evidence for predictability out of sample beyond that of simple random walk models was nil.

Later studies such as Brock, Lakonishok, and LeBaron,[20] LeBaron[46] for stock market indices, Antoniewicz[5] for individual stocks; and LeBaron[49] and Guillaume et al.[37] for foreign exchange showed that evidence for short-term out-of-sample predictability is good *provided* conditioning was done on the "right" information sets.

If opportunities for such predictability were rare enough over the whole sample, they would be hard to detect in bulk average measures of predictability over the whole sample. Tests such as the BDS test (Brock, Dechert, and Scheinkman[14]) are useful for exhibiting evidence for such "pockets of predictability" but more refined tools such as bootstrapping economically motivated quantities such as profits from different trading strategies under null models whose development was guided by structural economic modeling (e.g., Brock, Lakonishok, and LeBaron[20]) are required to locate such pockets and measure the "statistical and economic" significance of such conditional predictability. Blind application of nonlinear high technology is not likely to be successful.

As we said before, there is evidence of short-term predictability provided one conditions on the "right information sets." For example, LeBaron has shown,[46] for stock indices and for foreign exchange, that predictability increases when near-past volatility and volume decreases. He has also shown,[49] that certain technical trading rules have predictive power for foreign exchange. But this predictive power drops when central bank intervention periods are deleted.

Lagged volume has predictive power for near-future returns. Recall that volume and volatility measures are contemporaneously correlated for both indices and individual stocks (see Antoniewicz[4,5]) for documentation for a large collection of individual stocks). For indices LeBaron[46,47] showed (for several indices and for IBM stock returns) that decreases in lagged volatility (volume) are associated with increases in first-order autocorrelation of returns. To put it another way, price reversals tend to follow abnormally high volume.

Conrad et al.[24] find the following for weekly returns on individual securities: "returns autocovariances are negative only for last period's heavily traded securities; autocovariances in returns are positive if trading declined last week." Campbell, Grossman, and Wang[22] give a theoretical story for this type of finding and document it for indices and some data on individual firms.

Campbell, Grossman, and Wang[22] consider the regression

$$\text{(R)} \qquad r_t = \alpha_0 + (\alpha_1 + \beta_1^M V_{t-1}^M) r_{t-1} + \varepsilon_t \,,$$

where r, V denote returns and detrended turnover and find $\alpha_1 > 0, \beta_1^M < 0$ for the NYSE/AMEX value weighted index. While they also study some individual firms,

Antoniewicz[5] runs regressions of the form (R) for a large sample of individual firms, for more lags, and for individual firm-specific volume as well as market volume.

The theoretical story consists of two classes of mean variance investors, one type has constant risk aversion, the other type has random risk aversion. The constant risk aversion types serve as "liquidity providers" and they must be rewarded for this task. There is no heterogeneity of beliefs. Since the random risk aversion process is persistent, a burst of volume signals not only an innovation to the random risk aversion process but also an increase in expected returns which must revert to the "normal" level. This leads to a volume spike predicting a decrease in first-order autocorrelation of future returns. As we said before, Campbell, Grossman, and Wang[22] find evidence consistent with the theory using data on indices and individual firms.

Antoniewicz[5] has shown that "spikes in aggregate volume do *not* influence the serial correlation for large capitalized securities. For smaller securities, episodes of abnormally high firm-specific volume can induce a change from negative serial correlation to positive serial correlation in returns. The effects from market volume and firm-specific volume on the serial correlation in returns depend on firm size and the sign of the previous day's return." Her findings may be consistent with the presence of different types of "liquidity providers" for large securities than for small securities. The differential presence of derivative securities, such as put and call options across different size classes, may play a role in the apparently different volume/return dynamics for large capitalized securities.

SIMPLE HETEROGENEOUS BELIEF STORIES TO "EXPLAIN" VOLUME AND AUTOCORRELATION

A very simple story based upon sequences of two-period temporary equilibria under heterogeneous beliefs may help point the way toward building models that may help explain some of the Antoniewicz findings for smaller securities. Let one set of traders have belief-biases away from the fundamental where the bias process is persistent. The traders which have no bias serve as the "liquidity" providers. Volume increases when there is an exogenous shock to the bias process. Since this exogenous shock is persistent, this generates positive first-order serial correlation in returns even though fundamental returns have no serial correlation.

A simple story based upon three-period temporary equilibrium adds the new ingredient that expectations must be formed by the "fundamentalists" on how many biased traders will be around next period when they wish to trade. One can tie down the sequence of equilibria by having the fundamentalists believe that values will revert to fundamental values three periods from now. This kind of model can induce endogenous changes in the quantity of risk in the system especially when it is coupled with the modeling strategy laid out in Brock[12]—where the interaction between intensity of choice and the desire to choose strategies used by people most like yourself can create large movements in the strategy mix induced by small differences in the payoff measure to each strategy.

FACT 4: AUTOCORRELATION FUNCTIONS OF RETURNS, VOLATILITY OF RETURNS, VOLUME MEASURES, AND CROSS-CORRELATIONS OF THESE MEASURES HAVE SIMILAR SHAPES ACROSS DIFFERENT SECURITIES AND DIFFERENT INDICES.
Autocorrelation functions of returns are roughly zero at all leads and lags. This is a version of the *efficient markets hypothesis*. However, at high frequencies, autocorrelations of returns are slightly negative for individual securities and for foreign exchange at small lags. This may be due to bid/ask bounce.[62] For stock indices, autocorrelations of returns are slightly positive at small lags. This is, at least, partly due to nonsynchronous trading effects.[62] That is, stocks which have most recently traded have adjusted to the latest "news" whereas sluggishly traded stocks have not adjusted. Their slow adjustment leads to a move in the same direction later on. This "nonsynchronous trading" effect leads to positive autocorrelation in returns. Yet if one tried to trade on it, the typically wider bid/ask spread and higher liquidity costs of the sluggish securities may wipe out the apparent profits.

Autocorrelation functions of volatility measures are positive at all lags with slow (approximately hyperbolic) decay for stock indices and foreign exchange. The decay is more rapid for individual securities.

This is the famous stylized fact that lies behind the huge ARCH literature surveyed by Bollerslev, Engle, and Nelson.[11] Microeconomic explanations for this stylized fact, which we call the "ARCH" fact, are discussed by Brock and LeBaron.[21]

Autocorrelation functions of volume measures look similar to those of volatility measures. Cross autocorrelations of volatility with volume are contemporaneously large and fall off rapidly at all leads and lags. There is some asymmetry in the fall off. Antoniewicz[4,5] documents many of these facts, as well as others, for a large collection of individual NYSE/AMEX and NASDAQ stocks. She also surveys many other studies documenting similar facts for indices and individual assets.

SOME THEORETICAL STORIES FOR THE ARCH-FACT

Models of endogenous belief heterogeneity (where belief types are driven by profitability of trading on those beliefs) by Brock and LeBaron[21] for short-lived assets and multiple time scales, and de Fontnouvelle[31] for long-lived assets and one-time scale are shown to produce time series roughly consistent with Fact 4 which includes the ARCH-fact. The general phenomenon of volatility bursting in complex interconnected systems has been recently treated by Hogg et al.[41]

FACT 5: SEASONALITIES AND OTHER PREDICTABLE NONSTATIONARITIES; NON-PREDICTABLE NONSTATIONARITIES. Seasonalities include the widening of the bid/ask spread at daily open and daily close, increase in volatility and trading volume at daily open and daily close as well as spillover effects across major international markets as they open and close over the trading day (see Goodhart and O'Hara[36]). The major theories are asymmetric information and differential arrival rates of information to portfolio rebalancing and peak load pricing of trading services. Guillaume et al.[37] show that use of an appropriate rescaling of chronological time into "economic" time is very useful for removing many seasonalities. They also show how estimates of objects like autocorrelation functions are badly contaminated and how statistical inference is damaged if these seasonalities are not removed.

FACT 6: LEAD/LAG RELATIONSHIPS (LINEAR AND NONLINEAR). Lo and MacKinlay[50] have documented that large firms in the same industry tend to lead smaller ones in returns. Conrad et al.[24] references work documenting similar leading effects for conditional variances. Kleidon and Whaley[44] review evidence that futures markets on stock indices tend to lead cash markets on stock indices. Abhyankar[2] and Hiemstra and Jones[39] have recently used nonlinearity tests based upon U-statistics theory and correlation integrals to document lead-lag relationships over and above those documented by linear methods.

Let us digress here to explain the correlation integral-based tests of Hiemstra and Jones[39] and Abhyankar.[2] We do this by expositing a simple special case of the general method.

Let $\{X_t\}$, $\{Y_t\}$ be a pair of stationary stochastic processes, let "$Pr\{E\}$" denote the "probability of event E," and let it be desired to test the proposition $Pr\{X_t|Y_{t-1}\} = Pr\{X_t\}$, using data $\{X_t\}$, $\{Y_t\}$, i.e., X_t is independent of Y_{t-1}. While one can do this by constructing estimators of $Pr\{X_t, Y_{t-1}\}/Pr\{Y_{t-1}\}$, $Pr\{X_t\}$ and setting up a test of the null hypotheis: $Pr\{X_t|Y_{t-1}\} = Pr\{X_t\}$ using these estimators, one can alternatively set up a test based upon

$$H_0 : \frac{Pr\{|X_t - X_s s| < \varepsilon,\ |Y_{t-1} - Y_{s-1}| < \varepsilon\}}{Pr\{|Y_{t-1} - Y_{x-1}| < \varepsilon\}} = Pr\{|X_t - X_s| < \varepsilon\}\,.$$

Now any object of the form $Pr\{|Z_t - Z_s| < \varepsilon\}$ can be estimated by the correlation integral estimator,

$$U_T = \left(\frac{1}{T^2}\right) \sum\sum 1\{|Z_t - Z_s| < \varepsilon\}\,,$$

using data $\{Z_t, t = 1, 2, \ldots, T\}$, where $\sum\sum$ runs over $s, t = 1, 2, \ldots, T$. Here, $1\{A\}$ is 1 if event A occurs, zero otherwise.

For the univariate case Brock, Dechert, and Scheinkman[14] showed that statistics of the form U_T are U-statistics and tests for i.i.d. can be set up using such statistics. Brock, Dechert, Scheinkman, and LeBaron[14] later showed that such statistics had the same first-order asymptotic distribution on *estimated* residuals as on the true residuals of i.i.d.-driven estimated models on data. This made such statistics useful for model specification tests (cf. Bollerslev, Engle, and Nelson[11]).

Baek and Brock[9] extended this analysis for the multivariate case, including proofs of invariance theorems on estimated residuals. Hiemstra and Jones[39] showed how to test H by replacing each population value in H_0 by a corresponding correlation integral estimator like U_T , taking the difference and multiplying it by $T^{1/2}$, and using the statistician's Taylor series approximation method, which statisticians call "the delta method," and working out the null distribution. Hiemstra and Jones[39] faced the challenge of dealing with a long and complicated formula for the variance because, unlike Baek and Brock,[9] they did *not* have an i.i.d. maintained hypothesis.

Using this test for "nonlinear causality of Y for X" where the word "causality" here is used in the narrow sense of incremental predictive ability of past Y's given, above and beyond, that already contained in past X's, they found evidence consistent with the proposition that daily stock returns help predict trading volume and trading volume helps predict daily stock returns. That is, they found evidence consistent with bidirectional causality. However, Gallant, Rossi, and Tauchen's[35] impulse response analysis showed that returns help predict trading volume, they did *not* find that trading volume helped predict prices. Antoniewicz[5] using an adaptation of the Brock, Lakonishok, LeBaron[20] bootstrap test found evidence that trading volume *did* help predict prices. The efficient markets argument would lead one to expect that any test capable of picking up evidence (that presumably would have been erased by traders' actions) that trading volume helps to predict prices would have to be quite subtle.

FACT 7: NONLINEAR IMPULSE RESPONSE FUNCTIONS FOR VOLATILITY AND VOLUME. Gallant, Rossi, and Tauchen[35] show that a shock to current volatility leads to an immediate increase followed by a long-term decline in future trading volume. However, shocks to current volume have little effect on future volatility. They develop general nonlinear impulse response analysis and apply it to finance. Potter[60] develops nonlinear impulse response analysis and applies it to macroeconomic time series. Both sets of authors find strong evidence for nonlinear effects that would not be found by standard linear impulse response analysis.

de Fontnouvelle[30] has developed a heterogeneous agent asset pricing model with endogenous evolution of signal purchasing belief types versus passive investor types (who free ride on the signal purchasers as in Brock and LeBaron[21]) with infinite lived assets, unlike the finitely lived assets in Brock and LeBaron.[21] His model generates time series output for returns, volume, and volatility that is approximately consistent with some of Gallant, Rossi, and Tauchen's findings on impulse responses as well as Fact 3 on conditional predictability and Fact 4 on autocorrelation and cross-correlation structure on returns, volume, and volatility. His model is promising enough for estimation by methods such as the Method of Simulated Moments (cf. Altug and Labadie[3] for a nice exposition) and specification testing.

FACT 8: MARKET MICRO STRUCTURE EFFECTS ON INDEX RETURNS AND RETURNS ON INDIVIDUAL SECURITIES. Ross[62] discusses construction of different types of stock market indices and surveys the major indices in use around the world. He discusses the advantage of using value-weighted (usually capitalization weighted, i.e., weighted by the value of the capitalization of the company's common stock) indices over other types. The Standard and Poors 500 is a value weighted index. He points out that smaller stocks trade in a less liquid market and trade less frequently than larger stocks. This is called the nontrading effect and it induces positive serial correlation in daily index returns. This effect may loom larger in more broadly based indices. An offsetting effect is the "bid/ask" effect which induces negative serial correlation in the returns for individual stocks. This effect looms larger for smaller time intervals and for less frequently traded stocks. Ross points out that in indices where large liquid stocks (which tend to have small bid/ask spreads) dominate the index, the nontrading effect will generally dominate.

An extensive discussion of the impact of different market institutions upon the stochastic structure of returns, volatility, and volume is given in the surveys of Guillaume et al.[37] and Goodhart and O'Hara.[36]

FACT 9: EVIDENCE OF DEPARTURES FROM "STANDARD" MODELS. There is evidence from studies of technical trading rules that popular volatility models such as GARCH and EGARCH over predict volatility following buy signals, under predict returns following buy signals, and over predict returns following sell signals. Indeed there is some evidence that returns following sell signals are negative even though volatility following sell signals is substantial. Evidence for the Dow Jones Index daily data was given in Brock, Lakonishok, and LeBaron.[20] Evidence for individual firms was given by Antoniewicz.[4,5]

Trading rule specification tests may be one promising way of getting around the issue of possible "economic" irrelevance of formal statistical testing. Let us explain. First, any statistical model is an approximation to the true joint distribution. Hence, any approximation will be formally rejected if the sample size is large enough. The financial analyst needs to know "how wrong is it" and "in what direction should repair be made." Second, if one does diagnostic analysis on residuals of a particular specification and "accepts" the specification if one fails to reject at the 5% level, then this strategy may unduly throw the burden of proof against alternatives. It is easy to create examples where the null model is wrongly accepted by this strategy. Third, trading rule analysis allows one to calculate the potential profitability value of rejections.

The technique of bootstrapping was applied to measure the statistical and economic "significance" of the results. We digress here to give a brief explanation of bootstrapping. We believe that bootstrapping of economically motivated quantities, such as trading rule (both technical and fundamental) profits, returns, and volatilities, represents the emerging scientific trend of the future in designing specification tests and goodness of fit tests of predictive models. This methodology will tend to replace traditional "analytics" in statistics as computers become more powerful and bootstrapping theory is worked out for more general environments.

BOOTSTRAPPING: A BRIEF EXPLANATION

Bootstrapping is a technique of resampling from i.i.d. driven null models to approximate the null distribution of complicated statistics under such models. Surrogate data (cf. Theiler et al.[65]) is an adaptation that preserves the spectrum or autocorrelation function. Surrogate data techniques have gained some popularity in testing versions of the linearity hypothesis, but we shall concentrate on explanation of bootstrapping here.

We shall follow statistics jargon and call the parametric class of probabilistic data generating mechanisms whose "goodness of fit" we wish to test on our dataset under scrutiny, the "null" model. Let $\{y_t, I_t, t = 1, 2, \ldots, T\}$ be a T-sample of data. Let bold case letters denote random variables and/or vectors. Consider the following null model

$$\mathbf{y}_t = H(\mathbf{I}_t, \mathbf{e}_{t+1}; \mathrm{a}),$$

where $\{\mathbf{e}_t\}$ is independently and identically distributed with distribution function F, i.e., i.i.d.F, where F is the cumulative distribution function, $F(e) = \text{Prob}\{\mathbf{e}_t < e\}$. Here, a is a vector of parameters to be estimated. Estimate the null model to obtain the estimator â. Given â, insert the data into the null model and invert H to find $\{\hat{e}_{t+1}\}$. Place mass $(1/T)$ at each $\hat{e}_{t+1}$ to create the empirical distribution $\hat{F}_T$. Suppose we want to calculate the null distribution of some statistic $S_T(Z_T; F)$, $Z_T \equiv (\mathbf{y}_1, \mathbf{I}_1, \ldots, \mathbf{y}_T, \mathbf{I}_T)$. Consider set A and the quantity

$$Pr\{S_T(Z_T; \hat{F}_T) \varepsilon A\} \simeq \left(\frac{1}{B}\right) \sum 1[S_T(Z_T^b) \varepsilon A],$$

where each $Z_T^b, b = 1, 2, \ldots, B$ is a resampling of $\hat{F}_T$ generated by T i.i.d. draws *with replacement* from $\hat{F}_T$.

Bootstrap is a useful method if $Pr\{S_T(Z_T; \hat{F}_T)\varepsilon A\}$ is a good approximation to $Pr\{S_T(Z_T; F_T)\varepsilon A\}$, and if $Pr\{S_T(Z_T; F_T)\varepsilon A\}$ is a good approximation to $Pr\{S_T(Z_T; F)\varepsilon A\}$. Glivenko-Cantelli and Edgeworth expansion theorems give sufficient conditions for F_T to be a good approximation to F. While it is technical, one can show that for certain statistics, called "pivotal" statistics, bootstrap can give better approximations than conventional asymptotic expansions. In any event, even for cases where such "incremental accuracy" theorems are not available, bootstrap allows one to use the computer to approximate the distributions for intractible statistics. Standard results on consistent estimation are available in the formal statistics literature, which give conditions for $\hat{a} \to a$, $T \to \infty$, $T^{1/2}(\hat{a} - a) \to X$, where X is some random variable and "$\to$" denotes convergence in distribution. These results can be used to evaluate the quality of convergence of $\hat{F}_T$ to F_T.

See Maddala, Rao, and Vinod,[52] Maddala and Rao,[53] and Engle and McFadden,[33] for more on bootstrapping and consistent estimation. The paper by Maddala and Li on bootstrap-based specification tests in financial models in Maddala and Rao[53] is especially relevant.

I go into some detail on the theory of bootstrapping here because it is sometimes carelessly used in the literature without verifications of the sufficient conditions required for bootstrap to give useful approximations to the null distributions of interest. Results of *any* bootstrap-based method or analogue should be viewed with caution unless the sufficient conditions for convergence have been formally verified.

If formal verification is not possible then computer experiments can be done where one can estimate the null model on a sample of length T, generate artificial datasets from the estimated null model of length N, unleash "fake bootstrapping econometricians" on each length N artificial dataset, bootstrap the null distribution of statistic S for this N-dataset and simply observe whether the N-null distributions appear to converge as N increases. Then observe whether this convergence appears to have taken place for $N = T$.

While Brock, Lakonishok, and LeBaron[20] discuss convergence quality experiments, one should go beyond their efforts, now that faster computers are available.

FACT 10: EVIDENCE CONSISTENT WITH LONG-TERM DEPENDENCE AND LONG MEMORY APPEARS STRONG IN VARIANCE BUT WEAK IN MEAN FOR RETURNS. CARE MUST BE TAKEN NOT TO CONFUSE EVIDENCE FOR LONG MEMORY WITH EVIDENCE FOR STOCHASTIC SWITCHING MODELS AND OTHER "NONSTATIONARITIES." Brock and de Lima[15] survey a lot of work on testing for long memory in mean and variance, much of which was inspired by Mandelbrot's early work. A rough consensus seems to have emerged that evidence for long memory in mean is weak, at least for individual securities.

However, on the surface, evidence for long memory in variance is very strong. But de Lima has done Monte Carlo experiments (reported in Brock and de Lima[15]) with a class of Markov switching models which were fitted to financial data by Hamilton and Susmel. It was found that the standard tests for long memory gave too many false positives for the Hamilton/Susmel models, especially when they were fitted to daily data. Brock and LeBaron[21] show by example how simple Markov switching models, which have nothing to do with long memory, can give slow decay in autocorrelation functions in squared returns.

This suggests that extreme care be taken when interpreting evidence for long memory in variance. Apparently this caveat may apply even to vivid evidence such as hyperbolic decay of autocorrelations of objects such as squared returns. Since one common vision of financial markets is that they are infected with nonforecastible nonstationarity with, perhaps, different stochastic dynamical systems operating over time spans of random lengths, buffeted by outside shocks, but yet this might be a short-memory process overall, therefore one must proceed with caution in interpreting results from tests for long memory. This same caveat also applies to tests for other features such as nonlinearity and chaos.

Interestingly enough de Lima found that estimates of the Hurst exponent itself were fairly robust. Turn now to a theoretical argument on the applicability of fractional Brownian motion models to the stock market.

There has been much recent interest in fractal Brownian Motion models of the stock market. Hodges[40] has generalized an older argument of Mandelbrot to show that a fractal Brownian motion market with a Hurst exponent that departs very far from $1/2$ (the Hurst exponent for the standard Brownian motion) is extremely vulnerable to risk free arbitrage. If $H > 1/2$ there is positive (negative) serial correlation ($H < 1/2$), hence Hodges is able to exploit this serial correlation (and the self-similarity of the fractal Brownian motion) to calculate how many transactions are needed to generate large profits per unit of risk borne as a function of the Hurst exponent and transactions costs. He shows that for Hurst values reported in some recent literature, the potential for easily generated, essentially risk free profits, is just too large to be compatible with common sense given the actual costs of transacting.

FACT 11: "EXCESS" VOLATILITY, FAT TAILS, "INEXPLICABLE" ABRUPT CHANGES. Financial time series generally exhibit excess kurtosis, a fact that has been noticed at least since Mandelbrot's work in the sixties which also raised the issue of nonexistence of the variance. Brock and de Lima[15] contains a survey of studies of estimation of maximal moment exponents. It appears that the evidence is consistent with existence of the variance but not moments of order four and beyond. However, there is debate over the econometric methods. de Lima's studies which are surveyed in Brock and de Lima[15] show that many tests for nonlinearity are seriously biased because they require existence of too many moments to be consistent with heavy tailed financial data. This causes some tests to give "false positives" for nonlinearity.

Many studies measure excess volatility relative to the standard set by the discounted present value model

$$\text{(PDV)} \qquad Rp_t = E_t\{p_{t+1} + y_{t+1}\}\,,$$

where E_t denotes expectation conditioned on past prices $\{p\}$ and past earnings, $\{y\}$. More recent studies attempt to correct for possible temporal dependence in the discount factor "R" but the spirit is the same.

There has also been concern expressed in the media, government, and policy communities about "excess" volatility in stock prices. While much of this concern comes from the October 1987, 1989 crashes, some of the academic concern follows from the realization that the solution class of (PDV) includes solutions of the form $p_t = p_t^* + b_t$, for any process $\{b_t\}$, that satisfies $Rb_t = E_t b_{t+1}$, where p_t^* is the "fundamental" solution which is the expectation of the capitalized y's,

$$\text{(FS)} \qquad p_t^* = E_t\{y_{t+1}/R + y_{t+1}/R^2 + \ldots\}\,.$$

This has motivated many tests for "excess volatility" that test for deviations of actual prices from the solution predicted by (FS). Econometric implementation of such testing is nontrivial and has given rise to a large literature.

While we do not have space to discuss this literature here, we point out how the models of Brock and LeBaron[21] can shed light on the issue of excess volatility. Let $\Delta p_t \equiv p_{t+1} - p_t$ denote the change in price of the risky asset over periods t, $t+1$. Define price volatility V to be the variance of Δp_t. Then Brock and LeBaron show that price volatility increases as the market signal precison increases. Here market signal precision is a measure of the average precision of signals about the future earnings of the risky asset. Thus when price is doing a good job in tracking future earnings (high market precision), we have high price volatility. This finding illustrates a beneficial aspect of price volatility.

This shows that price volatility is not something to necessarily be concerned with unless one can identify that such volatility is coming from a harmful source rather than what the market is *supposed to be doing when it is working well*. Furthermore, this kind of model can easily be generalized to allow signals on a piece of earnings to become less precise the further away in the future the piece is, to allow individuals to purchase signals of higher precision for higher costs, and to allow traders to balance such prediction costs against profits generated by better information. This leads to a dynamic equilibrium of heterogeneously informed traders where the heterogeneity is endogenous.

Such models have been built by Brock and LeBaron[21] and de Fontnouvelle.[30,31] These kinds of models generate data that are roughly consistent with the autocorrelation and cross correlation Fact 4. de Fontnouvelle's[30] model also is consistent with LeBaron's version of conditional predictability in Fact 3. Brock and LeBaron[21] show how to use statistical models of interacting systems combined with statistical models of discrete choice over trading strategies to produce sudden jumps in asset market quantities such as returns, volatility, and volume in response to tiny changes in the environment.

Empirical work using such interacting systems models is just beginning. Altug and Labadie[3] contains an excellent discussion of structural modeling in finance and empirical estimation of more conventional models.

Before continuing, I wish to stress that many models have been built to attempt to understand the Facts listed above. Many of these are reviewed in the excellent survey by Goodhart and O'Hara.[36]

FACT 12: SCALING LAWS. There have been scaling laws reported in finance. The early work of Mandelbrot, which stressed the apparent self-similar structure across frequencies of financial time series, has stimulated a large literature on scaling laws of the form $\ln[Pr\{\mathbf{X} > x\}]A + B\ln(x)$, often called "Pareto-tail" scaling. The stylized fact is that estimates of B are similar across frequencies, across time, and replacment of the event $\{\mathbf{X} > x\}$, by the event $\{\mathbf{X} < -x\}$. This is illustrated in Mandelbrot's famous cotton plots. Loretan and Phillips[51] have developed formal statistical inference theory for these objects and have applied this theory to several financial datasets.

Scaling laws must be handled with care because many different stochastic processes with very different structure of time dependence can generate the *same* scaling law. For example, think of limit theorems as $N \to \infty$ for appropriately scaled sums of N random variables (the standard Central Limit Theorem with normal limiting distribution) and appropriately scaled maxima of N random variables (where the limiting distribution is one of the extreme value distributions) stated in the form of "scaling laws."

LeBaron has done computer experiments that show that GARCH processes fit to financial data can generate scaling "laws" of the form $\ln[Pr\{\mathbf{X} > x\}] = A + B\ln(x)$ that look very good to the human eyeball.

As early as 1940, Feller[34] (see p. 52 and his reference to his 1940 article) had pointed out that the scaling plots used to demonstrate the "universality" of the "logistic law of growth" were infected with statistical problems and may be useless for any kind of forecasting work. He pointed out that logistic extrapolations are unreliable and that other distributions such as the cauchy, normal, can be fitted to the "*same material with the same or better goodness of fit*," (Feller,[34] p. 53, italics are his).

A problem with scaling "laws" for financial analysis is this. In financial analysis we are interested in *conditional objects* such as conditional distributions, whereas scaling laws are *unconditional objects*. For example, nonlinear impulse response analysis and estimation of conditional objects as in, for example, Gallant, Rossi, and Tauchen[35] and Potter[60] is likely to be of more use for financial work, than simply presenting a scaling relationship. Of course, scaling "laws" are useful for disciplining the activity of model building and for stimulating the search for explanations.

Some nice examples of work on scaling are contained in Mantegna and Stanley[56] for high-frequency financial data and in Stanley et al.[64] for size distribution of business firms. A general discussion of scaling in economics is contained in Brock.[13]

This completes the list of facts we wish to concentrate the reader's attention upon. There are more in Guillaume et al.,[37] Goodhart and O'Hara,[36] and Brock and de Lima[15] as well as their references. In section 2, we shall briefly indicate how dynamic adaptive belief model building may contribute to understanding the economic mechanisms that give rise to these facts.

2. SOME BRIEF EXAMPLES OF ADAPTIVE BELIEF MODELING

2.1 A BASIC MEAN VARIANCE FRAMEWORK

It is useful to have a basic model of adaptive belief systems in order to organize discussion and to shed light upon the facts listed in section 1. Although we shall work in a discrete time context with step size $s = 1$, one should remember that the ultimate objective is to understand dynamical phenomena for different step sizes. We should also be aware of different behavior that can arise in synchronous adjustment setups versus asynchronous adjustment setups. We shall work with synchronous adjustment here.

Brock and LeBaron[21] added adaptive dynamics to received asymmetric information models in order to study the dynamics of the market's precision at tracking the underlying earnings stream that it was pricing. They showed that this kind of model produces dynamics that are roughly consistent with Facts 2, 4, and 11. de Fontnouvelle's models[30,31] produce dynamics that are consistent with Facts 2, 4, and 11 as well as Fact 3. Furthermore, because he has long-lived assets, he did not have to introduce two scales of time to generate dynamics consistent with these facts, unlike Brock and LeBaron's[21] short-lived assets model.

However, work by Cragg, Malkiel, Varian, and others discussed in Brock and LeBaron,[21] has stressed that measures of diversity of earnings forecasts are associated with risk in the financial markets. Risk, in turn, is associated with volatility, while volume is associated with diversity. We shall stress diversity of forecasts in the development we sketch here in order to show how complex systems methods can contribute to model building in this area. This complements the use of complex system techniques by Brock, de Fontnouvelle, and LeBaron.

Let p_t, y_t denote the price (ex dividend) and dividend of an asset at time t. We have

$$\mathbf{W}_{t+1} = RW_t + (\mathbf{p}_{t+1} + \mathbf{y}_{t+1} - Rp_t)z_t\,, \tag{2.1.1}$$

for the dynamics of wealth, where bold face type denotes random variables at date t and z_t denotes number of shares of the asset purchased at date t. In Eq. (2.1.1) we assume that there is a risk-free asset available with gross return R as well as the risky asset. Now let E_t, V_t denote conditional expectation and conditional variance based on a publicly available information set such as past prices and dividends, and let E_{ht}, V_{ht} denote conditional expectation and variance of investor type h. We shall sometimes call these conditional objects the "beliefs of h." Assume each investor type is a myopic mean variance maximizer so that demand for shares, z_{ht} solves

$$\max\{E_{ht}\mathbf{W}_{t+1} - \left(\frac{a}{2}\right) V_{ht}(\mathbf{W}_{t+1})\} \text{ i.e.}\,, \tag{2.1.2}$$

$$z_{ht} = \frac{E_{ht}(\mathbf{p}_{t+1} + \mathbf{y}_{t+1} - Rp_t)}{aV_{ht}(\mathbf{p}_{t+1} + \mathbf{y}_{t+1})} \tag{2.1.3}$$

Let z_{st}, n_{ht} denote the supply of shares per investor at date t and the fraction of investors of type h at date t. Then equilibrium of supply and demand implies,

$$\sum n_{ht} \left\{ \frac{E_{ht}(\mathbf{p}_{t+1} + \mathbf{y}_{t+1} - Rp_t)}{aV_{ht}(\mathbf{p}_{t+1} + \mathbf{y}_{t+1})} \right\} = z_{st} \,. \tag{2.1.4}$$

Hence, if there is only one type h, equilibration of supply and demand yields the pricing equation

$$Rp_t = E_{ht}(\mathbf{p}_{t+1} + \mathbf{y}_{t+1}) - aV_{ht}(\mathbf{p}_{t+1} + \mathbf{y}_{t+1}) z_{st} \,. \tag{2.1.5}$$

Given a precise sequence of information sets $\{\mathcal{F}_t\}$ (e.g., $\{\mathcal{F}_t\}$ is a filtration) we may use Eq. (2.1.5) to define a notion of *fundamental solution* by letting E_{ht}, V_{ht} denote conditional mean and variance upon $\mathcal{F}_t$.

If one makes specialized assumptions such as $\{\mathbf{y}_t\}$ i.i.d. or AutoRegressive of order 1 with i.i.d. innovations (i.e., $\mathbf{y}_{t+1} = a_1 + a_2\mathbf{y}_t + \mathbf{e}_{t+1}, \{\mathbf{e}_t\}$ i.i.d. with mean zero and finite variance), and assumes that $z_{st} = z_s$ independent of t, then an explicit solution of Eq. (2.1.5) may be given. As an especially tractible example, we examine the fundamental solution for the special case $z_{st} = 0$, for all t. Then we have

$$Rp_t = E\{(\mathbf{p}_{t+1} + \mathbf{y}_{t+1}) | \mathcal{F}_t\} \,. \tag{2.1.5'}$$

Note that Eq. (2.1.5) typically has infinitely many solutions but (for the "standard" case, $R > 1$) only one satisfies the "no bubbles" condition

$$\lim \left(\frac{E_1 p_t}{R^t} \right) = 0 \,, \tag{2.1.6}$$

where the limit is taken as $t \to \infty$.

Proceed now to the general case of Eq. (2.1.5). Suppose that a "no bubbles" solution, p_t^*, exists for $\{\mathcal{F}_t\}$ for Eq. (2.1.5). Replace $\mathbf{y}_{t+1}$ by $\mathbf{y}\#_{h,t+1} \equiv \mathbf{y}_{t+1} - aV_{ht}(\mathbf{p}_{t+1} + \mathbf{y}_{t+1}) z_{st}$ in Eq. (2.1.5). This move allows us to interpret $\mathbf{y}\#_{h,t+1}$ as a risk-adjusted dividend.

Bollerslev, Engle, and Nelson[11] review Nelson's work in continuous time settings which shows that conditional variances are much easier to estimate (especially from high-frequency data) than conditional means. Motivated by Nelson's work and analytic tractibility, we shall assume homogeneous conditional variances, i.e., $V_{ht}(\mathbf{p}_{t+1} + \mathbf{y}_{t+1}) \equiv V_t(\mathbf{p}_{t+1} + \mathbf{y}_{t+1})$, all h, t. Hence, $\mathbf{y}\#_{h,t+1} \equiv \mathbf{y}_{t+1} - aV_t(\mathbf{p}_{t+1} + \mathbf{y}_{t+1}) z_{st}$, for all h, t.

In order to proceed further, we must make some assumptions. We shall list the assumptions here and assume:

ASSUMPTIONS A.1–A.3 (i) $V_{ht}(\mathbf{p}_{t+1} + \mathbf{y}_{t+1}) \equiv V_t(\mathbf{p}_{t+1} + \mathbf{y}_{t+1})$ for all h, t; (ii) $E_{ht}(\mathbf{y}_{t+1}) = E_t\mathbf{y}_{t+1}$, all h, t where $E_t\mathbf{y}_{t+1}$ is the conditional expectation of $\mathbf{y}_{t+1}$ given past $\mathbf{y}$'s and past $\mathbf{p}$'s (e.g., if $\mathbf{y}_{t+1} = a_1 + a_2\mathbf{y}_t + \mathbf{e}_{t+1}$, $\{\mathbf{e}_t\}$ i.i.d. with mean zero and finite variance, then $E_t\mathbf{y}_{t+1} = a_1 + a_2\mathbf{y}_t$); (iii) all beliefs $E_{ht}\mathbf{p}_{t+1}$ are of the form $E_{ht}\mathbf{p}_{t+1} = E_t\mathbf{p}^*_{t+1} + f_h(x_{t-1}, x_{t-2}, \ldots)$, all t, h.

Notice Assumption A.2 amounts to homogeneity of beliefs on the one-step-ahead conditonal mean of earnings and Assumption A.3 restricts beliefs, $E_{ht}\mathbf{p}_{t+1}$ to time stationary functions of deviations from a commonly shared view of the fundamental. Notice that one can simply assume that all beliefs considered are of the form, $E_{ht}(\mathbf{p}_{t+1}+\mathbf{y}_{t+1}) = E_t\mathbf{p}^*_{t+1}+E_t\mathbf{y}_{t+1}+f_h(x_{t-1}, x_{t-2}, \ldots)$ all t, h; and all of the analysis below is still valid. It seemed clearer to separate the assumptions as we did above. In any case, Assumptions A.1–A.3 will allow us to derive a *deterministic dynamical system* for the equilibrium path of the system. This will be shown below.

Since we assume homogeneous conditional variances, we may use Assumptions A.2, and A.3 to write (putting $f_{ht} = f_h(x_{t-1}, x_{t-2}, \ldots)$)

$$E_{ht}(\mathbf{p}_{t+1} + \mathbf{y}^{\#}_{h,t+1}) = E(\mathbf{p}^*_{t+1} + \mathbf{y}^{\#}_{t+1}|\mathcal{F}_t) + f_{ht}\,. \tag{2.1.7}$$

Now use Eq. (2.1.4) to equate supply and demand and obtain the basic equilibrium equation in deviations form,

$$Rx_t = \sum n_{ht} f_{ht}\,, \tag{2.1.8}$$

were $x_t \equiv p_t - p^*_t$.

At this stage we shall be very general about the beliefs $\{f_{ht}\}$ which we have expressed in the form of deviations from the fundamental solution $\{p^*_t\}$. Notice that we may deal with a large class of $\{\mathbf{y}_t\}$ processes, even nonstationary ones, by working in the space of *deviations from the fundamental*.

This device allows us to dramatically widen the applicability of stationary dynamical systems analysis to financial modeling. Before we continue we wish to remind the reader that general financial models stress covariances among assets in large markets.

In these more general models (e.g., the Capital Asset Pricing Model (CAPM) of Sharpe, Lintner, Mossin, Black et al. and the Arbitrage Pricing Model (APT) of Ross) only *systematic* risk (risk that cannot be diversified away) gets priced. See Altug and Labadie[3] for a nice general exposition that gives a unified treatment of asset pricing theory and associated empirical methods.

In CAPM/APT models discounting for systematic risk gets built into the discount *rate* rather than into an adjustment to earnings. Notice that the adjustment to earnings depends on shares per trader, so more traders per share will lower this adjustment. Since we wish to stress the addition of belief dynamics, we shall work with the simpler setup given here. This is enough to suggest what a general development might look like.

Later on we shall make the basic working assumption that we can express the beliefs $\{f_{ht}\}$ as functions of $(x_{t-1}, x_{t-2}, \ldots, x_{t-L})$ and parameters. The parameters shall represent biases and coefficients on the lags x_{t-1} for special cases such as linear cases. We shall examine two types of systems: (i) systems with a small number of types; (ii) systems with a large number of types, where the large system limit is taken. But, before we can do any of this kind of analysis we must determine how the $n_{ht}, h = 1, 2, \ldots, H$ are formed.

In order to determine the fraction of each type of belief we introduce a discrete choice model of belief types as in Brock[12] and Brock and LeBaron.[21] Let

$$U_{ht} = U_{ht} + \mu e_{ht}\,, \tag{2.1.9}$$

where U_{ht} is a deterministic index of utility generated by belief h (e.g., a distributed lag of past trading profits on that belief system) and $\{e_{ht}\}$ is a collection of random variables which are mutually and serially independent extreme value distributed. For example, U_{ht} could be a distributed lag measure of past profits, past risk-adjusted profits, or be related to wealth in Eq. (2.1.1). Here μ is a parameter to scale the amount of uncertainty in choice. The parameter $\beta \equiv 1/\mu$ is called the *intensity of choice*.

de Fontnouvelle[30] builds on the discrete choice literature in econometrics to give a nice motivation for the presence of the $\{e_{ht}\}$ in Eq. (2.1.9). Four possible sources for the uncertainty in choice represented by the $\{e_{ht}\}$ are: (i) nonobservable characteristics (these are characteristics of the agents that the modeler cannot observe), (ii) measurement errors (agents make errors in constructing a "fitness" measure for each belief system), (iii) model approximation error (the functions used in the model are wrong, even if the characteristics of the agents are observed by the modeler), (iv) agents do not have the computational abilities assumed by the modeler (i.e., agents are *boundedly rational*).

Some of these reasons for uncertainty in choice motivate model building that endogenizes β rather than assuming that it is fixed. Brock[12] discussed a motivation for it by trading off the cost of making more precise choices against the gain from more precise choice. More will be said about this below.

However, one could imagine that (i) there is learning by doing in choice (the longer an agent has been doing this kind of choosing, the higher β becomes), (ii) some agents have a higher "choice IQ" (choice IQ is a characteristic which varies across agents), (iii) the longer the agent has been choosing strategies in similar settings the better it gets simply by the accumulation of "habits" (this is similar to learning by doing but emphasizes passage of time, rather than passage of volume of experience), (iv) technological change in choice machinery is occurring (faster computers and interlinked databases are making estimation of relevant quantities for more accurate choice possible).

de Fontnouvelle[30] argues, that in the context of financial modeling, one could imagine that the sources of the $\{e_{ht}\}$ could consist of many small nearly independent errors so the Central Limit Theorem would apply. Hence the $\{e_{ht}\}$ would be approximately normally distributed in such a situation. We use the extreme value distribution for tractibility, but many of the insights we draw are valid for more general distributions. The extreme value assumption leads to discrete choice "Gibbs" probabilities of the form,

$$n_{ht} \equiv \text{prob}\{\text{choose } h \text{ at date } t\} = \frac{\exp(\beta U_{ht})}{Z_t}, \tag{2.2.10}$$

$$Z_t \equiv \sum \exp(\beta U_{jt}), \tag{2.1.11}$$

where β is called the intensity of choice. See the book by Manski and McFadden[55] for a general discussion of discrete choice modeling.

In order to form the "fitness measure" U_{ht} we need a measure of profits generated by belief system f_{ht}. Let the last type have $f_{Ht} = 0$, i.e., the last type is a "fundamentalist." Measure profits relative to the fundamentalists and leave out risk adjustment for simplicity. We shall say more about risk adjustment below.

The conditional expectation of realized (excess) profits at date $t+1$ based upon past prices and earnings available at date t is given by

$$\pi_{h,t+1} = E_t\left\{\frac{(E_{ht}(\mathbf{p}_{t+1} + \mathbf{y}_{t+1}) - Rp_t)(\mathbf{p}_{t+1} + \mathbf{y}_{t+1} - Rp_t)}{aV_{ht}(\mathbf{p}_{t+1} + \mathbf{y}_{t+1})}\right\}. \tag{2.1.12}$$

We shall use $\pi_{h,t+1}$ as the fitness measure for h. This is a very idealized modeling of "learning" because it is not clear how agent h could ever "experience" such a "fitness" measure. We work with this measure of fitness here because it is analytically tractible and leave treatment of more realistic measures to future work. In future work one could introduce a "slow" scale of time for belief change and a "fast" scale of time for trading on beliefs. Then one could form sample analogs of $E_t(.)$ and use these to approximate the population counterparts. Or one could imagine an ergodic stationary setting where each information set is visited infinitely often so an analog estimator could be constructed for $E_t(.)$. Let us continue with the current development.

Assume homogeneous beliefs on conditional variances, abuse notation by continuing to let $\pi_{n,t+1}$ denote profit *differences* measured relative to profits of type H we obtain

$$\pi_{h,t+1} = E_t\left\{\frac{f_{ht}(\mathbf{p}_{t+1} + \mathbf{y}_{t+1} - Rp_t)}{aV_{ht}(\mathbf{p}_{t+1} + \mathbf{y}_{t+1})}\right\}. \tag{2.1.13}$$

Now put $\sigma_t^2 \equiv V_{ht}(\mathbf{p}_{t+1}+\mathbf{y}_{t+1})$ since we have assumed that beliefs are *homogeneous* on conditional variances. Note that

$$E_t\{f_{ht}(\mathbf{p}_{t+1} + \mathbf{y}_{t+1} - Rp_t)\} = f_{ht}E_t(\mathbf{x}_{t+1} - Rx_t) = f_{ht}E_t\mathbf{R}_{t+1} \equiv f_{ht}E_t\rho_{t+1} \tag{2.1.14}$$

and

$$n_{ht} = \frac{\exp[\beta\pi_{h,t-1}]}{Z_{t-1}} = \frac{\exp[(\beta/a\sigma_{t-1}^2)f_{h,t-2}\rho_{t-1}]}{Z_{t-1}}. \tag{2.1.15}$$

Here $\rho_{t+1} \equiv E_t R_{t+1} = E_t x_{t+1} - R x_t$, is just the conditional expectation of excess returns per share, conditioned upon past prices and past earnings.

We shall argue that, Assumptions A.1–A.3 imply that x_{t+1} is a deterministic function of $(x_t, x_{t+1}, \ldots)$ so that $E_t x_{t+1} = x_{t+1}$. Start the equilibrium Eq. (2.1.8) off at $t = 1$ using Eq. (2.1.15) with all x's with dates older than date 1 given by history. Then Eq. (2.1.8) implies that x_1 is given as a stationary function of $(x_0, x_{-1}, \ldots)$. Hence, it is deterministic at time $t = 0$, in particular $E_0 x_1 = x_1$. Continue the argument forward in this manner to conclude that for each t, x_{t+1} can be written as a deterministic function of $(x_t, x_{t-1}, \ldots)$. Equations (2.1.8) and (2.1.15) together give us an *equilibrium time stationary deterministic dynamical system.* We shall analyze some special cases below. Before we do that analysis we wish to make one remark.

In many applications of this kind of modeling, one can put $U_{h,t+1}$ equal to a distributed lag of the profit measures, $\pi_{h,t+1}$ and give the system more "memory." In many cases, the more memory the system has, the better it does at eliminating belief systems that deviate away from the fundamental. This can be seen by doing steady state analysis. Also if one writes down continuous time versions of our type of system, one will see a close relationship to replicator dynamics which are widely used in evolutionary game theory. A drawback of more memory (slower decay of past profit measures in the distributed lag) is sluggish adaptation to "regime changes."

2.2 SPECIAL CASES

Consider a special case where all beliefs on conditional variance are the same, are constant throughout time, and outside share supply is zero at all times. Recall that p denotes the fundamental and let $x_t = \bar{p}_t - p_t^*$, $\eta \equiv \beta / a\sigma^2$. Here $\bar{p}_t$ denotes equilibrium price. Consider expectations that are deterministic deviations from the fundamental. Then calculating profits generated by each belief system and using the discrete choice model to determine $\{n_{ht}\}$ as above, equilibrium is given by

$$R x_t = \sum n_{ht} f_{ht}, n_{ht} = \frac{\exp(\eta \rho_{t-1} f_{h,t-2})}{Z_{t-1}}, \rho_{t-1} \equiv x_{t-1} - R x_{t-2}\,. \tag{2.2.1}$$

Rewrite Eq. (2.2.1) as

$$R x_t = \frac{\{\sum f_{ht} \exp(\eta \rho_{t-1} f_{h,t-2}) / H\}}{\{\sum \exp(\eta \rho_{t-1} f_{h,t-2}) / H\}}\,. \tag{2.2.2}$$

This is a ratio of two "plug in" or analog estimators for the obvious moments on the R.H.S. of Eq. (2.2.2). Hence, once we specify distributions for the cross sectional variation in $\{f\}$ we can take the Large Type Limit as $H \to \infty$.

Let us be clear about the type of limits we are taking here. In order to get to Eq. (2.2.1) the number of traders has been taken to infinity, while *holding the number of types H fixed*, to get the discrete choice probabilities appearing on the R.H.S. of Eq. (2.2.1). After taking the number of traders to infinity first, we take the number of types to infinity second to get what we shall call the "Large Type Limit" (LTL). Before going on, let us say a few words about the notion of LTL.

One might think of the device of LTL as a way to get some analytic results for computer simulation experiments on complex adaptive systems such as those conducted at SFI. The idea is to ask what the typical behavior might be if a large number of types were drawn from a *fixed* type distribution.

Of course the answers are going to depend on how the parameters of the dynamics of each experiment are drawn from the space of possible experiments, as well as on the experimental design. Since financial markets typically have a large number of players and since equilibrium prices are a weighted average of demands it is not unreasonable to imagine the law of large numbers would operate to remove randomness at the individual level and leave only the bulk characteristics such as mean and variance of the type diversity distribution to effect equilibrium.

The notion of LTL goes even further, however. It recognizes that, at each point in time, financial equilibrium conditions such as Eq. (2.2.2) for example, can be typically be written in the "moment estimator" form even after the number of traders have been taken to infinity. This suggests that if type diversity is itself drawn from a type distribution, then large diversity systems should behave like the probability limit of Eq. (2.2.2) as $H \to \infty$. Hence if H is large the $H \to \infty$ limit should be a good approximation.

The research strategy carried out in Brock and Hommes[18,19] was to examine systems with a small number of types and try to catalogue them. The LTL approach is to analyze systems with a large number of types and approximate them with the LTL limit. This gives a pathway to analytic results for two polar classes of systems. We illustrate the usefulness of the LTL device with a couple of calculations.

LARGE TYPE LIMIT CALCULATIONS

EXAMPLE 1: Let $\{f_{ht}\}$ be given by $\{b_{ht}\}$ where b_{ht} is i.i.d. with mean zero and finite variance across time and "space." Then the LTL is given by

$$Rx_t = Ef_{ht}\frac{E(\exp(\eta\rho_{t-1}f_{h,t-2}))}{E(\exp(\eta\rho_{t-1}f_{h,t-2}))} = Ef_{ht} = Eb_{ht} = 0\,. \qquad (2.2.3)$$

In an LTL with pure bias one might expect $Eb_{ht} = 0$ since for each type that's biased up, one might expect a contrarian type that's biased down. In systems with a large number of such types it might be natural to expect an "averaging" out of the biases toward zero. Equation (2.2.3) says that $x_t = 0$, $t = 1, 2, \ldots$ for such systems. That is, the market sets the price of the asset at its fundamental value. Of course if $Eb_{ht} = \mu_b$, not zero, then $Rx_t = \mu_b$ for all t.

As an aside, note how the market can protect a biased trader from his own folly if he is part of a group of traders whose biases are "balanced" in the sense that they average out to zero over the set of types. Centralized market institutions can make it difficult for unbiased traders to prey on a set of biased traders provided they remain "balanced" at zero. Of course, in a pit trading situation, unbiased traders could learn which types are biased and simply take the opposite side of the trade. This is an example where a centralized trading institution like the New York Stock Exchange could "protect" biased traders, whereas in a pit trading institution, they could be eliminated.

Consider a modification of Example 1 where

$$f_{ht} = b_{ht} + w_{h,t-1}x_{t-1} + \ldots + w_{h,t-L}x_{t-L}\,, \tag{2.2.4}$$

where $(b_{ht}, w_{h,t-1}, \ldots, w_{h,t-L})$ is drawn multivariate i.i.d. over h, t.

That is, each type has a bias component and L extrapolative components based upon past deviations from the commonly understood fundamental value. Given a specific distribution such as multivariate normal, it is easy to use Eq. (2.2.3) and moment-generating functions to calculate the LTL and produce a deterministic dynamical system for x_t.

"QUENCHED" BELIEF SYSTEMS

Another useful way to generate LTL's is to draw $f_{ht} = b_{ht} + w_{h,t-1}x_{t-1} + \ldots + w_{h,t-L}x_{t-L}$ at date $t = 0$ and *fix it* for all $t = 1, 2, \ldots$. Call this draw $(b_h, w_{h,-1}, \ldots, w_{h,-L})$ and write

$$Rx_t = \frac{E\{f_{ht}\exp(\eta\rho_{t-1}f_{h,t-2})\}}{E(\exp(\eta\rho_{t-1}f_{ht-2}))} \tag{2.2.5}$$

and use moment-generating functions for specific distributions to calculate the LTL. This "quenching" of the form of $\{f_{ht}\}$ at time 0 and *then* letting time run while the belief types f_{ht} replicate according to the relative profits they generate probably gives a better representation of the "sea" of belief types evolving in artificial economic life settings like Arthur et al.[7] Before we do any calculations let us discuss some general issues.

First we must remark, that of course, we still do not yet capture the use of Holland's full blown adaptive complex systems approach to generating belief types as used in Arthur et al.[7] They introduce crossover and mutation of types as well as replication of types.

Second, consider a second stage attempt to capture the spirit of Holland in the artificial way of building a large set of $f_{ht} = f_h(\mathbf{b}, \mathbf{w}, \mathbf{X}_{t-1})$ where b, w denote a vector of bias parameters, a matrix of "loadings" on lagged x's, all drawn at the starting date $t = 0$, and $\mathbf{X}_{t-1}$ denotes a vector of lagged x's. Even though this does not yet capture Holland's "bottoms up" approach to building belief systems out of elementary building blocks, it goes part way, because one can generate a very large set of f's and let them compete against each other as in the above discrete choice approach where the utility of each f is the profits it generates. At each point in time the fraction of each f-type is determined by its relative profits.

An interesting recent paper of Dechert et al.[29] can be applied to make a conjecture about the dynamics of f-systems like the above. Dechert et al.[29] show that when dynamical systems of the form

$$x_t = g(\mathbf{X}_{t-1}; \theta)$$

are parameterized by a family of neural nets then *almost all such systems are chaotic* provided there are enough lags in $\mathbf{X}_{t-1}$. The probability of chaos is already high when $\mathbf{X}_{t-1}$ only has a modest number of lags.

This finding is suggestive of a possibility for the generic development of complexity as the intensity of choice β increases. Let us indicate a possible line of argument suggested by the work of Doyon et al.[32] Draw an f-system at random and increase β (i.e., increase η) for it. One might expect it to be "likely" that the "probability" that the first bifurcation is Hopf with an irrational angle between the real and imaginary part of the complex root pair to "converge" to unity as the number of lags in $\mathbf{X}_{t-1}$ increases. Speculating even more wildly, one might think that it would be quite likely that another Hopf bifurcation might appear after the first one. Furthermore, one might argue that for large enough lags "most" of the systems are chaotic. This suggests the possibility of construction of a successful argument that the "generic" route to chaos as η increases is quasi periodicity.

Of course, this is only suggestive of a research project to formalize and prove such assertions. One possible route might be to follow Doyon et al.,[32] who use Girko's Law on the distribution of eigenvalues of large matrices and computer simulations to argue that one might expect a quasi-periodic route to chaos. Recall that the measure of the set of real eigenvalues is zero under Girko's limit distribution, which is uniform over a circle in the complex plane.

Since the measure of the set of complex numbers with rational angle between the complex part and the real part is also zero under Girko's limit distribution, the Doyon et al.[32] argument is suggestive that the first bifurcation might be "generic Hopf" as the number of lags increases. A reading of Doyon et al.[32] suggests, due to difficulties raised by discrete time, that the argument that the first bifurcation is followed by another one which is generic Hopf is more problematic. For example, in discrete time, the first bifurcation can be followed "by a stable resonance due to discrete time occuring before the second Hopf bifurcation" (Doyon et al.,[32] p. 285).

However, the problems with this line of argument for our case are several. First, Girko's Law depends upon the explicit scaling of the system so that convergence of the eigenvalue spectrum of the sequence of Jacobians extracted from the linearizations at the steady state to Girko's limit occurs. This scaling may not make economic sense. Second, while one may have a chance getting the first bifurcation to be generic Hopf using the Doyon et al.[32] strategy and still making some economic sense out of it, the second bifurcation is complicated in discrete time settings. Third, the simulations of Doyon et al.[32] indicate the presence of a large quantity of real eigenvalues, even for large systems.

Another possible research route to exploration of the impact of adding more lags to the dynamics would be to explore Schur-type necessary and sufficient conditions for all eigenvalues of the linearization to lie inside the unit circle. One would expect the probability that a randomly drawn system's linearization to satisfy the Schur conditions to fall as the number of lags increases.

The safest thing that can be said at this stage is this. Exploration of "routes to complexity" of "randomly" drawn systems is a subject for future research. It would be interesting to formulate and prove a theorem on the quasi-periodic route to complex dynamics, like that suggested by the Doyon et al.[32] work, because it would drastically simplify the classification of pathways to complex dynamics in large systems.

Return now to explicit calculations. Consider Eq. (2.2.5) with quenched pure biases, $f_{ht} = b_h$, before the LTL is taken,

$$Rx_t = \frac{\sum\{b_h \exp(\eta\rho_{t-1}b_h)\}}{\sum(\exp(\eta\rho_{t-1}b_h)) \equiv \phi(\eta\rho_{t-1})}\,. \tag{2.2.6}$$

A direct computation shows that $\phi\prime > 0$. Hence, steady states x^*, $y^* = x^* - Rx^*$ must must satisfy, $Rx^* = \phi(-r\eta x^*)$, $r \equiv R - 1$. Now $\phi(0) = (1/H)\sum b_h$, which is a measure of the average bias. If this is zero, then the only steady state is $x^* = 0$, i.e., there can be no deviation from the fundamental. In general there can only be one steady state. Of course, since $\rho_{t-1} \equiv x_{t-1} - Rx_{t-2}$, the steady state may be unstable. Turn now to calculation of LTL's for some specific "quenched" cases.

Consider Eq. (2.2.6) but draw b i.i.d.N(μ_b, σ_b^2) across h. Let $\tau \equiv \eta\rho_{t-1}$. Note that the moment-generating function $E\{\exp(\tau b)\}$ and its derivative w.r.t. τ, $\{b\exp(\tau b)\}$ for the normal are given by

$$E\{\exp(\tau b)\} = \exp(\tau\mu_b + \left(\frac{1}{2}\right)\sigma_b^2\tau^2)\,, \tag{2.2.7a}$$

$$E\{b\exp(\tau b)\} = (\mu_b + \sigma_b^2\tau)\exp(\tau\mu_b + \left(\frac{1}{2}\right)\sigma_b^2\tau^2)\,. \tag{2.2.7b}$$

Hence, we have,

$$Rx_t = \frac{E\{b\exp(\tau b)\}}{E\{\exp(\tau b)\}} = \mu_b + \sigma_b^2\tau = \mu_b + \sigma_b^2\eta(x_{t-1} - Rx_{t-2})\,. \tag{2.2.8}$$

This is a linear second-order difference equation, which is trivial to analyze by linear methods. The case of economic interest is where R is near one. For R near one, it is easy to see that Eq. (2.2.8) becomes unstable as η, σ_b^2 increase. Let $\alpha \equiv \sigma_b^2\eta$.

Notice that the two roots of Eq. (2.2.8) are complex with modulus less than one for small α. As α increases the roots increase in modulus and pass out of the unit circle in the complex plane. Hence, the bifurcation of Eq. (2.2.8) under bifurcation parameter α is Hopf.

For a similar type of example, draw b i.i.d. with mass points −A,+A with probability 1/2 each. Calculate as above to obtain,

$$Rx_t = A \tanh[A\eta(x_{t-1} - Rx_{t-2})] . \qquad (2.2.9)$$

The analysis of the linearization of Eq. (2.2.9) is similar to the analysis of Eq. (2.2.8). The global analysis of system Eq. (2.2.9) may now be analyzed as in Brock and Hommes.[18,19] Note that the linearizations of Eq. (2.2.8) and Eq. (2.2.9) have an autoregressive structure of order 2 that is reminiscient of a common structure found in detrended macroaggregates such as detrended real Gross National Product.

CONCLUSION. In pure bias cases, instability tends to appear when η increases and when the variance of the b-distribution increases. Therefore, one expects an increase in heterogeneity dispersion to increase chances of instability in these pure bias cases. Depending on one's intuition one might expect an increase in diversity as measured by the variance of the type distribution to lead to an increase in stability. This is so because large diversity of types implies that for any biased type it is more likely that there's another type who would be willing to take the opposite side of the desired trade. Hence, an increase in diversity might be expected to lead to an increase in "liquidity" and, possibly, to an increase in "stability." However, when types interact with fitness and replication-like dynamics over time, an increase in diversity measured by an increase in σ_b^2 in Eq. (2.2.8) or an increase in A in Eq. (2.2.9) can lead to an increase in instability. This is so because large variance magnifies the system's response to last period's excess returns $x_{t-1} - Rx_{t-2}$.

The same type of calculations can be done to study the Large Type Limit dynamical system for general $f_h = b_h + w_{h,-1}x_{t-1} + \ldots + w_{h,t-L}x_{t-L} \equiv b_h + g(\mathbf{X}_{t-1}, \mathbf{w})$. One can say more for specific distributions such as multivariate normal where there is a closed-form expression for the moment-generating function. This kind of calculation allows one to explore the effects of changes in means and variances as well as covariances of bias and lag weights on stability of the Large System Limit dynamical system which can be explored by deterministic dynamical systems methods. In cases where the parameters in the g-part of f are independent of the bias term, and the g-part is small, then the primary bifurcation is likely to be Hopf as in Eq. (2.2.8). We refer to Brock and Hommes[19] for more on this.

Before continuing on to a treatment of social interaction in adaptive belief systems, we wish to make an extended remark concerning the use of risk-adjusted profit measures as fitness functions. Put $\mathbf{r}_{t+1} \equiv \mathbf{p}_{t+1} + \mathbf{y}_{t+1} - R\mathbf{p}_t$, let $\rho_{t+1} = E_t\mathbf{r}_{t+1}, \rho_{h,t+1} = E_{ht}\mathbf{r}_{t+1}$. Refer to Eqs. (2.1.1), (2.1.2), and (2.1.3) in the following short discussion.

Let $g(\rho_{t+1};\rho_{h,t+1})$ denote the conditional expectation of *actual* risk-adjusted profits experienced when choice of z_{ht} was made based upon $E_{ht}(.)$. Let $g(\rho_{t+1};\rho_{t+1})$ denote the conditional expectation of *actual* risk-adjusted profits when choice of z_{ht} was made upon the correct conditional expectation $E_t(.)$. It is easy to use Eqs. (2.1.2), (2.1.3), and Assumptions A.1–A.3 to show

$$g(\rho_{t+1};\rho_{h,t+1}) - g(\rho_{t+1};\rho_{t+1}) = -\left(\frac{1}{(2a\sigma_t^2)}\right)(E_{ht}(\mathbf{q}_{t+1}) - E_t(\mathbf{q}_{t+1}))^2\,, \quad (2.2.10)$$

where $q_{t+1} \equiv \mathbf{p}_{t+1} + \mathbf{y}_{t+1}$ and $\sigma_t^2 \equiv V_t(\mathbf{q}_{t+1})$.

Equation (2.2.10) allows us to make three points. First, if memory is introduced into the fitness function for type h, i.e.,

$$U_{ht} = dg_{h,t-1} + MU_{h,t-1}, dg_{h,t-1} \equiv g(\rho_{t-1};\rho_{h,t-1}) - g(\rho_{t-1};\rho_{t-1})\,, \quad (2.2.11)$$

then it may be shown that if $M = 1$ (or η goes to infinity in the above setting), rational expectations, $E_t\mathbf{q}_{t+1}$, eventually drives out all other beliefs. Intuitively, any belief that is consistently different from rational expectations eventually generates unbounded "value loss" from rational expectations and, hence, its weight, n_{ht} goes to zero for positive intensity of choice β. Brock and Hommes[19] develops this result and other results using risk-adjusted profit measures.

Second, the use of a profit measure that is risk-adjusted as above makes the profit measure used in the fitness function *compatible* with the goal function that generated the underlying demand functions. This consistency property is not only intellectually desirable but also allows unification of the evolutionary theory with the received theory of rational expectations. Brock and Hommes[19] show that LTL-type calculations are messier but still tractible. They also show that rational expectations tend to squeeze out all other beliefs when memory M is close to one, intensity of choice is close to infinity, risk aversion is near zero, or commonly held beliefs on conditional variance are very small. However, they also show that if there is a cost to obtaining rational expectations which must be renewed each period then this "creates niche space" for nonrational beliefs to compete evolutionarily. Hence, costly rational expectations can lead to complicated dynamics in the space of deviations from the fundamental.

Third, the fact that Eq. (2.2.10) always has negative R.H.S. induces a type of quasi norm on the space of belief measures for each h conditional on commonly held information at each time t. Since the goal function (2.1.2) is mean/variance therefore this quasi norm only can push conditional mean and conditional variance to their rational expectations values. One can obtain such convergence when, for example, memory M tends to one.

More general goal functions such as the standard Lucas tree model treated in Sandroni[63] may induce a more powerful quasi norm. The intuition is this: let $\mu_h(.)$ denote agent h's measure. Agent h makes decisions to maximize its goal (which is the conditional expectation of capitalized utility of consumption in Sandroni's case) and let $d(\mu_h)$ denote this decision. Now evaluate the goal function using the actual measure $\mu(.)$ but with the decision set at $d(\mu_h)$. Obviously agent h does better by using $d(\mu)$, *not* $d(\mu_h)$.

Hence, one can imagine developing an evolutionary approach in more general settings than the mean/variance setting which is exposited here. Provided that the goal function is "rich enough" one could imagine evolutionary forces for rational expectations. Recall that under regularity conditions characteristic functions determine measures and vice versa. In analogy with characteristic function theory one might imagine weak sufficient conditions on a class of goal functions that would deliver strong forces for convergence to rational expectations when memory is infinite (i.e., $M = 1$).

Notice that Araujo and Sandroni[6] and Sandroni[63] develop a theory that locates sufficient conditions for convergence to rational expectations in a setting where agents who do a better job of predicting the future achieve higher values of their goal functions. One could imagine constructing an evolutionary version of their more general theory parallel to the evolutionary mean/variance theory outlined here.

Finally, before turning to social interactions, we would like to state that Marimon's review[57] gives a broad survey of recent work on learning dynamics. We urge the reader to study this review in order to gain a proper perspective on the breadth of work that has been done on learning and evolution in economics.

2.3 SOCIAL INTERACTIONS IN ADAPTIVE BELIEF SYSTEMS

While there is not space here for an extensive treatment of social interactions, let us outline a brief treatment since social interactions are one way to build models consistent with the abrupt change Fact 11.

Specialize Eq. (2.1.9) to the case of two choices 1,2 which we recode as $\omega = -1, +1$ and consider the probability model for n choosers,

$$Pr\{(\omega_1, \ldots, \omega_n)\} = \frac{\exp[\beta(\sum(U(\omega_i) - C(\omega_i - \bar{\omega})^2)]}{Z}, \tag{2.3.1}$$

where $C > 0$ is a penalty for agent i to deviating from the average choice $\bar{\omega} \equiv (1/n) \sum \omega_i$ of the community of agents. Here Z is chosen so that the probabilities given in Eq. (2.3.1) add up to unity over the 2^n configurations $(\omega_1, \ldots, \omega_n)$. It is easy to rewrite Eq. (2.3.1) into the form of a standard statistical mechanics model, i.e., the Curie-Weiss model, and follow Brock[12] to show that $\bar{\omega}$ converges in distribution to the solution of

$$m = \tanh(\left(\frac{\beta}{2}\right) dU + \beta J m) \,, J = 2C \,, \tag{2.3.2}$$

where the root with the same sign as $dU \equiv U(+1) - U(-1)$ is picked. There are two roots when $\beta J > 1$.

Brock and Durlauf[16] have developed relatives of this type of model for application to macroeconomics and to modeling group interactions in social behaviors which are vulnerable to peer pressure. They have started the work of development of econometric theory for such models and have started the work of estimation of such models on data sets.

Let us pause here to take up the issue of endogeneity of the parameter β. In statistical mechanics models, β is the inverse "temperature." In the kind of models we treat here, β is the intensity of choice parameter which can be related to individual and population characteristics as in Manski and McFadden.[55]

There is another approach to β which is based upon ideas of E. T. Jaynes and was outlined in Brock.[12] Consider the maximum entropy problem

$$\text{Maximize } \{-\sum p_i \ln(p_i)\}, \text{ subject to} \tag{2.3.3}$$

$$\sum p_i = 1, \tag{2.3.4}$$

$$\sum p_i U_i = \bar{U}, \tag{2.3.5}$$

where $\bar{U}$ is a target level of utility. W.L.O.G. rank U's as $U_1 \leq U_2 \leq \ldots \leq U_H$. Substitute out Eq. (2.3.4), set up the Lagrangian $L = \sum p_i \ln(p_i) + \lambda(\bar{U} - \sum p_i U_i)$, write out the first-order conditions and solve them to obtain

$$p_i = \frac{\exp(\beta U_i)}{Z}, \beta \equiv -\lambda, Z = \sum \exp(\beta U_j), \tag{2.3.6}$$

and β is determined by Eq. (2.3.5).

Now introduce a "production function" of average utility of choice, call it $\bar{U}(e)$, and introduce "effort" e with "cost" q per unit effort. Choose effort level e to solve

$$\text{Maximize } \bar{U}(e) - qe, \tag{2.3.7}$$

let e^* solve Eq. (2.3.7), insert $\bar{U}(e^*)$ in place of $\bar{U}$ in Eq. (2.3.5), and obtain $\beta \equiv \beta^* \equiv -\lambda^*$. If one puts $\bar{U}(0) = (1/n)\sum U_i$, we see that $U_n - U_1 = 0$ implies $e^* = 0$ and $\beta^* = 0$. One can show that β^* weakly increases when cost q decreases and "choice gain" $U_n - U_1$ increases.

Even though it is trivial, this simple endogenization of β is useful as a reasoning tool when coupled with the modeling outlined above. For example, consider agent types where effort cost q is low (e.g., professional traders who have access to large databases, high-speed computers with highly trained support staff, as well as years of training and discipline). If there is anything in the reward system of such types that penalizes them from deviation from their group average choice, then Eq. (2.3.2) suggests that a small change in dU can lead to big changes in the limiting value of $\bar{\omega}$ in the large system limit.

As another example consider debates about application of interacting systems models to economics because the parameter β is "exogenously fixed" and corresponds to "inverse temperature" which has no meaning in economics. The maximum entropy story given above, trivial as it is, is enough to show that some of this debate may be misplaced. Turn now to a general discussion of adaptive belief modeling.

GENERAL THEORY OF ADAPTIVE EVOLUTION OF BELIEFS

During the last few years I have been working alone and with co-authors on developing models of diversity of beliefs and social interactions in belief formation (Brock,[12] Brock and LeBaron,[21] Brock and Hommes[18,19]) not only to help explain the facts listed in the introduction, but also to uncover general complexity producing mechanisms in the economy.

Let us discuss a most interesting pathway to complexity that appears very common in economic situations. I shall call it the Prediction Paradox.

PREDICTION PARADOX

Consider a situation where a collection of agents are predicting a variable (for example, a product price) whose value is determined by the joint aggregate of their decisions (their aggregate supply must equal aggregate demand, for example). Let prediction accuracy for each agent increase in cost. While the cost for some agents may be less than for others, cost increases for all as accuracy increases. If all agents invest a lot in prediction accuracy, then the gains for any one agent may not be enough to cover its cost given that the others are all investing. Vice versa, if no agent is investing in prediction, the gains to any one agent may cover the cost of investing in prediction. We shall show that a situation like this is a good source of complex dynamics.

There are many examples of this sort of situation in economics. A leading example, relevant to finance and to the real world is the "Index Paradox."

INDEX PARADOX. If everyone invests in index funds in the stock market, it will pay someone to do security analysis and invest in mispriced securities. If everyone does security analysis, then indexers can save on the costs of securities analysis. So it will pay to avoid the costs of security analysis and just put all of one's funds into an index fund.

de Fontnouvelle[30,31] develops adaptive belief models that address the Index Paradox.

We shall use a simple cobweb production setting to show how the prediction paradox can lead to complex dynamics. Think of a collection of firms with anticipated profit function $\pi^e = p^e q - c(q)$, p^e denotes price expectation which must be formed to guide the production decision q. Let each firm j make a prediction p_j^e

and maximize anticipated profits. Equilibrium of demand and supply determines actual price p and actual profits $\pi = pq_j - c(q_j)$ for each firm j.

In Brock and Hommes[18] predictions are made by choosing a predictor from a finite set P of predictor or expectations functions. These predictors are functions of past information. Each predictor has a performance measure attached to it which is publicly available to all agents. Agents use a discrete choice model along the lines of Manski and McFadden[55] to pick a predictor from P where the deterministic part of the utility of the predictor is the performance measure. This results in the *Adaptive Rational Equilibrium Dynamics* (A.R.E.D), a dynamics across predictor choice which is coupled to the dynamics of the endogenous variables.

The A.R.E.D. incorporates a very general mechanism, which can easily generate local instability of the equilibrium steady state and very complicated global equilibrium dynamics.

For example, under the assumption of unstable market dynamics when all agents use the cheapest predictor, it is shown that in the A.R.E.D. the "probability" of reaching the equilibrium steady state is zero when the intensity of choice to switch between predictors is large enough. This result holds under very general assumptions on the collection of predictors, $\mathcal{P}$. Of course, if there are predictors in $\mathcal{P}$ that are not too costly and are stabilizing, they will be activated if the system strays too far from the steady state. In this case the system just wanders around in a neighborhood of the steady state. We illustrate the typical behavior in a two-predictor case below.

When $\mathcal{P}$ contains only two predictors, much more can be said. We shall explain a typical result. Agents can either buy at small but positive information costs C a sophisticated predictor H_1 (for example rational expectations or long-memory predictors such as "Ljung-type" predictors or freely obtain another simple predictor H_2 (e.g., adaptive, short-memory or naïve expectations). Agents use a discrete choice model to make a boundedly rational decision between predictors and tend to choose the predictor which yields the smallest prediction error or the highest net profit.

Suppose that if all agents use the sophisticated predictor H_1 all time paths of the endogenous variables, say prices, would converge to a unique *stable* equilibrium steady state, whereas when all agents would use the simple predictor H_2 the same unique equilibrium steady state would occur, but this time it would be *unstable*. In the cobweb production setting, instability under naïve expectations tends to occur when the elasticity of supply is larger than the elasticity of demand.

Consider an initial state where prices are close to the steady state value and almost all agents use the simple predictor. Then prices will diverge from their steady state value and the prediction error from predictor H_2 will increase. As a result, the number of agents who are willing to pay some information costs to get the predictor H_1 increases. When the intensity of choice to switch between the two beliefs is high, as soon as the net profit associated to predictor H_1 is higher than the net profit associated to H_2, almost all agents will switch to H_1.

Prices are then pushed back toward their steady state value and remain there for a while. With prices close to their steady state value, the prediction error corresponding to predictor H_2 becomes small again whereas net profit corresponding to predictor H_2 becomes negative because of the information costs. When the intensity of choice is high, most agents will switch their beliefs to predictor H_2 again, and the story repeats.

There is thus one "centripetal force" of "far-from-equilibrium" negative feedback when most agents use the sophisticated predictor and another "centrifugal force" of "near-equilibrium" positive feedback when all agents use the simple predictor. The interaction between these two opposing forces results in a very complicated Adaptive Equilibrium Dynamics when the intensity of choice to switch beliefs is high. Local instability and irregular dynamics may thus be a feature of a fully rational notion of equilibrium.

The Brock and Hommes[18] paper makes this intuitive description rigorous and shows that the conflict between the "far-from-equilibrium stabilizing" and "near-equilibrium destabilizing" forces generates a *near homoclinic tangency*. A homoclinic tangency is created when the increase of a parameter, e.g., the intensity of choice β causes the unstable manifold of a steady state to pass from nonintersection, to tangency, to intersection of the stable manifold of that steady state. This situation is associated with complicated dynamical phenomena. Brock and Hommes[18] show that near homoclinic tangencies are typically generated if β is large enough. This generates phases where the market is close to equilibrium which are punctuated by bursts of instability which are quickly quashed by a mass of producers who now find it worthwhile to pay the costs of more accurate prediction.

Brock and Hommes[19] apply a similar methodology to adaptive belief evolution and has begun the daunting task of building a taxinomy of "universality classes" of "routes to complex dynamics" in such systems. One could view this type of modeling as the analytic "support" to the computer based artificial economic life work of Arthur et al.[7]

The discussion of the role of bounded rationality in creating complex dynamics has been theoretical. However, this kind of work suggests tractible econometric settings where traditional rational expectations can be nested within a model where some fraction of the agents choose rational expectations and the rest of the agents choose a form of boundedly rational expectations. In this setting, if one assumes the fraction of each type is constant, one can estimate the fraction of each type as well as other parameters of the model on data and set up a statistical test of the significance of departures from rational expectations. This is especially tractible in linear quadratic stochastic settings.

Brock and Hommes[18] suggested this type of research strategy. Baak[8] and Chavas[23] have carried out estimation and testing for cattle and pork data, respectively. They find evidence that purely rational expectations models are not adequate to describe the data. We hope to do some econometric work in finance that nests rational expectations models within heterogeneous expectations models in order to shed light on what kinds of belief systems are consistent with the data.

For example, a Ph.D. student of ours, Kim-Sung Sau, has done some preliminary work that suggests evidence consistent with the presence of a type of boundedly rational expectation whose emergence is stimulated by positive runs or negative runs in returns.

Turn now to a short summary and conclusion statement.

3. SUMMARY AND CONCLUSIONS

We started out this chapter with a list of facts that financial theorizing should attempt to explain. We surveyed some theoretical work which goes part way to explaining some of the facts, especially the ones concerning the typical shape of autocorrelation functions and cross autocorrelation functions of returns, volatility of returns, and trading volume. We cited surveys that cover theoretic work that contributes to explanation of some of the other facts. We discussed the facts in enough detail so that the reader can appreciate the caution one needs to display while interpreting evidence from financial datasets.

We sketched some examples of adaptive belief system modeling which we believe have promise into developing into theory which can shed light on the economic mechanisms lying behind the facts we have listed. We consider the approach to modeling that was sketched here to be in the style of complex systems approach of the Santa Fe Institute which stresses Complex Adaptive System modeling. While much remains to be done, a lot of progress has been made since 1988.

ACKNOWLEDGMENTS

William A. Brock is grateful to the NSF (SBR-9422670) and to the Vilas Trust for essential financial support. He is grateful to his friends Steven Durlauf and Blake LeBaron for many hours of discussions on the contents of this chapter, on science in general, and on things complex. He alone is responsible for the views expressed and for any errors in this paper.

REFERENCES

1. Abhyankar, A., L. Copeland, and W. Wong. "Nonlinear Dynamics in Real-Time Equity Market Indices: Evidence from the UK." *Econ. J.* (1995): forthcoming.
2. Abhyankar, A. "Linear and Nonlinear Granger Causality: Evidence from the FT-SE 100 Index Futures and Cash Markets." Discussion Paper No. 94/07, Department of Accountancy and Finance, University of Stirling, Scotland, January, 1994.
3. Altug, S., and P. Labadie. *Dynamic Choice and Asset Markets.* New York: Academic Press, 1994.
4. Antoniewicz, R. "A Causal Relationship Between Stock Prices and Volume." Division of Research and Statistics, Board of Governors, U.S. Federal Reserve System, Washington, DC, 1992.
5. Antoniewicz, R. "Relative Volume and Subsequent Stock Price Movements." Division of Research and Statistics, Board of Governors, U.S. Federal Reserve System, Washington, DC, 1993.
6. Araujo, A., and A. Sandroni. "On the Convergence to Rational Expectations when Markets are Complete." Instituto De Mathematica Pura E Aplicada, Estrada Dona Castorina, 110 22460, Rio de Janiero, RJ, Brazil, 1994.
7. Arthur, W. B., J. Holland, B. LeBaron, R. Palmer, and P. Taylor. "Artificial Economic Life: A Simple Model of a Stockmarket." *Physica D* **75** (1994): 264–274.
8. Baak, S. "Bounded Rationality: An Application to the Cattle Cycle." Department of Economics, University of Wisconsin, Madison, 1995.
9. Baek, E., and W. Brock. "A Nonparametric Test for Temporal Dependence in a Vector of Time Series." *Statistica Sinica* **42** (1992): 137–156.
10. Blume, L., and D. Easley. "Evolution and Market Behavior." *J. Econ. Theory* **58** (1992): 9–40.
11. Bollerslev, T., R. Engle, and D. Nelson. "ARCH Models." In *Handbook of Econometrics*, edited by R. Engle and D. McFadden, Vol. IV. Amsterdam: North Holland, 1994.
12. Brock, W. "Pathways to Randomness in the Economy: Emergent Nonlinearity and Chaos in Economics and Finance." *Estudios Economicos* **8(1)** January-June (1993): 3–55. Also, Working Paper 93-02-006, Santa Fe Institute, Santa Fe, NM, 1993, and Working Paper #9302, SSRI Department of Economics, University of Wisconsin, Madison, WI 1993.
13. Brock, W. "Scaling in Economics: A Reader's Guide." Department of Economics, University of Wisconsin, Madison, WI, 1993.
14. Brock, W., W. Dechert, and J. Scheinkman. "A Test for Independence Based Upon the Correlation Dimension." Revised version with LeBaron (1991). Department of Economics, University of Wisconsin, Madison; University of Houston; and University of Chicago, 1987. *Econ. Rev.* **15** (1996): 197–235.

15. Brock, W., and P. de Lima. "Nonlinear Time Series, Complexity Theory, and Finance." In *Handbook Of Statistics 12: Finance*, edited by G. S. Maddala, H. Rao, and H. Vinod. Amsterdam: North Holland, 1995.
16. Brock, W., and S. Durlauf. "Discrete Choice with Social Interactions." Department of Economics, University of Wisconsin, Madison, WI, 1995.
17. Brock, W., D. Hsieh, and B. LeBaron. *Nonlinear Dynamics, Chaos, and Instability: Statistical Theory and Economic Evidence.* Cambridge, MA: MIT Press, 1991.
18. Brock W., and C. Hommes. "Rational Routes to Randomness." SSRI Working Paper #9505 and Working Paper #95-03-029, Santa Fe Institute, Santa Fe, NM 1995.
19. Brock, W., and C. Hommes. "Rational Routes to Randomness: The Case of the Stock Market." University of Wisconsin, Madison, WI, 1991; and University of Amsterdam, 1995.
20. Brock, W., J. Lakonishok, and B. LeBaron. "Simple Technical Trading Rules and the Stochastic Properties of Stock Returns." *J. Finance* **47** (1992): 1731–1764.
21. Brock, W., and B. LeBaron. "A Dynamic Structural Model for Stock Return Volatility and Trading Volume." *Rev. Econ. & Stat.* **78** (1996): 94–110.
22. Campbell, J., S. Grossman, and J. Wang. "Trading Volume and Serial Correlation in Stock Returns." *Quart. J. Econ.* **108** (1993): 905–939.
23. Chavas, J. "On the Economic Rationality of Market Participants: The Case of Expectations in the U.S. Pork Market." Department of Agricultural Economics, University of Wisconsin, Madison, 1995.
24. Conrad, J., A. Hameed, and C. Niden. "Volume and Autocovariances in Short-Horizon Individual Security Returns." University of North Carolina Business School, Raleigh, NC and College of Business Administration, Notre Dame University, South Bend, IN, 1992.
25. Casti, J. *Searching For Certainty: What Scientists Can Know About the Future.* New York: William Morrow, 1990.
26. Corcella, K. "Market Prediction Turns the Tables on Random Walk." *Wall Street & Tech.* **12(13)** (1995): 37–40.
27. Creedy, J., and V. Martin, eds. *Chaos and Nonlinear Models in Economics: Theory and Applications.* Aldershot, UK: Edward Elgar, 1994.
28. Diebold, F., and J. Nason. "Nonparametric Exchange Rate Prediction?" *J. Intl. Econ.* **28** (1990): 315–332.
29. Dechert, W., J. Sprott, and D. Albers. "On the Frequency of Chaotic Dynamics in Large Economies," Department of Economics, University of Houston, TX, and Department of Physics, University of Wisconsin, Madison, WI, 1995.
30. de Fontnouvelle, P. "Stock Returns and Trading Volume: A Structural Explanation of the Empirical Regularities." Department of Economics, Iowa State University, Ames, Iowa, 1994.

31. de Fontnouvelle, P. "Informational Strategies in Financial Markets: The Implications for Volatility and Trading Volume Dynamics." Department of Economics, Iowa State University, Ames, Iowa, 1995.
32. Doyon, B., B. Cessac, M. Quoy, and M. Samuelides. "Control of the Transition to Chaos in Neural Networks with Random Connectivity." *Intl. J. Bif. & Chaos* **3(2)** (1993): 279–291.
33. Engle, R., and D. McFadden, eds. *Handbook of Econometrics*, Vol. IV. Amsterdam: North Holland, 1994.
34. Feller, W. *An Introduction to Probability Theory and Its Applications*, Vol. II, 2nd Ed. New York: Wiley, 1966.
35. Gallant, R., P. Rossi, and G. Tauchen. "Nonlinear Dynamic Structures." *Econometrica* **61** (1993): 871–907.
36. Goodhart, C., and M. O'Hara. "High Frequency Data in Financial Markets: Issues and Applications." London School of Economics, London, UK and Johnson Graduate School of Management, Cornell University, Ithaca, New York, 1995.
37. Guillaume, D., M. Dacorogna, R. Dave', U. Muller, R. Olson, and O. Pictet. "From the Bird's Eye to the Microscope: A Survey of New Stylized Facts of the Intra-Daily Foreign Exchange Markets." Zurich: Olsen & Assoc., 1994.
38. Hector, G. "What Makes Stock Prices Move?" *Fortune* (1988): 69–76.
39. Hiemstra, C., and J. Jones. "Testing for Linear and Nonlinear Granger Causality in the Stock-Volume Relation." *J. Fin.* **49** (1994): 1639–1664.
40. Hodges, S. "Arbitrage in a Fractal Brownian Motion Market." Financial Options Research Centre, University of Warwick, March 1995.
41. Hogg, T., B. Huberman, and M. Youssefmir. "The Instability of Markets." Dynamics of Computation Group, Xerox Palo Alto Research Center, 1995.
42. Hsieh, D. "Chaos and Nonlinear Dynamics: Applications to Financial Markets." *J. Fin.* **47(5)** (1991): 1145–1189.
43. Hsieh, D. "Using Nonlinear Methods to Search for Risk Premia in Currency Futures." *J. Intl. Econ.* **35** (1993): 113–132.
44. Kleidon, A., and R. Whaley. "One Market? Stocks, Futures, and Options During October 1987." *J. Fin.* **47** (1992): 851–877.
45. Kurz, M. "Asset Prices with Rational Beliefs." *Econ. Theor.* (1996): forthcoming. Also, Department of Economics, Stanford, Stanford, CA, 1992.
46. LeBaron, B. "Some Relations Between Volatility and Serial Correlations in Stock Returns." *J. Bus.* **65** (1992): 199–219.
47. LeBaron, B. "Persistence of the Dow Jones Index on Rising Volume." Working Paper #9201, SSRI Department of Economics, University of Wisconsin, Madison, WI 1992.
48. LeBaron, B. "Chaos and Nonlinear Forecastability in Economics and Finance." *Phil. Trans. Roy. Soc. London A* **348** (1994): 397–404.
49. LeBaron, B. "Technical Trading Rule Profitability and Foreign Exchange Intervention." Working Paper #9445, SSRI Department of Economics, University of Wisconsin, Madison, WI 1994.

50. Lo, A., and C. MacKinlay. "When are Contrarian Profits Due to Stock Market Overreaction?" *Rev. Fin. Stud.* **3(2)** (1990): 175–205.
51. Loretan, M., and P. Phillips. "Testing the Covariance Stationarity of Heavy-Tailed Time Series: An Overview of the Theory with Applications to Several Financial Datasets." *J. Emp. Fin.* **1** (1993): 211–248.
52. Maddala, G., C. Rao, and H. Vinod, eds. *Handbook of Statistics, Volume 11: Econometrics.* New York: North Holland, 1993.
53. Maddala, G., and C. Rao. *Handbook of Statistics, Volume 14: Statistical Methods in Finance.* New York: North Holland, 1996.
54. Manski, C. "Identification Problems in the Social Sciences." In *Sociological Methodology*, edited by P. Marsden, Vol. 23, Cambridge, UK: Basil Blackwell, 1993.
55. Manski, C., and D. McFadden. *Structural Analysis of Discrete Data With Econometric Applications.* Cambridge, MA: MIT Press, 1981.
56. Mantegna, R., and H. Stanley. "Scaling in the Dynamics of an Economic Index." *Nature* **376** July (1995): 46–49.
57. Marimon, R. "Learning from Learning in Economics." European University Institute and Universitat Pompeu Fabra, 1995.
58. Meese, R., and A. Rose. "Nonlinear, Nonparametric, Nonessential Exchange Rate Estimation." *Amer. Econ. Rev.* **80(2)** (1990): 192–196.
59. Nerlove, M., and I. Fornari. "Quasi-Rational Expectations, an Alternative to Fully Rational Expectations: An Application to U.S. Beef Cattle Supply." Department of Agricultural and Resource Economics, University of Maryland, College Park, MD, 1993.
60. Potter, S. "A Nonlinear Approach to U.S. GNP." *J. App. Econ.* **10** (1995): 109–125.
61. Ridley, M. "The Mathematics of Markets: A Survey of the Frontiers of Finance." *The London Economist*, October 9–15, 1993.
62. Ross, S. "Stock Market Indices." In *The New Palgrave Dictionary of Money and Finance*, edited by P. Newman, M. Milgate, and J. Eatwell, 582–588. London: MacMillan Press, Ltd. 1992.
63. Sandroni, A. "Do Markets Favor Agents Able to Make Accurate Predictions?" Department of Economics, University of Pennsylvania, 1995.
64. Stanley, M., L. Amaral, S. Buldyrev, S. Havlin, H. Leschhorn, P. Maass, M. Salinger, and H. Stanley. "Zipf Plots and the Size Distribution of Firms." *Econ. Lett.* **49** (1995): 453–457.
65. Theiler, J., S. Eubank, A. Longtin, B. Galdrikian, and J. Farmer. "Testing for Nonlinearity in Time Series: The Method of Surrogate Data." *Physica D* **58** (1992): 77–94.

Professor Lawrence E. Blume
Department of Economics, Uris Hall, Cornell University, Ithaca, NY 14853;
E-mail: LB19@CORNELL.EDU

Population Games

1. INTRODUCTION

In most economists' view, aggregate economic activity results from the interactions of many small economic agents pursuing diverse interests. To some, this statement is a normative proposition. It immediately brings to mind Adam Smith's famous metaphor of the invisible hand, an expression of the normative superiority of markets as a resource allocation device. But even among those economists who hold a more cautious view of the superiority of markets, methodological individualism—the emphasis on the primacy of individual agents—has had an enormous impact on research into aggregate economic phenomena. For instance, it has led to the wholesale scrapping of those parts of macroeconomics lacking adequate "microfoundations," and to the acceptance of the "new classical macroeconomics," search theories of macroeconomic activity and other microbased models. Nowadays a model of aggregate economic activity is only acceptable if it can be derived by aggregation from acceptable models of the behavior of individual agents.

Methodological individualism has reached a peculiar extreme in modern economic analysis. The great achievement of general equilibrium theory is that, given perfectly competitive behavior, equilibrium price levels can be determined without reference to the institutional details of market organization. Similarly, the inductive view of noncooperative game theory abstracts from all details of how players interact. In this view, equilibrium describes mutually consistent beliefs among players. The nature of this consistency is the subject of the common-knowledge literature. The paradigmatic strategic interaction has two undergraduates sitting in separate closed rooms, their only link a computer screen, and so forth. The analysis puts enormous emphasis on the decision problem, but employs a model of player interaction which is irrelevant to the study of social organization.

One significant consequence of the lack of specificity in the modeling of how agents interact with one another is the indeterminacy of equilibrium behavior. In general equilibrium analysis this indeterminacy is the subject of a celebrated result, the Debreu-Mantel-Mas Colell-Sonnenschein Theorem, which states that in an exchange economy with more people than goods, any closed set of positive price vectors can be an equilibrium price set for some specification of preferences. Indeterminacy is even more problematic in general equilibrium theory with incomplete markets. Here given preference and endowment data the models can predict a continuum of equilibria. And in game theory, practitioners know that most interesting games come with an embarrassingly large set of equilibria, even after all of one's favorite refinements have been exercised.

Recent developments in so-called evolutionary game theory offer a new methodological approach to understanding the relationship between individual and aggregate behavior. These models, drawing inspiration from similar models in population genetics, utilize only vague descriptions of individual behavior, but bring into sharp focus the aggregate population behavior. Attention shifts from the fine points of individual-level decision theory to dynamics of agent interaction. Consequently the analysis is fairly robust to descriptions of individual behavior, but depends crucially on the underlying institutions and norms that govern interactions. Collectively, these models have come to be known as *population games.*

The idea of population games is as old as Nash equilibrium itself. Literally! In his unpublished Ph.D. dissertation, John Nash writes:[1]

> It is unnecessary to assume that the participants have full knowledge of the total structure of the game, or the ability and inclination to go through any complex reasoning processes. But the participants are supposed to accumulate empirical information on the relative advantages of the various pure strategies at their disposal.
>
> To be more detailed, we assume that there is a population (in the sense of statistics) of participants for each position of the game. Let us also assume that the 'average playing' of the game involves n participants selected at random from the n populations, and that there is a stable average frequency frequency with which each pure strategy is employed by the 'average member' of the appropriate population.

[1]Nash,[26] pp. 21–23, quoted in Weibull.[31]

> ...
>
> Thus the assumption we made in this 'mass-action' interpretation lead to the conclusion that the mixed strategies representing the average behavior in each of the populations form an equilibrium point.
>
> ...
>
> Actually, of course, we can only expect some sort of approximate equilibrium, since the information, its utilization, and the stability of the average frequencies will be imperfect.

Nash's remarks suggest that he saw the mass-action justification as providing a rationale for mixed strategy equilibrium. Only recently have scholars attended to those processes which could serve as the source of the mass-action interpretation. Research has demonstrated the expected result, that the relationship of population games to Nash equilibrium is more complex than Nash's comments suggest.

This chapter will discuss that research which has attended to the dynamics of systems continually subject to random perturbations, or noise. Dean Foster and H. Peyton Young[16] introduced the study of these systems to evolutionary game theory, and also the concept of a *stochastically stable state.* A stochastically stable state is a set of states minimal with respect to the property that the fraction of time the system spends in the set is bounded away from zero as the random perturbations become vanishingly small. Both the Foster and Young paper and Drew Fudenberg and Christopher Harris[17] introduce noise to the aggregate process. David Canning,[9] Lawrence Blume,[4] Michihiro Kandori, George Mailath, and Raphael Rob[18] (KMR) and Young[32] (Young) introduce noise directly into individual decision-making.

The early papers that actually identify stochastically stable sets get their strongest results in two-by-two games. They demonstrate, under differing assumptions, the stochastic stability of the risk-dominant equilibria in coordination games. Kandori and Rob[19] extend the KMR analysis to K-by-K coordination games and supermodular games. Young[33] applies the techniques of his[32] paper to a "divide the dollar" bargaining game. He shows that for a particular class of learning processes, the unique stochastic outcome is a generalization of the Nash bargaining solution. Georg Noldeke and Larry Samuelson[25] apply related techniques to the problem of two populations of players from which players are matched to play a signaling game.

Many different models of player interaction, models of can be studied with population game techniques. The forerunner of the entire population games literature is Thomas Schellings's[29] analysis of segregation in housing markets. This chapter presents a "local interaction" model; that each player directly competes with only a few players in the population. Blume[4] developed the asymptotic analysis and selection results for local interaction models, and Glenn Ellison[13] compared the sample path behavior of local and global interaction models applied to two-by-two coordination games. Blume[5] studied the behavior of local interaction in particular classes of $K \times K$ games with pure best-response dynamics; that is, without noise.

More recently, other topics have been studied, including local interaction with endogenous partner choice, forward-looking behavior, and variable and endogenous population size. It is almost certainly too early for a comprehensive survey of the literature, and none will be attempted here. Instead, this paper surveys some of the important questions addressed in the literature, with the particular view of pushing toward economic applications.

Left out of this survey is the related literature on deterministic dynamics in a continuum population of players. These models are frequently justified as some kind of large-numbers limit of stochastic population games. But this connection, discussed in section 4.1, is frequently ill-founded.

2. THE BASIC MODEL

This section sets up a basic population model. Later sections will specialize this model in different ways. The motivation for the story comes from KMR and Young, but here a modeling strategy is introduced which makes the usual limit analysis transparent.

2.1 DESCRIPTION

Begin with a K-player normal-form game, where player k has I_k pure strategies. Let S denote the set of pure strategy profiles, and let $G_k : S \to \mathbf{R}$ denote player k's payoff function. Suppose that there is a population consisting of N players each of K different types, each type corresponding to a player in the original game.

A *strategy revision process* $\{p_t\}_{t\geq 0}$ tracks the number of players choosing each strategy. The state space for the process is

$$\mathcal{N} = \prod_{k=1}^{K} \left\{ (M_1, \ldots, M_{I_k}) : \sum_{i=1}^{I_k} m_i = N \right\}$$

The vector $n = (n_{11}, \ldots, n_{KI_K})$ lists the number of players playing each strategy. Each state encodes the distribution of strategies played by each type. A Nash equilibrium state, for instance, is a state in which the distribution of strategies played by the types is a (mixed strategy) Nash equilibrium.

At random times τ_{ni} a *strategy revision opportunity* arrives for player n. The stochastically independent times $\tau_{ni} - \tau_{n\,i-1}$ between her ith and i – 1st strategy revision opportunities are independently distributed exponentially with parameter σ. In between times, she may not revise her choice. So when player n makes a choice at time τ_{ni}, she is *locked in* to that choice for some random period until the next revision opportunity arrives.

When a strategy revision opportunity arrives, the lucky player will choose according to the immediate expected return given her expectations about play. It is through the process of expectation generation that players' behavior are coupled together. For instance, in the KMR model and throughout this paper except where stated, expectations are just the empirical distribution of current play. Young has a more complicated random expectation model. The underlying assumption in all these models is the *myopia* of players.players, myopia of Whatever their source, expectations are applied to the payoff matrix to compute immediate expected payoffs. That is, decisions are based on immediate returns and not on other, perhaps more long-run desiderata. When there are multiple best responses, the player is equally likely to choose any one of them.

Myopic best-response and lock-in do not completely describe the evolution of choice. The third component of the model is *noisy behavior*. We assume that at each strategy revision opportunity, with some probability, the player fails to realize her best response. A parameter ϵ will describe the probability that the player selects the best response. That is, at a strategy revision opportunity a player of type k will realize a choice drawn from the distribution $b_k(n,\epsilon)$, where $\lim_{\epsilon\to 0} b_k(n,\epsilon) = \mathrm{br}_k(n)$, the uniform distribution on the best responses of the player given her expectations in state N, and $\lim_{\epsilon\to\infty} b_k(n,\epsilon)$ is a given completely mixed distribution on the I_k strategies of type k. The choice distribution $b_k(n,\epsilon)$ is assumed to be continuous in ϵ for each n. Two such rules are

$$b_k(n,\epsilon) = (1-\epsilon)\mathrm{br}_k(n) + \epsilon q \qquad 0 \le \epsilon \le 1 \tag{2.1}$$

and

$$b_k(n,\epsilon) = \frac{1}{Z}\left(\exp \epsilon^{-1} \sum_{s_{-k}\in I_{-k}} G_k(s_{-k}, s_k)\,\mathrm{pr}_n(s_{-k})\right)_{s_k=1}^{I_k}, \tag{2.2}$$

where Z is a normalizing constant and pr_n is the empirical distribution of joint strategies of players of types other than k. The first rule is that studied by KMR and, with important modifications, Young. I shall refer to this as the "mistakes model" of choice, because players make mistakes with a fixed probability ϵ. The second rule, the log-linear choice rule, was introduced in Blume[4] for the study of local interaction models. This model of stochastic choice has a long history in psychology and economics, dating back to Thurstone.[30] Several arguments have been put forth to justify the assumption of random choice: Random experimentation, random preferences and, most important, unmodeled bounded rationality.

Because the probability that two or more players have a strategy revision opportunity at the same instant is zero, the state can advance or decline by at most 1 unit at any step. The process $\{p_t\}_{t\ge 0}$ is a K-type birth-death process, whose long-run behavior is easy to study. Birth-death processes are characterized by a set of parameters that describe the rate at which the process switches from one state to the next. For single-type birth-death processes these parameters are called birth- and death-rates. Let λ_{nm} denote the transition rate from state n to state m. For all positive ϵ the process is irreducible, hence ergodic, and the invariant distribution ρ^ϵ is characterized by the *balance conditions*

$$\sum_m \lambda_{nm}\rho^\epsilon(n) - \lambda_{mn}\rho^\epsilon(m) = 0\,. \tag{2.3}$$

For all single-type birth-death processes and for a special class of multitype birth-death processes known as *reversible* birth-death processes, the invariant distribution is characterized by the *detailed balance conditions*

$$\lambda_{nm}\rho^\epsilon(n) - \lambda_{mn}\rho^\epsilon(m) = 0 \tag{2.4}$$

which directly determine the odds-ratios of the invariant distribution.

Most of the λ_{nm} are 0. They are non-zero only for n-m pairs such that m can be reached from n by a player of some type, say k, switching from some strategy l to another strategy l'. In these cases the λ_{nm} can be computed as follows: Strategy revision opportunities arrive to each player at rate σ, so they arrive collectively to the group of type k players playing l at rate σn_{kl}. The probability that such a player will switch to l' is $b_k(n,\epsilon)(l')$. Consequently,

$$\lambda_{nm} = \sigma n_{kl} b_k(n,\epsilon)(l')\,.$$

It is hard to solve for ρ^ϵ analytically from the balance conditions (2.3), but ρ^ϵ is accessible by computation. The detailed balance conditions (2.4) can easily be solved for ρ^ϵ in those models where they apply.

Another version of this model has only one type, and the players in the single population play a symmetric game against each other. In this version, as before, the state space is the number of players playing each strategy. When players have only two choices, knowledge of the number of players playing one strategy is enough to infer the number of players playing the other, and so the state can be summarized by the number of players playing strategy 1. The resulting process will be a single-type birth-death process, and the invariant distribution is characterized by the detailed balance conditions (2.4).

2.2 BEST RESPONSE DYNAMICS

When $\epsilon = 0$, a transition from state n to state m where a k player switches from l to l' is possible only if l' is a best response for players of type k in state n. The $\epsilon = 0$ process is stochastic because of the random order of strategy revision opportunities, but it need not be irreducible. In particular, if a game has two or more strict Nash equilibria the process cannot be irreducible. In this case the state space can be partitioned into a finite number of sets $Q_0, Q_1, \ldots, Q_L$ such that Q_0 is the set of transient states, and the Q_l, $l \geq 1$ are the *communication classes* of the processes. That is, for any state in Q_0 the set of paths that remain in Q_0 has probability 0. For $l > 0$ there is a path with positive probability between any two states in each Q_l, and there are no paths out of any Q_l. The communication classes provide an ergodic decomposition of the state space $\mathcal{N}$. The process $\{p_t\}_{t\geq 0}$ with initial condition in any communication class Q_l is irreducible on Q_l, and hence, will have a unique invariant distribution ρ_l with support Q_l. The set of invariant distributions is the convex hull of the ρ_l. It follows immediately from the balance conditions (2.3) that if ρ is an invariant distribution for the $\epsilon = 0$ process and $\rho(n) > 0$, then $\rho(m) > 0$ for all other states m in the same communication class as n.

The important questions at this stage concern the relationship between the supports of the communication classes and the payoff structure of the game. In the following game, $K = 2$. This game has two Nash equilibrium Nash equilibria: The strict equilibrium (D, L) and the mixed equilibrium where the players play D and L, respectively, with probability 0, and put equal weight on the remaining two strategies.

	R	C	L
U	1,3	2,2	0,0
M	2,2	1,3	0,0
D	0,0	0,0	1,1

Example 1.

There are two communication classes. One is a singleton; the state in which all row players play D and all column players play L. The other is the set of states in which no row player plays D and no column player plays L. All states not in one of these two communication classes are transient. Notice that the first communication class describes a Nash equilibrium. This illustrates a connection between singleton communication classes and Nash equilibria. A *strict Nash equilibrium* is a Nash equilibrium in pure strategies wherein each player has a unique best response.

PROPOSITION 2.1: A state n is a singleton communication class if and only if it is a strict Nash equilibrium state.

PROOF 2.1 Suppose that n is a singleton communication class. All players of each type must be best-responding to the play of the other types; otherwise there is another state m with $\lambda_{nm} > 0$ and the communication class is not a singleton. Similarly, each type must have a unique best response. If players of a given type k are all choosing a given action i and action j is also a best response, there is a path from n to any other state in which the distribution of play for types other than k is that of n, some players of type k are choosing i and the remaining players are choosing j.

Conversely, suppose that $(i_1, \ldots, i_K)$ is a strict Nash equilibrium, and consider the state n such that $n_{ki_k} = N$. Any change of strategy for any player will leave that player worse off, so $\lambda_{nm} = 0$ for all $m \neq n$. Thus $\{n\}$ is a singleton communication class. □

It is possible for all communication classes to be singletons. The *best-reply graph* of a game G is a directed graph whose vertices are the pure strategy profiles. There is an arrow from s to s' if there is one player i such that $s'_i \neq s_i$ is a best-response for player i to s_{-i} and $s'_j = s_j$ for all other j. A *sink* of the graph is a state from which no arrow exits. Clearly a sink is a strict Nash equilibrium. The graph is *acyclic* if there are no paths that exit and later return to any state s. Then every path from any state s leads to a sink. The graph is *weakly acyclic* if from every state there is a path leading to a sink. Weakly acyclic games may have best-response cycles, but there is an alternative best response from some state that leads out of the any cycle.

THEOREM 2.1: (Young) If the best-reply graph of the game is weakly acyclic, then each sink is a singleton communication class, and these are the only communication classes.

Many games, including coordination games and ordinal potential games with only strict equilibria, are weakly acyclic. □

2.3 STOCHASTIC STABILITY

When $\epsilon > 0$, the random deviations from best-response choice make every state reachable from every other state, and so the process $\{p_t\}_{t \geq 0}$ is irreducible and recurrent. Consequently it will have a unique invariant distribution, and the invariant distribution will have full support. Nonetheless, as ϵ becomes small, distinctive behavior emerges. States that are transient when $\epsilon = 0$ can only be reached by stochastic perturbations when ϵ is positive, and these will be very unlikely. It will also be easier to reach some communication classes than others, and harder to leave them. This can be studied by looking at the limiting behavior of the invariant distributions as ϵ becomes small, and this limiting behavior is quite regular. Again let ρ^ϵ denote the invariant distribution for the process $\{p_t\}_{t \geq 0}$ with parameter value ϵ.

THEOREM 2.2: If the $b_k(n, \epsilon)$ are continuous in ϵ, then $\rho^* = \lim_{\epsilon \to 0} \rho$ exists and is a probability distribution on $\mathcal{N}$. The distribution ρ^* is an invariant distribution for the process $\{p_t\}_{t \geq 0}$ with parameter $\epsilon = 0$.

The proof of this follows immediately upon noticing that the solution set to the equation system (2.3) varies upper-hemicontinuously with ϵ. □

Those states in the limit distribution ρ^* are stable in the sense that they are likely to arise when small random perturbations of best-reply choice are introduced.

DEFINITION 2.1: A state $n \in \mathcal{N}$ is *stochastically stable* if $\rho^*(n) > 0$.
Stochastic stability is a property of communication classes. If a state n in communication class Q_l is stochastically stable, then all states m in Q_l are stochastically stable.

Stochastic stability is a means of identifying communication classes that are easier to reach and those that are harder. To the extent that it sorts among communication classes, stochastic stability serves the same role in evolutionary game theory that refinements do in eductive game theory. But the motivations are quite different. Where the trembles in the conventional refinements literature exist only in the heads of the players, the trembles of evolutionary game theory are real shocks to a dynamic system, which have an effect on the system's ability to reach particular equilibria.

KMR, Young, Ellison,[14] and Evans[12] provide powerful techniques that apply to very general stochastic perturbations of discrete time, countable state dynamical systems. One advantage of the birth-death formalism is that its special structure makes the identification of stochastically stable states easier. Identifying the stochastically stable states is straightforward when the detailed balance condition (2.4) is satisfied for all ϵ, and the invariant distributions can be computed.

When only the balance conditions (2.3) are satisfied, stable states can be identified by a technique which is related to Ellison's and Evans'. The following fact is easy to prove. Let $\{x_v\}_{v=0}^{\infty}$ be a discrete time, irreducible Markov process on a countable state space, and let $\{y_u\}_{u=0}^{\infty}$ be the process that results from recording the value of the x_v process every time it hits either state 0 or state 1. It is easily seen that the $\{y_u\}_{u=0}^{\infty}$ process is Markov and irreducible. Let ρ_x and ρ_y denote the invariant distributions for the two processes. Then

$$\frac{\rho_x(1)}{\rho_x(0)} = \frac{\rho_y(1)}{\rho_y(0)}.$$

The second process is much easier to study. The transition matrix of the $\{y_u\}_{u=0}^{\infty}$ process is

$$\begin{pmatrix} 1 - f_{01} & f_{10} \\ f_{01} & 1 - f_{10} \end{pmatrix}$$

where f_{ij} is the probability of transiting from i to j without passing again through i (a so-called *taboo probability*). Thus

$$\frac{\rho_x(1)}{\rho_x(0)} = \frac{f_{01}}{f_{10}}.$$

The connection to strategy revision processes $\{p_t\}_{t\geq 0}$ goes as follows. Sample $\{p_t\}_{t\geq 0}$ at every strategy revision opportunity. Let x_v denote the value of $\{p_t\}_{t\geq 0}$ at the time of arrival of the vth strategy revision opportunity. The process $\{x_v\}_{v=1}^{\infty}$ is a discrete time Markov chain, called the *embedded chain* of the process $\{p_t\}_{t\geq 0}$.

Moreover, because the arrival rate of strategy revision opportunities is unaffected by the value of $\{p_t\}_{t\geq 0}$, the invariant distributions of ρ_x are precisely those of $\{p_t\}_{t\geq 0}$.

The problem of identifying stable states now comes down to a minimization problem. Suppose there is a path that goes from 0 to 1 with h trembles, and that at least $h + k$ trembles are required to return from 1 to 0. If the probability of a tremble is ϵ, the ratio f_{01}/f_{10} will be of order ϵ^{-k}, which converges to ∞ as ϵ becomes small. Thus, conclude that 1 is more stable than 0. Maximal elements of this "more stable than" order are the stochastically stable states. This method extends in a straightforward manner to sets. It should be clear that if 1 and 0 are in the same communication class and 1 is more stable than 2, then 0 is more stable than 2. Therefore we can talk of the stability of communication classes.

An example will illustrate how this works. Consider the game

	a	b	c
A	3,2	2,3	0,0
B	2,3	3,2	0,0
C	0,0	0,0	x,x

Example 2.

There are two populations of size N playing against each other, and x is positive. There are two communication classes, $Q_1 = \{(0,0,N),(0,0,N)\}$ with support $\{C\} \times \{c\}$ and $Q_2 = \{(L, N-L, 0), (M, N-M, 0) : 0 \leq L, M \leq N\}$ with support $\{A, B\} \times \{a, b\}$. Dynamics are of the KMR variety: each player best responds with probability $1-\epsilon$ and trembles with probability ϵ to a given completely mixed distribution.

First, find the path from Q_1 to Q_2 with the minimum number of trembles. If enough column players deviate to b, then it will pay the row players to deviate to B. Once enough of them do so, it will be in the interests of the remaining column players to switch to b. A calculation shows that B becomes a best response when fraction $x/(3+x)$ or more of the population plays b, so the number of trembles required is, up to an integer, $Nx/(3+x)$. Clearly no path from Q_1 to Q_2 requires fewer trembles. On the other hand, the easiest way to move from Q_2 to Q_1 is to start from a state where all row players play B and all column players play a. (This state can be reached from any other state in the communication class using only best responses.) If enough row players deviate to C, then it pays the rest to do so. Once enough row players have switched to C, the column players will want to switch to c. Again, the number of trembles required is approximately $2N/(2+x)$.

The singleton set Q_1 will be stochastically stable when the number of trembles needed to leave exceeds the number needed to return. A computation shows that, for large enough N that the integer problem does not interfere, the critical value of x is $\sqrt{6}$. If $x > \sqrt{6}$, then for N large enough only Q_2 is stochastically stable. If $x < \sqrt{6}$, then only Q_1 is stochastically stable.

In a number of papers, stochastic stability appears to be a kind of equilibrium refinement, selecting, for instance, the risk-dominant equilibrium in a two-by-two coordination game. Samuelson[28] demonstrates that stochastically stable sets can include states where all players use weakly dominated strategies. And as the previous example demonstrates, stochastically stable sets may contain states which correspond to no Nash equilibria. The singleton set Q_2 contains a Nash equilibrium state. Assuming N is even, the state $\big((N/2, N/2, 0), (N/2, N/2, 0)\big) \in Q_1$ is a Nash equilibrium state. But no other state in Q_1 corresponds to a Nash equilibrium.

3. EXAMPLES

The canonical examples of this emerging literature are the results on the stochastic stability of risk-dominant play in coordination games, due to KMR and Young. This section works out a birth-death version of the KMR model, a selection model somewhat in the spirit of Young, a general analysis for potential games, and a local interaction model.

3.1 THE KMR MODEL

In the KMR model, expectations held by each player are given by the empirical distribution of play at the moment of the strategy revision opportunity, and players' choices follow the mistakes model (2.1). That is, if $p_t > p^*$ the player will choose a, if $p_t < p^*$ the player will choose b, and if $p_t = p^*$ the player will randomize in some given fashion. It summarizes the effects of a presumed *interaction model.* At discrete random moments pairs of players are briefly matched, and the payoff to each player from the match is determined by the two players' choices and the payoff matrix. All matches are equally likely, and players do not know with whom their next match will be.

Let M^* denote the largest integer m such that $M^*/N < p^*$. For all $M \leq M^*$ a player's best response is to choose a. It will soon be apparent that without any essential loss of generality we can assume that $M^*/N < p^* < M^* + 1/N$, so that the possibility of indifference never arises. For all $M > M^*$ a player's best response is to choose a. For the KMR model, the birth and death rates are

$$\lambda_M = \begin{cases} \sigma(N-M)\epsilon q_a & \text{if } M < M^*, \\ \sigma(N-M)\big((1-\epsilon) + \epsilon q_a\big) & \text{if } M \geq M^*, \end{cases}$$

$$\mu_M = \begin{cases} \sigma M\big((1-\epsilon) + \epsilon q_b\big) & \text{if } M \leq M^*, \\ \sigma M \epsilon q_b & \text{if } M > M^*. \end{cases}$$

To see this, consider the birth rate for $M < M^*$. A birth happens when a b player switches to a. There are $N - M$ such players, so the arrival rate of strategy revision opportunities to the collection of b players is $\sigma(N-M)$. When a player gets a strategy revision opportunity in state M, b is her best response. Consequently, she will switch to a only by error, which happens with probability ϵq_a. Thus, the birth rate $\sigma(N-M)\epsilon q_a$. The other cases are argued similarly.

The invariant distribution is

$$\log\frac{\rho^\epsilon(M)}{\rho^\epsilon(0)} = \begin{cases} \log\binom{N}{M} - M\log\dfrac{1-\epsilon+\epsilon q_b}{\epsilon q_a} & \text{if } M \le M^*, \\ \log\binom{N}{M} + (M-2M^*)\log\dfrac{1-\epsilon+\epsilon q_b}{\epsilon q_a} & \text{if } M > M^*. \end{cases}$$

When $\epsilon = 1$, the dynamics are governed entirely by the random term. On average, fraction q_a of the population will be choosing a at any point in time. Furthermore, the least likely states are the full coordination states.

As ϵ decreases, the invariant distribution changes in a systematic way. The hump in the middle begins to shrink and shift one way or the other, and the probabilities at the two ends begin to grow. At some point they become local modes of the distribution, and finally they are the only local modes. The probability function is U-shaped. I call this process *equilibrium emergence.*

Equilibrium selection describes the behavior of the invariant distribution for very small ϵ. As ϵ continues to shrink, one of the local modes grows at the expense of the other; and as ϵ converges to 0, the invariant distribution converges to point mass at one of the full coordination states. In the birth-death version of the KMR model, the odds ratio of the two states is

$$\log\frac{\rho^\epsilon(N)}{\rho^\epsilon(0)} \approx 2\left(\frac{1}{2}-p^*\right)\log\frac{1-\epsilon+\epsilon q_b}{\epsilon q_a},$$

which converges to $+\infty$ as $\epsilon \to 0$ since a is risk-dominant. Notice that the selection criterion is independent of the error distribution q. No matter how large q_b, so long as it is less than 1, full coordination at a is selected for. At $q_b = 1$ there is a dramatic change in behavior. When $q_b = 1$ the invariant distribution converges to point mass at state 0 as ϵ shrinks, because all states greater than p^* are transient.

Equilibrium emergence and selection is illustrated in the four graphs above (Figure 1(a)–(d)). In this game, a is risk dominant and the threshold probability is $p^* = 0.4$. When the tremble probability ϵ is 1, the invariant distribution is the binomial distribution with sample size n and success probability q_a. As ϵ falls, the valley with bottom at the threshold state p^*n quickly appears. When the tremble probability falls to 17%, the right and left hump are approximately the same height. By the time the tremble probability reaches 10%, the mode on the right, at 93, is 20,000 times more likely than the mode on the left, at 2. By the time the tremble probability is 1%, the local modes are 0 and 100, and the former occurs with infinitesimal probability.

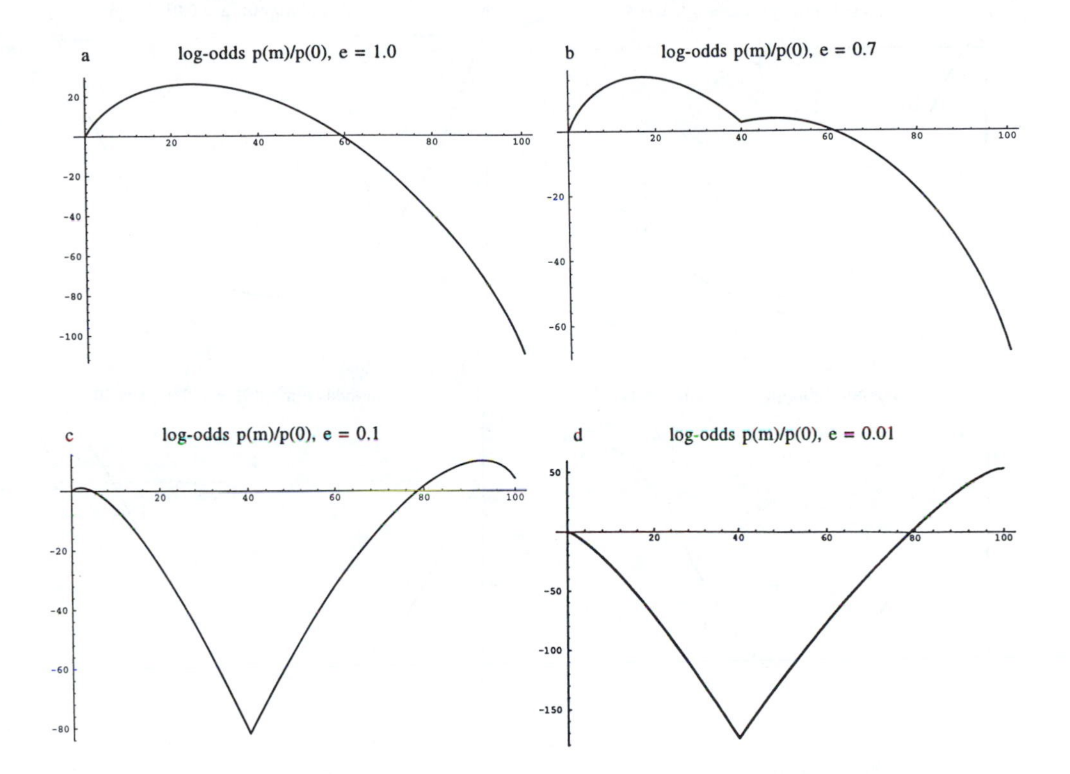

FIGURE 1 Equilibrium emergence and selection.

3.2 NOISY SELECTION MODELS

Young's model has a more sophisticated expectations generation mechanism. Each player has a memory of decisions from the immediate past. Each player samples that memory, and best responds to the empirical distribution of the sample. Thus, expectations are stochastic, because different samples can be drawn from the same memory. Binmore, Samuelson, and Vaughn[3] have called such models "noisy selection models." In the KMR model, in any state there is one action which is a best response, and the other action can arise only because of a tremble. In Young's model and other noisy selection models, any action can be a best response except in a very few absorbing states.

Developing that model here would take too much space. A simpler story in similar spirit goes as follows. A player receiving a decision opportunity when the state of the process is M draws a random sample of size K from a binomial distribution with probability M/N. If the number of successes exceeds p^*K she chooses

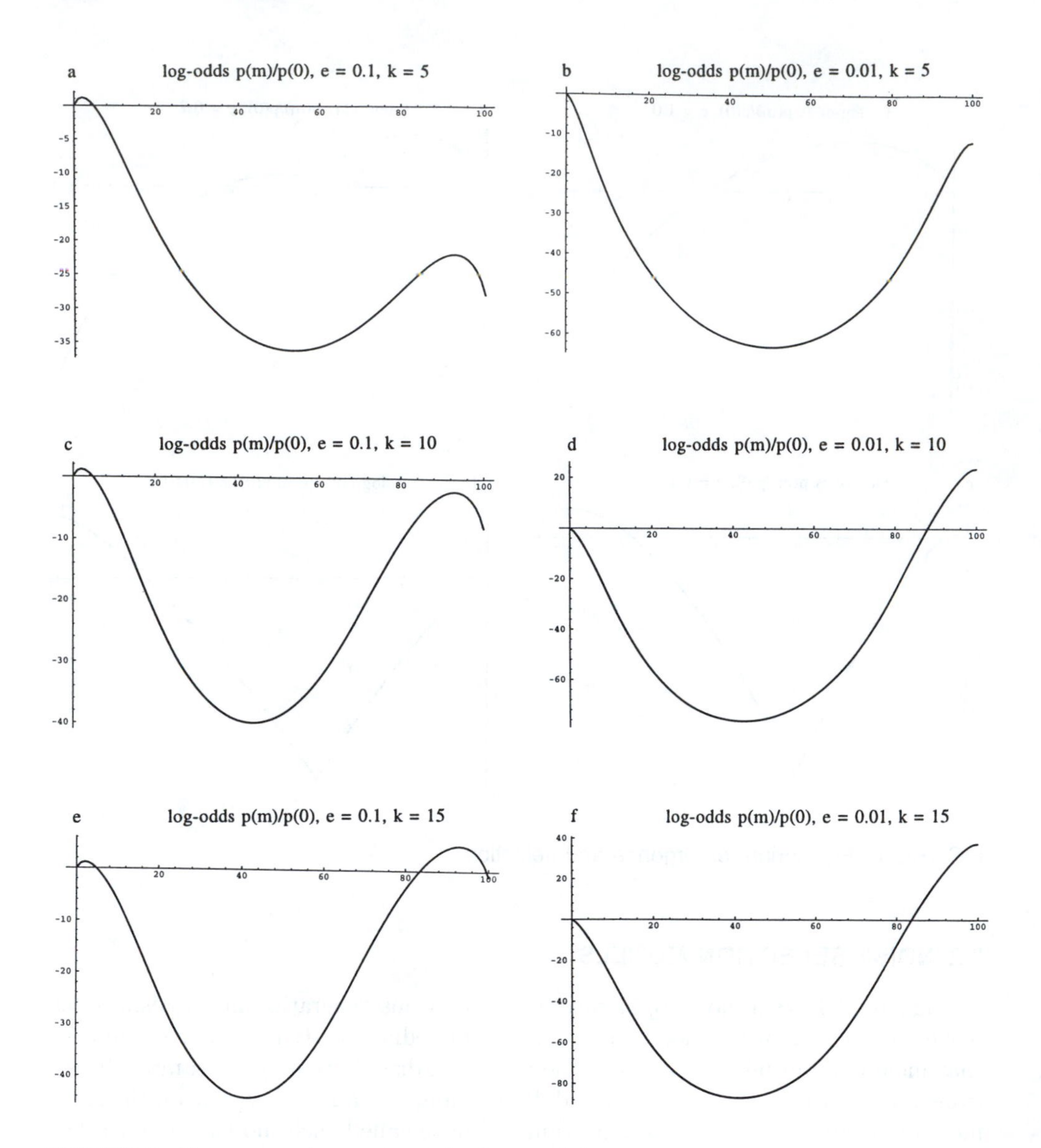

FIGURE 2 Emergence in a noisy selection model.

a, otherwise b. Thus, expectations are random, and the mean expectation is M/N. When K is large, this model behaves like the KMR model. But when K is small, the noise in expectations has a considerable effect on the long-run dynamics. This can be seen in the series of pictures above (Figure 2(a)–(f)). On the left $\epsilon = 0.1$, while on the right $\epsilon = 0.01$. The game and all other parameters are the same as before.

3.3 POTENTIAL GAMES

Potential games are games in which all payoff differences to all players can be described by first differences of a single real-valued function. Nash equilibria appear as local maxima of this potential function, and consequently the dynamics of adjustment to equilibrium are easier to study for potential games than they are for general games. Special as this structure might seem, it turns out that many important economic models have potentials. The analysis of potential games is due to Monderer and Shapley.[24]

Consider a K-player normal-form game with a set S of pure strategy profiles and payoff functions $G_k : S \to \mathbf{R}$.

DEFINITION 3.1 A *potential* for a K-player game is a function $P : S \to \mathbf{R}$ such that for each player k, strategy profiles s and strategies s'_k for player k,

$$G_k(s) - G_k(s'_k, s_{-k}) = P(s) - P(s'_k, s_{-k})\,.$$

Potential games have many special properties for particular dynamics. For instance, fictitious play converges for potential functions. Here we will see that the strategy revision process with the log-linear choice rule satisfies the detailed balance equations if and only if the game has a potential.

If the detailed balance conditions (2.4) are met, the ratios of the transition rates define odds ratios. That is, if the transition rates from n to m and m to n are positive,

$$\frac{\rho(m)}{\rho(n)} = \frac{\lambda_{nm}}{\lambda_{mn}}\,.$$

Imagine now a closed path of states $n, m_1, \ldots, m_J, n$, with positive transition rates between adjacent states in both directions. Since

$$\frac{\rho(m_1)}{\rho(n)}\frac{\rho(m_2)}{\rho(m_1)}\cdots\frac{\rho(n)}{\rho(m_J)} = 1\,.$$

It follows that if the detailed balance conditions are satisfied,

$$\frac{\lambda_{nm_1}}{\lambda_{m_1n}}\frac{\lambda_{m_1m_2}}{\lambda_{m_2m_1}}\cdots\frac{\lambda_{m_Jn}}{\lambda_{nm_J}} = 1\,.$$

Conversely, it can be shown that if this condition, known as Kolmogorov's condition, is satisfied, then Eq. (2.4) has a solution.

Checking Kolmogorov's condition for all closed paths would be impossible. But in fact it suffices to check it only for four-cycles, because all closed paths can be put together out of four-cycles. Let n denote a given state. A four-cycle is the following kind of closed path: (i) a player of type k switches from s_k to s'_k; (ii) a player of type l switches from s_l to s'_l; (iii) a type k player switches from s'_k back to s_k; (iv) a type l player switches from s'_l back to s_l. (Reversing the order of the return trip breaks up the four-cycle into two two-cycles, and the conditions are always satisfied for two-cycles.) For a four-cycle Kolmogorov's condition is

$$\frac{b_k(n,\epsilon)(s'_k)}{b_k(n-e_{ks_k}+e_{ks'_k},\epsilon)(s_k)}\,\frac{b_l(n-e_{ks_k}+e_{ks'_k},\epsilon)(s'_l)}{b_l(n-e_{ks_k}+e_{ks'_k}-e_{ls_l}+e_{ls'_l},\epsilon)(s_l)}$$
$$\frac{b_k(n-e_{ks_k}+e_{ks'_k}-e_{ls_l}+e_{ls'_l},\epsilon)(s'_k)}{b_k(n-e_{ls_l}+e_{ls'_l},\epsilon)(s_k)}\,\frac{b_l(n-e_{ls_l}+e_{ls'_l},\epsilon)(s_l)}{b_l(n,\epsilon)(s_l)}=1\,.$$

If players' choice behavior is described by the log-linear choice rule (2.2), a calculation shows that the left-hand side of the preceding equation is

$$\begin{aligned}\exp\frac{1}{e}\{P(n-e_{ks_k}+e_{ks'_k})-P(n)+P(n-e_{ks_k}+e_{ks'_k}-e_{ls_l}+e_{ls'_l})\\ -P(n-e_{ks_k}+e_{ks'_k})+P(n-e_{ls_l}+e_{ls'_l})\\ -P(n-e_{ks_k}+e_{ks'_k}-e_{ls_l}+e_{ls'_l})+P(n)\\ -P(n-e_{ls_l}+e_{ls'_l})\}=1\end{aligned}$$

so the condition is satisfied.

Since the invariant distribution is characterized by the detailed balance condition (2.4), it is straightforward to compute odds ratios of various states simply by chaining together the detailed balance conditions along a path. The general relationship is

$$\frac{\rho^\epsilon(m)}{\rho^\epsilon(n)}=C_{nm}\exp\frac{1}{\epsilon}\{P(m)-P(n)\}$$

where C_{nm} is a positive constant depending only on n and m, and not on ϵ. Fix a reference state n_0. The invariant distribution is of the form

$$\rho^\epsilon(n)=\frac{C_{n_0n}\exp\frac{1}{\epsilon}P(n)}{\sum_m C_{n_0m}\exp\frac{1}{\epsilon}P(m)}\,.$$

There is a general equilibrium selection result for potential games, corresponding to the risk-dominant equilibrium selection result for two-by-two coordination games. As ϵ becomes small, the mass of the invariant distribution concentrates on the set of global maxima of the potential function

$$\lim_{\epsilon\to0}\rho^\epsilon\{\mathrm{argmax}P(n)\}=1\,.$$

All symmetric two-by-two games are potential games. For coordination games, the potential maximum is at the state where the players coordinate on the risk-dominant equilibrium. Thus there is a connection between this result and those of KMR and Young. KMR and Young introduce trembles in the same way but use different best-response rules. This analysis of potential games uses the same best-response rule as KMR, but introduces trembles in a different fashion. Nonetheless, all three analyses find the same selection Theorem, that the unique stochastically stable set is the state in which all players play the risk-dominant strategy.

3.4 A LOCAL INTERACTION MODEL

The previous models have each player interacting with the entire population. Another important class of models has each player interacting only with a subset of the population—her neighbors. Although each player interacts directly with only a few others, a chain of direct interactions connects each player indirectly with every other player in the population. Imagine a population of players, each interacting with her neighbors in a coordination game. Initially, different geographic regions of the populace are playing different strategies. At the edge of these regions the different strategies compete against one another. Which strategies can survive in the long run? In the long run, can distinct regions play in distinct ways, or is heterogeneity unsustainable?

Suppose that players are interacting in a $K \times K$ symmetric game with payoff matrix G. Let I denote the set of K pure strategies. Given is a finite graph. Vertices represent players, and the edges connect neighbors. Let V denote the set of players (vertices), and V_v the set of neighbors of player v. A configuration of the population is a map $\phi : V \to I$ that describes the current choice of each player. If the play of the population is described by the configuration ϕ, then the instantaneous payoff to player v from adopting strategy s is $\sum_{u \in V_v} G(s, \phi(u))$.

A strategy revision process is a representation of the stochastic evolution of strategic choice, a process $\{\phi_t\}_{t \geq 0}$. The details of construction of the process are similar to those of the global interaction models. At random times τ_{v_i} player v has a strategy revision opportunity, and $\tau_{vi} - \tau_{vi-1}$, the ith interarrival time, is distributed exponentially with parameter σ and is independent of all other interarrival times. At a strategy revision opportunity, players will adopt a new strategy according to rules such as Eq. (2.1) or Eq. (2.2), but here, best responses are only to the play of v's neighbors.

One would think that the behavior of stochastic strategy revision processes for local interaction models would be profoundly affected by the geometry of the graph. As we shall see, this is certainly the case for some choice rules. But when the game G has a potential, the basic result of the previous section still applies. If G has a potential $P(s, s')$, then the "lattice game" has a potential

$$\tilde{P}(\phi) = \frac{1}{2} \sum_{v} \sum_{u \in V_v} P\big(\phi(v), \phi(u)\big).$$

To see that $\tilde{P}$ is indeed a potential for the game on the lattice, compute the potential difference between two configurations which differ only in the play of one player. Fix a configuration ϕ. Let ϕ_v^s denote the new configuration created by assigning s to v and letting everyone else play as specified by ϕ. Then

$$\tilde{P}(\phi_v^{s'}) - \tilde{P}(\phi_v^s) = \frac{1}{2} \sum_{u \in V_v} P\big(s', \phi(u)\big) - P\big(s, \phi(u)\big) + P\big(\phi(u), s'\big) - P\big(\phi(u), s\big)$$

$$= \sum_{u \in V_v} P(s', \phi(u)) - P(s, \phi(u))$$

$$= \sum_{u \in V_v} G(\phi(u), s') - G(\phi(u), s) ,$$

which is the payoff difference for v from switching from s to s'. The middle inequality follows from the symmetry of the two-person game.

If choice is described by the log-linear strategy rule (2.2), then as the parameter ϵ becomes small, the invariant distribution puts increasing weight on the configuration that maximizes the potential. The invariant distribution is

$$\rho^\epsilon(\phi) = \frac{\exp - \frac{1}{\epsilon} \tilde{P}(\phi)}{Z(\epsilon)}$$

where $Z(\epsilon)$ is a normalizing constant.

Two-person coordination games are potential games, and the strategy profile that maximizes the potential has both players playing the risk-dominant strategy. Consequently the configuration that maximizes $\tilde{P}$ has each player playing the risk-dominant strategy. More generally, for any symmetric K-by-K potential game, if there is a strategy t such that $\max_{u,v} P(u, v) = P(t, t)$, then the configuration $\phi(v) \equiv t$ maximizes $\tilde{P}$, and so strategy t is selected for as ϵ becomes small. This is true without regard to the geometry of the graph.

This independence result depends upon the game G. A two-by-two symmetric "anti-coordination game" has a potential, but it is maximized only when the two players are playing different strategies. Now the set of maxima of $\tilde{P}$ depends on the shape of the graph. Consider the following two graphs

and suppose coordinating pays off 0 and failing to coordinate pays off 1. In the first graph $\tilde{P}$ is maximized by a configuration which had the four corners playing identically, but different from that of the player in the center. In the second graph $\tilde{P}$ is maximized only by configurations that have the three corners alternating their play.

4. ANALYTICAL ASPECTS OF THE MODELS

Both the short-run and the long-run behavior of population models are important objects of study. Long-run behavior is characterized by the process' ergodic distribution. When the population is large, the short-run dynamics are nearly deterministic, and can be described by the differential equation which describes the evolution of mean behavior. A fundamental feature of these models, but all too often ignored in applications, is that the asymptotic behavior of the short-run deterministic approximation need have no connection to the asymptotic behavior of the stochastic population process. A classic example of this arises in the KMR model applied to a two-by-two coordination game.

4.1 THE SHORT-RUN DETERMINISTIC APPROXIMATION

Begin with the KMR model described above, and rescale time so that when the population is of size N, strategy revision opportunities arise at rate σ/N. The mean direction of motion of $\{p_t\}_{t\geq 0}$ depends upon the state. Suppose $p_t < p^*$. The process advances when a b player trembles in favor of a, and it declines when an a player either best responds or trembles in favor of b. Thus, the probability of an advance by $1/N$ is $(1-p)\epsilon q_a$, and that of a decline is $p(1-\epsilon+\epsilon q_b)$, and so the expected motion is $-p_t + \epsilon q_a$. Similar reasoning on the other side of p^* gives an expected motion of $1 - p_t + \epsilon q_b$. When N is large, this suggests the following differential equation for mean motion of the system.

$$\dot{p}_t = \begin{cases} -p_t + \epsilon q_a & \text{if } p_t < p^*, \\ 1 - p_t - \epsilon q_b & \text{if } p_t > p^*. \end{cases}$$

A simple strong-law argument suggests that this should be a good local approximation to the behavior of sample paths for large N. In fact, it can be made uniformly good over bounded time intervals. The following Theorem is well-known in population genetics (Ethier and Kurtz[15]). An elementary proof can be found in Binmore, Samuelson, and Vaughn.[3] Fix $p_0 = p(0)$ and let $p(t)$ denote the solution to the differential equation for the given boundary value.

THEOREM 4.1 For all $T < \infty$ and $\delta > 0$, if N is large enough then $\sup_{0\leq t\leq T} ||p(t) - p_t|| < \delta$ almost surely.

When $\epsilon = 1$, the equation has a unique globally stable steady state at $p = q_a$. But when ϵ is small, p^* divides the domain into basins of attraction for two stable steady states, at ϵq_a and $1 - \epsilon q_b$.□

Differential equations like this arise in the literature on deterministic population models with a continuum player population. Similar equations are derived for equilibrium search and sorting models. (See, for example, Diamond.[11]) But the problem with using these equations to study long-run behavior can be seen in this example. The asymptotic analysis of the differential equation does not approximate the long-run behavior of the stochastic process. For small ϵ, the process spends most of its time near only one equilibrium state—that where players coordinate on the risk-dominant equilibrium. In general, it is true that the support of the limit invariant distribution taken as $\epsilon \to 0$ is a subset of the set of stable steady states, but the two sets are not identical.

Other stochastic approximations have also been studied. Binmore, Samuelson, and Vaughn[3] construct a degenerate diffusion approximation by the discrete-time version of expanding the action of the birth-death generator on a function into a Taylor series and dropping the terms of order greater than 2. The resulting approximation is degenerate because as N grows the diffusion coefficient converges to 0. For all finite N, the diffusion approximation has a unique invariant distribution which converges to a point mass in the limit. Examples show, however, that this point mass can be concentrated on the wrong equilibrium state; that is, coordination on the risk-dominated equilibrium. They fail to predict the asymptotic behavior of the stochastic population process for the same reason the differential equations fail. Nonetheless, they can be useful for making probabilistic statements about the magnitude of the error in the short-run differential approximation.

4.2 HOW LONG IS THE LONG RUN?

The equilibrium selection theorems for population games are all asymptotic results. They describe properties of the long-run distribution of states. For economic analysis the long run may be too long. Thus, it is also important to have an idea of how long it might take to reach a particular state. First-passage time distributions are useful for describing the relevant aspects of sample path behavior. Consider first one-dimensional models such as the KMR model or the noisy selection model in a two-by-two coordination game with a single population. Suppose strategy a is risk-dominant. The equilibrium selection result says that when ϵ is small, the invariant distribution puts most of its mass on the state N in which all players use a. The first-passage time from state M to 1 is $\tau(M) = \inf_t\{p_t = N; p_0 = M\}$. The first moment $\nu^1(M)$ of the first-passage time distributions is given by the following difference equation:

$$\nu^1(M) = \frac{1 + \lambda(M)\nu^1(M+1) + \mu(M)\nu^1(M-1)}{\lambda(M) + \nu(M)}.$$

For a small population of 25, the log of the expected first-passage time to state 1 from initial state M (on the horizontal axis) is plotted in the following graphs for $\epsilon = 0.1$ and $\epsilon = 0.01$ (Figure 3(a) and (b), respectively). For these plots, $p^* = 0.4$, so the boundary between the two basins of attraction is at $M = 10$.

Graphs (a) and (b) show that once the process enters the basin of attraction of a given equilibria, it will move to that equilibrium rather quickly. But should it leave that basin of attraction, a long time will pass before the process returns. A better way of scaling time is by the number of strategy revision opportunities required for the transit. Strategy revision opportunities come at rate 1 for each individual—at

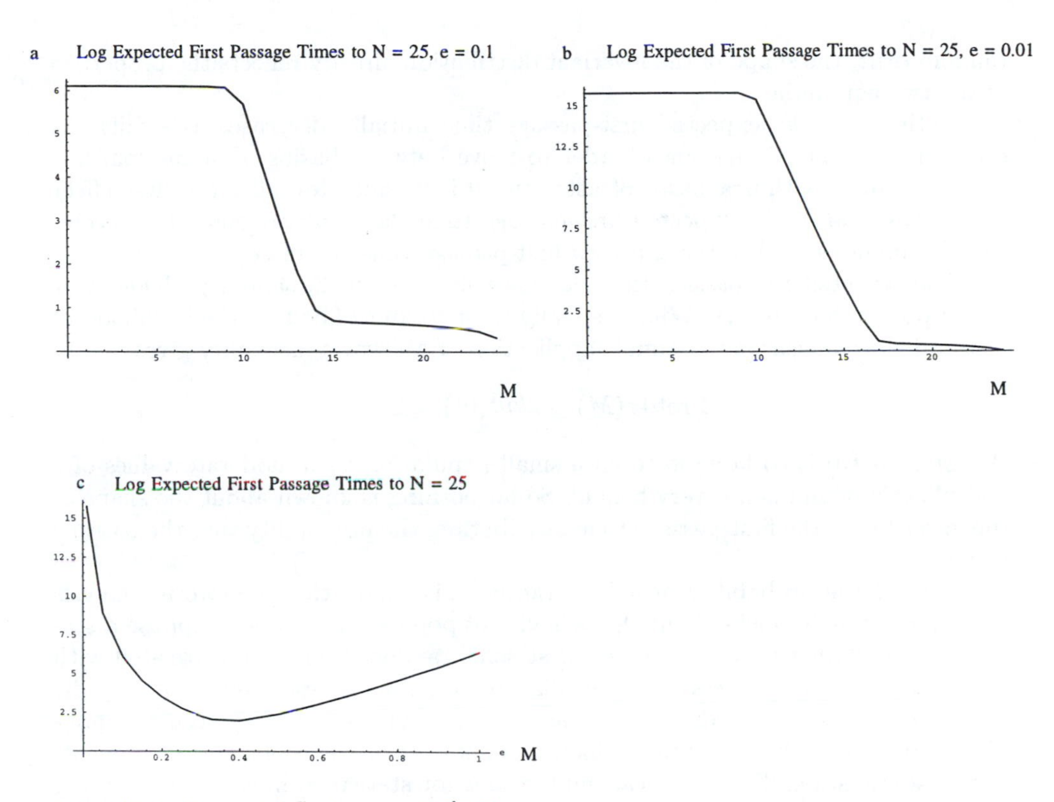

FIGURE 3 Expected first passage times.

rate 25 for the entire population. Thus, when $\epsilon = 0.1$, for the process to move from state 0 to state 25 requires, on average, on the order of 10^6 strategy revision opportunities for each individual in the population, or 10^7 revision opportunities for the entire population. These orders increase to 10^{15} and 10^{16}, respectively, when $\epsilon = 0.01$.

The usual complaint about stochastic perturbations is that when perturbations are small, the expected first-passage time from one equilibrium state to the other is so small that the asymptotic analysis, including the selection results, is irrelevant. Graph (c) above plots expected first-passage times from 0 to 25 as a function of the tremble probability ϵ. The striking fact illustrated in this graph is that, to the extent that the waiting times are viewed as being long, this is a function of the number of intermediate states that need to be traversed and not a consequence of the improbability of trembles. In this example, $\epsilon = 1$ represents pure random drift; at a strategy revision opportunity a move in either direction is equally likely. The expected waiting time for $\epsilon = 1$ is comparable to that of $\epsilon = 0.1$. For $N = 25$ and $\epsilon = 0.1$, the mode of the invariant distribution is at $p = 0.84$, and $\rho(25)/\rho(0) \approx 2400$, so at an ϵ value for which the expected first-passage time is comparable to that for random drift, the shape of the invariant distribution already reflects the properties of the two equilibria.

Notice that the expected first-passage time initially decreases as ϵ falls. As ϵ becomes smaller, it becomes harder to move between basins of attraction, but easier to move within a basin of attraction to its sink. Initially, the first effect dominates, and so the expected first-passage times fall. Subsequently, the second effect dominates, and so the expected first-passage times increase.

The expected first-passage time alone says little about the sample path behavior of the population process. What we really want to know about is the likelihood of fast and slow transitions. A simple application of Markov's inequality gives

$$\text{Prob}\{\tau(M) \geq 2k\nu^1(0)\} \leq 2^{-k}.$$

But since $\nu^1(0)$ is so large for even a small population with moderate values of ϵ and p^*, this bound is not very helpful. So far nothing is known about the shape of the lower tail of the first-passage time distribution; the probability that the passage is quick.

Even if the probability of a fast transition is small, the asymptotic analysis tells us meaningful facts about the behavior of population process. Suppose a stationary population process is observed at some random time τ, uncorrelated with the evolution of the process. Then the distribution of p_τ is described by the ergodic distribution. In other words, an econometrician studying a naturally occurring population process with small ϵ would have to assume he or she is observing a stochastically stable state. This has clear implications for statistical inference concerning the values of payoffs, the likelihood of trembles, and other important parameters of the model.

4.3 ROBUSTNESS TO THE PERTURBATION PROCESS

Population games are intentionally vague about the details of noisy choice. This is an acceptable modeling strategy only if these details do not significantly effect the results. This question has been investigated in two recent papers, by Bergin and Lipman[2] and Blume.[6]

The equilibrium selection results of KMR and Young are shown by their respective authors to satisfy some invariance to the specification of noise. If the updating rule is the mistakes model, then both KMR and Young demonstrate risk-dominant selection regardless of the distribution q from which choice is determined in the event of a tremble, so long as q places positive weight on all strategies. The log-linear model also leads to risk-dominant equilibrium selection, although it is distinctly different from the mistakes model. But Bergin and Lipman[2] show that it is possible to introduce noise to choice in a way that leads to risk-dominated equilibrium selection. Their analysis is really much more general than this, but the following example will illustrate their point. Consider again the KMR model, but with the following change: When b is a best response, b is chosen with probability

$1-\epsilon^2$, while when a is a best response, a is chosen with probability $1-\epsilon$. So trembling away from b is ϵ times as likely as trembling away from a. A calculation like that of section 3.1 shows that

$$\frac{\rho(N)}{\rho(0)} = C\left(\frac{\epsilon^2}{1-\epsilon^2}\right)^{M^*}\left(\frac{1-\epsilon}{\epsilon}\right)^{N-M^*}$$
$$= O\left(\epsilon^{3M^*-N}\right).$$

If strategy a is risk dominant, then $2M^* < N$. But in this model, selecting strategy a requires that $3M^* < N$, that $p^* < 1/3$. In fact, by making trembling away from b ϵ^k times as likely as trembling away from a for large k, the threshold probability p^* necessary to guarantee a-selection can be made arbitrarily close to 0.

Since it is possible to change the stochastically stable states by changing the noise process, it becomes important to identify the class of noise processes which give rise to the same stochastically stable states as do the mistakes model and the log-linear choice model. This question is addressed in Blume[6] for symmetric two-by-two games. Consider the class of choice rules where the log-odds of choosing a over b depend only on the payoff difference between the two states and on a multiplicative parameter β:

$$\log\frac{\text{pr}\{a\,|\,p\}}{\text{pr}\{b\,|\,p\}} = \beta g(\Delta(p))$$

where $\Delta(p)$ is the payoff difference between a and b when fraction p of the population plays a. This includes the log-linear model, where $g(x) = x$, and the mistakes model with $q_a = q_b = 1/2$, where $g(x) = \text{sgn}(x)$. (Take $\beta = \log\big((1-\epsilon)/\epsilon\big)$.)

Suppose the choice rule has the property that only payoff differences matter. That is, the odds of choosing a over b when the payoff difference between a and b is Δ, are the same as the odds of choosing b over a when the payoff difference between b and a is Δ. Then g will be skew-symmetric. All skew-symmetric g give rise to the same stochastically stable states. Furthermore, if a given g is not skew-symmetric, then there is a game for which this g gives stochastically stable states different from that given by any skew-symmetric g.

The skew-symmetry says that names of strategies do not matter—only their payoff advantage. Although it is a natural property within this framework, this framework does not encompass models such as the mistakes model where $q_a \neq q_b$. However, the results extend to the class of choice rules

$$\log\frac{\text{pr}\{a\,|\,p\}}{\text{pr}\{b\,|\,p\}} = \beta g(\Delta(p)) + \beta r(\Delta(p),\beta)$$

where g is skew-symmetric but r is not, so long as $\limsup_\beta \beta r(\Delta(p),\beta) < \infty$. This larger class of models includes all mistakes models. (In fact, the only requirement on

r is that $\beta||r(\Delta, \beta)||$ diverge slower than $\beta g(\Delta)$ as β grows.) The lesson of these two papers is that biases in the effect of deviations from best response on the identity of stochastically stable states matter, but only if the biases are unbounded as choice converges to pure best response.

These invariance results for global interaction models are encouraging. They suggest that, within broad limits, the actual specification of noise in choice does not matter. Unfortunately the situation is not so good for local interaction models. Consider the graph on the left:

Take any two-by-two coordination game with a unique risk-dominant strategy and $p^* > 1/3$, and consider the two configurations which have all players playing identically. With either the mistakes model (2.1) or the log-linear model (2.2) no other configuration can be stochastically stable (or part of a stochastically stable set). With the mistakes model, both fully coordinated configurations are stochastically stable (singleton) sets. It takes two mutations to move from all play a to all play b, and two to move back. In the log-linear model, only the configuration in which everyone plays the risk-dominant strategy is stochastically stable, as it alone maximizes the value of the potential function $\tilde{P}(\phi)$. This is essentially an integer problem, that only a majority of ones' neighbors playing the risk-dominant strategy can make the risk-dominant strategy a best response. Links can be added to this graph to preserve the phenomenon for an arbitrarily large population of players. For the graph on the right, suppose that $p^* > 2/5$. Then again, it takes two mutations to move from one coordinated configuration to the other, so both are stochastically stable. Notice that there is only one way of moving from the dominated state to the dominant state, but eight ways of moving from the dominated state to the dominant state. Thus, the limit distribution will put more weight on "all play the dominant strategy" than on "all play the dominated strategy," but both will be present. Again the potential function $\tilde{P}(\phi)$ is maximized only by the configuration "all play the dominant strategy." How special these examples are is unknown, so the adequacy of not modeling noise is not yet determined. Notice, too, that these examples demonstrate that the invariance of the selected configuration to the geometry of the graph is not a universal property of nice selection rules. In the graphs above, the "all play the dominated strategy" configuration is stochastically stable under the mistakes model, but if more edges were added to the graphs this would no longer be the case.

5. ECONOMIC APPLICATIONS

There are four barriers to the application of population game techniques to serious economic models. The treatment of time is inadequate, different interaction models are needed, heterogeneity among agents must be considered, and the population game paradigm must be extended beyond two-player games. In this section I discuss some recent attempts to bring population game models closer to economic problems.

5.1 TIME

In the models discussed so far, decision-making is essentially static. Players form expectations about the current state of the process, and respond to those expectations, subject to noise in choice, so as to maximize their instantaneous return. This separation between choice and dynamics makes the analysis of population games models particularly simple. But economic decision-makers are typically concerned about the future as well as the present. Evaluating the future requires that each trader try to forecast the strategy revision process, and take account of these forecasts when searching for the best response at a strategy revision opportunity. If there is any connection between the forecasts and the actual behavior of the strategy revision process, such as the hypothesis that expectations are rational, then the dynamic behavior of the strategy revision process cannot be simply computed from the choice rule.

The substantive effects of allowing decision makers to consider the future are important. Some have argued that in coordination games where the risk-dominant strategy is not payoff-dominant, risk-dominant selection is an artifact of myopia. That is, players who put sufficient emphasis on the future will tend to coordinate on the payoff-dominant choice when choice is not too noisy. This question was addressed by Blume,[8] who found quite the opposite. If players are sufficiently patient, then when noise is sufficiently small the only equilibrium is to select the risk-dominant strategy at every outcome, regardless of the state of the process. The equilibrium concept employed here is called a *rational population equilibrium,* where all players correctly identify the stochastic process that describes the evolution of their opponents' play. Given this knowledge, each player's choice problem is a dynamic program.

Consider again the basic model of section 2 where, at random intervals, strategy revision opportunities arrive. That model simply specified a rule of behavior for each player; a choice of action in each state. Now we need to derive such a rule. In order to do so, we need to be specific about how players interact. The story—told, but not formally modeled—in most population games papers is the "billiard ball model" of player interaction. At random intervals, two players (billiard balls) meet (collide), and receive payoffs depending on the strategies they are currently using as described by the payoff matrix G. This can be described formally by assigning to each pair of players i, j a sequence of times $\{\tau_{ijn}\}_{n=0}^{\infty}$ such that $\tau_{ij0} = 0$, and the random variables $\tau_{ij\,n} - \tau_{ij\,n-1}$ are distributed independent of each other and of the strategy revision opportunity interarrival times, exponentially with parameter 1. The time τ_{ijn} is the time of the nth meeting of players i and j.

We shall suppose that the game is symmetric; that is, there is only one player type. Let S denote the set of K pure strategies for the game. Player's actions depend upon what their opponents are doing. Let $\Delta_{N-1} = \{M_1, \ldots, M_K : M_k \geq 0$ and $\sum_k M_k = N-1\}$. This is the state space for each player's decision problem. A *policy* is a map $\pi : \Delta_{N-1} \to P(S)$, the set of all probability distributions on S. If all opponents are employing policy π, their collective play, the state of the decision problem, evolves as a multitype birth-death process. Suppose the decision maker is currently employing strategy k. The transition rate from state m to state $m' = m - e_j + e_l$, requiring that a j player switches to l, is

$$\lambda^k_{mm'} = \sigma m_j \big((1-\epsilon)\pi_l(m - \epsilon_j + \epsilon_k) + \epsilon q_l\big) .$$

Since m_j opponents are playing j, σm_j is the arrival rate of strategy revision opportunities to this group. If our player is playing k, then the state of one of her opponents playing j is $m - e_j + e_k$. Thus $\pi_l(m - e_j + e_k)$ is the probability that the opponent playing j will want to choose l, and $(1-\epsilon)\pi_l(m - \epsilon_j + \epsilon_k) + \epsilon q_l$ is the probability that the opponent will end up playing l. Let $l^k_m = \sum_{m'} \lambda^k_{mm'}$.

Suppose that players discount the future at rate r. If all opponents are playing according to policy π, then the optimal choice for a player whose opponent process is in state m is the solution to a dynamic programming problem. That problem has the Bellman equation

$$\begin{aligned} V(k,m) =& E\Big\{ e^{-r\tilde{\tau}} \sum_{m' \neq m} \frac{\lambda^k_{mm'}}{\lambda^k_m + \sigma + 1} V(k,m') + \frac{1}{\lambda^k_m + \sigma + 1}\big(v(k,m) + V(k,m)\big) \\ &+ \frac{\sigma}{\lambda^k_a + \sigma + 1} \max\nolimits_{\pi \in P(S)} \sum_l \big((1-\epsilon)\pi_l + \epsilon q_l\big) V(l,m) \Big\} \\ =& \sum_{m' \neq m} \frac{\lambda^k_{mm'}}{r + \lambda^k_m + \sigma} V(k,m') + \frac{1}{r + \lambda^k_m + \sigma} v(k,m) \\ &+ \frac{\sigma}{r + \lambda^k_m + \sigma} \max\nolimits_{\pi \in P(S)} \sum_l \big((1-\epsilon)\pi_l + \epsilon q_l\big) V(l,m) \end{aligned} \tag{5.1}$$

where $v(k,m) = (N-1)^{-1} \sum_{l \in S} m_l G(k,l)$ is the expected value of a match with a randomly chosen opponent. A *rational population equilibrium* is a policy π such that $\pi_k(m) > 0$ implies $k \in \operatorname{argmax} V(k,m)$ when all opponents are using policy π. Among other things, Blume[8] shows that when r is large, the optimal policy is to maximize the expected payoff flow from the current state—the KMR model. More surprising, in a two-by-two coordination game with unique risk-dominant strategy a, if r and ϵ are sufficiently small, the only rational population equilibrium is $\pi_a(m) \equiv 1$.

A virtue of the birth-death formalism is that, for a given conjectured equilibrium the dynamic program is easily solved and so the equilibrium can be checked. This is really the main point of the paper, that intertemporal equilibrium analysis can be conducted within the population games framework.

5.2 INTERACTION MODELS

A criticism frequently leveled at the existing population games literature is that the "billiard ball" model of player interaction is not useful for modeling any interesting economic phenomena.[2] Fortunately, other kinds of player interaction are equally amenable to analysis. Strictly within the population game framework, Blume[6] also considers a model where players receive a continuous payoff flow from their current match. At random moments, players are matched with new opponents, and so forth. At any moment of time, each player is matched with an opponent. Consequently, the myopic behavior of this model is not the KMR model. Instead, the myopic player maximizes the return against the play of her current opponent. This leads to different asymptotic behavior. In a two-by-two coordination game there are two stochastically stable sets, each one a singleton consisting of a state in which all players coordinate. Patient play in this interaction model mimics that of billiard-ball interaction.

Still other models are within easy reach of contemporary technique. For instance, consider a model where from time to time players are matched, and later these matches break up. New matches can only be made from among the unmatched population. This matching model can be set up by assigning to each player pair two Poisson processes. When the first one advances, the player pair forms a match if both are unmatched; otherwise nothing happens. When the second one advances, if the players are matched they break up; otherwise nothing happens. Now different experiments can be performed with strategy revision opportunities by assuming that strategy revision opportunities arrive at different rates for the paired and unpaired players. Suppose they arrive at rate 0 for the unpaired players. For myopic players this hypothesis generates continuous payoff flow dynamics. On the other hand, suppose they arrive at rate 0 for paired players. This leads to KMR dynamics. Most interesting (and not yet done) is to study the intermediate regimes in order to understand the transition between voter model and KMR dynamics. I conjecture that in two-by-two coordination games all intermediate regimes have only one stochastically stable set, consisting only of the state in which all players coordinate on the risk-dominant strategy.

[2]The Menger-Kiotaki-Wright model of monetary exchange (Menger[23] and Kiyotaki and Wright[20]) and some versions of Diamond's search equilibrium model (Diamond[11]) are exceptions.

5.3 BEYOND TWO-PLAYER GAMES

More crucial than how players meet is the assumption that the interaction, however it occurs, takes the form of a single two-person game. This section will exhibit two models which depart from the conventional paradigm in different ways. The first model introduces player heterogeneity along with two-person interactions. In the second model, players choose which of two groups to join, and payoffs are determined by the size of the groups.

5.3.1 MATSUYAMA'S FASHION MODEL. The model, due to Kiminori Matsuyama,[21] provides an elegant study of the evolution of fashion. Refer to the original publication for a discussion of the motivations, modeling strategy, and implications. Matsuyama presented what can be interpreted as the large-numbers limit analysis. Here the model will simply be recast as a population model. The simple story is that there are two types of clothes, red and blue, and two types of people. Types 1s are conformists and type 2s are nonconformists. Conformists want to look like the rest of the population, and therefore, will match the dress of the majority of the population. Nonconformists want to look different, and so will match the dress of the minority of the population. The number of type i people wearing red is m_i. The total population size is n, and the size of the subpopulation of type i is n_i.

This leads to the following specification of dynamics. Transition rates are:

$$m_1 \to m_1 + 1 \text{ at rate } \sigma(n_1 - m_1)\begin{cases} \epsilon q_r & \text{if } m_1 + m_2 < n/2 \\ 1 - \epsilon + \epsilon q_r & \text{if } m_1 + m_2 > n/2 \end{cases},$$

$$m_1 \to m_1 - 1 \text{ at rate } \sigma m_1 \begin{cases} 1 - \epsilon + \epsilon q_b & \text{if } m_1 + m_2 < n/2 \\ \epsilon q_b & \text{if } m_1 + m_2 > n/2 \end{cases},$$

$$m_2 \to m_2 + 1 \text{ at rate } \sigma(n_2 - m_2)\begin{cases} 1 - \epsilon + \epsilon q_r & \text{if } m_1 + m_2 < n/2 \\ \epsilon q_r & \text{if } m_1 + m_2 > n/2 \end{cases},$$

$$m_2 \to m_2 - 1 \text{ at rate } \sigma m_2 \begin{cases} \epsilon q_b & \text{if } m_1 + m_2 < n/2 \\ 1 - \epsilon + \epsilon q_b & \text{if } m_1 + m_2 > n/2 \end{cases}.$$

Take $\sigma = 1/n$, and let θ denote the fraction of conformists. The differential equation system describing the short-run dynamics is

$$\begin{aligned} \dot{p}_1 &= -\theta(p_1 - \epsilon q_r) & &\text{if } p_1\theta + p_2(1-\theta) < 1/2, \\ \dot{p}_2 &= (1-\theta)(1 - p_2 - \epsilon q_b) & & \\ \dot{p}_1 &= \theta(1 - p_1 - \epsilon q_b) & &\text{if } p_1\theta + p_2(1-\theta) > 1/2, \\ \dot{p}_2 &= -(1-\theta)(p_2 - \epsilon q_r) & & \end{aligned}$$

where p_i is the fraction of red-dressers in the type-i population. Above the boundary line $p_1\theta + p_2(1-\theta) = 1/2$ the conformists switch to red and the nonconformists to blue. Below the boundary line they switch the other way.

When $\theta < 1/2$, evaluation of the differential equation system shows that for small ϵ the system dives into the boundary line. The slope of the two solutions at the boundary are such that the escape from the boundary cannot occur. When $\theta > 1/2$, the natural sinks near $(0, 1)$ for the regime where conformists switch to blue and the nonconformists to red, and $(0, 1)$ for the other regime, are below and above the line, respectively. (They are only "near" because of the tremble.) Thus, curves never hit the line, and the flow is toward the sinks. The two cases are illustrated in Figure 4(a) and (b). In both cases, the curves in the lower part of the boxes are moving up and to the left, and those in the upper boxes are moving down and to the right.

The asypmtotic behavior is interesting, and hinted at by the analysis of the short-run behavior. Suppose first that $\theta < 1/2$. The sinks near $(n_1, 0)$ and $(0, n_2)$ are on the "wrong" side of the line. When $\epsilon = 0$, it is easy to see that states (n_1, n_2) where $n_2 > 1/2(1 - \theta)$ and $n_2 < (1 - \theta)^{-1}(1/2 - \theta)$ are transient, and all other states are recurrent. Figure 5(a) shows the invariant distribution when $n1 = 8$ and $n2 = 24$. The probability of the mode is about 0.11.

When $\theta > 1/2$ the two sinks are absorbing states for the $\epsilon = 0$ process, and neither is favored by selection. To see this, note that the minimum number of trembles necessary to move from $(n_1, 0)$ to $(0, n_2)$ is the same as the number necessary to move from $(0, n_2)$ back to $(n_1, 0)$. This is illustrated in Figure 5(b), for $\epsilon = 0.1$. It is easy to introduce asymmetries that will lead to a selection result simply by putting in a bias in favor of one color or the other. This has the effect of shifting the line to the left or the right, increasing the number of trembles needed to move in one direction and decreasing the number of trembles needed to move in the other.

Matsuyama has also varied the interaction technology by allowing players to meet other players of the same and the other type at different rates. This increases the possibilities for interesting short-run dynamics. For some specifications of the model, fashion cycles arise. That is, the differential equations describing the large-numbers short-run behavior of the system have cycles as solutions. This illustrates

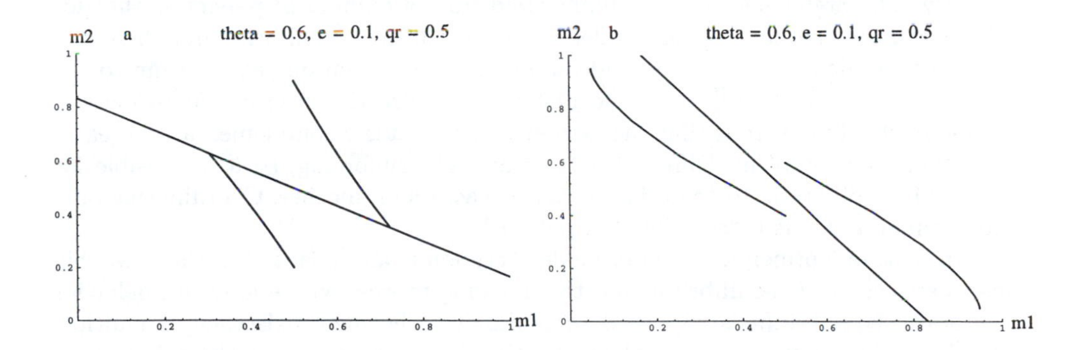

FIGURE 4 Short-run mean dynamics for the fashion model.

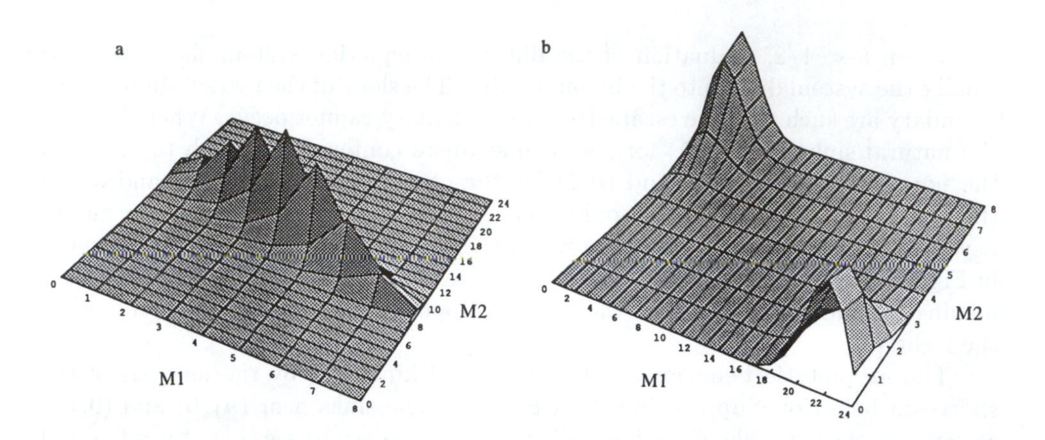

FIGURE 5 Invariant distributions for the fashion model.

once again how details of the interaction technology effect the qualitative performance of the system.

5.3.2 KEYNESIAN COORDINATION FAILURE. The study of strategic complementarities has become important throughout economics. Macroeconomics, industrial organization, and development economics are only some of the fields where problems posed in terms of strategic complementarities are drawing significant attention. Arthur's[1] contribution to the first Santa Fe volume on the economy as a complex system surveys some of the dynamic phenomena that can arise from systems exhibiting strategic complementarities.

In the macroeconomic literature on Keynesian coordination failure, the existence of strategic complementarities has been claimed as a source of cyclic economic activity. The argument, crudely summarized from a number of papers, is that for one reason or another a given model will have multiple equilibria, and therefore, cyclic economic activity can be understood as a drift from one equilibrium to the next. Cooper and John[10] were the first to recognize that the common analytic theme in this literature is that the existence of a strategic complementarity creates a coordination problem. Different actions are self-reinforcing, so it is possible for players in equilibrium to "coordinate" on one action or another. Coordinating on a suboptimal action is a "coordination failure."

In fact, one principle lesson of modern economic analysis is that there is a big leap between static equilibrium and the dynamic process whose long-run behavior the static equilibrium is supposed to represent. The mere existence of multiple equilibria due to strategic complementarities does not guarantee the existence of interesting economic dynamics. So Blume[7] investigates a population game with a strategic complementarity to see if it can generate cyclic investment behavior. The model is in the spirit of Diamond's[11] search equilibrium paper, where the strategic complementarity is due to an externality from market participation. When more traders participate in a market, the returns to market participation are higher.

Imagine, then, a population of N potential investors. Those who choose not to invest receive a fixed payoff flow. Those who invest in a project receive an expected payoff flow whose value depends upon the number of other investment projects currently active. The net return function is $W(M)$. Details of the market organization determine $W(M)$, but one can write down a number of models where $W(M)$ is initially increasing in M due to strategic complementarities in market participation. If there is an M^* such that $W(M) < 0$ for $1 \leq M < M^*$ and $W(M^*) > 0$, there will be at least two Nash equilibria of the market participation game, one where no one participates and one where some number $M \geq M^*$ invest. Decision opportunities arrive for each player at the jump times of a Poisson process, and the process for each player is independent of all other processes. When a nonparticipant has a decision opportunity she may choose to enter the market, while an investor with a decision opportunity may choose to leave the market. For the moment, suppose that the scrap value of an investment project is 0, and that there are no fixed costs to investment. Finally, suppose that decisions are based on the current return to investment; that is, investors' discount rates are sufficiently high relative to the arrival rate of investment decisions that their investment decision is determined by the return to investment given the current size of the market. Randomness in decision-making is a consequence of shocks to expectations. I assume that the resultant random utility choice model has a logit representation. In other words, the probability of investing or not (disinvesting or continuing to invest) at a decision opportunity is given by the log-linear choice rule (2.2).

This model tells a very Keynesian story of the trade cycle. Keynes viewed investment as perhaps interest-inelastic, but very responsive to expectations, and made much of the "instability" of the marginal efficiency of investment on this account. One consequence of this expectations-driven investment analysis is that different kinds of expectations, optimistic and pessimistic, can support different levels of investment of which only one corresponds to full employment. In the present model, shocks to expectations may cumulate to drive investment from one basin of attraction to the next.

The investment model departs from the previous two-player game models in that the net gain to participation is not linear in the number of market participants. This nonlinearity has no consequences for the analytical paradigm outlined above. The process describing the number of active investors is a birth-death process, and its properties can be examined by studying the birth and death rates. In particular, the asymptotic behavior of the economy is characterized by a potential function. Let

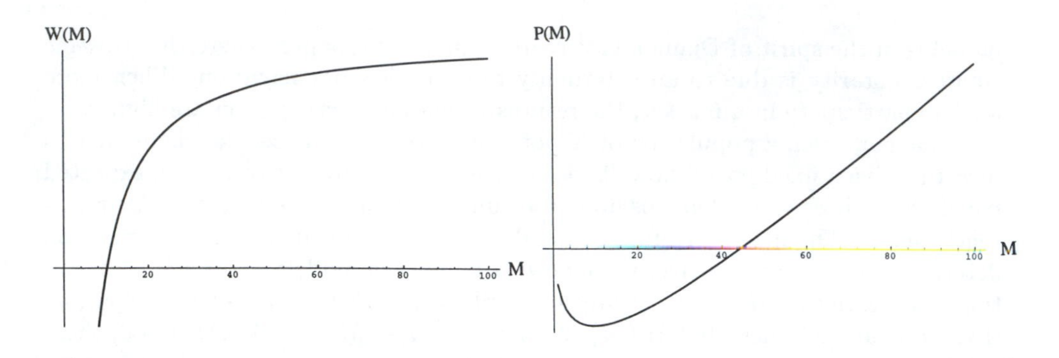

FIGURE 6 Keynesian Coordination: expected returns and potential.

$P(0) = 0$ and $P(M) = P(M-1)+W(M)$. A computation from the detailed balance conditions (2.4) shows that the invariant distribution satisfies the relationship

$$\log \frac{\rho(M)}{\rho(0)} = C(M,N) + \frac{1}{\epsilon} P(M)$$

where $1/\epsilon$ is the proportionality constant in the log-linear choice rule.

When ϵ is small, the i.i.d. investment shocks are small. The invariant distribution will be concentrated near the global maximum of the potential function. The model is illustrated in Figure 6(a) and (b). On the left (Figure 6(a)) is the expected return function $W(M)$, and the potential function $P(M)$ is on the right (Figure 6(b)). The expected return $W(M)$ exceeds 0 for all $M > 10$. Thus, viewing the model as a static investment model, if $N > 10$ there are two equilibrium states: A low-investment equilibrium at $M = 0$, and a high-investment equilibrium at $M = N$. If the i.i.d. expectations shocks are not large, the economy will spend a large fraction of time near the equilibrium of greatest potential value. The potential always has a local maximum at $M = 0$. If $N <= 10$ this is the only local maximum, and indeed there is only one equilibrium, with no active investors. When $N > 10$, there is a local maximum at $M = N$, corresponding to the high-investment equilibrium. But, for $N < 45$ the global maximum is at $M = 0$, and so the economy never really gets off the ground. Investment may occasionally spurt up, but ultimately it falls back to the ground state $M = 0$, and stays near there most of the time. For $N > 45$ the global maximum moves to $M = N$, and high-investment levels predominate.

The model can be elaborated in several ways. Consider adding a fixed cost to investment. With a fixed cost, investments that generate small positive returns will not be undertaken, but active investors in such a state will not (on average) close out their investments. Consequently, new equilibria are created, just where $W(M)$ goes positive; in this case at $M = 11$, $M = 12$, etc., depending on the magnitude of the fixed cost. These new equilibria have no significant dynamic consequences.

They are not local maxima of the appropriate potential function, and so they can never be favored as ϵ becomes small.

This model exhibits two interesting properties. First, the high productivity equilibrium can only be sustained in the long run if the pool of potential investors is sufficiently large. If the economy has not achieved a critical mass, attempts to jump-start the economy by short-run programs to stimulate investment are doomed to fail. The second interesting property speaks directly to the original purpose of the model. The mere existence of multiple equilibria is not enough to conclude that significant cyclic activity will arise. The "cyclic" behavior of the model when expectation shocks are small is nothing like that observed in the actual economy. Dynamics here look as follows: The economy spends a negligible amount of time away from the equilibrium states, and it spends vastly more time in one equilibrium state than the other. Even without going through a calibration exercise it is clear that this description of dynamics does not correspond with any current picture of business cycle activity.

6. CONCLUSION

Two views of the invisible hand appear repeatedly in economic thought: Adam Smith's positive and benign invisible hand leads the merchant "to promote an end which is no part of his intention"; an end which represents the "interests of society." Austrian economists, most notably Menger and Hayek, see that institutions themselves evolve as a result of the flux of individual actions. "What Adam Smith and even those of his followers who have most successfully developed political economy can actually be charged with is ... their defective understanding of the unintentionally created social institutions and their significance for economy."[3] There is no argument that the "spontaneous order" of the Austrians is necessarily beneficent. The invisible hand could equally well be Shakespeare's bloody and invisible hand of night as Adam Smith's hand of Pangloss.

Unfortunately, it is hard to separate the notion of spontaneous order from that of a beneficent invisible hand. Robert Nozick[27] coined the phrase "invisible hand explanation" to refer to exactly the spontaneous generation of order of which the Austrians wrote. "An invisible hand explanation explains what looks to be the product of someone's intentional design, as not being brought about by anyone's intentions."[4] This explanation suggests some sort of optimization. But his leading example of invisible hand explanations are evolutionary explanations of the traits of organisms and populations. Few biologists claim that evolution is a global optimization device.

[3] Menger,[22] p. 172.

[4] Nozick,[27] p. 19.

The population games literature clarifies how the spontaneous evolution of order can arise as the unintended social consequence of many private actions. The resultant social order is frequently consistent with some kind of optimization hypothesis—the optimization of a potential. But the optimization of a potential does not imply optimization of social welfare. As coordination games demonstrate, the potential maximum may be the Pareto inferior equilibrium.

Invisible hand phenomena of the Austrian variety abound in social systems. Although population games are a useful tool for modeling invisible hand phenomena, more work is needed to get the most from them. A variety of methods exist for uncovering stochastically stable sets, but less is known about the asymptotic behavior of these models when ϵ is not small—when significant residual noise remains in the system. Little is known about the sample-path behavior of these processes. I argued above that the mean of the first-passage time distribution is not informative about short-run behavior. Finally, application of these models to problems more complex than the choice of driving on the left or the right will require the development of computational techniques for calibration.

Further conceptual development is also needed. Population games are viewed as a branch of evolutionary game theory. This is true, to the extent that evolutionary game theory is that which is not eductive. But there is no real evolution going on in these models. The global environment established by the game and the interaction model is fixed, and players are equipped with a fixed set of rules. Some rules are better than others, and the better rules tend to be adopted. Consequently, the local environment, determined by the rules in play, changes. The population game analysis tracks the interaction between individual play and the local environment. The appropriate biological analogy is an ecological model. In population games there is no innovation. The set of rules is fixed at the start. More important, there is no variation in the global environment. It would seem that an appropriate topic of a truly evolutionary game theory is evolution in the rules of the game. Such evolution is the ultimate source of the spontaneous order to which invisible hand explanations refer.

REFERENCES

1. Arthur, W. B. "Self-Reinforcing Mechanisms in Economics." In *The Economy as an Evolving Complex System*, edited by P. W. Anderson, K. Arrow, and D. Pines, 9–31. Santa Fe Institute Studies in the Sciences of Complexity, Proc. Vol. V. Redwood City, CA: Addison-Wesley, 1988.
2. Bergin, J., and B. Lipman. "Evolution with State-Dependent Mutations." Unpublished, Queens University, 1994.
3. Binmore, K., L. Samuelson, and R. Vaughn. "Musical Chairs: Noisy Equilibrium Selection." Unpublished, University of Wisconsin, 1993.
4. Blume, L. "The Statistical Mechanics of Strategic Interaction." *Games & Econ. Behav.* **5** (1993): 387–426.
5. Blume, L. "The Statistical Mechanics of Best Response Strategy Revision." *Games & Econ. Behav.*, forthcoming. Economics Working Paper Archive ewp-game/9307001, 1993.
6. Blume, L. "How Noise Matters." Unpublished, Cornell University, Economics Working Paper Archive ewp-game/9410001, 1994.
7. Blume, L. "An Evolutionary Analysis of Keynesian Coordination Failure." Unpublished, Cornell University, 1994.
8. Blume, L. "Evolutionary Equilibrium with Forward-Looking Players." Unpublished, Cornell University, 1995.
9. Canning, D. "Average Behavior in Learning Models." *J. Econ. Theory* **57** (1992): 442–472.
10. Cooper, R., and A. John. "Coordinating Coordination Failures in Keynesian Models." *Quart. J. Econ.* **103** (1988): 441–63.
11. Diamond, P. "Aggregate Demand Management in Search Equilibrium." *J. Pol. Econ.* **90** (1982): 881–94.
12. Evans, R. "Observability, Imitation, and Cooperation in the Repeated Prisoner's Dilemma." Unpublished, Cambridge University, 1993.
13. Ellison, G. "Learning, Local Interaction, and Coordination." *Econometrica* **61** (1993): 1047–1072.
14. Ellison, G. "Basins of Attraction and Long Run Equilibria." Unpublished, MIT, Cambridge, MA, 1995.
15. Ethier, S., and T. Kurtz. *Markov Processes.* New York: Wiley, 1986.
16. Foster, D., and H. P. Young. "Stochastic Evolutionary Game Dynamics." *Theor. Pop. Biol.* **38** (1990): 219–232.
17. Fudenberg, D., and C. Harris. "Evolutionary Dynamics with Aggregate Shocks." *J. Econ. Theory* **57** (1992): 420–441.
18. Kandori, M., G. Mailath, and R. Rob. "Learning, Mutation, and Long Run Equilibria in Games." *Econometrica* **61** (1993): 29–56.
19. Kandori, M., and R. Rob. "Evolution of Equilibria in the Long Run: A General Theory and Applications." *J. Econ. Theory* **65** (1995): 383–414.

20. Kiyotaki, N., and R. Wright. "On Money as a Medium of Exchange." *J. Pol. Econ.* **92** (1989): 927–54.
21. Matsuyama, K. "Custom Versus Fashion: Path-Dependence and Limit Cycles in a Random Matching Game." Unpublished, Hoover Institution, Stanford University, Stanford, CA, 1992.
22. Menger C. *Untersuchungen über die Methode der Soczialwissenschaften und der Politischen Oekonomie inbesondere.* Leipzig: Duncker & Humboldt, 1883; English Translation, *Investigations into the Method of the Social Sciences with Special Reference to Economics.* New York: New York University Press, 1985.
23. Menger, C. "On the Origin of Money." *Econ. J.* **2** (1892): 239–255.
24. Monderer, D., and L. Shapley. "Potential Games." Unpublished, The Technion, 1993.
25. Noldeke, G., and L. Samuelson. "Learning to Signal in Markets." Unpublished, University of Wisconsin, Economics Working Paper Archive ewp-game/9410001, 1994.
26. Nash, J. "Non-Cooperative Games." Unpublished Ph.D. Thesis, Mathematics Department, Princeton University, Princeton, NJ, 1950.
27. Nozick, R. *Anarchy, State, and Utopia.* New York: Basic Books, 1974.
28. Samuelson, L. "Stochastic Stability in Games with Alternative Best Replies." *J. Econ. Theory* **64** (1994): 35–65.
29. Schelling, T. "Dynamic Models of Segregation." *J. Math. Sociol.* **1** (1971): 143–186.
30. Thurstone, L. "A Law of Comparative Judgment." *Psychol. Rev.* **34** (1927): 273–286.
31. Weibull, J. "The Mass-Action Interpretation of Nash Equilibrium." Unpublished, The Industrial Institute for Economic and Social Research, Stockholm, 1994.
32. Young, H. P. "The Evolution of Conventions." *Econometrica* **61** (1993): 57–84.
33. Young, H. P. "An Evolutionary Model of Bargaining." *J. Econ. Theor.* **59** (1993): 145–168.

Ken Kollman,* John H. Miller,† and Scott E. Page‡
*Department of Political Science and Center for Political Studies, University of Michigan, Ann Arbor, MI 48109
†Department of Social and Decision Sciences, Carnegie Mellon University, Pittsburgh, PA 15213
‡Division of Humanities and Social Sciences, California Institute of Technology 228-77, Pasadena, CA 91125

Computational Political Economy

In this chapter, we address the use of adaptive computational modeling techniques in the field of political economy. The introduction considers the advantages of computational methods. The bulk of the chapter describes two computational models: a spatial model of electoral competition and a Tiebout model. At the end of the chapter, we discuss what the future may hold for these new techniques.

1. INTRODUCTION

There has been growing interest in the use of computer experiments in scientific inquiry in recent years. Some advocates believe that computational modeling will lead to breakthroughs and fundamentally alter how we understand both the physical and social worlds.[31,32] Such bold claims are met with varying degrees of skepticism. In our view, there are reasons to be optimistic about the use of computational modeling in building theory in the social sciences, particularly in the field of political field of political economy. The potential contributions alone deserve attention: the construction of flexible theoretical models generating rich empirical predictions; the inclusion of dynamic analyses of social systems, including a theory of patterns; and the comparison of institutional structures in complex social environments. Several early, provocative accomplishments have reinforced our optimism and have led us to include these techniques in much of our research.[1] Certainly the grand promises of changes in the nature of science have not yet been fulfilled, and perhaps never will be. Nonetheless, we believe that the current state of the techniques calls for a tempered advocacy.

In political economy, computational techniques are especially valuable because they can complement and extend current theory. "Theory" is often associated with mathematics in the social sciences (especially economics), but it is important to separate the notions of what are useful theories from what are the tools, such as mathematics, that allow us to develop such theories. Useful theories generate accurate and testable predictions about interesting phenomena. Tools have various strengths and weaknesses, but they should be evaluated based on how well they help us construct useful theories. It is possible to use multiple tools.

The interplay between mathematical and computational theoretical investigations promises to be a growing research area.[32] Though often seen as substitutes, the two approaches overlap on many dimensions. A good example is that both approaches place high value on formally stated assumptions. Also, they often produce similar types of predictions. For example, in our spatial voting models (reported below), we generate predictions about final candidate positions in much the same fashion as mathematical models of electoral competition. Our predictions regarding end states are similar in certain respects to those of the current mathematical theory, even under different conditions than those previously modeled. However, this does not mean that computational and mathematical theory must have identical predictions. In our case, we find several contrasting results, which we then investigate empirically.

As a complement to mathematical theory, computational experiments in general are hardly controversial.[42] Few social scientists dispute the usefulness of generating examples and counterexamples, testing counterfactuals, and relaxing assumptions. But whether computational modeling alone can solve theoretical problems is another story. Apparent breakthroughs may turn out to be peculiarities of a technique, assumption, or algorithm which initially appears innocuous.[2] The risk of any one computational model being "a mere example" unfortunately exists. For example, Huberman and Glance[29] have found that many of the interesting spatial patterns in cellular automata models disappear when updating is asynchronous, and Page[55] has shown that when the timing of updating is incentive based, the dynamics change dramatically. Unless those using computational techniques set high standards for robustness, theoretical findings will be greeted with skepticism.

[1] See especially the work of Thomas Schelling[60] and Robert Axelrod.[3]

[2] Of course, mathematical theories run similar risks.

Nevertheless, the advantages of a purely computational approach to theory are extensive. Holland and Miller[26] argue in favor of computational modeling as a good middle ground. Prior to the advent of computational techniques, social science theory relied on one of two methodologies: mathematical or verbal. The strength of mathematical analysis stems from its rigor. However, constructing mathematical models is a slow, challenging pursuit, where assumptions must often be guided by tractability rather than realism. Verbal analysis offers more flexibility and realism at the cost of reduced certainty. James Madison and Karl Marx were unconstrained by the requirements of mathematical consistency imposed upon ideas by twentieth century theorists such as Arrow and Debreu. Accordingly, the scope of analysis and the specificity of explanations in qualitative theory are far greater. An ideal tool for social science inquiry would combine the flexibility of qualitative theory with the rigor of mathematics. But flexibility comes at the cost of rigor. Holland and Miller believe that for computational models this cost is relatively low given that the computer programs guarantee that a logically consistent analysis proceeds from the encoded assumptions. They also state that the attendant gain in flexibility is large, as computer programs can encode a wide-range of behaviors. Alternative behavioral and environmental assumptions can be included quickly and at low cost in a computational model. Assumptions need not be dictated by the abilities of the researcher to prove formal claims.

Our own recent work[39] on federal systems of government exemplifies the flexibility of computational approaches. We began to develop a theory of using states as policy laboratories by highlighting four basic characteristics of the problem: the difficulty of the policy function, the relative abilities of states and the federal government to search for innovative policies, the heterogeneity of preferences across states, and whether states could adopt new policies instantaneously or gradually.[3] None of these characteristics fits easily into a dynamic programming framework, the standard mathematical approach to modeling search for innovation.[14] Yet, in a computational model, we were able to include these features. We found unexpected interaction effects between the difficulty of policy functions and whether policies could be implemented by states instantaneously.

The second area in which computational modeling has advantages concerns dynamics. Many social phenomena are dynamic in nature. The best way to understand some social systems is to watch them unfold rather than to compare end states or equilibria (which may or may not exist). An array of computational techniques have been developed to explore and capture dynamical phenomena. In many instances, mathematical techniques are often strained, and computational techniques serve as neither complement nor substitute: they are the only alternative. Returning to our voting model example, mathematical models typically ignore the process of candidate position-taking over time, instead focusing on the final positions.[16]

[3] By difficult here, we mean a static, hard to solve problem. See Page[54] for a more complete discussion of the difference between difficult and complex.

Our computational techniques enable us to make predictions about paths taken by candidates toward equilibria.

One way to gauge the contribution of computational models to theory in political economy is to compare them with rational choice and game theory models, methodologies which have been applied for a longer period of time. Though initial findings evoked interest, game theory and rational choice modeling survived a long period of indifference by many social scientists. Yet, despite recent attacks,[22] the rational choice/game theoretic approach has to be considered a big success. It has allowed researchers to generate theoretical insights, to guide empirical investigations, and to assist in the evaluation of potential political institutions. Computational modeling has also generated provocative insights[3,4,35,53] yet presently endures a mild response. Whether it will make similar or even larger contributions to social science than rational choice modeling, or whether it will collapse under its own weight in the manner of catastrophe theory, remains to be seen. The point is that methodological movements take time. One should not expect the discipline to embrace these new techniques, or to agree upon their contributions at such an early stage.

In the next section, we describe some of our work in computational political economy focusing on two models: a spatial model of electoral competition and a Tiebout model. We conclude with a discussion of the future of computational modeling in political economy. The description of the two models demonstrates some of the advantages of computational modeling. In the spatial electoral model, we see how computational methods can extend previous mathematical results and develop new testable hypotheses of party position-taking in two-party elections. We find strong support for these hypotheses in data from American presidential elections.[4]

In the Tiebout model, we demonstrate how computational models can lead to new insights, in this case an unexpected change in institutional performance when conditions change. When citizens can relocate to other jurisdictions, unstable electoral institutions, which perform poorly in single jurisdiction models, may outperform more stable institutions. This counterintuitive finding can be explained by borrowing an idea from computational physics: annealing. In an annealing algorithm, the level of noise is decreased, or "cooled," over time to help a complex dynamical system settle into a good equilibrium. In the Tiebout model, instability together with citizen relocations function as noise. The level of noise cools over time because the districts become more homogeneous, reducing the institutional instability and the incentives to relocate.

[4] For a detailed overview of political methodology which includes a discussion of path dependency see Jackson.[31]

2. A COMPUTATIONAL AGENDA

In our own computational modeling research, we try to balance the generation of new theoretical insights with an attempt to understand better the methodology itself. The research described below on spatial voting models extends and challenges well-known theoretical models and leads to new, testable hypotheses. In developing these and other computational models, the most pressing methodological issues have concerned the modeling of adaptive, nonoptimizing behavior. We shall discuss our approach and the motivations which led to it in the following subsection.

2.0.1 MODELING ADAPTIVE PARTIES. The spatial voting models and the Tiebout model described below incorporate what we call *adaptive political parties* in place of the fully rational parties used in nearly all mathematical theories of political competition. Modeling parties as adaptive captures the many limitations on parties when they attempt to change platforms to appeal to voters. Standard rational choice models assume parties can maneuver virtually without limit in an issue space.[16,17,41] In contrast, adaptive parties are limited informationally, computationally, and spatially. We assume that adaptive parties' information about voters' preferences comes entirely from polling data. Parties do not know individual voters' utility functions. Nor do adaptive parties respond to their information optimally. They do not begin with Bayesian priors from which they update. Instead, they rely on heuristics, or rules of thumb, to navigate around the issue space in search of votes. Finally, adaptive parties are restricted in how far they can move in the policy space.

The depiction of party behavior as adaptive and incremental has several justifications. Here we consider two. First, parties, like other organizations, need to maintain credibility and keep diverse supporters aboard, so wild policy swings or rapid changes are unlikely.[7,12,63] Second, uncertainty over whether a policy change will lead to an increase or decrease in vote totals—the uncertainty might be a product of sampling or measurement error, or just the difficulty of the decision—makes parties tentative in changing their issue positions. Granted, if a party interprets polls as suggesting that a particular shift in a policy position will lead to a larger percentage of the vote, then the party will probably adopt the change. However, because parties do not know actual voters' preferences, when they change positions toward the median voter on one issue with the rest of the platform unchanged, they cannot be certain this will lead to an increase in vote totals. This uncertainty stems not so much from the fact that parties do not know the median position, but rather from the imperfect information and multidimensional nature of the issue space.

Precisely how to model formally the behavior of adaptive organizations is not straightforward. Undoubtedly the biggest criticism of computational models is that findings are sensitive to particular heuristics and parameters. While optimal choices are often unique, there are typically many choices that lead to improvement. For example, a political party may be adapting in an environment in which many platforms generate higher vote totals, though typically only one will yield a maximal number of votes. This potential multitude of good choices creates the possibility of the search rule determining the predictions a given adaptive model will generate. Findings from a particular rule of thumb, such as a steepest or first ascent method, may not extend to other rules of thumb.

To guard against nonrobust findings, we consider several decision heuristics and consider a wide range of parameters in our research. If we find similar results under all rules and many parameter values, we can be more secure that we are reporting general phenomena and not anomalies. Besides ensuring robustness, our search heuristics also satisfy two criteria. The first is that the rules are crudely accurate as descriptions of the behavior of actual parties both in the information they acquire and how they process it. More specifically, the amount, type, and quality of information should correspond to that which parties may actually obtain. To assume that parties rely on polling data is reasonable. To assume that they take gradients of utility functions is not. In addition, the parties should process their information in a realistic fashion: trying new policies to gather more information and updating their beliefs about voters. The second criterion is that the rules be widely used search algorithms with known strengths and weaknesses. In addition to having a class of problems for which it performs well, every search heuristic has an Achilles heel. Comparing performance across algorithms, we can try to discover which algorithms perform well in which environments. Also, if we amend an algorithm, such as appending an operator onto a genetic algorithm, we may introduce behavior which contaminates our findings.

2.0.2 THREE HEURISTICS. In our research, we have relied for the most part on three algorithms to model adaptive party behavior: a hill-climbing algorithm, a random search algorithm, and a genetic algorithm. A hill-climbing algorithm is a sequential search algorithm. A single platform is chosen randomly in the first election as the party's status quo point. In subsequent elections, the party's platform from the previous election becomes the status quo. The algorithm proceeds in two steps. A platform is chosen in a neighborhood of the status quo, and a poll is taken comparing this platform against the opponent. If it receives more votes than the status quo, it becomes the new status quo. Otherwise, the status quo remains. This process continues for a fixed number of iterations and then an election is held. A great deal is known about hill-climbing algorithms. For example, they perform well on functions with low levels of difficulty and poorly on very difficult functions.[54] Metaphorically, we can interpret the hill-climbing algorithm as a party that selects a candidate and then alters the candidate's platforms during the course of the campaign in response to polls and focus groups.

A random search algorithm differs from a hill-climbing algorithm in two ways. First, rather than pick one random neighboring platform, many are chosen. The new status quo is simply the the best from the group of randomly chosen platforms and the old status quo. The random algorithm is only run for one generation. Random algorithms perform moderately well on functions of varying degrees of difficulty. They outperform hill-climbing on difficult functions but are less effective on easy functions. They are also not deceived easily. An algorithm is deceived if, during the course of search, it is systematically led away from good regions of the domain. A random search algorithm represents a party that chooses a candidate from a collection of volunteers. Once the candidate is selected, her platform is assumed to be immutable.

A genetic algorithm (GA) is a population-based search algorithm.[24] A GA begins with a population of platforms, all near the party's status quo. A GA proceeds in two steps: *reproduction* and *modification*. Each step captures a characteristic of biological evolution. In the reproduction step, more fit platforms—those that attract more votes—are more likely to be reproduced than less fit platforms. In the modification stage, platforms exchange blocks of positions on issues (crossover), and randomly change some positions on individual issues (mutation). Each application of reproduction and modification is called a generation. A GA is run for several generations and then the best platform in the final generation becomes the party's new platform. GAs perform well for many types of functions, usually far better than random search. Hill-climbing can outperform a GA on simple functions. Unlike random search though, a GA can be deceived. GAs perform poorly on some functions. Metaphorically, the GA represents competition and deliberation within the party. Candidates compete with the better surviving (reproduction), they share ideas (crossover), and they experiment with novel policy changes (mutation).

In each of the models we describe, our GA parties outperformed, on average, the other two types of parties. The differences in performance, however, were not large. The results suggest that the vote function, while not separable across issues, has sufficient regularity, that is, moves toward the median tend to be good, to enable simple search algorithms to perform well. In the abbreviated presentations which follow, we restrict attention to the hill-climbing algorithm. Qualitatively similar results hold for the other two types. For a comparison of algorithms we refer readers to the original papers.

2.1 SPATIAL ELECTIONS

The spatial election model is a building block of formal political theory. In a series of four papers, we examine the spatial model from a computational perspective. Our research in this area has combined mathematical and computational theories with empirical testing. In our first paper,[35] we construct a computational model of adaptive parties in two-party elections. We find that parties tend toward moderate positions, but that they do not converge to a single platform. In a second paper,[36] we vary voters' preferences and discover correlations between voter characteristics and the separation or divergence of parties in the issue space. A mathematical paper[37] demonstrates why these correlations occur. Finally,[38] we test our results empirically and find support for our claims about party separation.

We begin by describing a spatial model of electoral competition. In the model we consider, each voter attaches both a strength and an ideal position to each of n issues. Voter j's strength on issue i, $s_{ji} \in [0,1]$, measures the issue's relative importance to the voter. For example, a voter considers an issue of strength zero irrelevant. The ideal position of voter j on issue i, $x_{ji} \in \Re$, denotes the voter's preferred position on the issue. The utility to voter j from a party's platform, $y \in \Re^n$, equals the negative of the squared, weighted Euclidean distance between the vector of j's ideal positions and the party's platform weighted by the voter's strengths:

$$u_j(y) = -\sum_{i=1}^{n} s_{ji} \cdot (x_{ji} - y_i)^2 .$$

Voter j computes the personal utility from each party's platform and casts a ballot for the party whose platform yields the higher utility. Parties care only about winning elections, and they compete for votes by adapting their platforms.[5] That is, they "move," or adapt, about the multidimensional issue space over the course of campaigns. Each election campaign begins with two parties, one of which is the incumbent. The incumbent's platform remains fixed during the campaign while the challenger party adapts its platform.[6]

The challenger party uses information about its current popularity and applies decision rules to change its platform. Polling information during the campaign comes in the form of mock elections, indicating the percentage of votes the party would receive if an election was held at the time of the poll. A party can try a platform change and observe whether the change will lead to an increase or decrease in vote percentages. An important feature is that the polls provide noisy, or imperfect, signals of the popularity of proposed platform alterations. At the completion of the campaign, the challenger party selects a platform and the two parties stand for election with the winning party becoming the new fixed incumbent (at the winning platform) and the losing party becoming the challenger.

The movements of the parties in the issue space over the course of several elections can be monitored, thus giving a picture of the trajectory of party policies over time. An intuitive way to conceive of the process being modeled is as parties adapting on an *electoral landscape*. Each possible platform is perceived as a physical location and its corresponding vote total against the incumbent's platform is perceived as an elevation.[7]

[5] In Kollman, Miller, and Page (KMP),[35] we also consider ideological parties that have preferences over platforms.

[6] Initially, both parties begin with a random platform.

[7] The landscape metaphor is common in models of adaptive search. Applications to political science include Axelrod and Bennett.[4]

2.1.1 A BASIC COMPUTATIONAL SPATIAL MODEL. In Kollman, Miller, and Page (KMP),[35] we analyze the behavior of adaptive parties in a two-party computational model. Our main findings are threefold. First, as previously mentioned, parties tend to converge to similar platforms that yield high aggregate utility. This appears robust to wide ranges of parameter values and methods of adaptive search. The fact that these platforms have high aggregate utility suggests that parties reliant on adaptive search rules processing limited information tend to adopt moderate platforms. They do not wander to the extremes of the issue space, even though such an outcome is mathematically possible. Instead, predictable, moderate, consensual (though distinct) platforms evolve, a state of affairs familiar to observers of American party politics. Second, even though an incumbent party can always be defeated, they often remain in office. In this, computational and mathematical models lead to opposite conclusions.[8] In our computational experiments, incumbents remain in office because their adaptive challengers are unable to locate winning platforms. Third, even though parties tend to converge, the rates of convergence differ systematically.

2.1.2 PREFERENCES, LANDSCAPES, AND OUTCOMES. Our computational spatial voting model[35] suggests a relationship between voters' preferences, electoral landscapes, and outcomes. The intuition behind electoral landscapes is straightforward: parties in search of more votes try to find points of higher elevation on the landscape. Landscapes may be rugged with many local optima, or they may be smooth with one large hill. On rugged landscapes, if parties have only limited and imperfect information, and can only move incrementally, they may have difficulty finding winning platforms because local optima may lead them away from winning platforms. Furthermore, because information about a landscape is imperfect, the slope of the landscape becomes paramount. To see this, note that if a party wants to move to a higher elevation, a steep slope in one direction is easy to recognize and even imperfect information will tend to lead the party in that direction. In contrast, if a party faces a rugged, gradual slope, small polling errors can lead to big mistakes. With gradual slopes, parties may not be able to recognize which direction will lead to enough votes to win.

To learn how voters' preferences influence electoral landscapes and platform changes, in KMP[36] we alter the distribution of preferences in empirically relevant ways, and then compare both the shape of landscapes and the resulting party behavior across distributions. Which characteristics of voters' preferences might make an electoral landscape complicated? Our model highlights one plausible set of characteristics: strengths across issues.

We introduce three types of correlations between ideal points and strengths in the model: *centrist*—voters place more weight on issues on which they have moderate views, *extremist*—voters place more weight on issues on which they have

[8] Mathematical models often rely on an exogenous incumbency advantage to find that incumbents win reelection.[6,9]

extreme views, and *uniform*—voters place equal weight on every issue. To simplify exposition, let ideal positions, x_{ji}, belong to $[-1,1]$. The second column in Table 1 indicates how strengths are distributed for the various preference types.

The slope of a landscape is determined by how many new voters are attracted by a small position change. Suppose voters are centrist about a social insurance program. Only those who have moderate or centrist views care a lot about the issue, while those with extreme views do not make voting decisions based on this issue. Moves toward the center by a party on social insurance will likely win a party a lot of votes without losing many votes. This is because voters with ideal points near the center care a lot about the issue, while voters away from the center do not. The large number of votes to be won by moving to the center will be reflected in a large, steep hill in the center of the electoral landscape. Now, say voters are extremist about abortion policies. Only those who have extreme views care a lot about the issue. Small moves toward the center on abortion will not win many votes, because voters with ideal points near the center do not care much about abortion policy. No single steep hill forms in the center of the landscape, but instead, there may be several local peaks. Each local peak corresponds to a stable policy position, and the hill supporting a local peak corresponds to portions of the platform space in which the party has incentive to move away from the center.[9] One obvious intuition about adaptive search on landscapes is that steep slopes should be relatively easy to climb. In the context of our model, a party confronting a steep slope should recognize improving platforms. These platforms will tend to lie near the center of the issue space, especially with centrist voters. Therefore, with

TABLE 1 The Impact of Various Preference Types

Type of Preference	Strength	Slope of Landscape	Separation of Platforms
Centrist	$s_{ji} = (1- \mid x_{ji} \mid)$	steep	least
Extremist	$s_{ji} = \mid x_{ji} \mid$	gradual	most
Uniform	$s_{ji} = 1/over2$	mid-range	mid-range

[9]Recall that voters in our model have quadratic utility functions, which amplifies the effect of change with distance. Small moves by nearby parties matter less among voters than small moves by far away parties. If voters are centrist, and parties are relatively far from the center of the issue space, small moves toward the center of the space by parties will win parties a lot of centrist votes. If voters are extremist, and parties are near the center, moves by parties toward the center will win few votes, while moves by parties away from the center will win a lot of extremist votes.

TABLE 2 Computational Results on Preference Types

preference type	platform separation
uniform	20.23 (0.54)
centrist	12.24 (0.38)
extremist	30.11 (0.62)

centrist voters, we should expect the parties to be "closer" together ideologically than with extremist voters. The fourth column of Table 1 lists the predictions of party platform separation. The more centrist voters predominate on an issue, the more likely parties will have similar positions on the issue.

The connection between landscape slope and adaptive party behavior works just as predicted. Table 2 shows data from the sixth election of computational experiments using hill-climbing parties. The results shown are means and standard errors from 500 trials in a computational experiment using 2501 voters and ten issues. In the algorithms, parties polled 251 randomly selected voters, and parties tested forty platform alterations. We performed difference of means tests on the computer-generated data and found strong support for our conjecture about party behavior: extremist preferences lead to the most platform separation followed by uniform preferences and then centrist preferences. To supplement these findings, we also measured the slope of the actual landscape and found that it varies in the expected way. According to several measures of slope, centrist landscapes have significantly greater slope than uniform landscapes, which in turn have significantly greater slopes than extremist landscapes.

2.1.3 A MATHEMATICAL ANALYSIS OF SLOPE. In KMP,[37] we found a mathematical relationship between the distribution of voters' strengths and the slope of electoral landscapes. For reasons of tractability, KMP[37] considers a simplified version of our model in which platform changes occur on a single issue of a multidimensional issue space. By restricting our analysis to a single dimension, we can isolate the effect of variations in preferences on landscapes. We make three assumptions about voters' ideal points. First, voter ideal points are uniformly distributed on issue 1 on $[-1,1]$. Second, voter strengths were as in Table 1. Third, for voters at each ideal point in $[-1,1]$ on issue 1, the utility difference between the challenger and the incumbent on the other $n-1$ issues is uniformly distributed on $[-b, b]$ where $b \geq 1$. These assumptions enable us to calculate the vote total that the challenger party receives as a function of its position on issue 1, y; the incumbent's position, z;

and the divergence of opinion on other issues, parameterized by b. The challenger's vote total equals the measure of the agents whose votes he receives. We can then compute the change in vote total as a function of the candidates' position and we can state the following three claims.

CLAIM 2.1.1: For any (y, z, b) with $| y | > 0$, the slope of a landscape formed by centrist preferences is strictly steeper than the slope of a landscape formed by extremist preferences.

CLAIM 2.1.2: For any (y, z, b) with $| y | > 0$, the slope of a landscape formed by centrist preferences is strictly steeper than the slope of a landscape formed by uniform preferences.

CLAIM 2.1.3: For any (y, z, b) with $| y | > 0$, the slope of a landscape formed by uniform preferences is strictly steeper than the slope of a landscape formed by extremist preferences.

As these claims show, the results from the mathematical analysis agree with the findings from the computational analysis.

2.1.4 AN EMPIRICAL INVESTIGATION. In KMP,[38] we find empirical support for the computationally generated hypotheses about the relationship between voter preferences and party separation suggested by KMP.[36,37] We focus on presidential elections in the United States. Recall that the models predict that the more extremist voters (in our sense of the term extremist) predominate on an issue, the more parties will diverge or separate on that issue. The more centrist voters predominate, the more the parties will have similar positions.

Measuring the policy positions of parties at a given time is notoriously challenging. Scholars have suggested various methods of data collection and estimation, each having strengths and shortcomings depending on the purposes of research.[1,8,17,20,21,43,52] Since we are concerned with how citizens influence party behavior and how parties appeal to citizens, (that is, since perceptions of ideological distance matter every bit as much as "true" ideological distance between and among parties and voters), to investigate our model we rely on polling data to measure citizens' perceptions of party issue positions.

In the American National Election Surveys (ANES), which are extensive, mass surveys of Americans prior to major elections, respondents are asked to place themselves, candidates, and parties on scales (usually seven-point scales) referring to ideologies and controversial political issues. In results reported in KMP[38] we use the mean of respondents' perceptions of a party's position on an issue to estimate the true position. The use of seven-point scales and the use of means of respondents' perceptions raise troubling methodological issues, and we address these issues in detail in KMP.[38] To summarize that discussion, we agree with Brady and Sniderman[8] that voters have reasonably accurate and stable *aggregate* perceptions of party positions on issues. Moreover, we use several different sample groups to estimate party positions, and our findings are robust over these different measures.[10] Here we summarize only the results using the difference between the mean evaluations of each party among all respondents.

The ANES also contains data on respondents' views on the most important problems facing the country. Open-ended responses were coded according to nine categories, such as foreign policy, the economy, social welfare, and racial issues. We use these data to measure strengths and how strengths relate to policy preferences.

Our measure of extremism uses three components.[11] The first component is *perim*, the percent of respondents who considered that issue the most important problem facing the country. Second, z^{sub} is the average of the absolute values of the z-scores of responses on the seven-point scale for the subpopulation considering the issue important. Note that the z-scores used in z^{sub} are calculated according to the entire set of respondents on the seven-point scale and the most important problem question. As z-scores give a measure of the standardized deviation of respondents from the mean, the average of the absolute values of the z-scores will offer a single measure of the aggregate outlier status of respondents on an issue. Third, z^{tot} is the average of the absolute values of the z-scores of responses on the seven-point scale for the entire population of respondents. Putting these three components together we have

$$extremism_{ie} = \frac{perim_{ie} \cdot z_{ie}^{sub}}{z_{ie}^{tot}},$$

where $perim_{ie}$ is the percentage of voters who felt issue i was the most important issue in election e. The measure, $extremism_{ie}$, captures how the distribution of the subpopulation which cares about issue i in election e differs from the distribution of the entire population of respondents, weighted by the percentage of the population that cares about the issue. The measure will be high when extremist voters weigh heavily on the issue and low if centrist voters weigh heavily on the issue. For example, if the subpopulation is 25% more extreme than the total population, and 25% of the population thought the issue was most important, then $extremism = (1.25) \cdot (.25) = .313$.

Our model predicts that parties' divergence on a given issue will depend on the distribution of voters' strengths on that issue. An assumption of linearity between *separation* and *extremism* is a good approximation of the results of our model, and we reproduce a scatter plot that reveals a positive linear relationship between the two measures. Each point on the scatter plot is an issue, so the correlation ($r = .51$, $p = .001$) between the two measures offers initial support for our hypothesis.

[10]The literature on scaling party positions is quite large. See, in particular, Aldrich and McKelvey.[1]

[11]We ran all tests with several measures of extremism. All results are congruent with those summarized here.

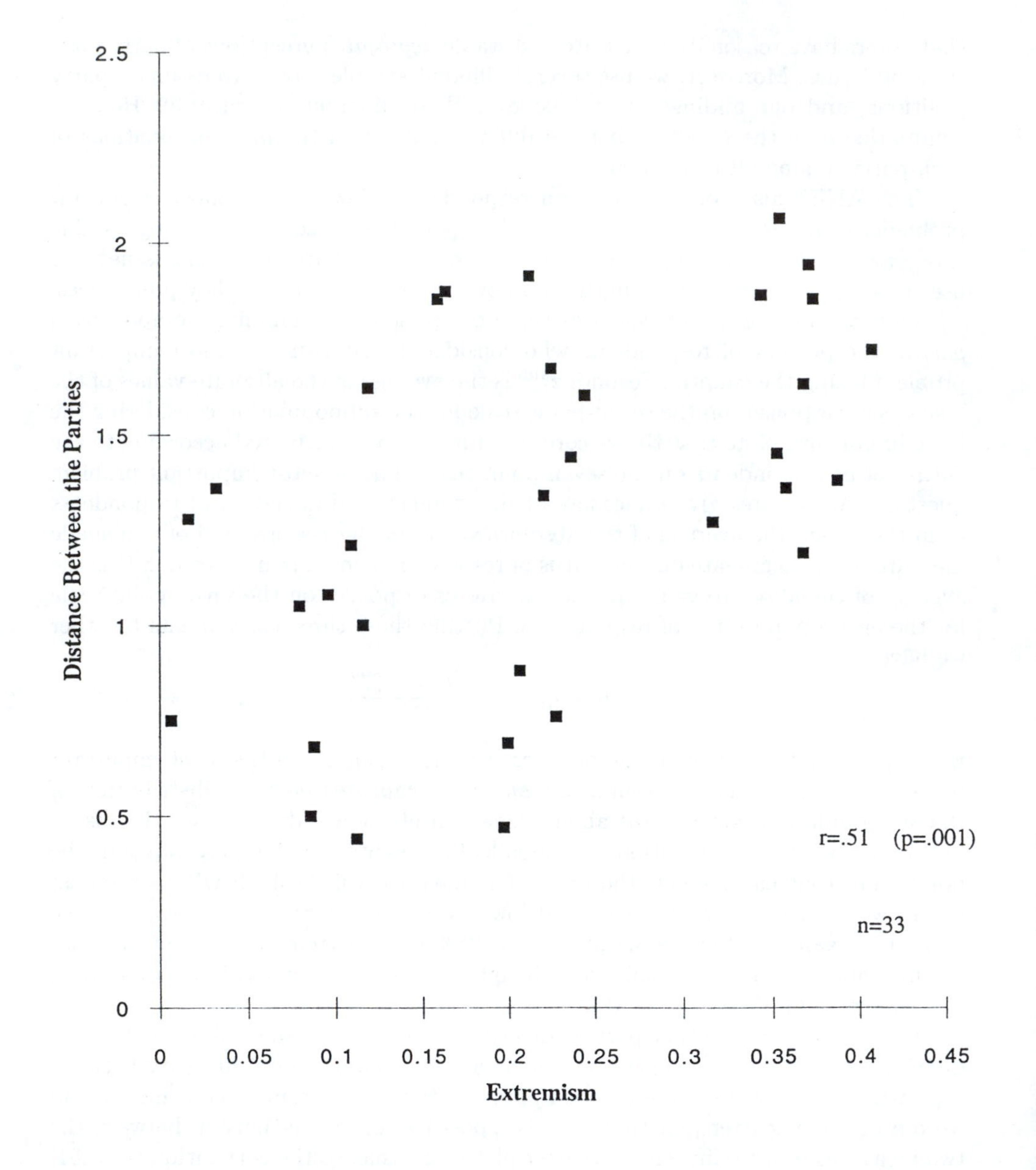

FIGURE 1 Policy Issues (1972–1992).

The more extremist the population is on an issue, the more the parties will separate on the issue. Using the candidate measure (the difference between the means of party candidate evaluations), the correlation is still fairly strong ($r = .31$, $p = .05$).

TABLE 3

Independent Variable	Party Separation (OLS)	Party Robustness (OLS)	Party Separation LSDV (OLS)	Party Separation Random Effects (REML)
extremism	1.97 (.46)		1.27 (.49)	.86 (.45)
z^{sub}/z^{tot}		1.30 (.70)		
perim		2.12 (.46)		
Cov. Par. Estimates				
issue				.09 (.06)
year				.05 (.04)
R^2	.26	.29	.20	
Adj. R^2	.24	.24	.20	
SE	.41	.41	.35	
n	33	33	33	33
Stan. Dev. Estimate				.24
REML log likelihood				-13.13
Akaike's Information Criterion				-16.13

[1] For each entry, the first row is the OLS or maximum likelihood coefficient, and the second row is the robust standard error.[65]

The linear relationship between party separation and extremism is detailed further in ordinary least squares (OLS) coefficients. The first column of Table 3 shows the coefficients for a simple bivariate regression of *separation* on *extremism*. The coefficient is significant by standard measures and in the expected direction. Moreover, to make sure only one component of the extremism measure is not causing the correlation with *separation*, we regress *separation* on the two parts of the extremism measure. The second column of Table 3 shows OLS coefficients for *perim* and the ratio of z^{sub} and z^{tot}. These data indicate that both components are important in predicting values of *separation*. Furthermore, the two component variables are not significantly correlated ($r = -.28$, $p = .11$). Candidate separation is not presented, though for all results presented here and in KMP,[38] using the measure of candidate separation leads to similar coefficients and errors.

These OLS coefficients may be biased, however, because the error terms among the issues are certainly correlated. The cases are issues that overlap in time (there are, for example, seven issues from the 1976 survey) and issues measured over time (the question on aid to minorities was asked for all six election surveys). In fact, our data on issues resemble panel data, which have both cross-sectional and time-series properties. Suitable procedures for analyzing such types of data (especially in cases involving missing data) are discussed in Baltagi[5] and Kmenta[34] (Ch. 12).

Two methods for estimating panel data coefficients, essentially controlling for time and cross-section effects, are commonly discussed. A random effects model allows the intercept to vary by each subset of data, while a fixed effects model, otherwise known as a least squares dummy variable model (LSDV), sets an intercept for each subset of the data.

In Table 3 the third and fourth columns show coefficients for these two methods. As is commonly noted in the econometrics literature (Hsiao,[30] Ch. 3), coefficients between a random effects and fixed effects model vary, sometimes dramatically. However, the coefficients of *extremism* for both models are positive and significant by standard measures. Even separating the two characteristics of the extremism measure works well with the data (shown in KMP[38]), indicating robustness of the results.

It is important to note that our main purpose in the empirical work in KMP[38] is to demonstrate that party separation and extremism covary, controlling for the effects of time. A more fully specified model, incorporating other pressures on parties to change—for example, pressures from activists and donors—would add further explanation for why parties take certain policy positions. We leave such questions for future research.

There is always the possibility of the results being highly sensitive to particular measures or to the use of survey data. In KMP,[38] we show that our results based on voters' preferences are robust when accounting for different measures, and the possible contamination of results due to biased perceptions by survey respondents.

2.1.5 DISCUSSION. Our research offers encouraging evidence in support of computational models. The original computational model[35] generated findings in accord with empirical evidence: challengers unable to locate winning platforms, and moderate, though distinct, platforms in two-party systems. It also generated a puzzle: why do parties appear to differ more on some issues rather than other issues? This led to a more detailed computational investigation in KMP,[36] which spawned both theoretical[37] and empirical investigations.[38] We also find that more formal mathematical and empirical analyses inspired by the initial computational results, support the insights from the computational models.

2.2 A COMPUTATIONAL TIEBOUT MODEL

In KMP,[40] we extend our computational spatial voting model to a Tiebout setting. In our Tiebout model, the polity is broken up into multiple jurisdictions within which political competition or referenda are used to determine policy. Once policies have been chosen, citizens relocate to jurisdictions with the most favorable policies for them. Ideally, as citizens sort themselves among jurisdictions according to their preferences, total utility increases. Tiebout's[64] original formulation was an attempt to disprove Samuelson's[59] conjecture that public goods could not be allocated efficiently. The core "Tiebout hypothesis" has since been extended to include additional propositions. Prominent among them is that Tiebout competition, as a result of enforcing efficiency, renders local politics unimportant: a political institution able to attract and retain citizens cannot waste resources, that is, it must be efficient.[28] This argument does not preclude the possibility that political institutions may differ in their ability to sort citizens according to preferences, which is the focus of our model.

Our computational model allows a direct comparison of political institutions as sorting devices. Guiding our analysis is the idea that the performance of a political or economic institution depends upon its ability to structure microlevel incentives to be in agreement with macrolevel goals.[60] In many important situations, microlevel incentives are consistent with multiple equilibria,[4,15] and one role of institutions may be to steer agents toward the best configuration, or at least bias outcomes toward better configurations.

Through computational experiments, we have found that some political institutions lead to a natural annealing process which improves sorting. Annealing is a concept unfamiliar to most social scientists, so we explain it in depth below. We find that political instability, such as voting cycles, can improve outcomes in a multijurisdiction environment when the degree of instability correlates positively with the heterogeneity of voters' preferences. Or, alternatively stated, institutions whose stability negatively correlates with the goodness of sort outperform more stable institutions. These findings occur under a variety of conditions and parameters in the model.

In KMP,[40] we consider three institutions: two-party competition, democratic referenda, and proportional representation. Here, we describe only the first two; they are sufficient to provide the primary intuition. In the formal model, we assume a set of N_a agents, each of whom resides in one of N_j possible jurisdictions. Within any jurisdiction, the local government takes positions on a set of N_i local public issues. For simplicity all such positions are binary. Examples of such issues include the presence (or absence) of a public good (say a community swimming pool) or policy (like smoking in public buildings). Let $p_{ji} \in \{Y, N\}$ give the position of jurisdiction j on issue i, and let a *platform*, $P_j \in \{Y, N\}^{N_i}$, give the vector of decisions p_{ji} across all N_i issues in jurisdiction j. Finally, define a *configuration* as a mapping of agents to jurisdictions.

Agents have linearly separable preferences on the issues and their per unit value for each issue lies in the interval $[-400/N_i, 400/N_i]$ distributed uniformly. Let ν_{ai} give agent a's per unit utility for issue i. Thus, agent a's utility from P_j is given by

$$u_a(P_j) = \sum_{i=1}^{N_i} \nu_{ai} \cdot \delta(p_{ji}),$$

where $\delta(Y) = 1$ and $\delta(N) = 0$. A straightforward calculation verifies that the expected value to an agent of an arbitrary platform equals zero, and the expected value of her optimal platform equals one hundred.

We model democratic referenda as majority rule on each issue.[12] The assumption that there are no external effects between projects implies that sincere voting is a dominant strategy for all agents. The outcome of democratic referenda is the *median platform*, p_j^m, for the jurisdiction, where $p_{ji}^m = Y$ if the number of agents in j with $\nu_{ai} > 0$ exceeds the number of agents in j with $\nu_{ai} < 0$; otherwise $p_{ji}^m = N$.

The platform P_j maximizes utility at jurisdiction j given a configuration, if and only if, on every issue the mean agent value and the median agent value have identical signs. Generally speaking, democratic referenda locate a policy of high aggregate utility given a configuration. A perceived advantage of democratic referenda is its stability: the policy prediction is unique and an individual agent migrating into or out of a jurisdiction rarely changes the median platform. We show, however, that this stability stifles sorting in multiple jurisdiction environments.

We model two-party competition using the previously discussed adaptive party model. Adaptive parties advocate policy platforms, and each agent votes for the party proposing the platform which yields her higher utility. Two-party competition is not as stable as democratic referenda, which, as we have just discussed, produces a unique outcome equal to the median voter's preference on each issue. Even with the linearly separable preferences considered here, policy predictions cannot be guaranteed to be unique without severe restrictions on preferences.[56] In fact, the top-cycle set, a commonly used solution concept that assumes any platform that

[12] In the case of a tie, we assume that the policy on the issue is N.

could be victorious over any other possible platform via some sequence of pairwise elections is a potential solution, can encompass the entire space.

2.2.1 ANNEALING AND INSTABILITY. The central insight in our Tiebout paper[64] is that the instability inherent in two-party competition may be beneficial in a Tiebout model, provided that instability and utility have a negative correlation. Our argument hinges on the relationship between the level of political instability within jurisdictions and the degree of homogeneity of preferences at each jurisdiction. At this point we should clarify that at present we do not have formal proofs for many of the ideas we put forth. What follows is an informal explanation of our results and findings from computational experiments.

We begin with an example that shows how the Tiebout equilibria with respect to two-party competition may be preferred to the Tiebout equilibria with respect to democratic referenda in a multiple jurisdiction model. Prior to describing this example, we must clarify what we mean when we say that a configuration is a *Tiebout Equilibrium with respect to an institution.* Given an institution, we first need a rule for the set of policies in each jurisdiction, which can result from each configuration. For democratic referenda, this rule consists of the median policy, p_j^m. For two-party competition, we assume it consists of all platforms such that each platform belongs to the top-cycle set for its jurisdiction. Finally, a configuration is a Tiebout equilibrium with respect to an institution if for any set of policies in the jurisdictions, no agent wants to relocate.

2.2.2 EXAMPLE: IMPROVED SORTING: There are two jurisdictions: α and β, eight agents: a, b, c, d, e, f, g, and h, and three issues: 1, 2, and 3. Define preferences as follows:

	Preferences		
Agent	Issue 1	Issue 2	Issue 3
a	+1	+1	+1
b	+1	−1	+0.5
c	+1	+0.5	−1
d	+1	−1	−1
e	+1	−1	−1
f	−1	−0.5	−1
g	−1	+0.5	−1
h	−1	+0.5	−1

Assume the following configuration of agents to jurisdictions: α contains agents a, b, and c; and β contains agents d, e, f, g, and h. If democratic referenda is the political institution, then the policy platform in α is YYY and in β the policy platform is NNN. It is easy to show that no agent wants to relocate. Therefore, the configuration of agents and the platforms form a Tiebout equilibrium with respect to democratic referenda. A simple calculation shows that the aggregate utility equals 4.0. We now show that two-party competition does not support this configuration of agents to jurisdictions as a Tiebout equilibrium. In two-party competition the platform YYY in α can be defeated by YNN.[13]

Note that democratic referenda are preferred to two-party competition on utilitarian grounds. The sum of the utilities to agents a, b, and c from the platform YYY equals 4.0, while the sums of their utilities of the other platforms in the top-cycle set YNN, YNY, and YYN equal 3.0, 3.5, and 3.5, respectively. If we assume for the moment that the set of policies resulting from two-party competition is randomly chosen from this set, then the lack of a unique outcome from two-party competition yields lower aggregate utility.

With multiple jurisdictions, the lack of a unique outcome from two-party competition may induce sorting and increase aggregate utility. Including jurisdiction β in the analysis, we find that two-party competition does not support this configuration of agents as an equilibrium. More importantly, this configuration lies in the basin of attraction of an equilibrium configuration which has a higher aggregate utility.[14] First, notice that if the agents in α elect a candidate advocating a platform from the set $\{YYY, YYN, YNY\}$, they will not create a strict incentive for any agents from β to move to α. If, however, a candidate wins election in jurisdiction α advocating the platform YNN, then agents d and e strictly prefer jurisdiction α to jurisdiction β, and they will relocate. Jurisdiction α would then contain agents a, b, c, d, and e, and YNN would be the unique member of the top-cycle set (as well as the median platform). Jurisdiction β would now contain agents f, g, and h and have the median platform NYN as the unique member of the top-cycle set. It is easy to show that this configuration of agents is a Tiebout equilibrium with respect to two-party competition. It is also a Tiebout equilibrium with respect to democratic referenda. More importantly, the aggregate utility of this configuration equals 5.5, which exceeds the aggregate utility of 4.0 from the previous configuration.

To provide deeper understanding of why this example is representative of a general phenomenon and not an anomaly, we can rely on insights from research applying *simulated annealing* to nonlinear optimization problems. Simulated annealing is a sequential search algorithm applied to a real valued function, f. A simulated annealing algorithm begins by assuming a neighborhood structure over the function's domain. Each point, x, in f's domain belongs to a neighborhood, $N(x)$, which contains at least one point different from x. Simulated annealing works much like hill climbing: at each step a new point in the neighborhood of the status quo point is tested and becomes the status quo point if it has a higher value under f. Where

[13] YNN can be be defeated by either YYN or YNY. Thus, these three platforms along with YYY form the top-cycle set.

[14] By basin of attraction, we mean the set of connected configurations that lead to higher aggregate utility than the status quo.

simulated annealing differs from hill climbing is that it also accepts points which have lower values with some probability. This probability depends on the difference in the function values, $\Delta(x, \hat{x}) = f(x) - f(\hat{x})$, and a *temperature*, $T(t)$, which is a decreasing function of the time spent searching, t. Formally, if x is the status quo point and $\hat{x}$ is the lower-valued neighboring point, the probability of acceptance is written:

$$p(x, \hat{x}, t) = e^{\frac{-\Delta(x,\hat{x})}{T(t)}} .$$

Notice that if the difference in function values, $\Delta(x, \hat{x})$, is large relative to the temperature, $T(t)$, then the probability of acceptance is low. The temperature can be interpreted as the degree of leniency. A high temperature allows for almost any new point to be accepted. A low temperature allows for a new point to be accepted only if the difference in function values is small. The temperature decreases to zero according to an *annealing schedule.* When the temperature nears zero, search converges to a local optimum with respect to the neighborhood structure.

A substantial body of theory exists to explain the performance of simulated annealing.[2] Essentially, simulated annealing locates good local optima because mistakes bias search away from local optima with low values and onto local optima with high values. This is more likely to occur when optima with low (high) values have small (large) basins of attraction. In such spaces, simulated annealing performs "as if" it can recognize whether a local optimum's value was relatively high or low, and escapes lower valued local optima. We propose that two-party competition performs similarly in a Tiebout model: *two-party competition makes more (fewer) mistakes in relatively low (high) valued configurations, enabling it to act "as if" it recognizes the value of a local optima and to escape it (remain in it) if it has a low (high) value.*

Two characteristics of Tiebout competition generate this fortuitous bias in error-making. First, aggregate utility for a configuration of agents is positively correlated to the homogeneity of preferences at each jurisdiction. Second, more homogeneous preferences result in more stable outcomes from two-party competition.[15] Combining these two effects, if agents in a configuration are not very homogeneous at each jurisdiction, their aggregate utility will be low and two-party competition may tend to roam about the space of platforms rendering the configuration unstable. If the agents in a configuration are homogeneous at each jurisdiction, their aggregate utility will be high, and two-party competition will select from a small set of platforms. This second configuration is more likely to be stable.

[15] An aspect of this second characteristic has been addressed formally by McKelvey,[46] who showed that the size of the uncovered set decreases with the level of symmetry of preferences.

2.2.3 COMPUTATIONAL FINDINGS. The computational model begins with a procedure in which agents' preferences are created and in which agents are assigned randomly to jurisdictions, with each agent being assigned to each jurisdiction with equal probability. We next begin a series of relocation decisions by agents and platform decisions by the political institutions until agents have had ten opportunities to relocate. For some political institutions, an equilibrium will have been attained at this point.

Following standard Tiebout models we assume that an agent moves to the jurisdiction providing her the highest utility. As mentioned previously, mathematical theory does not provide a unique platform prediction for the two-party model, except in special cases. Therefore, we rely on the adaptive party model described earlier in this chapter. The findings below are from hill-climbing parties which take forty polls each of the entire population of their district. The total number of citizens equals 1000.

We first consider a single-jurisdiction model. The findings are not surprising. Democratic referenda outperforms two-party competition. Democratic referenda produces an outcome of the median on each issue, which produces nearly maximal utility given our assumptions on the distribution of preferences. Democratic referenda had an aggregate utility of 2.49, while two-party competition had an aggregate utility of 1.84.[16] The differences are statistically significant at a 99% level.

With multiple jurisdictions, agents can sort according to their preferences. Tiebout's theory predicts that aggregate utility will increase with the number of jurisdictions. We present findings from models with three, seven, and eleven jurisdictions, and for both institutions considered, aggregate utility increases with the number of jurisdictions. The increases from the one-jurisdiction model to the three-jurisdiction model are most dramatic. More importantly, the *performance of the political institutions reverses*: two-party competition now outperforms democratic referenda. These differences in aggregate utility (standard errors of the distributions in parentheses) are also statistically significant at a 99% level.

Earlier we described how, with multiple jurisdictions, two-party competition might induce more sorting of agents. Configurations of agents that had relatively low aggregate utility would be less likely to be stable with respect to two-party competition. A small policy change might lead to a small migration, which in turn might lead to yet another policy change and still more migration, until eventually a stable configuration has been located. One approach to capturing the extent of sorting is to keep track of the total number of agent relocations for each political institution. Two-party competition results in significantly more relocations than democratic referenda with three, seven, and eleven jurisdictions.

[16]The numerical findings we present here are from 50 trials with 1000 agents and eleven issues. Utilities have been normalized so that the expected maximal utility to an agent equals 100 and the expected utility to an agent of a randomly selected platform equals zero. These findings appear to be robust to large variations in parameter values and search heuristics.

TABLE 4 Multiple Jurisdictions: Utility

Institution	3 Loc's	(s.d.)	7 Loc's	(s.d)	11 Loc's	(s.d.)
Democratic Ref.	34.40	(2.31)	48.51	(2.04)	55.64	(2.05)
Two-party Comp.	34.74	(1.83)	49.35	(1.74)	56.64	(1.96)

TABLE 5 Number of Agent Relocations

Institution	3 Loc's	(s.d.)	7 Loc's	(s.d)	11 Loc's	(s.d.)
Democratic Ref.	864.16	(121.90)	863.2	(73.55)	887.3	(37.22)
Two-party Comp.	915.68	(103.71)	1162.5	(61.03)	1293.7	(50.43)

Our argument as to why two-party competition outperforms democratic referenda boils down to a conjecture that instability is positively correlated with the degree of heterogeneity of voters' preferences. To test whether this is true, we create populations of agents of varying degrees of heterogeneity. We then measure the number of changes in platforms over time and verify whether the number of platform changes increases with the heterogeneity of preferences. We manufacture populations of agents with specified levels of heterogeneity as follows: each agent's ideal point was made a convex combination of a *base preference* and an *individual preference*. Both an agent's individual preference and the common base preference were drawn from the same distribution as agents' preferences in the basic model. We can write an agent's preferences as $(1 - \theta) \cdot base + \theta \cdot individual$. We refer to θ as the *degree of heterogeneity*. We examine a one jurisdiction model with 250 agents. The restriction to one jurisdiction guarantees that none of the changes in platforms are attributable to the shifting population. We vary θ between 0.5 and 1.0.[17] We define the variable *flips* to be the number of policies which change in each election. For democratic referenda, the number of flips equals zero for all θ. This occurs because with a fixed population, democratic referenda results in a unique outcome, and in this scenario no agents move into or out of the jurisdiction. For two-party competition, a different picture emerges: at $\theta = 0.5$, the number of flips is approximately one-quarter (0.24), at $\theta = 0.8$, the number of flips increases to over one-half (0.53), at $\theta = 0.95$, the number of flips exceeds two and a one-half

[17]Values of θ lower than this create agents whose preferences are too homogeneous: almost all have the same preferred platform.

(2.65), and finally, at $\theta = 1$, the number of flips is nearly three and a half (3.36).[18] As expected, the average number of flips increases with the degree of heterogeneity. These findings strongly support our assumption that as preferences in a jurisdiction become more homogeneous the propensity for platforms to wander decreases.

2.2.4 DISCUSSION. In constructing a computational Tiebout model, we find that institutional instability can generate better sorting provided that the instability is correlated negatively with the homogeneity of preferences. This result, though counterintuitive to social scientists, may not be surprising to those who study nonlinear systems. The institutions we model mimic a class of annealing procedures which have been found to be effective in finding good solutions to highly nonlinear problems. Thus, the computational model reveals a new insight (an insight, for which other computational results provide an explanation). Tiebout models are just one example of a much broader class of models in which agents sort among alternative configurations. These include models of coalition formation and economic organization. Under decentralized sorting mechanisms, these systems can become trapped in suboptimal configurations. If, however, there are means by which these poor configurations can be annealed in an appropriate manner, then the global system can escape these configurations and achieve superior outcomes. Whether other institutions also naturally anneal, or whether institutions with built-in annealing features can be constructed, is a potential research area within computational social science.

3. THE FUTURE OF COMPUTATIONAL MODELS IN POLITICAL ECONOMY

Adaptive computational models in political economy will likely become more prevalent because they offer several advantages to researchers. The models provide a flexible, yet precise, means of capturing a variety of phenomena. In particular, they allow the creation of dynamic models of adaptive agents. The ability to analyze adaptive behavior opens up important territories for research, permitting the creation of new classes of models that cannot be analyzed with previous tools. Moreover, such models also have the potential to elucidate and enhance existing theoretical tools. For example, computational techniques can be used to define the limits of rational choice approaches by demonstrating where static equilibrium models and dynamic adaptation models diverge. Irrespective of these advantages, the use of computational models is likely to increase if for no other reason than the changing relative cost advantages of computational versus mathematical modeling.

[18]The findings presented are from fifty series of ten elections. The differences reported are significant by standard statistical measures.

In this chapter we have summarized our use of computational methods in the context of two basic models in political economy. Our decision to focus on well-known problems is not coincidental. Only by returning to the basic models that define our disciplines and challenging the fundamental assumptions of these models will computational approaches lead to real breakthroughs. By studying what we think we already know, we reveal the strengths and weaknesses of both computational and mathematical approaches. Computational approaches can change the way social scientists understand social phenomena by forcing them to reconsider core assumptions.

While the advantages of computational models make a compelling case for their implementation, practitioners must address important methodological issues. The scientific validity of such models can be evaluated only if methodological norms are established. Computational modeling is no different than any other scientific enterprise; there are instances of brittle results being driven by both ignorance and political agendas. With time, the ability of the scientific community to evaluate these models will be enhanced.

In our own work, we have handled potential shortcomings with a computational approach in a variety of ways. As discussed, we have incorporated multiple adaptive algorithms to establish the robustness of our results. We have analyzed the effect of issues such as timing on updating rules.[53] The fact that such changes can have an impact on theoretical predictions is an important lesson for both computational and traditional modeling efforts. (In particular, traditional modeling often trivializes issues of timing and procedure, for example, by assuming instantaneous information updating.) We have also concentrated our modeling efforts on the creation of simple computational worlds that provide a framework within which the adaptive agents can "fill in" the details. For example, in Miller[47] rather than defining a set of "black-box" strategies for the agents, a genetic algorithm was used to allow agents to adaptively select their strategies from an enormous set of potential strategies. Finally, we have examined the importance of heterogeneity in strategy encodings and in the heuristics used to locate strategies.[26] Our efforts and those of others are far from comprehensive. At present, little consensus exists as to what makes a robust computational model different than just "a bunch of examples." In the future, the ability to confront and solve methodological problems will be as important to the acceptance of computational models as the ability to generate substantive theoretical breakthroughs.

We are neither advocating nor predicting a dramatic change in the social sciences, where computational techniques replace mathematical techniques. Instead, we foresee the future of computational modeling as a steady development of new methods that extend and occasionally overturn existing theoretical insights. Perhaps the greatest contribution of computational methods will be an indirect one. Such models allow theorists to interact with a convenient "artificial world." Such interaction extends a theorist's intuition and allows the computer to serve as a collaborator, in both the inspiration and refinement of theoretical ideas. By introducing new principles and conceptual tools, computational modeling will lead social scientists in valuable new directions.

ACKNOWLEDGMENTS

The authors would like to thank R. Michael Alvarez, Jennifer Bednar, John Ledyard, Richard McKelvey, and Charlie Plott for their help with this chapter. A portion of this work was funded by NSF grants SBR 94-09602, SBR 94-11025, and SBR 94-10948. Data from the American National Election Studies were made available by the Inter-University Consortium for Political and Social Research, which bears no responsibility for analysis or interpretation presented here.

REFERENCES

1. Aldrich, J., and R. McKelvey. "A Method of Scaling with Applications to the 1968 and 1972 Presidential Elections." *Am. Pol. Sci. Rev.* **71** (1977): 111–130.
2. Aldrich, J., and R. McKelvey. *Algorithmica* (A Special Issue on Simulated Annealing) **6** (1991).
3. Axelrod, R. "An Evolutionary Approach to Norms." *Am. Pol. Sci. Rev.* **80** (1986): 1095–1111.
4. Axelrod, R., and S. Bennett. "A Landscape Theory of Aggregation." *Brit. J. Pol. Sci.* **23** (1993): 211–233.
5. Baltagi, B. *Econometric Analysis of Panel Data.* New York: Wiley, 1995.
6. Baron, D. "Electoral Competition with Informed and Uninformed Voters." *Am. Pol. Sci. Rev.* **88** (1994): 33–47.
7. Bendor, J., and T. Hammond. "Rethinking Allison's Models." *Am. Pol. Sci. Rev.* **86** (1992): 301–322.
8. Brady, H., and P. Sniderman. "Attitude Attribution: A Group Basis for Political Reasoning." *Am. Pol. Sci. Rev.* **79** (1985): 1061–1078.
9. Calvert, R. "Robustness of the Multidimensional Voting Model: Candidate Motivations, Uncertainty, and Convergence." *Am. J. Pol. Sci.* **29** (1985): 69–95.
10. Chappell, H., and W. Keech. "Policy Motivation and Party Differences in a Dynamic Spatial Model of Party Competition." *Am. Pol. Sci. Rev.* **80** (1986): 881–899.
11. Coughlin, P. "Candidate Uncertainty and Electoral Equilibria." In *Advances in the Spatial Theory of Voting*, edited by J. Enelow and M. Hinich. New York: Cambridge University Press, 1990.
12. Dahl, R., and C. E. Lindblom. *Politics, Economics, and Welfare.* New York: Harper & Brothers, 1953.
13. Davis, O., M. Hinich, and P. Ordeshook. "An Expository Development of a Mathematical Model of the Electoral Process." *Am. Pol. Sci. Rev.* **64** (1970): 426–448.
14. Dearden, J., B. Ickes, and L. Samuelson. "To Innovate or Not to Innovate: Incentives and Innovation in Hierarchies." *Amer. Econ. Rev.* **80** (1990): 1105–1124.
15. DeVany, A. "Hard Cores, Soft Cores, and the Emergence of Self-Organized Coalitions." Working Paper, University of California, Irvine, CA, 1994.
16. Downs, A. *An Economic Theory of Democracy.* New York: Harper & Row, 1957.
17. Enelow, J., and M. Hinich. *The Spatial Theory of Voting.* New York: Cambridge University Press, 1984.
18. Enelow, J., and M. Hinich. "A General Probabilistic Spatial Theory of Elections." *Public Choice* **61** (1989): 101–113.

19. Enelow, J., and M. Hinich. *Advances in the Spatial Theory of Voting.* New York: Cambridge University Press, 1990.
20. Enelow, J., N. Mandel, and S. Ramesh. "A Comparison of Two Distance Metrics Through Regression Diagnostics of a Model of Relative Candidate Evaluation." *J. Pol.* **50** (1988): 1057–1071.
21. Granberg, D., and T. Brown. "The Perception of Ideological Distance." *West. Pol. Quart.* **45** (1992): 727–750.
22. Green, D., and I. Shapiro. *Pathologies of Rational Choice Theory.* New Haven, CT: Yale University Press, 1994.
23. Hinich, M. J., and P. C. Ordeshook. "Plurality Maximization Versus Vote Maximization: A Spatial Analysis with Variable Participation." *Am. Pol. Sci. Rev.* **64** (1970): 722–791.
24. Holland, J. H. *Adaptation in Natural and Artificial Systems.* Ann Arbor. MI, University of Michigan Press, 1975.
25. Holland, J. H., and J. H. Miller. "Artificial Adaptive Agents and Economic Theory." *Amer. Econ. Rev., Papers & Proc.* **81** (1991): 365–370.
26. Hong, L., and S. E. Page. "A Model of Heterogeneous Problem Solvers." Working Paper, California Institute of Technology, 1995.
27. Hoyt, W. H. "Local Government Inefficiency and the Tiebout Hypothesis: Does Competition among Municipalities Limit Local Government Inefficiency." *Southern Econ. J.* (1990): 481–496.
28. Huberman, B. A., and N. S. Glance. "Evolutionary Games and Computer Simulations." *Proc. Natl. Acad. Sci.* **90** (1993): 7716–7718.
29. Hsiao, C. *Analysis of Panel Data.* New York: Cambridge University Press, 1986.
30. Jackson, J. "Political Methodology: An Overview." In *A New Handbook of Political Science*, edited by R. E. Goodin and H. Klingemann. Oxford, UK: Oxford University Press, 1996.
31. Judd, K. "Computational Approaches to Theory: Complement or Substitute?" Paper presented at the Computational Economics Conference, Austin, Texas, 1995.
32. Kauffman, S. A. *At Home in the Universe: The Search for Laws of Self-Organization and Complexity.* New York: Oxford University Press, 1995.
33. Kmenta, J. *Elements of Econometrics.* New York: Macmillan, 1986.
34. Kollman, K., J. H. Miller, and S. E. Page. "Adaptive Parties in Spatial Elections." *Am. Pol. Sci. Rev.* **86** (1992): 929–937.
35. Kollman, K., J. H. Miller, and S. E. Page. "Political Parties and Electoral Landscapes." Working Paper, 1994.
36. Kollman, K., J. H. Miller, and S. E. Page. "An Investigation of Adaptive Parties' Platform Convergence." Working Paper, 1994.
37. Kollman, K., J. H. Miller, and S. E. Page. "Policy Position-Taking in Two-Party Elections." Working Paper, 1995.
38. Kollman, K., J. H. Miller, and S. E. Page. "On States as Policy Laboratories." Working Paper, 1995.

39. Kollman, K., J. H. Miller, and S. E. Page. "Political Institutions and Sorting in a Tiebout Model." Working Paper, 1996.
40. Kramer, G. "A Dynamical Model of Political Equilibrium." *J. Econ. Theory* **15** (1977): 310–334.
41. Kydland, F., and E. Prescott. "The Computational Experiment." *J. Econ. Perspectives* **10** (1996): 69–85.
42. Laver, M., and N. Schofield. *Multiparty Government.* New York: Oxford University Press, 1990.
43. March, J., and J. Olsen. "The New Institutionalism: Organizational Factors in Political Life." *Am. Pol. Sci. Rev.* **78** 734–749.
44. McKelvey, R. D. "Intransitivities in Multidimensional Voting Models and Some Implications for Agenda Control." *J. Econ. Theory* **12** (1976): 472–482.
45. McKelvey, R. "Covering, Dominance, and the Institution-Free Properties of Social Choice." *Am. J. Pol. Sci.* **30** (1986): 283–314.
46. McKelvey, R. "The Evolution of Automata in the Repeated Prisoner's Dilemma." In *Two Essays on the Economics of Imperfect Information*, Ph.D. dissertation, University of Michigan, 1988.
47. Miller, J. H. "The Coevolution of Automata in the Repeated Prisoner's Dilemma." *J. Econ. Behav. & Org.* **29** (1996): 87–112.
48. Ordeshook, P. C. *Game Theory and Political Theory.* New York: Cambridge University Press, 1986.
49. Page, B. *Choices and Echoes in Presidential Elections.* Chicago: University of Chicago Press, 1978.
50. Page, B., and C. Jones. "Reciprocal Effects of Policy Preferences, Party Loyalties, and the Vote." *Am. Pol. Sci. Rev.* **73** (1979): 1071–1089.
51. Page, B., and R. Shapiro. *The Rational Public.* Chicago: University of Chicago Press, 1992.
52. Page, S. "Two Measures of Difficulty." *Econ. Theory* (1995): forthcoming.
53. Page, S. "A Note on Incentive Based Asynchronous Updating in Cellular Automata." *Comp. Econ.* (1996): forthcoming.
54. Plott, C. "A Notion of Equilibrium and Its Possibility Under Majority Rule." *Amer. Econ. Rev.* **79** (1967): 787–806.
55. Rabinowitz, G., and S. E. McDonald. "A Directional Theory of Issue Voting." *Am. Pol. Sci. Rev.* **83** (1989): 93–123.
56. Riker, W. *Liberalism Against Populism.* New York: Freeman, 1982.
57. Samuelson, P. "The Pure Theory of Public Expenditure." *Rev. Econ. & Stat.* **36** (1954): 387–389.
58. Schelling, T. *Micromotives and Macrobehavior.* New York: Norton, 1978.
59. Schofield, N. "Instability of Simple Dynamic Games." *Rev. Econ. Studies* **45** (1978): 575–594.
60. Simon, H. *Administrative Behavior: A Study of Decision-Making Processes in Administrative Organization.* New York: Macmillan, 1957.
61. Stokes, D. "Spatial Models of Party Competition." *Am. Pol. Sci. Rev.* **57** (1963): 368–377.

62. Simon, H. *Administrative Behavior: A Study of Decision-Making Processes in Administrative Organization.* New York: Macmillan, 1957.
63. Stokes, D. "Spatial Models of Party Competition." *Am. Pol. Sci. Rev.* **57** (1963): 368–377.
64. Tiebout, C. M. "A Pure Theory of Local Expenditures." *J. Pol. Econ.* **64** (1956): 416–424.
65. White, H. "A Heteroscedasticity–Consistent Covariance Matrix Estimator and a Direct Test for Heteroscedasticity." *Econometrica* **48** (1980): 817–838.

Alan P. Kirman
G.R.E.Q.A.M., E.H.E.S.S. and Université d'Aix-Marseille III, Institut Universitaire de France, 2 Rue de la Charite, 13002 Marseille, France; E-mail: kirman@ehess.cnrs-mrs.fr

The Economy as an Interactive System

In standard economic models agents only interact through the price system. In fact, the state that agents find themselves in depends on the states of the other agents in the economy. In game theoretical models each agent interacts with every other agent and is aware of the consequences of doing so. This paper discusses economic models in which agents interact directly with each other but use relatively simple rules and learn from their experience. In particular, the models considered include not only ones with global interaction in which all agents can interact with each other but also ones in which agents can only interact with their immediate neighbors. In both cases, static and dynamic models are considered, and the latter includes the class of evolutionary economic models. To study models with local interaction it is necessary to specify the network in which the agents are situated. In general this network is given, but here, models in which communication networks evolve are also discussed.

INTRODUCTION

Interaction in standard "competitive" economic models is through market signals. Agents are thought of as receiving signals, such as prices, and as reacting to them in isolation. The coordination of the agents' choices is achieved without the agents' having to be directly aware of each others' existence. Models in which agents consciously interact with each other are referred to as ones of "imperfect competition." The implication is that the competitive model is a bench-mark and that models of economic behavior should be thought of in terms of the ways in which they deviate from that underlying model. One justification for this is that the competitive model is described as "decentralized" and is, therefore, appropriate to study modern economies which are, by any standard, decentralized. The solution to the allocation problem occurs when the signals induce the individuals to make choices which clear the markets. This "competitive equilibrium" is also considered as a bench mark by which to judge outcomes of any allocation process. The model and its solution leave a number of questions unanswered. Who decides on the market signals, and how are the trades necessary to achieve the equilibrium outcome organized? Without an answer to the first question, one has to create a fictitious agent who transmits and adjusts the signals which decentralize the decision-making process. This is the role of the Walrasian auctioneer. However, although this allows one to overcome the difficulty, the result is a communication structure that is highly centralized.

Even if the appropriate signals were forthcoming, there has to be some mechanism by which the transactions that effectively clear the markets take place. Again, what one would hope to define is something akin to real markets in which individuals do, in fact, make bilateral trades with each other, communicate with each other, and give information to and obtain information from other agents. There have been a number of attempts to tackle the sort of problems I have outlined. One of the simplest approaches is to model the economy as if there was only one individual in a sector. Thus, the aggregate behavior of consumers is modeled as though it were the behavior of one, maximizing "representative individual." This essentially assumes that the direct interaction between individuals has no consequence and that micro- and macrobehavior are of the same type. This approach has both logical and technical problems (see Jerison,[70] Summers,[104] Kirman,[75] and Stoker[103]), and indeed, the basic idea of this paper is to suggest that the understanding of aggregate economic phenomena requires an understanding both of individual behavior and the way in which individuals interact. For the same reason that as a neurologist would not model thought processes by considering the average neuron, and the entomologist would not model the behavior of an ant colony by looking at a representative ant, the economist should not model market behavior by modeling markets as individuals. In models in which individuals interact directly, and possibly randomly, as is well known from statistical mechanics for example, macroregularities may appear which cannot be thought of as corresponding to average individual behavior.

However, once one moves away from the vision of an economy or market as a collection of individuals reacting, in isolation, to market signals, an important problem arises. The notion of equilibrium in the normal economic model is simple and well defined. Agents make choices as a function of market signals. An equilibrium signal is one which engenders choices that satisfy some rule of consistency. In the market framework, the signal is the price vector and the consistency condition is that markets clear. In a system where there is interaction, this idea may need to be modified.

To see this consider the opposite extreme to the model with isolated decision makers, that in which every individual reacts to every other individual. This is the situation in a noncooperative game. If one thinks of the communication in an economy as represented by a graph, then the standard, or Walrasian model can be thought of as a star with each individual linked to the source of the signals. In the game theoretic situation, on the other hand, there would be a complete graph with a link between every two individuals in the game or economy. In the full-blown game theoretic model, every individual takes account of the strategies of every other player and what is more, knows that the other does so and knows that he knows and so forth. Thus, not only is the structure of the communication network very different from that of the standard economic model, but the type of reasoning attributed to the individuals is also of a very different nature. However, despite the complexity of the reasoning imputed to the individuals, the basic equilibrium notion, the Nash equilibrium also clear and well defined in this case. It is a choice of strategy for each player that cannot be improved upon, given the strategies of the others. It is clear that market equilibrium can be recast in this way, no agent, given the market rules, could make a better choice (see Debreu[32]).

Nash equilibrium and the full-blown game theoretic model share fundamental defects with the market model. Such equilibria are not, in general, unique and there is no obvious way in which they would be arrived at. Furthermore, the complexity of these games and the reasoning involved means that they pose problems of logical consistency (see Binmore[22]) and that only very simple examples are analytically tractable.

Given these difficulties, I suggest that the appropriate route to follow is one which lies between the two extremes I have just described. My purpose here, therefore, is to describe models in which agents communicate with some, but not necessarily all, other agents, and in which their reasoning is less complex than that envisaged in the game theoretic approach. Within the general class of models I shall describe, there are several different categories. On the one hand, agents may be capable of communicating with all the other agents even though they may actually interact with only a limited sample. This can be thought of as "global interaction," and one can either think of static models in which the interaction only takes place once, or of models in which the interaction between agents is repeated and one considers the dynamic evolution of the system. At the other extreme are models in which a neighborhood structure is given and agents can only interact with those near to them. This I shall refer to as "local interaction." Here also, one can analyze static or dynamic models. In the latter case, one can investigate the diffusion of the consequences of local actions across the system. These consequences may take some time to have an effect on agents with whom the agents who take the actions are not directly linked.

There are a number of ways of moving from considering a static equilibrium notion to that of studying the dynamic evolution of the systems I have in mind. One approach is to think of a repeated series of markets or games, each of which is the same as its predecessor and in which the payoffs of any specific action in the current game are unaffected by the players' actions in previous rounds. This does not, of course, mean that the strategies involved consist of a simple choice of action in each game. Those choices will, in fact, be conditioned by the evolution of the game or market. Typical examples of such situations are markets for perishable goods, where the same actors meet day after day to trade the same goods, but where the only link between successive days is the experience of the traders, since no inventories are held. In such a case one might look for a stationary state that is just a sequence of identical choices for each round. There are two ways in which such a situation can arise: Either economic agents think about the nature of the repeated game they are playing and work out the appropriate choice of strategy, or agents have simply found from previous experience which actions have performed better. In the first case, even if current payoffs only reflect current strategies, the latter will, and should, take into account histories of play to date. This simple consideration leads to the "folk theorem" type of results with a large class of equilibria. Here, an equilibrium would correspond to the occurrence of one outcome in each of the sequences of identical games. This happens because individuals will have strategies that involve a number of choices which are conditional on the history observed. The fact that other choices are not observed is a property of the equilibrium. These choices are never observed because of the other players responses. Two difficulties are apparent here. The reasoning involved is intricate and sophisticated but, perhaps more importantly, observation of the choices made reveals very little about the strategies.

An interesting problem, therefore, and one which has been widely investigated, is under what circumstances players will, by simply learning from previous experience, converge to some particular state which would correspond to the equilibrium that would have been obtained by much more complicated game theoretic reasoning. The difference between the two approaches, which Binmore[21] has described as "eductive" and "evolutive" should not be underestimated. The way in which agents behave is very different in the two cases and the outcome will depend crucially on what information the actors have at each point. Another issue in the context here, will be whether agents always interact with the same partners and, if not, who they play against in each period.

In repeated games, therefore, it is clear that the future payoffs from an action may vary depending on the choice of actions of other agents. Yet the situation is even more complicated in general economic models, where current actions influence future outcomes and, in particular, the payoff from future actions. In other words, even if all the actions taken by the actors are the same at period t as they were at time $t - k$ the payoffs may be different depending on the actions taken at other periods. When agents take account of the consequences of their own actions and those of other agents for current and future payoffs the situation becomes highly complex. There are two ways of dealing with this. Either one can try to solve the full-blown equilibrium problem *ab initio* or, alternatively, one might ask whether players would learn or adjust from one period to the other and whether such behavior would converge to any specific outcome. Then one can compare the limit point with an equilibrium that might have occurred if all the actors had solved the problem at the outset. The idea that individuals learn in relatively simple ways from their own experience and that of others is more persuasive than the alternative idea that they solve highly complicated maximization problems involving calculations as to their own course of action and that of others. Thus, the models I describe will, in general, be of this type.

Pursuing this line of argument, one might argue that in global interaction models, even though agents may have the chance to interact with all other individuals, they will not necessarily learn from the experience of all other agents but might, for example, learn from the experience of a randomly drawn subset. Thus, individuals might interact with some randomly drawn partners and react to the results of this interaction alone. Another way of looking at this problem is to make a radical simplification of the overly complex, general repeated-game model and to deprive the players of any rationality at all. This can be done by identifying each agent with a strategy and allowing Darwinian selection to operate—agents with less successful strategies will die off and new agents with more successful strategies will replace them. This is the approach adopted by evolutionary game theory. It provides an interesting counterbalance to the full rationality situation, and the question that arises immediately is whether such a mechanical type of selection can lead to a situation which would have been achieved by the "eductive" approach. If this is the case, many economists have argued that one might as well take the short-cut of assuming complete rationality and studying the equilibria that would arise. This is the position explicitly adopted by Lucas[82] when he says,

> In general we view or model an individual as a collection of decision rules (rules that dictate the action to be taken in given situations) and a set of preferences used to evaluate the outcomes arising from particular situation-action combinations. These decision rules are continuously under review and revision: new decisions are tried and tested against experience, and rules that produce desirable outcomes supplant those that do not. I use the term 'adaptive' to refer to this trial-and-error process through which our modes of behavior are determined.

However, Lucas then goes on to argue that we can safely ignore the dynamics of this process since,

> Technically, I think of economics as studying decision rules that are steady states of some adaptive process, decision rules that are found to work over a range of situations and hence are no longer revised appreciably as more experience accumulates.

Thus, the difference between the approach of many economists who rely on standard theory and many of the authors who have contributed to this book is, to a considerable extent, one concerning the stability of the adjustment to equilibrium and the speed of that adjustment. The basic tenet of those who concentrate on equilibrium is that individuals learn relatively rapidly to behave optimally and that the economic environment changes sufficiently slowly so that in the resulting situation they have no need to continue to learn. The contention of authors who, like myself, hold the view that the economy is a complex adaptive system is that the very process of learning and adaptation, and the feedback from the consequences of that adaptation generate highly complicated dynamics which may well not converge to any standard economic equilibrium.

Furthermore, once one takes account of the fact that individuals only interact with a limited collection of other agents, the evolution of the system becomes yet more complicated. One is then led to construct local interaction models in which one is interested in knowing how the configuration of states, of the individuals who are related to each other through a specific network, behaves over time, whether this settles to any stationary situation and whether this is characterized by local "clusters" of individuals in a particular state.

Perhaps the most interesting problem in this context, is to model how the links between individuals develop and hence how market structure itself evolves. This can be thought of in two ways. Firstly, one can think of some physical, say spatial, framework in which locations could be thought of as nodes of a lattice structure. In this case agents might choose where to situate themselves in the lattice and one could observe the evolution of activity and of resources over time (see, e.g., Durlauf[36]). Alternatively, one could think of individuals as interacting with each other in different ways, trading, communicating, and reinforcing their links with those people with whom interaction has proved to be more profitable in the past. For example, as I have mentioned, in analyzing games in an evolutionary context, players are typically matched with each other with a uniform probability. It is of interest to see what happens when players can refuse to play with partners with whom they have had bad experiences in the past (see Stanley et al.[102]).

For those who are in the habit of working within the complete and elegant framework of general equilibrium analysis, many of the models described below will appear as simple and disjointed examples. Yet each captures some aspect of economic reality and should be thought of as giving some insight into a particular phenomenon. We are far from having a general model that captures all the different types of economic interaction one can observe. For the moment we must content ourselves with looking at simple lattice-like models when trying to get insight into spatial problems, random matching models for the labor market and some goods markets, sequential decision models for looking at some aspects of financial markets, and so forth.

When we allow for the full feedback from observed outcomes to current behavior and for the evolution of new types of behavior, even the examples I shall present display considerable analytical problems. We are far from having an analytical framework for examining the economy as a whole as a complex adaptive interactive system, but the models outlined here represent small steps in this direction. My particular emphasis here will be on the development of models that will allow us to analyze how the graphs which link agents together in their pursuit of economic activities develop, but this is, of course, only one of the aspects of the general problem of establishing the relationship between microeconomic interaction and aggregate phenomena.

SOME BASIC MODELS OF INTERACTION: RANDOM MATCHING

Before examining models in which the structure of the network of links between agents is, or can be, specified, I shall briefly consider those in which agents do interact directly but in which there is no specific structure determining who interacts with whom. In other words all agents can, or are equally likely to, meet each other. I will first look at those for which the equilibrium notion is static, which include many of the standard random matching models. There is a substantial literature in economics in which the role of bilateral encounters and trading has been analyzed. Many of the founders of the general equilibrium model, such as Walras and Pareto and their contemporaries Edgeworth and Wicksteed, actually described trading processes which might arrive at an equilibrium. Indeed, Wicksteed anticipated recent work by outlining a bilateral trading structure and by specifying the terms at which trade would take place. Yet as the full model became more formal and more abstract, attention was focused on the existence of equilibrium and not how it was attained. Thus, no trading mechanism is present or envisaged in the Arrow-Debreu model. However, the influence of the equilibrium concept of the latter is strong and it is often invoked as a potential solution or reference point for models in which a market and trading are explicitly specified. The basic aim of Rubinstein and Wolinsky's[93] contribution, for example, is to examine whether a process of bilateral bargaining and transaction will converge to a competitive outcome. Another earlier type of model is the "search" model, (see, e.g., Diamond[35]) in which trading for an indivisible good takes place between many anonymous buyers and sellers. Buyers use some rule such as a "reservation price" to decide at what price to buy and, given the reservation prices, the sellers choose which price to set. Although this seems to specify direct interaction between agents this is largely illusory. Agents are unaware of the identity of those with whom they interact, and no direct mechanism is given for the adjustment of sellers' prices nor for the reservation prices of buyers. Attention is concentrated on the equilibrium state, which ceases to be necessarily a uniform price for units of the same commodity. Instead, it is a distribution of prices. Since agents are supposed to know the distribution of prices in the market at any point, an equilibrium will be a distribution—of prices for sellers and of reservation prices for buyers—which regenerates itself. Thus one is once again reduced to looking for a fixed point of a particular transition mapping as an equilibrium.

Two points should be made here. First, in the Diamond model (see e.g., Diamond[35]) an individual buyer is still confronted by a market signal, the current price distribution, and responds individually to that by establishing his reservation price; while the seller, knowing the distribution of reservation prices, will set his price accordingly. Most important in this connection is the fact that nowhere is it specified how the central market signal concerning the distribution of prices is generated. Thus, just as in the general equilibrium model, we are left depending on some fictitious agent who makes the distribution known to all the agents.

Secondly, although transactions are pairwise and, thus, there is apparent individual interaction, the equilibrium notion is essentially a static one corresponding closely to the standard notion of equilibrium. The term "search equilibrium" conjures up an idea of active interaction, but although this type of model captures another dimension of the nature of market signals, it does not really involve active interaction between economic agents. Such a criticism also holds for other models that involve random matching, such as those analyzed by Roth and Sotomayor.[92] Although the solution concepts involved differ from those of the standard market in many cases, they can be sustained as competitive solutions of a market, as Gale[52] has shown. Although algorithms for finding such solutions are known, they also depend, in general, on some central organizer and so cast little light on the market evolution which I described at the outset.

An alternative approach to studying an adjustment process that leads to a Pareto efficient outcome but which involves individual pairwise transactions and dispenses with any central price signal, is that adopted by Feldman.[47] He, as in the previous models, allows for random matching and requires that each pair, if they can, make a Pareto improving trade. Intuitively it might seem clear that such a process should converge to a Pareto-efficient allocation. However, once the notion of any central coordinating signal is removed, new problems arise. Pairwise interaction or trading can easily get into a cul de sac, simply because there are possible improvements for larger groups but not for pairs. This is Edgeworth's problem of the lack of "double coincidence of wants." Coordination must be achieved in some other way. One simple solution is to require that there should be one good which every agent holds in sufficiently large quantities. This good then plays the role of money and overcomes the basic problem. However, although this type of work suggests a step forward to a genuinely decentralized mechanism it is still concerned with processes that will yield a static equilibrium of the classic sort.

A natural development of this approach is to consider the nature and role of money in general, since it becomes needed as soon as transactions are not centrally organized. A basic model in which exchange takes place through voluntary bilateral trade and money plays the role of medium of exchange was introduced by Kyotaki and Wright.[79] The step from the use of a given fiat money to the emergence of a particular good as money is studied, for example, by Gintis[54] with agents who adopt trading strategies according to the distribution of inventories in the population, which in turn, result from the previous strategy choices of the agents. This approach is, of course, open to the same criticism that I have made of other models—how do agents become aware of the distribution of inventories in the population? Nevertheless, although this is not the place to go into the foundations of a theory of money, it is revealing that money, as a medium for transactions, is a natural concern in the sort of decentralized, interactive system I am concentrating on here, while it does not fit comfortably into the general equilibrium framework![1]

A closely related literature, which allows agents to learn about opportunities from their own experience but gives rise to other problems, is that concerned with sequential bargaining between pairs of individuals who are unaware of each others' valuations of the indivisible good over which they are bargaining. As the process evolves, individuals update their estimates of their partners' valuations. Myerson and Satterthwaite[90] have shown that there is no mechanism or rule which guarantees that the outcome of bargaining between two agents will give rise to *ex post* efficient outcomes. Goyal[55] shows, however, that if there are several agents who all know at each stage the strategies of previous traders, then bargaining can converge to *ex post* efficient outcomes even if the players may not come to know the valuations of the others. Once again we are basically concerned with convergence to a static equilibrium, but agents are no longer reacting anonymously to some central signal and this represents a possible avenue for the study of institutions such as double auctions or posted price markets.

A BASIC MODEL OF GLOBAL INTERDEPENDENCE

The interaction that I have discussed up to this point involves the fact that agents can be thought of as encountering other agents but, of course, interaction may involve reaction to the behavior or characteristics of others without any meeting in the normal sense taking place. Thus, the general notion of interaction is expressed by the idea that the state that an agent is in is influenced by the states in which others find themselves. "State" here may have many meanings, it could be some action that an agent has chosen, such as adopting a technology, choosing a place to live or work, or playing some strategy; it could refer to some characteristic of an agent, such as his preferences, one agent's preferences being influenced by those of others. The influence may be deterministic or stochastic.

A model which allows us to examine this sort of problem was introduced into economics by Föllmer[48] and has been more recently taken up again by Brock[26] and Durlauf.[38] The basic framework for the analysis, consists of a set S of possible

[1] Of course, other roles for money can be studied as Grandmont[56] has done in a temporary general equilibrium setting.

"states" in which the agents can be, and a conditional probability law $\pi_a(\cdot|\eta)$ for a in A, where A is the set of agents and where η is the environment, i.e., the specification of the states of the other agents. Of course, if we wish to consider deterministic dependence, then the conditional probability of an agent being in a certain state will be one or zero depending on the states of the others.

Now if we denote by $w : A \to S$ the state of the economy, by Ω the set of all possible such states, and by $\Im$ the σ field on Ω generated by the individual states, one would like to think of our simple model as a measure μ on $(\Omega, \Im)$. In other words, it would specify the probability that the model or economy will be in any particular subset of states. More particularly, it will tell us which states for individual agents are consistent with each other. (Of course, if *both* A and S are finite then matters are greatly simplified from a formal point of view.) An important observation should be made at this point, for it will recur in much of what follows. A state w of the economy is a very full description because it specifies precisely which individual is in precisely which of the possible individual states. For many applications such detailed information will not be necessary, and agents may only be influenced by some summary variables such as "how many agents are in each state." In this case, there will be equivalence classes of states and this enables one to use alternative mathematical approaches which, in the dynamic case in particular, can simplify matters. When I come to discuss the use of this sort of model to analyze the transition from one state of the economy to another, the importance of this will become evident.

To return to the situation in which the full specification of the states is used as Föllmer did, the problem is that once we allow for interdependence there may be none or many global measures which are consistent with the individual specifications of the conditional probability laws for each agent. A simple relationship or even any consistency between microeconomic characteristics and macroeconomic characteristics may be lost. Recalling that the microeconomic characteristics of the agent a are given by the probability law $\pi_a(\cdot|\eta)$, then a measure μ on $(\Omega, \Im)$ is compatible with the underlying microeconomic characteristics if

$$\mu[w(a) = s|\eta] = \pi_a(s|\eta) \quad \mu \quad \text{almost surely} \quad (a \in A, s \in S).$$

Such a measure is called a *macroeconomic phase* of the economy, and individual laws are *consistent* if they admit at least one macroeconomic phase. To see the sort of problem that may arise, even when one does have consistency, consider the following example due to Allen.[1] She discusses the adoption of one of a finite number of technologies by firms. Thus, the state of a firm can be described by the technology that it has chosen. By making the assumption that all agents have a positive probability of being in every state, regardless of the choices of the others, and by assuming that there is a finite number of agents, she is able to use a standard theorem (see Spitzer[100]), which guarantees that there is a unique global phase or, put alternatively, that there is a unique distribution of adoptions of technologies that is consistent with the underlying microcharacteristics. She gives a simple example in the one technology case to show why her positive probability condition is necessary. Let 1 indicate the state of having adopted the technology and 0 represent not having adopted it. Let there be two agents, i.e.,

$$A = \{1, 2\}$$

and consider the following local characteristics

$$\pi_a(1|0) = \pi_a(0|1) \quad \text{for} \quad a = 1, 2\,.$$

Then μ such that

$$\mu(1,1) = q \quad \text{and} \quad \mu(0,0) = 1 - q$$

is consistent with π for q in $[0, 1]$, and thus, there are an infinite number of global phases. Thus, even though one has a precise specification of the conditional probabilities of being in each state for each individual firm, one has no idea of what the probability of any aggregate state is. It is clear that the problem here is produced by the fact that there is a total interdependence of the agents. In the case with a finite number of states for the agents and a finite number of agents, any relaxation of this dependence restores a simple relationship between individual and aggregate probability laws.

Föllmer's[48] aim was to study a problem and his work received a considerable amount of attention. If individuals in a simple exchange economy have characteristics, i.e., preferences and endowments which are stochastic rather than deterministic, will it be true that, if the economy is large enough the set of equilibrium prices will be the same regardless of the particular characteristics "drawn" by the individuals? Previous authors such as Hildenbrand,[61] Malinvaud,[84] and Bhattacharya and Majumdar[19] also studied random economies and looked for equilibrium prices when these economies were large enough. In each case some law of large numbers prevailed so that the link from micro to macro did not produce any difficulties. The importance of Föllmer's contribution was to show how important interdependence had to be before the sort of law of large numbers, invoked by the authors mentioned, broke down. Thus, the essential idea here is to understand under what circumstances we can effectively ignore the interaction between the agents in the economy and treat the economy as behaving as it would, loosely speaking, "on average" or "in the mean." Föllmer's particular question is whether this will be true if there are "enough" agents. It was precisely the fact that he was dealing with economies in which the number of agents tended to infinity that meant that the standard result, invoked above, could not be used. However, in many applications that involve interacting agents, the assumption of an arbitrarily large number of agents is made, and without it no law of large numbers could be used. Paradoxically then, from a mathematical point of view, since the question asked by Föllmer necessitates an infinite set of agents, it makes things more complicated. In particular, it does not rule out the case of multiple phases and of there being no price vector which equilibrates all of them.

One way of bringing back finiteness of dependence and, thus, to restore the unique relationship between individual and aggregate probability laws is by assuming that agents are only influenced by a finite number of "neighbors"; but this requires the specification of a graph-like structure on the agents and the study of local interaction to which I will come back later.

Though FöllmerFöllmerH. considered a static problem, most of the recent work on interaction and interdependence using his random field approach has been in the context of dynamic stochastic processes, and I will come back to this. And it will become clear that the problem of the nonuniqueness of the global phase plays an important role in characterizing the nature of aggregate behavior.

SOME DYNAMIC MODELS OF GLOBAL INTERACTION

Up to this point I have done no more than to consider the relation between individual but interactive behavior and aggregate outcomes as reflected in some static equilibrium notion. I would now like to consider models, which while retaining the global or uniform communication structure just described, consider the dynamic evolution of the aggregate behavior resulting from individual interaction. In so doing, they capture some important features of the aggregate phenomena one would like to analyze. In each of the three classes of models I shall consider, the interaction among individuals may result in quite complicated aggregate dynamics. In some cases there will be convergence to an equilibrium, in the sense of a solution, which will then remain unchanged over time. In others, the state at any time will change continually over time, and the appropriate equilibrium notion will be some sort of limit distribution of the process itself. The difference between the two sometimes rests on apparently small changes in the structure of the underlying model. The three categories of model I will treat are firstly, those which involve "herd" or "epidemic" behavior such as those of Banerjee,[16] Bikhchandani et al.,[20] Sharfstein and Stein,[98] and Kirman,[76] and which are often thought of as being particularly applicable to financial markets. Here an important feature will be the fact that individuals infer information from the actions of others.

Secondly, there is the type of model developed by Arthur[10] and David[31] to explain the adoption of new technologies when the profitability of a certain technology for one firm depends on the number of firms that have already adopted it. Thus the significant feature, in this case, is the existence of an externality deriving from the adoption of technologies by other firms. This can also be thought of as the existence of increasing returns at the industry level.

Thirdly, there is the literature on the evolution of populations in which players are identified with strategies and are randomly matched against each other, and play a game such as Prisoner's Dilemma. The distribution of strategies then evolves according to their relative success (see, e.g., Axelrod,[14] Young and Foster,[113] and Lindgren,[81] and the survey by Blume[24] pp. 425-460 in this volume).

The very notion of "herd behavior" suggests that there is some tendency for individuals to follow the choices made by other members of the population. This tendency may be for no special reason other than fad or fashion, or it may be that agents deduce from the actions of others that they have some information concerning the value of those actions. An example is given by Banerjee.[16] In his model, individuals sequentially choose options from those indexed on a line segment. One of these options is profitable, the others are not. Each player receives a signal with probability a, and this signal is correct with probability β. People choose their options sequentially. Thus, observing individuals' previous choices may reveal information about the signals that they have had. Banerjee looks for a Bayesian Nash equilibrium and finds that the equilibrium outcome will be inefficient from a welfare point of view, if the population is large enough. Furthermore, he shows that the probability that none of the N players will choose the "superior" option is bounded away from zero and that the equilibrium pattern of choices is highly volatile across different plays of the same game.

A very simple example explains the origin of the Banerjee problem. There are two restaurants A and B. Assume that B is, by some objective standard, better. A public signal, which has 51% probability of being accurate, and which is observed by everyone, says that restaurant A is better. However, 100 potential clients receive private signals, which are known to be correct with 99% probability. Of the 100 clients, 99 receive a signal that B is better and 1 that A is better. Thus, the aggregate information given by all the private signals together, were it to be revealed, indicates that B is practically certainly better. Yet all the clients may be led to make the wrong choice. To see this, suppose that the player who received signal A chooses first. He will choose A since his private signal reinforces the public signal. The second player observes signal B. Since all private signals are of equal quality, and he knows from the first player's behavior that he received signal A, the two private signals cancel out. He chooses A since he has only the public signal to go by, and this indicates the choice of restaurant A. The third player is left in a situation where he can infer nothing from the second player's choice, and is, therefore, in the same position as that player was and will choose A. By the same reasoning the unsatisfactory result occurs that all clients end up at the worse restaurant. Paradoxically, both welfare loss and instability would be reduced by preventing some people from using other than their own private information. Thus, reducing interaction would be better for all involved. The instability that occurs when the Banerjee market is repeated is strongly related to the feedback between players, and what is of particular interest is the crucial role played by the sequential structure of the moves. Bikhchandani et al.[20] emphasize the fact that after a sufficient time the cumulated actions of other actors contain so much information that an individual will have an incentive to ignore his own information and a "cascade" will start. In their model, as a cascade develops, it becomes virtually impossible for it to be overturned by an individual making a choice based on his own information rather than on following the crowd.

In Kirman,[76] the evolution of two opinions over time in a population is discussed. The basic idea was stimulated by the observed behavior of ants who, when faced with two apparently equally productive food sources concentrate largely on one and then focus their attention on the other. This is due to the recruiting process which is such that the more ants are being fed the stronger the trail to the source and the higher the probability that an ant leaving the nest will go to that source. Using a simple stochastic model developed jointly with Hans Föllmer, it is shown that provided there is a minimal amount of "noise" in the system the proportion of ants feeding at each source will stay close to 1 or to 0 for a long time and then switch to the other extreme.

The feedback involved can either be thought of as a stronger trail or if recruiting is of the tandem type, of a higher probability of meeting a successful forager from the food source that is currently most frequented. The appropriate equilibrium notion here is then not some fixed proportion but rather a *limit distribution* of the underlying stochastic process. Thus, thinking of the state of the system as k/N, the proportion of ants at the first source, we can write $f(k/N)$ for the limit distribution, and this should be viewed as the proportion of time that the system spends in any state k/N. Föllmer[2] shows that if one lets N become large, and approximates f by a continuous distribution $f(x)$ where x takes on values between 0 and 1, then this distribution will be a symmetric beta distribution, i.e., of the form

$$f(X) = X(1 - x)\,,$$

and with appropriate assumptions on the parameters of the original model the distribution will be concentrated in the tails. Thus, the process will indeed spend time at each limit and occasionally switch between the two.

In a symmetric situation with a priori symmetric agents, aggregate behavior displays violent stochastic swings.

This sort of model can be applied to a market for a financial asset as is done in Kirman,[74] the abruptness of the change where is amplified by including into agents' observations a signal concerning which opinion is held by the majority. Observing the behavior of the majority provides additional information over and above that obtained by direct encounters. One has, of course, to model the market in such a way that agents do actually gain by acting with the majority.

This model is related to that developed by Topol,[106] for example, who refers to "mimetic contagion." Ellison and Fudenberg[43] also develop a model in which both how well a certain choice has done but also its "popularity" are considered. They apply their model to the choice of technologies, and I will turn to this in the next section.

[2] Private communication.

TECHNOLOGICAL CHOICE

It has long been argued that as a result of externalities the value of choosing a technology may be enhanced by the fact that other firms have previously chosen it. This idea has been formalized by Arthur.[10] He gives very simple examples of how the presence of an externality due to the number of firms that have already adopted a technology may lead a whole industry to become "locked in" to an inferior technology. Although each individual acts in his own interests the system does not generate the best collective solution. Examples of such situations are the adoption of the QWERTY keyboard for typewriters, analyzed by David,[31] or the construction of light-water nuclear reactors in the U.S. in the 1950s (see Cowan[30]).

The formal model, developed by Arthur et al.[11,12] is very simple and bears a family resemblance to the Föllmer-Kirman process mentioned earlier.

There are K technologies, x is the K vector of current shares of firms using each technology. *Initial conditions* are given by the vector

$$y = (y_1, \ldots, y_k)$$

and the total

$$w = \sum_{i=1}^{k} y_i \, .$$

Consider the vector

$$y_n = (y_{n1}, \ldots, y_{nk}) \, ,$$

which is the vector of firms attributed to each technology after $n - 1$ additional units have been added to the w initial ones.

Before we can discuss the evolution of this process, we must specify how the additional units are allocated. Let p define the vector of probabilities for attracting a new firm and denote by x_i the proportion of firms using technology i, then we have

$$p = (p_1(x), p_2(x), \ldots, p_k(x))$$

where $p_i(x)$ is the probability that a new firm will adopt technology i given the proportions that have currently adopted each of the technologies.

Note that, as Arthur[9] remarks, one cannot invoke a standard Strong Law of Large Numbers in order to make statements about long-run proportions, since the increments are not added independently. However, it is clear what a long run equilibrium of this system, if it exists, should be. The mapping p takes the k simplex S^k into itself. Therefore, an equilibrium will be a fixed point of this mapping.

The evolution of the process of the proportions x_n will be described by

$$x_{n+1} = x_n + \frac{1}{(n+w)}(b(x_n) - x_n) \qquad x_1 = \frac{y}{w}$$

where $b(x_n)$ is the jth unit vector with probability $p_j(x_n)$. Rewriting the above expression we get

$$x_{n+1} = x_n + \frac{1}{n+w}(p(x_n) - x_n) + \frac{1}{n+w}\varepsilon(x_n)$$

where ε is a random vector given by

$$\varepsilon(x_n) = b(x_n) - p(x_n)\,.$$

The point of this is that we can decompose the system into a deterministic and a random part. Since the conditional expectation of ε given x_n is zero, we can derive the expected notion of the shares as

$$E(x_{n+1}|x_n) - x_n = \frac{1}{n+w}(p(x_n) - x_n)$$

and we can think of the *deterministic* component of the system or the *equivalent deterministic system* as

$$x_{n+1} = x_n + \frac{1}{n+w}(p(x_n) - x_n)\,.$$

Now the question: does this system converge and if so to what? Let p be continuous[3] and consider the set B of fixed points of p. Arthur[9] gives the following results which are generalized in Arthur et al.[13]

THEOREM. (Arthur[9]) If $p : S^k \to S^k$ is continuous and such that the equivalent deterministic system of x possesses a Lyapounov function v whose motion is negative outside $B = \{x|p(x) = x\}$, then x_n converges with probability one to a point z in B.

Suppose p maps the interior of S^k into itself and that z is a stable point, then the process of x_n converges to z with positive probability.

If z is a nonvertex unstable point of B then x_n does not converge to z with positive probability.

The intriguing point of this analysis, which contrasts with that given in Kirman[76] is that the industry *does* settle down to fixed proportions. This is due to two things. Firstly and importantly, the number of firms is continually increasing which gives qualitatively different results from the fixed population urn model used in the Föllmer-Kirman type of process. Secondly, the restriction that the equivalent deterministic system be a gradient system is not an innocent one and rules

[3] p can actually be time variant. In other words at time n we have p_n. However, although this has the advantage of generating more initial fluctuations, Arthur et al.[8] require that the sequence $\{p_n\}$ converges "relatively rapidly" to a limit p, at which point the analysis reduces to that given above.

out cycles for example. Under certain assumptions the Arthur process will actually converge to a vertex point, in which case one technology dominates the whole industry.

In modeling technological change, it is by no means clear whether the assumption of a fixed number of firms adopting different technologies or being replaced by new firms is better than that of an arbitrarily growing number of firms. Which model is chosen certainly has consequences for the nature of the results. Lastly, it should, of course, be said that the set of K technologies available should expand over time to take account of the arrival of new candidates. This, however, would add considerably to the analytical complications. Again, it is worth insisting on the fact that the nature of the equilibrium depends crucially on these assumptions.

Returning to the Ellison and Fudenberg[43] model mentioned earlier, they show that using the information given by the popularity of a technology can lead to the general adoption of the most efficient technology. The idea is simple. If agents are faced with payoffs from one of the two technologies from which they can choose, which are subject to shocks, but not all agents modify their choice at each time, then watching the evolution of the choices of others reveals information about the success of those choices. If those agents who can change simply choose the best technology for the last period without looking at other firms, the result will be a limit distribution over the states of the system x_t when x is the fraction of individuals who choose technology 1. Moreover, the average of x reflects the probability in any one period that technology 1 will give a payoff superior to that of technology 2.

Now, however, if players also choose by attaching some weight to the popularity of a technology, then Ellison and Fudenberg show that if the popularity weight is chosen appropriately the system will converge with probability one to a state in which everyone uses the best technology.

In a different model, Ellison and Fudenberg[44] look at the consequence of sampling other agents to obtain information about their experience. Surprisingly perhaps, the social learning is most efficient when communication is limited, i.e., sample size is small. Fluctuations may occur perpetually if too large samples are taken. This is, of course, related to the fact that in their earlier model there was an optimum weight to put on what amounted to a complete sampling of the population.

Once again, the extent to which agents interact has a crucial bearing on the type of evolution that may occur. Two features of the models discussed are also common to many models with interaction. Firstly, the important aspect of many of them is that there is some summary statistic for the state which conveys all the agents need to know as information. For example, in the models of technological adoption this might be the proportion of firms having adopted each technology. Such a variable or vector is often referred to as a "field variable" (see Aoki[5]). Thus, a full description of the state of the system is not necessary. This often considerably simplifies the analysis. Secondly, the underlying formal structure implicit in the study of the evolution of a field variable, such as the vector of proportions of agents adopting different technologies or strategies, involves the study of a mapping from a simplex into itself. As will become apparent, this is the basis of much of the dynamic analysis in the models I will describe. The differences between them lie essentially in the particular specifications of the mappings involved.

EVOLUTION IN GAMES

Rather than attribute the full capacity to undertake game theoretic reasoning to players, an alternative approach has been to ask whether strategies playing each other at random will "evolve" toward an equilibrium. A substantial literature addressing this question has developed recently. The basic idea is to identify each player in a large population with a particular strategy. The players are then randomly matched with each other and play a particular one-shot game. Successful strategies reproduce faster than unsuccessful ones and one studies the evolution of strategies, (see, e.g., Hofbauer et al.,[66] Zeeman,[114] Schuster and Sigmund,[97] Bomze,[25] Hofbauer and Sigmund,[65] Foster and Young,[49] Lindgren,[81] and, for a survey, Blume[24]).

As I have said, the basic notion involved in this literature is that the proportion of players playing a particular strategy should grow according to the relative success of that strategy. Therefore, an equilibrium will consist either of a point in the n simplex where n is the number of available strategies or of a distribution over that simplex. One might hope, for example, that the equilibrium, if it did not consist of a unique point in the simplex, would consist of a distribution with all the mass concentrated on some of the Nash equilibria of the underlying game. Note that we are once again studying a mapping from an n simplex into itself.

Since the models involved are either based on, or invoke, biological examples, it is worth mentioning two basic differences between biological and economic models, which should be kept in mind in the following discussion. Firstly, the notion of "fitness," which is the biologist's criterion for success, is not always well defined in economic models. Why should the payoffs from the game reflect fitness? Since payoffs are measured in terms of the utilities of the players, for otherwise, taking expectations would make no sense, what does it mean for all players to be playing "the same game." This can only make sense if the population consists of identical players, but this is highly restrictive and raises the spectre of the "representative agent" again. Secondly, removing any possibility for players to reason may be excessive. One interpretation is that players and their associated strategies do not die out but rather that players with unsuccessful strategies learn to use more successful ones. Furthermore, human players may try to calculate the future consequences of their actions and, thus, the typical myopic behavior attributed to them may not be appropriate. An a priori awareness of the goal and the desire to head in that direction may not only accelerate the evolutionary process but may also modify it.

(A discussion of this sort of problem may be found in Banerjee and Weibull.[17]) However, the relationship between individual rationality, however sophisticated, and fitness is tenuous (see, e.g., Samuelson[94]).

Several approaches to modeling the general problem of the evolution of strategies as a function of the results from interaction with other players, have been suggested. One approach is to consider a deterministic rule, which maps proportions of players using each strategy at time t into proportions at time $t+1$. Then one defines the "replicator" dynamics and examines the basins of attraction of the dynamic process. This, of course, is a special case of the general framework mentioned earlier. A second approach is to loosen the identification of individuals with strategies as suggested above, to take into account the experience and information of players and to allow them to adjust their strategies individually (see, e.g., Nachbar,[91] Friedman,[50] Kandori et al.,[72] Samuelson and Zhang[95]). A last approach is to introduce stochastic elements into the replicator dynamics. Axelrod's[14] Repeated Prisoner's Dilemma tournament has been adapted, for example, by Foster and Young,[49,113] to allow for uncertainty and this significantly changes the result that cooperation emerges over time. They consider three strategies, "cooperate," "defect," and "tit-for-tat." Once again using the evolution of the proportions as represented in the simplex in R^3, they show that most of the mass of the limit distribution is near the apex corresponding to all players "defecting." This is more pronounced when noise is small, and this casts doubt on the notion that "cooperative" behavior naturally involves in such games.

Let me first briefly examine the case of deterministic replicator dynamics. Consider strategies in a normal form game with a large population N of players, and let each of them be identified with a particular strategy. Now, the outcome at each point will depend on the proportion of individuals who have chosen each strategy. In game theory, as I have said, the usual idea is that individuals meet each other at random and then play a game with the particular opponent they draw. In large populations the result of this will coincide with the expected payoffs at the outset. Consider a matrix A, the elements a_{ij} of which give the payoff to the player using strategy i when his opponent uses strategy j. Denoting the ith row of A by A_i and the n vector of the current proportions of the population using each strategy as p then the expected payoff of strategy i is given by

$$s_i = A_i P \,.$$

Denoting by s the average expected payoff over the n strategies, then the change in the proportions is assumed to be given by

$$dp_i(t) = s_i - s \,.$$

One can then look for the rest points of the system and examine their stability.

This brings us to an idea introduced by Maynard Smith.[86] He suggested that a situation could be called "evolutionarily stable" if any small proportion ε of invading mutants is subsequently eliminated from the population. In a two strategy game the idea of evolutionary stability has a natural interpretation. A strategy $\hat{p}$ is evolutionarily stable if $(\hat{p}, \hat{p})$ is a Nash equilibrium, that is if $\hat{p}$ is a best reply to itself *and* if for any other best reply $\tilde{p}$ to $\hat{p}$, the latter must be a better reply to $\tilde{p}$ than $\tilde{p}$ itself.

As Binmore[22] explains, evolutionary stability was thought of as a useful concept since it would mean that the way in which the population evolved over time would not have to be studied in detail. Thus, if we consider the n strategy situation, what can be shown is that if p is evolutionarily stable then it is an asymptotic attractor of the replicator dynamics. Thus, checking for evolutionary stability permits us to identify the attractors of the dynamics of the system. In particular, when there are two players, Nash equilibria play a special role. As Bomze[25] showed, stable points in the replicator dynamics are Nash equilibria and Nash equilibria are stationary in the replicator dynamics. Yet, when more than two strategies or technologies are available there may be no evolutionarily stable situation.

However, we do know something about how rational strategies have to be for them to survive, Weibull[108] shows that any strictly dominated strategy will be eliminated along a convergent path of the replicator dynamics. Samuelson and Zhang[95] show that even if the dynamic path does not converge, any nonrationalizable strategy will be eliminated from the population. Unfortunately, as Dekel and Scotchmer[33] have shown, this is not true in discrete time models.

Nevertheless, the idea of looking at the evolution of strategies over time in this way in a large population leads to the examination of time paths that are characterized by populations which are polymorphous, at least as far as their strategies are concerned, that is where heterogeneity persists. Furthermore, it is already clear that the evolution of the aggregate system, even in these simple models with random matching, is sensitive to the precise specification of the model.

As I mentioned above a great deal of attention has been paid to the Repeated Prisoner's Dilemma and much of this stems from Axelrod's famous tournament in which various strategies were matched against each other. The winner in that tournament, and in a subsequent one, was "tit-for-tat," which starts out by cooperating and then imitates the last action of the opponent. This led to the idea that cooperation would emerge as the title of Axelrod's[14] book indicates.

However, the work initiated by Foster and Young, (see Foster and Young[49,115] and Young and Foster[113]) shows how vulnerable such a conclusion is to the presence of a small amount of uncertainty. They introduce this uncertainty through several channels. They make the length of the playing time for each encounter genuinely stochastic and allow for mutations and immigration which always maintain the presence of all strategies. They then examine the evolutionary dynamics and define the set S of "stochastically stable equilibria." The basic idea being that the system will almost certainly be in every open set containing S as the noise goes to zero. This notion is important since, unlike the idea of an evolutionarily stable strategy, it takes into account the fact that perturbations are happening regularly and are not just isolated events. In a repeated game they show how each strategy in turn will come to dominate for a period before being superseded by the next.

All of this is done within the context of a fixed finite number of strategies. However, Lindgren[81] allows for the possible introduction of new strategies as a result of evolution. Any finite memory strategy in the repeated Prisoner's Dilemma framework can be written as a binary sequence which specifies the history of the actions taken by both the players and the action to be taken. This "genetic code" can then be modified by point mutations, which change one of the elements in the sequence; by split mutations, which divide the sequence in half and retain one or the other half; and by gene duplication, which duplicates the sequence. Thus, although the system starts out with strategies of a fixed memory length other strategies with different memory lengths enter the population. The evolution of the system is complicated, with periods in which strategies of increasing memory length dominate and then periods of extinction. There are also periods in which the average payoff is low as a new mutant exploits the existing species. Such an open-ended system can exhibit a number of interesting features, but it is worth noting that the evolutionarily stable strategy still plays an important role.

MARKET EVOLUTION: A BRIDGE TO LOCAL INTERACTION

A further line of research involving the evolution of the strategies of a population is that which has been pursued by Vriend.[107] He constructs a very simple model of a market in which stores can sell units of a perishable good to consumers, each of whom requires one unit. The price is fixed exogenously. Stores can choose how many units to stock and how many random "signals" to send out to consumers. Buyers can choose to buy at the store at which they were previously satisfied, to go to a store from which they received a signal or to search randomly. Individuals do not reflect strategically as to where they should shop but rather simply learn from experience. They update the "value" of each choice using a classifier system and, in addition, firms use a genetic algorithm to develop new rules. The results are striking. The distribution of firm sizes becomes rather dispersed, consumers have different behavior and the market almost clears, 95% of consumers being satisfied. Two features are of particular interest. Firstly, although the size distribution of firms is stable, individual firms do move around within that distribution. Secondly, although of no intrinsic value, signals persist and even increase in quantity. This is reminiscent of the role of "cheap talk" in games. In fact, here the problem is more severe, since producing signals uses resources. Thus, if a stable situation, in which buyers were sure to be served and sellers knew how many buyers they would have in each period, could be achieved without signals it would be more efficient than any arrangement with signals. Unfortunately, buyers learn to attach value to signals they receive, and sellers learn that buyers are influenced by signals. This situation is self-reinforcing and produces a feature of the market which is stable but inefficient.

Thus, Vriend's paper makes several important contributions. First, it shows how asymmetric behavior can result from an initially perfectly symmetric situation. Although not game theoretic in the normal sense, once again one can think in terms of the persistence of a variety of strategies. Second, it shows how agents can learn to attach importance to intrinsically unimportant phenomena, and this is reminiscent of the literature on learning to believe in "sunspots," (see Woodford[112]). Last, it has one additional feature which is that although the interaction is global at the outset it becomes, at least partially, local, as certain consumers become tied to certain firms. This leads naturally to the next class of models.

LOCAL INTERACTION: SOME STATIC NETWORK MODELS

The thrust of this paper is that it is important to consider economic models in which the "states" of individuals depend on the states of others. Up to now I have only looked at situations in which individuals are affected by all the other members of the population. This influence could, as in the model proposed by Föllmer, be based on the full description of the states of all the agents or, as in many models, on some "field variable" that summarizes the distribution of the states of the individuals. Yet, in many situations agents are only in contact with, or influenced by a limited number of other individuals. Thus, agents will only be influenced by a subset of the others and these might be thought of, in some sense, as those who are their "neighbors."

To look at this problem, I shall now turn to models in which there are links between specific agents, that is there is a graph of which the agents are the nodes and a notion of "nearness" can be defined. In this context, each agent is only influenced by a limited (finite) number of other agents who are near to him and this "influence structure" is defined a priori. I will come back to discuss the forms that this structure might take a little later, but it is worthwhile here looking again at Föllmer's[48] model and his use of neighborhoods and, in particular, his introduction of the notion of a "Markov random field" which has been widely readopted recently.

In order to examine the consequences of interaction, one has to specify the nature of the communication graph linking agents. A typical assumption as to the graph and one which is frequently used in spatial models, and which was used by Föllmer, is to think of the links as forming a lattice structure. In this case the neighborhood $N(a)$ of agent a might consist of those individuals who were directly linked to agent a. More generally, however, the idea is that once the graph is specified, one assumes that the agent a is influenced only by his neighbors in $N(a)$ where $N(a)$ is defined as the set of agents less than a certain "distance" from a. Influence is used in the following sense

$$\pi_a(\cdot|\eta) = \pi_a(\cdot|\eta') \quad \text{if } \eta \text{ coincides with } \eta' \text{ on } N(a)\,. \tag{1}$$

That is, individuals' probabilities of being in a certain state are conditioned only on the states of their neighbors. Now consider the class of economies with local probability laws, which are consistent (i.e., such that there is an aggregate probability law consistent with the local ones) and satisfy (1). These are called Markov economies. Furthermore, assume that the economy $\mathcal{E}$ is homogeneous, by which we mean that π is *translation invariant*, i.e., that every agent reacts in the same way to the states of the other agents. The phase μ is said to be homogeneous if it is a translation invariant measure. Now as Föllmer explains, two things can happen. There may be more than one homogeneous phase, and there may also be phases which are not homogeneous. The latter case in which the conditional global probabilities may vary from agent to agent, is called a symmetry breakdown.

The importance of Föllmer's contribution is that he shows that even with his completely symmetric local structure, nonhomogeneities may arise at the aggregate level. His question was, if there are many agents in an economy and they all draw their characteristics in the same way, can one find prices which will equilibrate the market, independently of which agent draws which characteristics. What he shows is that this is not the case and it may not be possible to equilibrate markets. The point here is that no *heterogeneity* is required at the individual level to produce these difficulties if interaction is strong enough. Put another way, if agents are sufficiently "outer directed," that is their characteristics depend sufficiently on their neighbors, one can have the impossibility of equilibrium prices. Föllmer gives, albeit extremely stylized, examples of Ising economies in which this occurs.

Let me reiterate the essential point at this stage which is that strong local random interaction amongst agents who are *a priori identical* may prevent the existence of equilibrium. This is in stark contrast to the determinate model without this type of interaction.

THE NEIGHBORHOOD STRUCTURE OF AN ECONOMY

As soon as we introduce, as Föllmer does, a structure for the neighborhoods of the agents in an economy, we are obliged to answer two questions. Firstly, what is the appropriate relationship to consider? Its geographical distance as in locational models (see Gabszewicz and Thisse,[51] for example), closeness in characteristics (see Gilles and Ruys[53]) or, if one is considering transactions, the probability of interaction as being related to the potential gains from bilateral trade?

In any event, to discuss local interaction we must effectively impose some graph-like structure on the space of agents. One approach introduced by Kirman[73] and developed by Kirman, Oddou, and Weber[77] and Ioannides[67] is to consider the graph structure itself as random. Once the communication network is established, then one has to define a corresponding equilibrium notion and study the characteristics of the equilibrium that arises from the particular structure one has chosen. The difference between this and the approach adopted by Föllmer and those such as Durlauf[36] and Benabou[18] is that while they impose an a priori locational relationship, the stochastic graph approach can allow for communication between agents who are "arbitrarily far" in the underlying structure. In the papers cited above, the basic idea is to consider a set of agents A and impose a probabilistic structure on the links between them. Consider p_{ab} as the probability that individual a "communicates" with agent b. In graph terms, this is the probability that an arc exists between nodes a and b. The graph is taken to be undirected, i.e., the existence of the arc ab implies the existence of ba and thus one way communication is ruled out.

In the case of a finite set A this is easy to envisage and the resulting stochastic graph can be denoted

$$\Gamma(p_{ab}) .$$

If there is no obvious underlying topological structure, then one could consider $p_{ab} = p$, that is the probability of interaction is the same regardless of "who" or "where" the individuals are. On the other hand the structure can, of course, be reduced to the sort of local interaction mentioned earlier if one has an underlying "geographical" structure (see also Gabszewicz et al.[51]) and one conditions p_{ab} on the proximity of a and b. The extreme case in which an integer lattice structure exists and agents communicate with their nearest neighbors is a special case of the stochastic graph with $p_{ab} = 1$ if distance $[a, b] \leq 1$.

Gilles and Ruys,[53] working within the deterministic approach developed by Myerson[89] and Kalai, Postlewaite, and Roberts,[71] adopt a more ambitious approach and use the topological structure of the underlying characteristics space to define the notion of distance. Thus, agents nearer in characteristics communicate more. To an economist interested in how trading relationships are established this seems a little perverse. In general, the possibility of mutually profitable trade increases with increasing differences in characteristics. Identical agents have no trading possibilities. However, there are other economic problems, the formation of unions and cartels for example, for which this approach may be suitable. The important point is that the relational structure is linked to, but not identical with, the topological characteristics structure.

Haller[60] links the Gilles and Ruys approach to the stochastic one by basing communicational links directly on the topological structure of the attribute space.

The basic aim of all of these models of interaction through communication is to look at equilibria which depend on the structure of the "connected components" of the graphs involved. "Connected" here means simply that a graph Γ_A is connected if for every a and b in A there exist arcs $(a, c)(c, d) \ldots (d.b)$ linking a and b. When looking at finite graphs, the idea of the distance between a and b is simply the number of arcs in the shortest path linking the two. The *diameter* of a graph Γ_A, i.e., a graph with the set of nodes A, is given by

$$D(\Gamma_A) \equiv \text{Max} \quad \text{distance}\,(a, b) \quad (a, b) \in A .$$

If individuals interact with a certain probability, and these probabilities are independent, it is easy to see that the appropriate tool for analysis is the random graph $\Gamma(p_{ab})$ mentioned earlier where p_{ab} specifies the possibility of an arc between two elements, a and b, of the underlying set of agents A existing. Thus, a particular realization of the random graph would be a deterministic graph, undirected if a communicates with b implies b communicates with a, and directed if one is possible without the other.

The sort of results that can be obtained using this approach involve, for example, linking the allowable coalitions to their connectivity as measured by something like the diameter. Typical examples are those obtained by Kirman, Oddou, and Weber[77] and Ioannides,[67] and in a survey by Ioannides.[68] The first of these contributions considers the problem of the core and its relation with the competitive equilibrium. Recall that the core of an economy is a cooperative notion and is the set of allocations to which no coalition has an objection. A coalition is said to have an objection to an allocation if it could, with the resources of its members, make all those members better off than they were in the proposed allocation. One typically assumes that all coalitions can form. Suppose however, that one only allows those coalitions to form that have a certain diameter, in particular diameter = 1 or diameter = 2. In the first case everyone in a coalition must communicate with everyone else, and in the second, every pair of individuals in a coalition must have a common friend in that coalition. Now consider a simple economy $\mathcal{E}$ with agents A. For each realization of the random graph on the agents A, the core of $\mathcal{E}$ is well defined. Thus, the core $C_i(\mathcal{E}, \Gamma(p_{ab}))$ is a random variable which depends on the underlying economy, on the random communication network, and on the criterion for the permissible coalitions. The choice of the latter is denoted by i which takes on two values, 1 and 2, depending on which diameter is required. Now, a classic result says that in large economies core allocations are essentially competitive, or Walrasian allocations. If we consider a sequence of economies $\mathcal{E}_n$ and define an appropriate "distance" between the core $C(\mathcal{E}_n)$ and the set of Walrasian equilibrium allocations $W(\mathcal{E}_n)$ which is independent of the graph structure, then a statement such as the distance between the core and the set of Walrasian equilibria of the economies goes to zero as n becomes large, can be proved.

In our case, the distance depends on the realization of the core but, for any particular economy $\mathcal{E}$ in the sequence, the expression

$$p(dist(C_i(\mathcal{E}, \Gamma(p_{ab}))W(\mathcal{E})) \leq \mathcal{E} \qquad i = 1, 2$$

is well defined.

What we would like to show is that limiting interaction or communication, in the way suggested, does not prevent the core from "shrinking" to the competitive equilibrium.

Consider again the sequence of expanding economies $\mathcal{E}_n$ (see Hildenbrand and Kirman[64]) and for each n let

$$p_{abn} = p_n, a, b \in A_n\,.$$

Now, Kirman, Oddou, and Weber[77] obtain the following two results, stated rather imprecisely,

DIAMETER ONE COALITIONS (STRONG CONDITION). For any $\varepsilon > 0$

$$\lim_{n\to\infty} p(dist(C_1(\mathcal{E}_n, \Gamma(p_n), W(\mathcal{E}_n)) \leq \varepsilon)) = 1$$

if $p_n \leq 1/\log N$ where $N = \#A_n$.

DIAMETER TWO COALITIONS (WEAKER CONDITION).

$$\lim_{n\to\infty} p(dist(C_2(\mathcal{E}_n, \Gamma(p_n), w(\mathcal{E}_n)) \leq \varepsilon)) = 1$$

if $p_n \leq 1/\sqrt{N}$ where $N = \#A_n$.

In the diameter 2 case the essential result is one due to Bollobas, which shows that provided the probability that two agents communicate does not go down faster than one over root N, then any large enough coalition will have almost certainly a diameter of 2. Thus, in contrast with the usual observation, it is clear that the probability of any two individuals having an acquaintance in common increases as the number of agents increases. Hence, one should remark that "It is a big world" rather than the contrary. Obviously then, the strategy in the diameter 2 case is to use large coalitions, of which there are enough. However, in the diameter 1 case we have to rely on small coalitions, but there are plenty of these.

Ioannides'[67] contribution is an examination of the evolution of the random graph describing the communication network in his economy as the number of agents becomes large. In his model, the graph becomes one with one large connected component and several smaller disjoint connected ones. He considers that Walrasian equilibria are only meaningful for economies in which the agents are linked. As a consequence, a situation persists in which one price prevails in most of the economy whereas there are small islands in which the equilibria are characterized by other prices. Thus, here again, the aggregate equilibrium outcome is directly affected by the nature of the stochastic interaction between the individual agents, though here, in contrast to the previous model, connectedness is not restored as the economy becomes large since there are always isolated "islands" of individuals.

LOCAL INTERACTION: DYNAMIC MODELS

Here, I will consider some examples in which there is local interaction and in which the dynamic evolution of the aggregate behavior and the existence of an appropriately defined equilibrium is studied. The classic approach derived from physics to this problem is to use the Markov random field approach mentioned already in the contest of Föllmer's[48] model.

Indeed it might appear that the Markov random field approach might apply to any local interaction process that might be encountered in economics. This is, unfortunately, not the case without some additional assumptions and, as was already mentioned, in the case of Allen's[1] contribution a typical problem is one which arises when not all switches of state are possible in all configurations of neighbors' states.

To see the sort of problem that can occur in a local contagion model, consider the following example. There is an integer lattice in which each site is "infected" (state 1) or not (state 0). The infection is transmitted stochastically but irreversibly to the nearest four neighbors, and the probability $\Pi_a(1)$ that agent a is in state 1 is positive if and only if at least one neighbor is in state 1. Suppose that the process is initiated by setting the two sites 0,0 and 0,1 to state 1. Clearly an event such as that shown in Figure 1 has probability 0.

But now consider the situation in Figure 2. If we are to have consistency then the site u, v must have probability 0 of not being infected, otherwise we have the event in Figure 1 with positive probability. However, in general

$$\prod_{(u,v)} (0) > 0$$

for some configuration of the states of the neighbors. The problem here is that while the event in question must have probability 0, all the individual components of the event have positive probability. Thus, this sort of process which could, for example, easily be thought of as one of technological adoption cannot be handled through the Markov random field approach. As I have mentioned, the difficulty that arose here was that there was a zero probability of moving to one state from another given certain configurations of the neighbors. Imposing a positivity condition

$$\begin{matrix} & 0 & \\ 0 & 1 & 0 \\ & 0 & \end{matrix}$$

FIGURE 1 An event with zero probability.

$$\begin{matrix} & 0 & S(u-1,v) & S(u-1,v+1) \\ 0 & 1 & S(u,v) & S(u,v+1) \\ & 0 & S(u+1,v) & S(u+1,v+1) \end{matrix}$$

FIGURE 2 $S(u, v)$ will not be zero with probability zero.

which assigns a positive probability to changing state whatever the configuration of the neighbors eliminates this problem, but does mean that, for some economic situations, the approach is more restrictive than it might at first appear.

A typical model which does use this approach is that of Blume[23] who considers a countable infinity of sites, each of which is occupied by one player who is directly connected to a finite set of neighbors. Each firm can then be thought of as adopting a strategy and then receiving a payoff, depending on the strategies adopted by his neighbors. If the set of sites is S and of strategies W, then a configuration will be $\phi : S \rightarrow W$ and the payoff to s of choosing w when his neighbors choose according to ϕ can be written $G_s(w, \phi(V_s))$ where V_s is the neighborhood of s, i.e., all sites with distance less than k from s. A stochastic revision process could then be defined, such as

$$log \frac{p_s(v|\phi(V_s))}{p_s(w|\phi(V_s))} = \beta G_s(v, \phi(V_s)) - G_s(w, \phi(V_s))$$

hence

$$p_s(v|\phi(V_s)) = \left(\sum_{w \in w} \exp \beta G_s(w, \phi(V_s)) - G_s(v, \phi(V_s)) \right)^{-1} .$$

This stochastic strategy revision process is a continuous-time Markov process. The particular class of revision processes has a long history in economics and is discussed in detail in Anderson et al.[4] and in Brock and Durlauf,[28] and I will come back to it when discussing the evolution of networks. Without entering into the details, the problem is to look at the limit behavior of this process. If β is large, individuals are very sensitive to the behavior of their opponents and, in effect, they will put all the weight on their current best strategy. In this case the process may settle down to any one of a number of equilibrium configurations. However, if β is small, then the process is ergodic, that is there is an *invariant measure* which describes the limit behavior of the process. Once again we see the distinction between a process which will converge to one of a number of particular "equilibrium" configurations, that is where the structure of the population is stabilized at certain distributions of strategies and that of a process which wanders through different configurations and which has, as an equilibrium, a distribution over states.

The key issue is how responsive is the probability of the choice of strategy to that of the neighbors or rather to the increase in payoff to be obtained by changing strategy given the choice of the neighbors and, as I will explain later, exactly the same issue arises when one examines the evolution of loyalty in markets.

As I have suggested, the limit when $\beta \rightarrow \infty$ is the "best response process." In this case, if there is a Nash configuration, that is a configuration in which every firm chooses the best technology or randomizes choice over best technologies if they are not unique, then one would like the concentrations of all weight on each of the points corresponding to Nash configurations to be invariant measures of stochastic revision processes. This is unfortunately not quite true. However, if in the Nash configuration the firm's best response is unique, this will be the case.

In the model just described, Blume examines the situation in which players play a game against each of their neighbors, and receive the average payoff. However, it is easy to consider more general payoff schemes to players for adopting strategies given those adopted by their neighbors.

The important thing, once again, is to see that variety can persist either through a stable distribution of strategies, which remains fixed over time or through a situation in which the system spends a fixed proportion of its time in each set of configurations. Note also that the interaction is local and that the time that an equilibrium will take to establish itself may be very large.

This may not necessarily be the case, however, as is shown by Ellison[42] who also emphasizes the importance of local interaction as opposed to global uniform matching. He considers two polar cases. First, he takes a simple two-by-two coordination game and considers a situation in which all of the N players are matched with uniform probability, i.e.,

$$P_{ij} = \frac{1}{N-1}\,.$$

On the other hand, he considers the case where the players are located around a circle and only allows players to be matched with their immediate neighbors, i.e.,

$$P_{ij} = \begin{cases} \frac{1}{2}, & \text{if } i-j \equiv \pm 1 (\text{mod} N); \\ 0, & \text{otherwise,} \end{cases}$$

or they could be matched with any of their $2k$ nearest neighbors, i.e.,

$$P_{ij} = \begin{cases} \frac{1}{2}k, & \text{if } i-j \equiv \pm 1 (\text{mod} N); \\ 0, & \text{otherwise.} \end{cases}$$

Many intermediate cases could be constructed with the probability of matching directly depending on distance on a lattice.

This idea, in effect, can be thought of in terms of the stochastic graph notion mentioned earlier. The rule for choice of strategy adopted by Ellison[116] is very similar to that used by Kandori, Mailath, and Rob.[72] A player chooses that strategy that would do best against those employed by his possible opponents in the previous period. To look at what happens consider the simple game used by Ellison.

TABLE 1 Ellison's game.

	A	B
A	2,2	0,0
B	0,0	1,1

In this game it is easy to see that a player will play strategy A if at least $1/3$ of his possible opponents did so in the last period. There are two steady states (a state is the number of people playing strategy A) 0, and N. Both have large basins of attraction and although, if noise is introduced, the system will eventually move from one steady state to the other, it will do so very infrequently.

Now, however, if we look at local interaction in the same game and consider a situation in which players may be matched with any of their nearest eight neighbors, it is clear that players will choose A if at least three of these neighbors did so last time. The point here is that if there is a small cluster of players playing A it will rapidly expand and take over the whole population. Consider the case when the whole group is playing B except for players 1 through 4 who play A. Thus we have

$$(A, A, A, A, B, ..., B).$$

Clearly at time $t+1$ players 5 and 6 will change to A as will players N and $N-1$. This experience will continue until the state with A for all players is reached. The important observation here is that if a small amount of noise (i.e., a small probability of self-change) is added, then it is sufficient that four adjacent players become A for the whole system to drift toward A. This is much more likely than the $(n-1)/3$ simultaneous mutations that would be necessary for the uniform matching model to shift from all B to the basin of attraction of A.

Perhaps counterintuitively, convergence to the equilibrium, that is to the steady state distribution which puts even less weight on all B in the local matching case than in the uniform matching case, is much more rapid in the local matching situation. Thus, restricting interaction leads to more rather than less rapid convergence to equilibrium.

In a related model, Ellison and Fudenberg[43] consider a situation in which individuals vary according to a parameter. They might vary in their geographical location or some basic characteristics. Players now learn from those with parameters close to their own. What is shown is that the system settles to a situation around a unique steady state "cut off value" of the parameter. Those with values above this use one technology, those with values below it use the other. There is a limit distribution, but its variance depends on the interval of parameters over which agents make their observations. Small intervals mean low variance but clearly also slow convergence. Thus, a more stable limit situation is offset by a slower movement to that situation.

Another contribution in this vein is that of Bala and Goyal,[15] who consider agents who learn from those with whom they are connected. If connectivity is high enough and individuals have identical preferences then, in the long run, every agent has the same utility. However, this may be a result of their having conformed to a suboptimal action. When preferences are heterogeneous, however, society may not conform at all. This again depends crucially on who is connected to whom. Thus, the nature of the communication network assumed plays an essential role.

In concluding, it is worth noticing that in some situations learning from local experience may not be socially efficient. Convergence to a uniform situation may not occur and "stratification" may persist (see, e.g., Benabou,[18] Durlauf,[39] and Anderlini and Ianni[3]).[4] Yet in all of this the connections between individuals are given.

To see what might occur if individuals could choose where to situate themselves rather than being fixed with relation to the other individuals, suppose, for example, that people in a neighborhood who become wealthy because of their education then move out. Observing those who remain would give wrong information about the value of investing in education.[5] This suggests that the notion of who is a neighbor depends on experience and this means, in terms of the earlier discussion, that the graph representing neighborhood structures, itself, evolves over time. Durlauf[39] endogenizes the formation of neighborhoods and looks at neighborhood structures which are in the core of all possible coalition structures. This is an intermediate step to a fully endogenous communication graph and this is the subject of the next section.

EVOLVING NETWORKS

It seems clear that the obvious way in which to proceed is to specify models in which the links between agents are reinforced over time by the gain derived from those links. Thus, longstanding economic relationships would be derived endogenously from the agent's experience. Instead of thinking of only the individuals learning, one could also think of the economy as learning and the graph representing the economy as evolving.[6] In the paper by Vriend[107] discussed earlier, relationships between traders do evolve over time and a number of stable bilateral arrangements emerge.

To see how the approach just suggested translates into more concrete form, consider a model of a simple market similar to that of Vriend and which has been extended by Kirman and Vriend[78] and Weisbuch et al.[110] Consider a wholesale market in which buyers update their probability of visiting sellers on the basis of the profit that they obtained in the past from those sellers. Denote by $J_{ij}(t)$ the cumulated profit up to period t, discounted by a factor γ that buyer i has obtained from trading with seller j, that is,

$$J_{ij}(t) = J_{ij}(t-l)(1-\gamma) + \pi_{ij}(\tau)$$

where $\pi_{ij}(\tau)$ is current profit.

[4] For results in the other direction see An and Kiefer.[2]

[5] This observation on Ellison and Fudenberg's results was made by R. Benabou.

[6] For a discussion of the formal problem of evolving networks see Weisbuch[109]

The probability p that i will visit j in that period is then given by,

$$P = \frac{\exp(\beta J_{ij})}{\sum j \exp(\beta J_{ij})}$$

where β is a reinforcement parameter, which describes how sensitive the individual is to past profits. This nonlinear updating rule is, of course, that discussed by Anderson et al.,[4] is familiar from the models described earlier by Blume[23] and Brock and Durlauf,[27] for example, and is widely used in statistical physics. Given this rule, one can envisage the case of three sellers, for example as corresponding to the simplex in Figure 3 below. Each buyer has certain probabilities of visiting each of the sellers and, thus, can be thought of as a point in the simplex. If he is equally likely to visit each of the three sellers, then he can be represented as a point in the center of the triangle. If, on the other hand, he visits one of the sellers with probability one then he can be shown as a point at one of the apexes of the triangle.

Thus, at any one point in time, the market is described by a cloud of points in the triangle and the question is how will this cloud evolve? If buyers all become loyal to particular sellers then the result will be be that all the points, corresponding to the buyers will be at the apexes of the triangle as in Figure 3(b), this might be thought of as a situation in which the market is "ordered." On the other hand, if buyers learn to search randomly amongst the sellers, then the result will be a cluster of points at the center of the triangle, as in Figure 3(a). What Weisbuch et al.[110] show, is that which of these situations develops depends crucially on the parameters β, γ, and the profit per transaction. The stronger the reinforcement, the slower the individual forgets, and the higher the profit, the more likely is it that order will emerge. In particular, the transition from disorder to order is very sharp with a

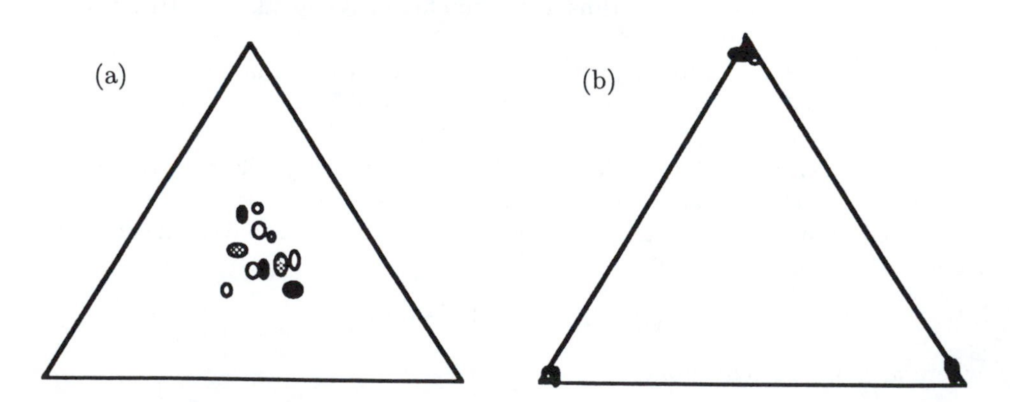

FIGURE 3

change in β. Order, in the sense mentioned, is clearly Pareto superior to disorder since there will be no sellers with unsold stocks and no unsatisfied buyers.[7] This recalls an earlier model by Whittle[111] where there are two sorts of individuals, farmers and traders, and under certain conditions markets may emerge where previously there were only itinerant traders. In the Weisbuch et al. model, this sort of "phase transition" is derived using the "mean-field" approach. The latter is open to the objection that random variables are replaced by their means and, in consequence, the process derived is only an approximation. The alternative is to consider the full stochastic process, but this is often not tractable, and one can then resort to simulations to see whether the theoretical results from the approximation capture the features of the simulated stochastic process, which is the case in Weisbuch et al.[110]

Stanley et al.[102] consider a Repeated Prisoner's Dilemma model in which players can choose and refuse to play against other players on the basis of updated expected payoffs. Tesfastion[105] applies this to a trading game. The introduction of choice of partners can lead to the emergence and persistence of multiple payoff bands. Another contribution is that of Mailath et al.[83] who look at the contribution to efficiency that can be made when agents can endogenously influence who their partners are.

In concluding this section, I shall suggest a simple approach to the modeling of evolving graphs. The basic idea is to model an evolving graph for a fixed population of size n by considering the set of all possible graphs, that is the 2^{n2} $n \times n$ incidence matrices and to define a probability distribution over them. Thus, a random directed graph is nothing other than a point in the unit simplex S in R^k where $k = 2^{n2}$ with the appropriate reduction in dimension for an undirected graph since the matrix is then symmetric. The evolution of the random graph is then described by a mapping from S into S and the dynamics will then be determined by the particular form of learning used to update the probabilities attached to the links. A vertex of the simplex corresponds to a deterministic network whilst the barycenter corresponds to the uniform matching model. Careful specification of the updating mechanism should lead to interesting and potentially testable conclusions about the form of the resulting network. Thus, one should be able to observe the evolution of trading groups and partnerships in markets and the development of groups playing certain strategies amongst themselves in repeated game models. An obvious extension of this is to consider agents as having several different types of functional links, for example, they might be linked within a firm, as trading partners, or as members of a household. However, the analysis of this sort of multilayered graph seems to be much less tractable.

[7]For an analysis of ordered and disordered markets in Renaissance Florence, but with a very different interpretation see McClean and Padgett.[87]

CONCLUSION

Agents in economic situations interact with each other in many ways other than through the market mechanism. By incorporating a consideration of how agents interact into our models we not only make them more realistic but we also enrich the types of aggregate behavior that can occur. One alternative to the standard model is to consider the economy or some particular market as a noncooperative game. The reasoning imposed on the individuals is extremely sophisticated. What I have suggested in this paper is a view of the economy as an interactive system in which agents, rather than fully optimizing, learn from experience. In particular, I have emphasized the importance of models in which agents only interact with a limited subset of other agents, their neighbors. I also suggest that the neighborhood structure itself should be considered as endogenous.

However, as soon as we introduce this sort of evolving interaction pattern the notion of equilibrium has to be reconsidered, and this is particularly true if we allow for stochastic interaction and study the ensuing dynamics. The relationship between the evolution of the system as a whole and that of its microcomponents may be both more complicated and different in nature to the type of aggregation with which economists are familiar. Indeed there may be no convergence at all to an equilibrium in any standard sense, and one is faced with analyzing a constantly evolving open ended system. Nevertheless, the sort of work discussed in this paper may represent a step in the direction of thinking of the evolution of the economy as that of a self-organizing system,[8] and, in particular, one can think of the economy itself as learning its way to a certain pattern of communication between agents. This, in turn will allow us to capture many features of market organization which are simply ignored in many standard economic models.

[8]See, e.g., Lesourne.[80]

REFERENCES

1. Allen, B. "Some Stochastic Processes of Interdependent Demand and Technological Diffusion of an Innovation Exhibiting Externalities Among Adopters." *Intl. Econ. Rev.* **23(3)** (1982): 595–608.
2. An, M. Y., and N. M. Kiefer. "Evolution and Equilibria Selection of Repeated Lattice Games." CAE Working Paper 92-11, Department of Economics, Cornell University, Ithaca, NY, 1992.
3. Anderlini and Ianni. "Path Dependence and Learning From Neighbors." *Games & Econ. Behav.* **13(2)** (1996): 141–177.
4. Anderson, S., A. de Palma, and J.-F. Thisse. *Discrete Choice Theory of Product Differentiation.* Cambridge, MA: MIT Press, 1992.
5. Aoki, M. "Economic Fluctuations with Interactive Agents: Dynamic and Stochastic Externalities." *The Japanese Econ. Rev.* **46** (1995): 148–165.
6. Aoki, M. *New Methods of Macroeconometric Modelling.* Cambridge: Cambridge University Press, 1996.
7. Arthur, W. B. "Competing Technologies and Lock-in by Historical Events: The Dynamics of Allocation Under Increasing Returns." I.I.A.S.A. Paper WP-83-90, Laxenburg, Austria. Revised as C.E.P.R. Paper 43, Stanford University, Stanford, CA, 1983.
8. Arthur, W. B. "Industry Location and the Importance of History." C.E.P.R. Paper 84, Stanford University, Stanford, CA, 1986.
9. Arthur W. B. "Self-Reinforcing Mechanisms in Economics." In *The Economy as an Evolving Complex System*, edited by P. W. Anderson, K. J. Arrow, and D. Pines, Santa Fe Institute Studies in the Sciences of Complexity, Proceedings Vol. V. Redwood City, CA: Addison-Wesley, 1988.
10. Arthur W. B. "Competing Technologies, Increasing Returns, and Lock-in by Historical Events." *Econ. J.* **IC** (1989): 116–31.
11. Arthur W. B., Y. Ermoliev, and Y. Kaniovski "Generalised Unproblem and Its Applications." *Kibernetika* **1** (1983): 49–56.
12. Arthur W. B., Y. Ermoliev, and Y. Kaniovski "Strong Laws for a Class of Path-Dependent Stochastic Processes, with Applications." In *Proceedings of the Conference on Stochastic Optimisation, Kiev, 1984*, edited by Arkin, Shiryayev, and Wets, Springer Lecture Notes in Control and Information Sciences. Berlin: Springer-Verlag, 1984.
13. Arthur, W. B., Y. Ermoliev, and Y. Kaniovski. "Path-Dependent Processes and the Emergence of Macro-Structure." *European J. Operational Res.* **30** (1987): 294–303.
14. Axelrod, R. *The Evolution of Cooperation.* New York: Basic Books, 1984.
15. Bala, V., and S. Goyal. "Learning from Neighbors." Mimeo, Department of Economics, McGill University, Montreal, Canada, 1996.
16. Banerjee, A. "A Simple Model of Herd Behavior." *Quart. J. Econ.* **108** (1992): 797–817.

17. Banerjee, A., and J. W. Weibull. "Evolution and Rationality: Some Recent Game-Theoretic Results." Mimeo, Institute for International Economic Studies, Stockholm University, Stockholm, 1992.
18. Benabou, R. "Heterogeneity, Stratification, and Growth." Mimeo, MIT, Cambridge, MA, 1992.
19. Bhattacharya, R. N., and M. Majumdar. "Random Exchange Economies." *J. Econ. Theory* **6** (1973): 36–67.
20. Bikhchandani, S., D. Hirschleifer, and I. Welch "A Theory of Fads, Fashion, Custom, and Cultural Change as Informational Cascades." *J. Pol. Econ.* **100** (1992): 992–1026.
21. Binmore, K. *Essays on the Foundations of Game Theory.* Oxford: Basil Blackwell, 1990.
22. Binmore, K. *Fun and Games: A Text on Game Theory.* Lexington, MA: D.C. Heath & Co., 1992.
23. Blume, L. "The Statistical Mechanics of Strategic Interaction." *Games & Econ. Behav.* **5** (1993): 387–424.
24. Blume, L. "Population Games." In *The Economy as an Evolving, Complex System, II,* edited by S. Durlauf, D. Lane, and W. B. Arthur, Santa Fe Institute Studies in the Sciences of Complexity, Proceedings Vol. XXVII. Reading, MA: Addison-Wesley, 1997.
25. Bomze, I. M. "Non-cooperative Two-Person Games in Biology: A Classification." *Intl. J. Game Theory* **15** (1986): 31–57.
26. Brock, W. A. "Pathways to Randomness in the Economy: Emergent Nonlinearity and Chaos in Economics and Finance." Working Paper 93-02-006, Santa Fe Institute, Santa Fe, NM, 1993.
27. Brock, W. A., and S. Durlauf. "Discrete Choice with Social Interactions I: Theory." Working Paper 95-10-084, Santa Fe Institute, Santa Fe, NM, 1995.
28. Brock, W. A., and S. Durlauf. "Discrete Choice with Social Interactions I: Theory." SSRI Working Paper 9521, University of Wisconsin, Madison, WI, 1995.
29. Cournot, A. A. *Exposition de la Theorie des Chances et des Probabilites.* Paris: Hachette, 1843.
30. Cowan, R. "Backing the Wrong Horse: Sequential Technology Choice Under Increasing Returns." Ph.D. dissertation, Stanford University, Stanford, CA, 1987.
31. David, P. "Clio and the Economics of QWERTY." *Amer. Econ. Rev. Proc.* **75** (1985): 332–337.
32. Debreu, G. "A Social Equilibrium Existence Theorem." *Proc. Natl. Acad. Sci., USA* **38** (1952): 886–893.
33. Dekel, E., and S. Scotchmer. "On the Evolution of Optimizing Behavior." *J. Econ. Theory* **57** (1992): 392–406.
34. Diamond, P. "Aggregate Demand in Search Equilibrium." *J. Pol. Econ.* **XV** (1982): 881–894.

35. Diamond, P. "Search Theory." In *The New Palgrave: A Dictionary of Economics*, edited by J. Eatwell, M. Milgate, and P. Newman, 273–279. London: Macmillan, 1987.
36. Durlauf, S. "Locally Interacting Systems, Coordination Failure, and the Behavior of Aggregate Activity." Mimeo, Stanford University, Stanford, CA, 1990.
37. Durlauf, S. "Multiple Equilibria and Persistence in Aggregate Fluctuations." *Amer. Econ. Rev.* **81** (1991): 70–74.
38. Durlauf, S. "Nonergodic Economic Growth." *Rev. Econ. Studies* **60(2)** (1993): 349–366.
39. Durlauf, S. "Neighborhood Feedbacks, Endogenous Stratification, and Income Inequality." Working Paper 95-07-061, Santa Fe Institute, Santa Fe, NM, 1995.
40. Durlauf, S. "Statistical Mechanics Approaches to Socioeconomic Behavior." Mimeo, Departement of Economics, University of Wisconsin at Madison, WI, 1996.
41. Edgeworth, F. Y. *Mathematical Psychics.* London: Kegan Paul, 1881.
42. Ellison, G. "Learning, Local Interaction, and Coordination." *Econometrica* **61(5)** (1993): 1047–1071.
43. Ellison, G., and D. Fudenberg. "Rules of Thumb for Social Learning." *J. Pol. Econ.* **101(41)** (1993): 612–643.
44. Ellison, G., and D. Fudenberg. "Word of Mouth Communication and Social Learning." *Quart. J. Econ.* **440** (1995): 93–125.
45. Evstigneev, I., and M. Taskar. "Stochastic Equilibria on Graphs, 1." *J. Math. Econ.* **23** (1994): 401–433.
46. Evstigneev, I., and M. Taskar. "Stochastic Equilibria on Graphs, 2." *J. Math. Econ.* **24** (1995): 401–433.
47. Feldman, A. "Bilateral Trading Processes, Pairwise Optimality, and Pareto Optimality." *Rev. Econ. Studies* **40** (1973): 463–73.
48. Föllmer, H. "Random Economies with Many Interacting Agents." *J. Math. Econ.* **1(1)** (1974): 51–62.
49. Foster, D., and P. Young. "Stochastic Evolutionary Game Dynamics." *Theor. Pop. Biol.* **38** (1990): 219–232.
50. Friedman, D. "Evolutionary Games in Economics." *Econometrica* **59** (1991): 637–666.
51. Gabszewicz, J. J., J. F. Thisse, M. Fujita, and U. Schweizer. "Spatial Competition and the Location of Firms." In *Location Theory: Fundamentals of Pure and Applied Economics*, edited by J. Lesourne and H. Sonnesche. London: Harwood Academic Publishers, 1986.
52. Gale, D. "Price Setting and Competition in a Simple Duopoly Model." Universities of Pittsburgh and Pennsylvania, first draft, 1987.
53. Gilles, R. P., and P. H. M. Ruys. "Relational Constraints in Coalition Formation." Department of Economics Research Memorandum, FEW 371, Tilburg University, 1989.

54. Gintis, H. "Money and Trade in a Primitive Economy: Theory and Artificial Life Simulation." Mimeo, Department of Economics, University of Massachusetts, Amherst, MA, 1996.
55. Goyal, S. "On the Possibility of Efficient Bilateral Trade." Mimeo, Econometric Institute, Erasmus University, Rotterdam, 1994.
56. Grandmont, J.-M. *Money and Value, a Reconsideration of Classical and Neoclassical Monetary Theories.* Paris: Cambridge University Press, Editions de la Maison des Sciences de l'Homme, 1983.
57. Grandmont, J.-M. "Distributions of Preferences and the 'Law of Demand.'" *Econometrica* **55(1)** (1987): 155–61.
58. Grandmont, J.-M. "Transformations of the Commodity Space, Behavioral Heterogeneity, and the Aggregation Problem." *J. Econ. Theory* **57** (1992): 1–35.
59. Guriev, S., and M. Shakhova. "Self-Organization of Trade Networks in an Economy with Imperfect Infrastructure." Mimeo, Computing Center of Russian Academy of Sciences, Moscow, 1995.
60. Haller, H. "Large Random Graphs in Pseudo-Metric Spaces." *Math. Soc. Sci.* **20** (1990): 147–164.
61. Hildenbrand, W. "On Random Preferences and Equilibrium Analysis." *J. Econ. Theory* **3** (1971): 414–429.
62. Hildenbrand, W. "Facts and Ideas in Microeconomic Theory." *European Econ. Rev.* **33** (1989): 251–276.
63. Hildenbrand, W. *Market Demand: Theory and Empirical Evidence.* Princeton, NJ: Princeton University Press, 1994.
64. Hildenbrand, W., and A. P. Kirman. *Equilibrium Analysis.* Amsterdam: North Holland, 1988.
65. Hofbauer, J., and K. Sigmund. "The Theory of Evolution and Dynamical Systems." London Mathematical Society Students' Texts, Vol. 7. Cambridge, MA: Cambridge University Press, 1988.
66. Hofbauer, J., P. Schuster, and K. Sigmund. "A Note on Evolutionary Stable Strategies and Game Dynamics." *J. Theor. Biol.* **81** (1979), 609–612.
67. Ioannides, Y. M. "Trading Uncertainty and Market Form." *Intl. Econ. Rev.* **31(3)** (1990): 619–638.
68. Ioannides, Y. M. "Evolution of Trading Structures." Mimeo, Dept. of Economics, Tufts University, Medford, MA, 1995.
69. Jackson, M., and A.Wolinsky. "A Strategic Model of Social and Economic Networks." Mimeo, Kellogg School of Management, Northwestern University, Evanston, IL, 1995.
70. Jerison, M. "The Representative Consumer and the Weak Axiom when the Distribution of Income Is Fixed." Working Paper 150, SUNY, Albany, NY, 1990.
71. Kalai, E., A. Postlewaite, and J. Roberts. "Barriers to Trade and Disadvantageous Middlemen: Nonmonotonicity of the Core." *J. Econ. Theory* **19** (1978): 200–209.

72. Kandori, M., G. Mailath, and R. Rob. "Learning, Mutation, and Long Run Equilibria in Games." *Econometrica* **61(1)** (1993): 29–56.
73. Kirman, A. P. "Communication in Markets: A Suggested Approach." *Econ. Lett.* **12(1)** (1983): 101–108.
74. Kirman, A. P. "Epidemics of Opinion and Speculative Bubbles in Financial Markets." In *Money and Financial Markets*, edited by M. Taylor, Chap. 17, 354–368. London: Macmillan, 1991.
75. Kirman, A. P. "Whom or What Does the Representative Individual Represent?" *J. Econ. Perspectives* **6(2)** (1992): 117–136.
76. Kirman, A. P. "Ants, Rationality, and Recruitment." *Quart. J. Econ.* **CVIII** (1993): 137–156.
77. Kirman, A. P., C. Oddou, and S. Weber. "Stochastic Communication and Coalition Formation." *Econometrica* **54(1)** (1986): 129–138.
78. Kirman, A. P., and N. Vriend. "Evolving Market Structure: A Model of Price Dispersion and Loyalty." Mimeo, Dept. of Economics, Virginia Polytechnic Institute, Blacksburg, VI, 1995.
79. Kyotaki, N., and R. Wright. "On Money as a Medium of Exchange." *J. Pol. Econ.* **94(4)** (1989): 927–954.
80. Lesourne, J. *The Economics of Order and Disorder.* Oxford: Clarendon Press, 1992.
81. Lindgren, K. "Evolutionary Phenomena in Simple Dynamics." In *Artificial Life II*, edited by C. G. Langton, C. Taylor, J. D. Farmer, and S. Rasmussen, Santa Fe Institute Studies in the Sciences of Complexity, Proc. Vol. X. Redwood City, CA: Addison-Wesley, 1991.
82. Lucas, R. "Adaptive Behavior and Economic Theory." *J. Bus.* **59** (1988): 5401–5426.
83. Mailath, G., L. Samuelson, and A. Shaked. "Evolution and Endogenous Interactions." CARESS Working Paper No. 94-13, University of Pennsylvania, Philadelphia, PA, 1994.
84. Malinvaud, E. "The Allocation of Individual Risks in Large Markets." *J. Econ. Theory* **4** (1972): 312–328.
85. Mas-Colell, A. *The Theory of General Economic Equilibrium: A Pifferentiable Approach.* Cambridge, MA: Cambridge University Press, 1985.
86. Maynard Smith, J. *Evolution and the Theory of Games.* Cambridge, MA: Cambridge University Press, 1982.
87. Mclean, P. D., and J. F. Padgett. "Was Florence a Perfectly Competitive Market?: Transactional Evidence from the Renaissance." *Theory and Society*, forthcoming.
88. Moscarini, G., and M. Ottaviani. "Social Learning in a Changing World." Mimeo, Department of Economics, MIT, Cambridge, MA, 1995.
89. Myerson, R. B. "Graphs and Cooperation in Games." *Math. Oper. Res.* **2** (1977): 225–229.
90. Myerson, R. B. and Satterthwaite "Efficient Mechanisms for Bilateral Trade." *J. Econ. Theory* **29** (1983): 265–281.

91. Nachbar, J. "Evolutionary Selection Dynamics in Games: Convergence and Limit Properties." *Intl. J. Game Theor.* **19** (1990): 59–89.
92. Roth, A., and M. Sotomayor. *Two Sided Matching: A Study in Game Theoretic Modelling and Analysis.* New York: Cambridge University Press, 1990.
93. Rubinstein, A., and A. Wolinsky. "Decentralized Trading, Strategic Behavior, and the Walrasian Outcome." *Rev. Econ. Studies* **57** (1990): 63–78.
94. Samuelson, L. "Recent Advances in Evolutionary Economics: Comments." *Econ. Lett.* **42** (1993): 313–319.
95. Samuelson, L., and J. Zhang. "Evolutionary Stability in Asymmetric Games." *J. Econ. Theory* **57(2)** (1992): 363–391.
96. Samuelson, P. A. "Problems of Methodology: Discussion." *Amer. Econ. Rev.* **54** (1963): 231–236.
97. Schuster, P., and K. Sigmund. "Replicator Dynamics." *J. Theor. Biol.* **100** (1983): 533–538.
98. Sharfstein, D. S., and J. C. Stein. "Herd Behaviour and Investment." *Amer. Econ. Rev.* **80(3)** (1990): 465–479.
99. Smith, L., and P. Sorensen. "Pathological Models of Observational Learning." Mimeo, Department of Economics, MIT, Cambridge, MA, 1994.
100. Spitzer, F. "Markov Random Fields and Gibbs Ensembles." *Am. Math. Monthly* **78** (1971): 142–154.
101. Stanley, E. A., D. Ashlock, and L. Tesfatsion. "Iterated Prisoner's Dilemma with Choice and Refusal of Partners." Economic Report No. 30, Iowa State University, Ames, IA, 1993.
102. Stanley, E. A., D. Ashlock, and L. Tesfatsion. "Iterated Prisoner's Dilemma with Choice and Refusal of Partners." In *Artificial Life III*, edited by C. G. Langton, Santa Fe Institute Studies in the Sciences of Complexity, Proc. Vol. XVII, Reading, MA: Addison-Wesley, 1994.
103. Stoker, T. "Empirical Approaches to the Problem of Aggregation over Individuals." *J. Econ. Lit.* **XXXI** 1827–1874.
104. Summers, L. H. "The Scientific Illusion in Empirical Macroeconomics." *Scandinavian J. Econ.* **93(2)** (1991): 129–148.
105. Tesfastion, L. "A Trade Coalition Game with Preferential Partner Selection." Working Paper, Iowa State University, Ames, IA, 1995.
106. Topol, R. "Bubbles and Volatility of Stock Prices: Effects of Mimetic Contagion." *Econ. J.* **101** (1991): 786–800.
107. Vriend, N. "Self-Organized Markets in a Decentralized Economy." *J. Evol. Econ.* (1995).
108. Weibull, J. W. "An Introduction to Evolutionary Game Theory." Mimeo, Department of Economics, Stockholm University, 1992.
109. Weisbuch, G. *Complex System Dynamics.* Santa Fe Institute Studies in the Sciences of Complexity, Lecture Notes Vol. II. Redwood City, CA: Addison-Wesley, 1991.

110. Weisbuch, G., A. Kirman, and D. Herreiner. "Market Organisation." Mimeo, Laboratoire de Physique Statistique de l'Ecole Normale Superieure, Paris, 1995.
111. Whittle, P. *Systems in Stochastic Equilibrium.* New York: Wiley, 1986.
112. Woodford, M. "Learning to Believe in Sunspots." *Econometrica* **58** (1990): 277–307.
113. Young, H. P., and D. Foster. "Cooperation in the Short and in the Long Run." *Games & Econ. Behav.* **3** (1991): 145–156.
114. Zeeman, E. C. *Population Dynamics from Game Theory.* Lecture Notes in Mathematics 819. New York: Springer-Verlag, 1980.

Leigh Tesfatsion
Department of Economics, Heady Hall 260, Iowa State University, Ames, IA 50011-1070;
E-mail: tesfatsi@iastate.edu; http://www.econ.iastate.edu/tesfatsi/

How Economists Can Get ALife

This chapter presents a summary overview of the fast-developing field of artificial life, stressing aspects especially relevant for the study of decentralized market economies. In particular, a recently developed trade network game (TNG) is used to illustrate how the basic artificial life paradigm might be specialized to economics. The TNG traders choose and refuse trade partners on the basis of continually updated expected utility and evolve their trade behavior over time. Analytical and simulation work is reported to indicate how the TNG is currently being used to study the evolutionary implications of alternative market structures at three different levels: individual trade behavior; trade network formation; and social welfare.

1. INTRODUCTION

Artificial life (alife) is the bottom-up study of basic phenomena commonly associated with living organisms, such as self-replication, evolution, adaptation, self-organization, parasitism, competition, and cooperation. Alife complements the traditional biological and social sciences concerned with the analytical, laboratory, and field study of living organisms by attempting to simulate or synthesize these phenomena within computers, robots, and other artificial media. One goal is to enhance the understanding of actual and potential life processes. A second goal is to use nature as an inspiration for the development of solution algorithms for difficult optimization problems characterized by high-dimensional search domains, nonlinearities, and multiple local optima.[1]

Alife researchers study complex adaptive systems sharing many of the following characteristics.[22] Most importantly, each such system typically consists of many dispersed units acting in parallel with no global controller responsible for the behavior of all units. Rather, the actions of each unit depend upon the states and actions of a limited number of other units, and the overall direction of the system is determined by competition and coordination among the units, subject to structural constraints. The complexity of the system thus tends to arise more from the interactions among the units than from any complexity inherent in the individual units per se. Moreover, the local interaction networks connecting individual units are continuously recombined and revised. In particular, niches that can be exploited by particular adaptations are continuously created, and their exploitation in turn leads to new niche creations, so that perpetual novelty exists.

Briefly put, then, alife research tends to focus on continually evolving systems whose global behavior arises from the local interactions of distributed units; this is the sense in which alife research is said to be bottom up. Although the units comprising the systems might be bit strings, molecules, or robotic insects, the abstract description of how the unit interactions result in global behavior is clearly reminiscent of a Schumpeterian economy, only filtered through an unfamiliar terminology.

The study of evolutionary economies has, of course, been pursued by many researchers in addition to Schumpeter. For example, one has Armen Alchian's work on uncertainty and evolution in economic systems, the work of W. Brian Arthur on economies incorporating positive feedbacks, the work by Richard Day on dynamic economies characterized by complex phase transitions, the work by John Foster on an evolutionary approach to macroeconomics, Ron Heiner's work on the origins of predictable behavior, Jack Hirshleifer's work on evolutionary models in economics and law, and Richard Nelson and Sidney Winter's work on an evolutionary theory of

[1]As detailed in the entertaining monographs by Levy[30] and Sigmund,[43] the roots of alife go at least as far back as the work of John von Neumann in the 1940s on self-replicating automata. The establishment of alife as a distinct field of inquiry, however, must be traced to the first alife conference, organized in 1987 by Christopher Langton at the Los Alamos National Laboratory; see Langton.[29]

economic change. These and numerous other related studies are reviewed by Witt[50] and Nelson.[39] In addition, as detailed in Friedman,[16] a number of researchers have recently been focusing on the potential economic applicability of evolutionary game theory in which game strategies distributed over a fixed number of strategy types reproduce over time in direct proportion to their relative fitness.

Exploiting the recent advent of object-oriented programming languages such as C++ and ObjectiveC, economists making use of the basic alife paradigm[2] have been able to extend this previous evolutionary work in several directions. First, much greater attention is generally focused on the endogenous determination of agent interactions. Second, a broader range of interactions is typically considered, with cooperative and predatory associations increasingly taking center stage along with price and quantity relationships. Third, agent actions and interactions are represented with a greater degree of abstraction, permitting generalizations across specific system applications. Fourth, the evolutionary process is generally expressed by means of genetic (recombination and/or mutation) operations acting directly on agent characteristics.[3] These evolutionary selection pressures result in the continual creation of new modes of behavior and an ever-changing network of agent interactions.

The central problem for economic alife researchers is to understand the apparently spontaneous appearance of regularity in economic processes, such as the unplanned coordination of trading activities in decentralized market economies that economists associate with Adam Smith's invisible hand. The challenge is to explain how these global regularities arise from the local interactions of autonomous agents channeled through actual or potential economic institutions rather than through fictitious coordinating mechanisms such as a single representative consumer. In line with this challenge, rationality is generally viewed as a testable hypothesis, or at least as a debatable methodological assumption, rather than as an unquestioned axiom of individual behavior.

Several economic studies that focus on key alife issues have either appeared or are in the pipeline. See, for example, Anderson et al.,[1] Arifovic,[2] Arthur,[3] Bell,[7]

[2]After the February 1996 UCLA Economic Simulation Conference organized by Axel Leijonhufvud, participants suggested that the name "agent-based economics" (ACE) be adopted for this line of work, roughly characterized as the computational study of evolutionary economies modeled as decentralized systems of adaptive agents.

[3]For example, the basic *genetic algorithm* used in many economic alife studies evolves a new population of agents from an existing population of agents using the following four steps: (1) *evaluation*, in which a fitness score is assigned to each agent in the population; (2) *selection for reproduction*, in which a subset of the existing population of agents is selected for reproduction, with selection biased in favor of fitness; (3) *recombination*, in which offspring (new ideas) are generated by combining the genetic material (structural characteristics) of pairs of parents chosen from among the most fit agents in the population; and (4) *mutation*, in which additional variations are introduced into the population by mutating the structural characteristics of each offspring with some small probability. See Goldberg[18] and Mitchell and Forrest.[38]

Birchenhall,[8] Bosch and Sunder,[9] Bullard and Duffy,[11] De Vany,[12] Durlauf,[13] Epstein and Axtell,[14] Holland and Miller,[23] Kirman,[25] Lane,[28] Mailath et al.,[32] Marimon et al.,[33] Marks,[34] Miller,[37] Routledge,[41] Sargent,[42] Tesfatsion,[48] and Vriend.[49]

To illustrate more concretely the potential usefulness of the alife approach for economics, as well as the hurdles that remain to be cleared, the following two sections take up a facet of alife work that appears to be particularly relevant for the modeling of decentralized market economies. Section 2 describes recent attempts to combine evolutionary game theory with preferential partner selection.[4,44,45] Section 3 discusses how a modified version of this framework is being used to study the endogenous formation and evolution of trade networks.[36,48] Concluding comments are given in section 4.

2. EVOLUTIONARY IPD WITH CHOICE AND REFUSAL

Following the seminal work of Axelrod,[5,6] the Iterated Prisoner's Dilemma (IPD) game has been extensively used by economists and other researchers to explore the potential emergence of mutually cooperative behavior among nonaltruistic agents. As detailed in Kirman[26] and Lindgren and Nordahl,[31] these studies have typically assumed that individual players have no control over whom they play. Rather, game partners are generally determined by an extraneous matching mechanism such as a roulette wheel, a neighborhood grid, or a round-robin tournament. The general conclusion reached by these studies has been that mutually cooperative behavior tends to emerge if the number of game iterations is either unknown or infinite, the frequency of mutually cooperative play in initial game iterations is sufficiently large, and the perceived probability of future interactions with any given current partner is sufficiently high.

In actuality, however, socioeconomic interactions are often characterized by the preferential choice and refusal of partners. The question then arises whether the emergence and long-run viability of cooperative behavior in the IPD game would be enhanced if players were more realistically allowed to choose and refuse their potential game partners.

This question is taken up in Stanley et al.[45] The traditional IPD game is extended to an IPD/CR game in which players choose and refuse partners on the basis of continually updated expected payoffs.[4]

[4]Other game theory studies that have allowed players to avoid unwanted interactions, or more generally to affect the probability of interaction with other players through their own actions, include Fogel,[15] Guriev and Shakhova,[19] Hirshleifer and Rasmusen,[21] Kitcher,[27] Mailath et al.,[32] and Orbell and Dawes.[40] See Stanley et al.[45] (section 2) and Ashlock et al.[4] (section 1) for more detailed discussions of related game theory work. There is also a growing body of work on multiagent systems with endogenous interactions in which the decision (or state) of an agent

The introduction of partner choice and refusal fundamentally modifies the ways in which players interact in the IPD game and the characteristics that result in high payoff scores. Choice allows players to increase their chances of encountering other cooperative players, refusal gives players a way to protect themselves from defections without having to defect themselves, and ostracism of defectors occurs endogenously as an increasing number of players individually refuse the defectors' game offers. On the other hand, choice and refusal also permit opportunistic players to home in quickly on exploitable players and form parasitic relationships.

The analytical and simulation findings reported for the IPD/CR game in Stanley et al.,[45] and in the subsequent studies by Smucker et al.,[44] Ashlock et al.,[4] and Hauk,[20] indicate that the overall emergence of cooperation is accelerated in evolutionary IPD games by the introduction of choice and refusal. Nevertheless, the underlying player interaction patterns induced by choice and refusal can be complex and time varying, even when expressed play behavior is largely cooperative. Consequently, it has proven to be extremely difficult to get an analytical handle on the mapping from parameter configurations to evolutionary IPD/CR outcomes.

A reasonable next step, then, is to focus on more concrete problem settings which impose natural constraints on the range of feasible player interactions. In the next section it is shown how a modified version of the IPD/CR game is being used to examine the endogenous formation and evolution of trade networks among resource-constrained traders.

3. A TRADE NETWORK GAME WITH CHOICE AND REFUSAL

The *trade network game* (TNG) developed by Tesfatsion[48] consists of successive generations of resource-constrained traders who choose and refuse trade partners on the basis of continually updated expected payoffs, engage in risky trades modeled as two person games, and evolve their trade strategies over time. The TNG has been implemented by McFadzean and Tesfatsion[36] with the support of a general C++ evolutionary simulation framework, *SimBioSys*, developed by McFadzean.[35]

The TNG facilitates the general study of trade from a bottom-up perspective in three key ways. First, the TNG traders are instantiated as autonomous endogenously interacting software agents (*tradebots*) with internal behavioral functions and with internally stored information that includes addresses for other tradebots. The tradebots can, therefore, display anticipatory behavior (expectation formation); and they can communicate with each other at event-triggered times, a feature not present in standard economic models. Second, the modular design of the TNG permits experimentation with alternative specifications for market structure,

depends on the decision (or state) of certain neighboring agents, where these neighbors may change over time. See, for example, Brock and Durlauf,[10] De Vany,[12] Ioannides,[24] and Young.[51]

trade partner matching, trading, expectation formation, and trade behavior evolution. All of these specifications can potentially be grounded in tradebot-initiated actions. Third, the evolutionary implications of alternative module specifications can be studied at three different levels: individual tradebot characteristics, trade network formation; and social welfare as measured by descriptive statistics such as average tradebot fitness.

Section 3.1 sets out and motivates the general TNG framework. To gain insight into the subtle interplay between game play and the choice and refusal of partners in the TNG, section 3.2 presents a detailed analytical study of an illustrative TNG with five tradebots. In particular, it is shown that the parameter space for this illustrative TNG partitions into economically interpretable regions corresponding to qualitatively distinct trade network formations. Section 3.3 reports on some illustrative TNG computer experiments for two alternative market structures: buyer-seller markets in which all tradebots can both make and receive trade offers; and two-sided markets in which a subset of buyers makes trade offers to a disjoint subset of sellers.

3.1 THE BASIC TRADE NETWORK GAME

The trade network game (TNG) consists of a collection of tradebots that evolves over time. As depicted in Table 1, this evolution is implemented through a hierarchy of cycle loops.

Each tradebot in the initial tradebot generation is assigned a random trade strategy and is configured with a prior expected payoff for each of his potential trade partners. The tradebots then engage in a *trade cycle loop* consisting of a fixed number of trade cycles. In each trade cycle the tradebots undertake three activities: the determination of trade partners, given current expected payoffs; the carrying out of potentially risky trades; and the updating of expected payoffs based on any new payoffs received during trade partner determination and trading. At the end of the trade cycle loop the tradebots enter into an *environmental cycle* during which the *fitness score* of each tradebot is calculated as the total sum of his payoffs divided by the total number of his payoffs and the current tradebot generation is sorted by fitness scores. At the end of the environmental cycle, a *generation cycle* commences during which evolutionary selection pressures are applied to the current tradebot generation to obtain a new tradebot generation with evolved trade strategies. This new tradebot generation is then configured, and another trade cycle loop commences.

TABLE 1 Pseudo-Code for the TNG

```
int main () {
  Init() ;                        // Construct initial tradebot generation
                                  //     with random trade strategies.
  For (G = 0,...,GMAX-1) {        // Enter the generation cycle loop.
    If (G > 0) {
      EvolveGen();                // Generation cycle: Evolve a
                                  //     new tradebot generation.
    }
    InitGen();                    // Configure tradebots with user-supplied
                                  //     parameter values (prior expected
                                  //     payoffs, quotas,...).
    For (I = 0,...,IMAX-1) {      // Enter the trade cycle loop.
      MatchTraders();             // Determine trade partners
                                  //     given expected payoffs,
                                  //     and record refusal and
                                  //     wallflower payoffs.
      Trade();                    // Implement trades and
                                  //     record trade payoffs.
      UpdateExp();                // Update expected payoffs
                                  //     using newly recorded payoffs.
    }
    AssessFitness();              // Record fitness scores.
  }
  Return 0 ;
}
```

The TNG currently uses the particular specifications for market structure, trade partner determination, trade, expectation updating, and trade behavior evolution detailed in Tesfatsion.[48] For completeness, these specifications are reviewed below.

Alternative market structures are currently imposed in the TNG through the prespecification of buyers and sellers and through the prespecification of quotas on offer submissions and acceptances. More precisely, the set of players for the TNG is the union $V = B \cup S$ of a nonempty subset B of *buyer tradebots* who can submit trade offers and a nonempty subset S of *seller tradebots* who can receive trade offers, where B and S may be disjoint, overlapping, or coincident. In each trade cycle, each buyer m can submit up to O_m trade offers to sellers and each seller n can accept up to A_n trade offers from buyers, where the offer quota O_m and the acceptance quota A_n can be any positive integers.

Although highly simplified, these parametric specifications permit the TNG to encompass two-sided markets, markets with intermediaries, and markets in which all traders engage in both buying and selling activities. For example, the buyers and sellers might represent customers and retail store owners, workers and firms, borrowers and lenders, or barter traders. The offer quota O_m indicates that buyer m has a limited amount of resources (credit, labor time, collateral, apples,...) to offer, and the acceptance quota A_n indicates that seller n has a limited amount of resources (goods, job openings, loans, oranges,...) to provide in return.

Three illustrations are sketched below.

CASE 1: A LABOR MARKET WITH ENDOGENOUS LAYOFFS AND QUITS. The set B consists of M workers and the set S consists of N employers, where B and S are disjoint. Each worker m can make work offers to a maximum of O_m employers, or he can choose to be unemployed. Each employer n can hire up to A_n workers, and employers can refuse work offers. Once matched, workers choose on-the-job effort levels and employers choose monitoring and penalty levels. An employer fires one of its current workers by refusing future work offers from this worker, and a worker quits his current employer by ceasing to direct work offers to this employer. This TNG special case thus extends the standard treatment of labor markets as assignment problems by incorporating subsequent strategic efficiency wage interactions between matched pairs of workers and employers and by having these interactions iterated over time.

CASE 2: INTERMEDIATION WITH CHOICE AND REFUSAL. The buyer subset B and the seller subset S overlap but do not coincide. The pure buyers in $V - S$ are the depositors (lenders), the buyer-sellers in $B \cap S$ are the intermediaries (banks), and the pure sellers in $V - B$ are the capital investors (borrowers). The depositors offer funds to the intermediaries in return for deposit accounts, and the intermediaries offer loan contracts to the capital investors in return for a share of earnings. The degree to which an accepted offer results in satisfactory payoffs for the participants is determined by the degree to which the deposit or loan contract obligations are fulfilled.

CASE 3: A LABOR MARKET WITH ENDOGENOUSLY DETERMINED WORKERS AND EMPLOYERS. The subsets B and S coincide, implying that each tradebot can both make and receive trade offers. Each tradebot v can make up to O_v work offers to tradebots at other work sites and receive up to A_v work offers at his own work site. As in Case 1, the degree to which any accepted work offer results in satisfactory payoffs for the participant tradebots is determined by subsequent work site interactions. Ex post, four pure types of tradebots can emerge: (1) pure workers, who work at the sites of other tradebots but have no tradebots working for them at their own sites; (2) pure employers, who have tradebots working for them at their own sites but who do not work at the sites of other tradebots; (3) unemployed tradebots, who make at least one work offer to a tradebot at another site, but who end up neither working at other sites nor having tradebots working for them at their own sites; and (4) inactive (out of the work force) tradebots, who neither make nor accept any work offers.

The determination of trade partners in the TNG is currently implemented using a modified version of the well-known Gale-Shapley[17] deferred acceptance mechanism. This modified mechanism, hereafter referred to as the *deferred choice and refusal* (DCR) mechanism, presumes that each buyer and seller currently associates an expected payoff with each potential trade partner. Also, each buyer and seller is presumed to have an exogenously given *minimum tolerance level*, in the sense that he will not trade with anyone whose expected payoff lies below this level.

The DCR mechanism proceeds as follows. Each buyer m first makes trade offers to a maximum of O_m most-preferred sellers he finds tolerable, with at most one offer going to any one seller. Each seller n, in turn, forms a waiting list consisting of a maximum of A_n of the most preferred trade offers he has received to date from tolerable buyers; all other trade offers are refused. For both buyers and sellers, selection among equally preferred options is settled by a random draw. A buyer that has a trade offer refused receives a nonpositive *refusal payoff*, R; the seller who does the refusing is not penalized. A refused buyer immediately submits a replacement trade offer to any tolerable next-most-preferred seller that has not yet refused him. A seller receiving a new trade offer that dominates a trade offer currently on his waiting list substitutes this new trade offer in place of the dominated trade offer, which is then refused. A buyer ceases making trade offers when either he has no further trade offers refused or all tolerable sellers have refused him. When all trade offers cease, each seller accepts all trade offers currently on his waiting list. A tradebot that neither submits nor accepts trade offers during this matching process receives a *wallflower payoff*, W.[5]

The buyer-seller matching outcomes generated by the DCR mechanism exhibit the usual static optimality properties associated with Gale-Shapley type matching mechanisms. First, any such matching outcome is core stable, in the sense that no subset of tradebots has an incentive to block the matching outcome by engaging in a feasible rearrangement of trade partners among themselves (Tesfatsion,[48] Proposition 4.2). Second, define a matching outcome to be B-optimal if it is core stable and if each buyer matched under the matching outcome is at least as well off as he would be under any other core stable matching outcome. Then, in each TNG trade cycle, the DCR mechanism yields the unique B-optimal matching outcome as long as each tradebot has a strict preference order over the potential trade partners he

[5]The DCR mechanism always stops in finitely many steps (Tesfatsion,[48] Proposition 4.1). A particularly interesting aspect of the DCR mechanism is that it requires the tradebots to pass messages back and forth to each other at event-triggered times, a requirement that is easily handled by a C++ implementation; see McFadzean and Tesfatsion.[36]

finds tolerable (Tesfatsion,[48] Proposition 4.3). As indicated by the computer experiments reported in section 3.3 below, however, these static optimality properties do not appear to be adequate measures of optimality from an evolutionary perspective.

TABLE 2 Payoff Matrix for the Prisoner's Dilemma Game

		Player 2	
		c	d
Player 1	c	(C,C)	(L,H)
	d	(H,L)	(D,D)

Trades are currently modeled in the TNG as Prisoner's Dilemma (PD) games. For example, a trade may involve the exchange of a good or service of a certain promised quality in return for a loan or wage contract entailing various payment obligations. A buyer participating in a trade may either cooperate (fulfill his trade obligations) or defect (renege on his trade obligations), and similarly for a seller. The range of possible payoffs is the same for each trade in each trade cycle: namely, L (the sucker payoff) is the lowest possible payoff, received by a cooperative tradebot whose trade partner defects; D is the payoff received by a defecting tradebot whose trade partner also defects; C is the payoff received by a cooperative tradebot whose trade partner also cooperates; and H (the temptation payoff) is the highest possible payoff, received by a defecting tradebot whose trade partner cooperates. More precisely, the payoffs are assumed to satisfy $L < D < 0 < C < H$, with $(L + H)/2 < C$. The payoff matrix for the PD game is depicted in Table 2.

The TNG tradebots are currently assumed to use a simple form of criterion filter[6] to update their expected payoffs on the basis of new payoff information. Specifically, whenever a tradebot v receives a trade or refusal payoff P from an interaction with a potential trade partner k, tradebot v forms an updated expected payoff for k by taking a convex combination of this new payoff P and his previous expected payoff for k. The inverse of the weight on the new payoff P is 1 plus v's current payoff count with k. As explained in Tesfatsion,[48] this updating procedure guarantees that the expected payoff tradebot v associates with k converges to the true average payoff v attains from interactions with k as the number of interactions between v and k becomes arbitrarily large.

[6] As detailed in Tesfatsion,[46] a criterion filter is an algorithm for the direct updating of an expected return function on the basis of past return outcomes without recourse to the usual interim updating of probabilities via Bayes' rule.

The trade behavior of each tradebot, whether he is a pure buyer in $V - S$, a buyer-seller in $B \cap S$, or a pure seller in $V - B$, is currently characterized by a finite-memory pure strategy for playing a PD game with an arbitrary partner an indefinite number of times, hereafter referred to as a *trade strategy*. Each tradebot thus has a distinct trading personality, even if he engages in both buying and selling activities. At the commencement of each trade cycle loop, tradebots have no information about the trade strategies of other tradebots; they can only learn about these strategies by engaging other tradebots in repeated trades and observing the payoff histories that ensue. Moreover, each tradebot's choice of an action in a current trade with a potential trade partner is determined entirely on the basis of the payoffs obtained in past trades with this same partner. Thus, each tradebot keeps separate track of the particular state he is in with regard to each of his potential trade partners.

In the current implementation of the TNG, the only aspect of a tradebot that evolves over time is his trade strategy. The evolution of the tradebots in each generation cycle is thus meant to reflect the formation and transmission of new ideas rather than biological reproduction.

More precisely, each tradebot's trade strategy is represented as a finite state machine (FSM) with a fixed starting state. The FSMs for two illustrative trade strategies are depicted in Figure 1: namely, a nice trade strategy, Tit-for-Two-Tats, that only defects if defected against twice in a row; and an opportunistic trade strategy, Rip-Off, that evolved in an experiment with an initial population of Tit-for-Two-Tats to take perfect advantage of the latter strategy by defecting every other time.[7] As is more carefully explained in McFadzean and Tesfatsion,[36] in each generation cycle the trade strategies (FSMs) associated with the current tradebot generation are evolved by means of a standardly specified genetic algorithm involving mutation, recombination, and elitism operations applied to bit string encodings for the strategies. The effect of these operations is that successful trade strategies are mimicked and unsuccessful trade strategies are replaced by variants of more successful strategies.

3.2 AN ILLUSTRATIVE 5-TRADEBOT TNG

Consider a TNG for which the player set contains a total of five buyer-seller tradebots who can both make and receive trade offers.[8] Each tradebot v has the same minimum tolerance level, 0. Also, each tradebot v has the same offer quota, $O_v = 1$, implying that he can have at most one trade offer outstanding at any given time and can receive at most four trade offers from other tradebots at any given time. Each

[7]The experimental discovery of Rip-Off was made by Daniel Ashlock of Iowa State University during the preparation of Stanley et al.[45]

[8]The specifications for this illustrative TNG were chosen to permit comparisons with a 5-player IPD/CR game analyzed in Stanley et al.[45] (pp. 153–156). See Tesfatsion[48] (section 7.3) for the details of this comparison.

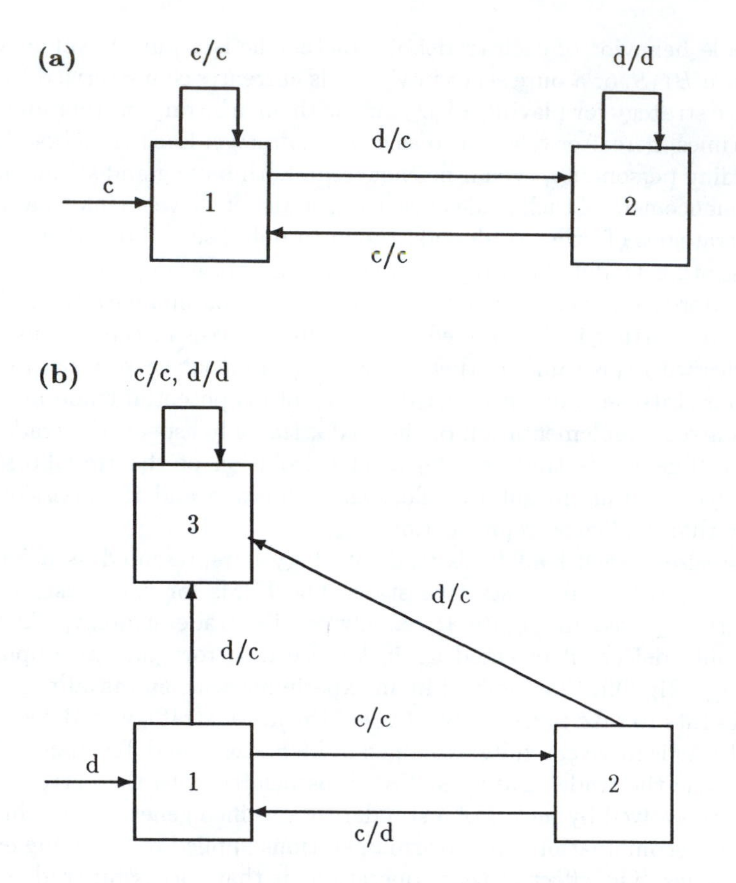

FIGURE 1 The FSM Representations for Two Illustrative Trade Strategies. (a) A nice trade strategy that starts by cooperating and only defects if defected against twice in a row; (b) An opportunistic trade strategy that starts by defecting and defects every other time unless defected against. An arrow label x/y means that y is the next move to be taken in response to x, the last move of one's opponent.

tradebot v is also assumed to have the same acceptance quota, $A_v = 4$, which then implies that each tradebot is effectively unconstrained with regard to the number of trade offers he can have on his waiting list at any given time. The refusal payoff, R, is assumed to be strictly negative, and the wallflower payoff, W, is assumed to be 0.

With regard to trade strategies, three of the tradebots are Tit-for-Two-Tats (TFTTs) and the remaining two tradebots are Rip-Offs (Rips); see Figure 1. A TFTT receives a payoff sequence $(C, C, C, \ldots)$ in repeated trades with another TFTT, and a payoff sequence $(L, C, L, C, \ldots)$ in repeated trades with a Rip, and a Rip receives a payoff sequence $(H, C, H, C, \ldots)$ in repeated trades with a TFTT, and a payoff sequence $(D, C, C, \ldots)$ in repeated trades with another Rip. Note that a Rip never triggers defection in a TFTT since a Rip never defects twice in a row. Consequently, in any match-up between a TFTT and a Rip, the Rip would definitely attain a higher fitness score than the TFTT if the TFTT were not permitted to refuse the Rip's trade offers.

One key factor affecting the emergence of trade networks in the TNG is the specification of the tradebots' prior expected payoffs. Low prior expected payoffs encourage tradebots to latch on to the first trade partner from whom they receive even a modestly high payoff. On the other hand, high prior expected payoffs encourage repeated experimentation with new trade partners in the face of continually disappointing payoffs from current trade partners. As will be seen below for the TFTTs, the combination of high prior expected payoffs and wide-spread experimentation can be detrimental to nice tradebots since it increases their chances of encountering opportunistically defecting trade partners whose defections are infrequent enough to avoid triggering refusals.

Two alternative benchmark assumptions will be considered for the tradebots' prior expected payoffs for their potential trade partners. The first assumption, a common prior, is the assumption made in Stanley et al.,[45] Smucker et al.,[44] and Ashlock et al.[4] for the IPD/CR game. The second assumption, long-run expectations, sets the prior expected payoffs that any two tradebots have for each other equal to the true long-run average payoffs that would result if these two tradebots were to engage in infinitely many trades.

Assumption (CP): Common Prior. Each tradebot associates the same prior expected payoff, U^0, with each other tradebot, where U^0 lies in the open interval from 0 (the minimum tolerance level) to H (the highest possible trade payoff).

Assumption (LR): Long-Run Expectations. Each TFTT associates a prior expected payoff C with each other TFTT and a prior expected payoff $(L + C)/2$ with each Rip; and each Rip associates a prior expected payoff $(H+C)/2$ with each TFTT and a prior expected payoff C with each other Rip.

Another key factor affecting the types of trade networks that can emerge in the TNG is the extent to which the benefits and costs associated with each potential trade partner balance out over time, either triggering eventual refusal or permitting long-term partnership. It is, therefore, useful to examine the various possible 2-tradebot match-ups for the illustrative 5-tradebot TNG before entering into an analysis of the full-blown model.

Given the form of the criterion filter that the tradebots use to update their expectations, a TFTT never judges another TFTT to be intolerable; for the expected payoff of a TFTT for another TFTT is always nonnegative. Similarly, a Rip never judges a TFTT to be intolerable. However, a Rip can become intolerable for another Rip or for a TFTT. In particular, as established in Tesfatsion[48] (section 7), under assumption (CP) one obtains the following four behavioral regions as the prior expected payoff U^0 ranges from low to high values:

(CP.1) $0 < U^0 < -D$: a TFTT finds a Rip intolerable after only one trade, and a Rip finds another Rip intolerable after only one trade;

(CP.2) $-D \leq U^0 < -L$: a TFTT finds a Rip intolerable after only one trade, and a Rip never finds another Rip intolerable;

(CP.3) $-L \leq U^0$ and $C < -L$: a TFTT finds a Rip intolerable after a finite odd number of trades, and a Rip never finds another Rip intolerable;

(CP.4) $-L \leq U^0$ and $-L \leq C$: a TFTT never finds a Rip intolerable, and a Rip never finds another Rip intolerable.

In contrast, under (LR) one obtains four regions characterized by somewhat different transient behaviors as the mutual cooperation payoff C ranges from low to high values:

(LR.1). $0 < C < -D$: a TFTT finds a Rip intolerable prior to trade, and a Rip finds another Rip intolerable after only one trade;

(LR.2). $-D \leq C < -L$: a TFTT finds a Rip intolerable prior to trade, and a Rip never finds another Rip intolerable;

(LR.3). $-L \leq C < -3L$: a TFTT finds a Rip intolerable after only one trade, and a Rip never finds another Rip intolerable;

(LR.4). $-3L \leq C$: a TFTT never finds a Rip intolerable, and a Rip never finds another Rip intolerable.

What, then, are the possible trade networks that can emerge in this 5-tradebot TNG over the course of a single trade cycle loop, assuming either (CP) or (LR) is in effect?

POSSIBLE TRADE NETWORKS UNDER ASSUMPTION (CP). At the beginning of the initial trade cycle, each tradebot judges each other tradebot to be equally tolerable, and he uses random selection to submit a trade offer to one of these tradebots. In turn, each tradebot places all received trade orders on his current waiting list, with no refusals. In accordance with the DCR mechanism, each tradebot then accepts all trade offers on his current waiting list.

Suppose that payoffs and prior expected payoffs are configured as in (CP.1). In this case, even though a Rip receives the highest possible payoff, H, from an initial trade with a TFTT, which encourages him to submit another trade offer to this TFTT in the next trade cycle, the TFTT neither submits trade offers to, nor accept trade offers from, this Rip after their first trade. Moreover, the two Rips cease all trade activity with each other after their first trade. Under the DCR mechanism, a tradebot receiving a refusal payoff from a refused trade offer during the course of a trade cycle immediately submits a replacement trade offer to any next-most-preferred tolerable tradebot who has not yet refused him. Consequently, by the end of the first four trade cycles, a Rip has triggered refusal in every one of his potential trade partners. Thereafter, the Rip submits trade offers only to the TFTTS, receiving only negative refusal payoffs in return, until the expected payoff he associates with each TFTT finally drops below zero and he turns into a wallflower.

In summary, by the end of the fourth trade cycle, the only trade networks that are viable for case (CP.1) involve trades among the three TFTTs, with both Rips ostracized and eventually reduced to wallflowers; cf. Figure 2(a). The fitness score of each TFTT thus tends toward the mutually cooperative payoff, C, whereas the fitness score of each Rip tends toward the wallflower payoff, 0. Whether or not the Rips survive and prosper in the generation cycle at the end of the trade cycle loop then depends on the length of this loop. Specifically, in order for a Rip to end up with a higher fitness score than the TFTTs, the loop must be short enough so that the H payoffs received by the Rip from his initial successful defections against the three TFTTs sufficiently outweigh the mutual defection payoff, D, that he receives from his one Rip-Rip trade, any refusal payoffs, R, that he receives from subsequent refused attempts to trade with the TFTTs, and any wallflower payoffs, 0, that he receives after ceasing all trade activity.

Cases (CP.2) and (CP.3) are similar to case (CP.1), except that the two Rips end up trading cooperatively with each other rather than as wallflowers; cf. Figure 2(b). Also, in case (CP.3) the TFTTs may take longer to reject the opportunistically defecting Rips. The fitness scores of all tradebots thus tend toward the mutually cooperative payoff, C, but it is now more likely that the Rips will have a higher fitness score than the TFTTs at the end of any given trade cycle loop and, hence, a reproductive advantage in the subsequent generation cycle.

Suppose, now, that case (CP.4) holds with $-L \leq U^0 < (H+C)/2$. As established in Tesfatsion[48] (Proposition 7.1), the only long-run trade networks viable in this case consist of the three TFTTs engaged in mutually cooperative trades

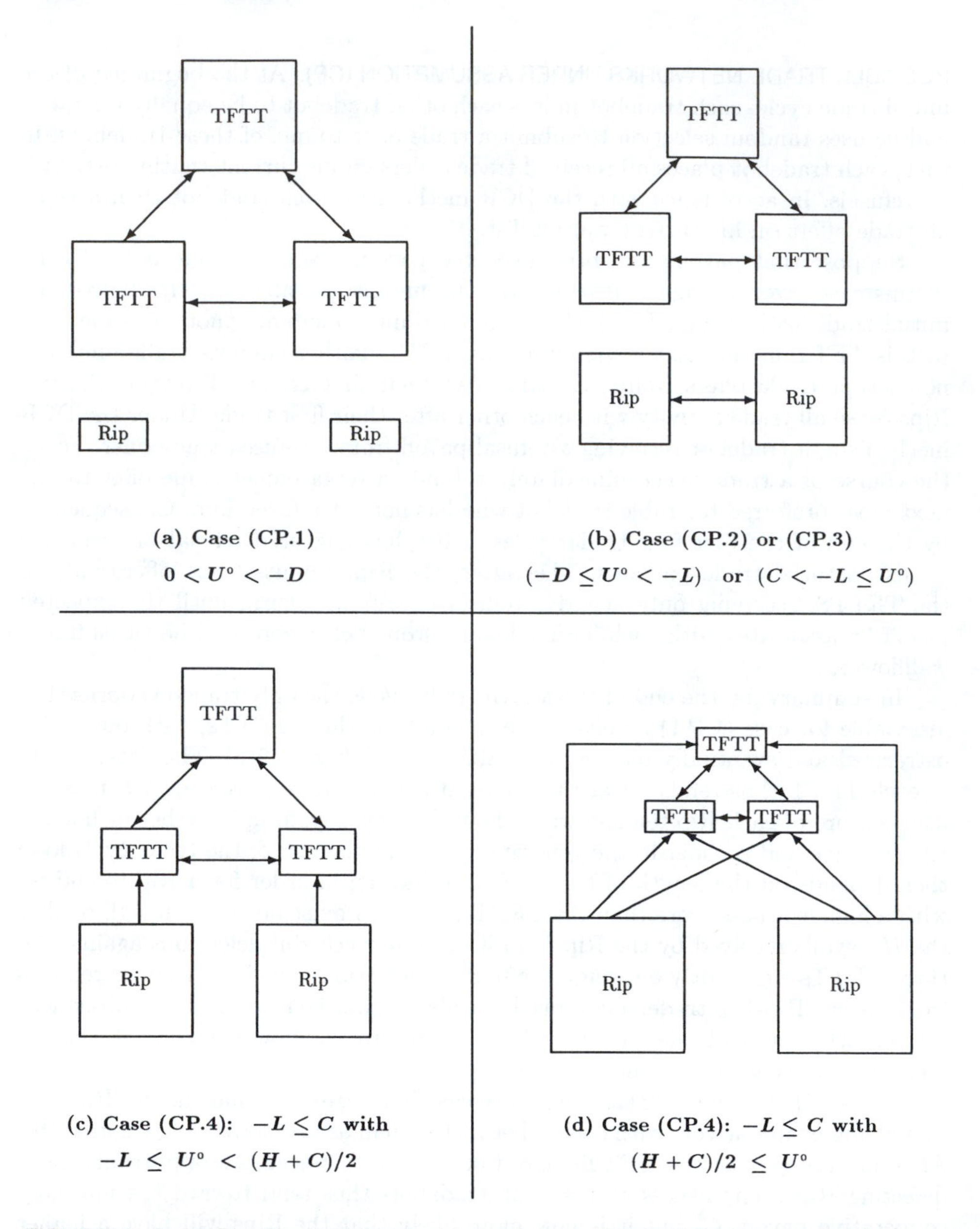

FIGURE 2 Long-Run Trade Networks Under Assumption (CP) for the Illustrative 5-Tradebot TNG. A relatively larger box indicates a definitely higher fitness score for a sufficiently long trade cycle loop. In case (d), the Rip-TFTT interactions are stochastic if $(H + C)/2 = U^0$ and deterministic if $(H + C)/2 < U^0$.

with other TFTTs, with each Rip latched on to one randomly determined TFTT. Figure 2(c) depicts the case in which each Rip happens to be latched on to a different TFTT.

In this case, then, each TFTT risks becoming a full-time host for a parasitical Rip. Prior expected payoffs are low enough to encourage latching behavior on the part of the Rips, who are delighted with their unexpectedly high payoffs from the TFTTs, but not low enough to induce the TFTTs to refuse the Rips. Although the average payoff, $(L+C)/2$, that a TFTT receives from repeated trades with a Rip is nonnegative and possibly positive, it is not as high as the average payoff accruing to a Rip from such trades, $(H+C)/2$, nor as high as the average payoff, C, that accrues to a TFTT in repeated trades with another TFTT. Hence, the relative fitness of a TFTT is lowered by interactions with a Rip, and this puts him at a reproductive disadvantage in the generation cycle.

It is interesting to note that at least one TFTT always avoids becoming parasitized by a Rip in case (CP.4) with $-L \leq U^0 < (H+C)/2$. The fitness score of any such TFTT tends toward C, whereas the fitness score of each parasitized TFTT is uniformly bounded below C. The structurally identical TFTT tradebots thus end up, by chance, with different fitness scores. Nevertheless, given a sufficiently long trade cycle loop, each Rip exits the loop with a higher fitness score than each TFTT; for the fitness score of each Rip tends toward $(H+C)/2$.

Next, consider case (CP.4) with $U^0 = (H+C)/2$. In this case, as established in Tesfatsion[48] (Proposition 7.2) and depicted in Figure 2(d), each Rip stochastically switches his trade offers back and forth among the three TFTTs for the duration of the trade cycle loop. Hence, each TFTT always has a positive probability of being parasitized by each Rip. The reason for the formation of this type of trade network is that a Rip is indifferent between any two TFTTs with whom he has traded an even number of times.

Finally, consider case (CP.4) with $(H+C)/2 < U^0 < H$. As established in Tesfatsion[48] (Proposition 7.3) and depicted in Figure 2(d), each TFTT is now a recurrent host for each of the parasitical Rips by the end of the fourth trade cycle. The intuitive reason for the formation of this type of trade network is that each Rip's prior expected payoff for a TFTT is so high that he essentially always prefers the TFTT with whom he has currently traded the least, and this leads him to repeatedly cycle his trade offers among the three TFTTs.

As in the previous (CP.4) cases, the fitness score of each Rip in this final (CP.4) case tends toward $(H+C)/2$. In contrast to the previous (CP.4) cases, however, no TFTT now has any chance of escaping parasitization by all three Rips. Consequently, the fitness score of all three TFTTs is uniformly bounded below C for all sufficiently long trade cycle loops. Here, then, is an example where optimistic prior expectations, leading to increased experimentation, turn out to be detrimental for the nicer tradebots.

POSSIBLE TRADE NETWORKS UNDER ASSUMPTION (LR). Comparing the behavioral regions under (LR) with the behavioral regions under (CP), one sees that the TFTTs tend to behave more cautiously under (LR) because their prior expected payoffs are less optimistic. In particular, a TFTT's prior expected payoff for a Rip is bounded strictly below C under (LR) and may even be negative. Consequently, no TFTT ever submits a trade offer to a Rip. Moreover, a TFTT will not accept an initial trade offer from a Rip under (LR) unless the benefit, C, from a mutual cooperation is at least as great as the cost, $-L$, that is incurred when the Rip successfully defects against him.

Consider, now, the trade networks that can emerge under (LR). In the initial trade cycle, each TFTT uses random selection to submit a trade offer to one particular TFTT, and all such trade offers are accepted. Each Rip likewise uses random selection to submit a trade offer to one particular TFTT; but whether or not these trade offers are accepted depends on the behavioral region.

In case (LR.1), a TFTT refuses all trade offers from a Rip prior to any trades taking place. Under the DCR mechanism, a tradebot who has a trade offer refused immediately submits a replacement trade offer to any next-most-preferred tolerable tradebot who has not yet refused him. Thus, by the end of the initial trade cycle, each Rip has made one trade offer to each TFTT which was refused, and one trade offer to the other Rip which was accepted. Nevertheless, after this one trade, the Rips find each other intolerable and never submit trade offers to each other again. In subsequent trade cycles each Rip only submits trade offers to the TFTTs, collecting negative refusal payoffs until, finally, his expected payoff for each TFTT drops below zero. Thereafter, each Rip subsides into wallflowerdom; cf. Figure 3(a).

Case (LR.2) differs from case (LR.1) in only one respect—the Rips never find each other intolerable. Thus, in the first few trade cycles, each Rip uses sequential random selection to submit a trade offer to each TFTT in turn, who refuses the offer, and then to the other Rip, who accepts the offer. Since each refusal results in a negative refusal payoff, R, the expected payoff that each Rip associates with each TFTT eventually drops below the expected payoff that each Rip has for the other Rip, which is always nonnegative. In each subsequent trade cycle the Rips submit trade offers only to each other; cf. Figure 3(b).

The interesting aspect of both (LR.1) and (LR.2) is that each TFTT is able to use refusal to protect himself completely from the Rips, so that he *never* sustains any low L payoffs. The fitness score of each TFTT at the end of the trade cycle loop is thus C, because C is the only payoff he ever experiences. In contrast, each Rip sustains negative refusal payoffs as well as at least one negative defection payoff before finally settling down either to wallflowerdom in case (LR.1) or to mutually cooperative trades with the other Rip in case (LR.2). Consequently, the fitness score of each Rip at the end of any trade cycle loop is definitely below C and may even be negative. It follows that each TFTT has a higher fitness score than each Rip at the end of any trade cycle loop and, hence, has a reproductive advantage over each Rip in the subsequent generation cycle.

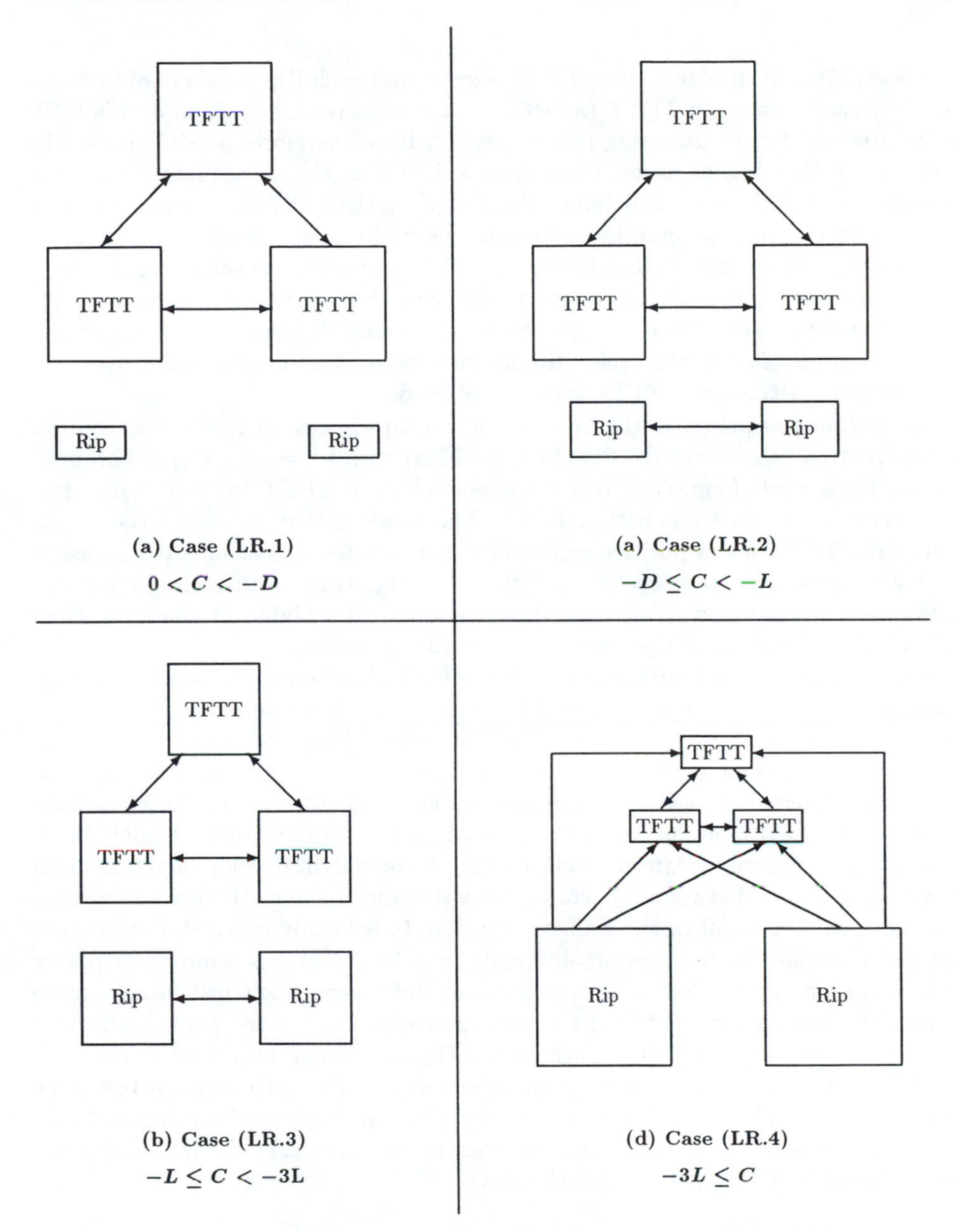

FIGURE 3 Long-Run Trade Networks Under Assumption (LR) for the Illustrative 5-Tradebot TNG. A relatively larger box indicates a definitely higher fitness score for a sufficiently long trade cycle loop. In case (d), the Rip-TFTT interactions are stochastic.

Case (LR.3) is similar to case (LR.2), except that each Rip is able to obtain one high H payoff from each TFTT (inflicting a low negative L payoff on each TFTT in the process) before collecting refusal payoffs. It is, therefore, possible for a Rip to end up with a higher fitness score than a TFTT in the subsequent generation cycle. In general, then, as depicted in Figure 3(c), neither the TFTTs nor the Rips have a definite long-run reproductive advantage under case (LR.3).

Finally, suppose case (LR.4) holds. The TFTTs continue to submit trade offers only to each other in each successive trade cycle; but, unlike the previous (LR) cases, they never refuse trade offers received from a Rip. Consequently, a Rip never submits a trade offer to the other Rip; for the trade offers he submits to the persistently more attractive TFTTs are never refused.

Indeed, not surprisingly, the behavior of the Rips in case (LR.4) is very similar to the eventual behavior of the Rips in case (CP.4) with $U^0 = (H+C)/2$. Throughout the trade cycle loop, each Rip randomly selects a TFTT to trade with after every even-numbered trade with a TFTT. The result is that, in every other trade cycle, each TFTT has a positive probability of becoming a host for each parasitic Rip for the next two trade cycles, and the Rips never trade with each other at all; cf. Figure 3(d). It follows that each Rip ends up with a higher fitness score than each TFTT, regardless of the length of the trade cycle loop.

Comparing Figure 3 with Figure 2, the TFTTs have an easier time protecting themselves against the Rips in case (LR), where they have an accurate prior understanding of the payoffs they can expect to obtain from each type of tradebot. While this amount of information may be excessive, it does seem reasonable to suppose that, based on past bad experiences, nice tradebots develop more cautious priors for untested trade partners than do street-wise opportunistic tradebots on the look-out for chumps. Alternatively, nice tradebots might develop high minimum tolerance levels, so that refusal occurs quickly if trades go sour. Having a minimum tolerance level set equal to the wallflower payoff, 0, is locally rational, in the sense that positive payoffs, however small, result in a better fitness score for a player than no payoffs at all. Yet a myopic focus on increasing one's own fitness score in absolute terms might not lead to reproductive success in the generation cycle if other opportunistic tradebots such as the Rips are doing even better; cf. cases (CP.4) and (LR.4). Ideally, then, prior expected payoffs and minimum tolerance levels should be allowed to evolve conjointly with the tradebots' strategies. Some preliminary simulation work along these lines in the context of the IPD/CR game can be found in Ashlock et al.[4] (section 5.4).

3.3 ILLUSTRATIVE TNG COMPUTER EXPERIMENTS

Two types of TNG computer experiments are reported in this section: (a) buyer-seller market experiments, in which each tradebot is both a buyer and a seller; and (b) two-sided market experiments, in which a subset of buyer tradebots makes offers to a disjoint subset of seller tradebots. All experimental findings were obtained using the C++ implementation of the TNG developed by McFadzean and Tesfatsion.[36]

For each experiment, multiple runs from different initial random seeds are described. The following features are set commonly across all of these experimental runs.

The wallflower payoff W is set at 0, the refusal payoff R is set at -0.6, the PD trade payoffs are set at $L = -1.6$, $D = -0.6$, $C = 1.4$, and $H = 3.4$, and each tradebot's minimum tolerance level is set at 0. Each tradebot assigns the same prior expected payoff, 1.4, to each other tradebot, implying that he is initially indifferent concerning which trade partners he interacts with and is fairly optimistic about the payoffs he will receive; and each tradebot assigns a negative prior expected payoff to himself, thus ensuring that he never trades with himself.

The total number of tradebots is set at 24, the number of generations is set at 50, and the number of trade cycles in each trade cycle loop is set at 150. Trade strategies are represented as 16-state finite state machines (FSMs) with fixed starting states and with memory 1, where the memory is the number of bits used to encode the past actions of an opponent that can be recalled in any given internal FSM state. For implementation of the genetic algorithm, the number of elite trade strategies retained unchanged for the next generation is set at 16, where the elite trade strategies are the trade strategies that have attained the highest relative fitness scores. Also, the probability of crossover is set at 1.0 and the probability of a bit mutation is set at 0.005.

As outlined in Table 1, each experimental run has the same dynamic structure. Each of the 24 tradebots in the initial tradebot generation is assigned a trade strategy encoded as a randomly generated bit string. The tradebots are then configured in accordance with user-supplied parameter values, and a trade cycle loop commences. At the end of the trade cycle loop, the randomly generated bit strings that encode the trade strategies of the current tradebot generation are evolved by means of a genetic algorithm employing two-point crossover, bit mutation, and elitism. The evolved set of 24 bit strings decodes to an evolved set of 24 trade strategies (FSMs) that are reinitialized to their fixed starting states and assigned to a new tradebot generation. This new tradebot generation is then configured with user-supplied parameter values, and another trade cycle loop commences.

BUYER-SELLER MARKETS. Each tradebot in these experiments was both a buyer and a seller, implying that he could both make and receive trade offers.

In the first batch of buyer-seller experiments, the acceptance quota of each tradebot was set equal to the total number of tradebots, 24, and the offer quota of each tradebot was set equal to 1. The tradebots were thus effectively unconstrained with regard to the number of trade offers they could have on their waiting lists at any given time.

As a benchmark, experiments were first run with random partner matching in place of the DCR matching mechanism. Random partner matching was effected by preventing the updating of the prior expected payoff 1.4 that each tradebot initially assigned to each potential trade partner, so that all tradebots remained indifferent concerning their potential trade partners and matching was accomplished by the default mechanism of a random draw. Although occasionally the average fitness score achieved by the tradebots under random matching rose to the mutual cooperation level, 1.4, a more typical outcome was a steady decline[9] to the mutual defection level, −0.6; see Figure 4. The size of the refusal payoff is irrelevant for this finding, since refusals never occur in TNG experiments with random matching and nonbinding acceptance quotas.

When the DCR matching mechanism was then restored, the average fitness score achieved by the tradebots typically evolved to the mutual cooperation level 1.4; see Figure 5. These TNG experiments reinforce the previous IPD/CR findings of Stanley et al.[45] and Ashlock et al.[4] that a preference-based matching mechanism tends to accelerate the emergence of mutual cooperation in the IPD when each agent is permitted both to make and to refuse game offers, is unconstrained with regard to the number of received offers he can accept, and is permitted to have at most one offer outstanding at any given time.

In the second batch of buyer-seller experiments, all parameter settings were retained unchanged except that the acceptance quotas were reduced from 24 to 1. Each tradebot could thus retain at most one trade offer on his waiting list at any one time; all other received trade offers had to be refused. Under random partner matching, the typical outcome was again the emergence of an average fitness score close to the mutual defection payoff level, −0.6. This same outcome obtained even when refusal payoffs were omitted from fitness scores, implying that refusal payoffs resulting from limited waiting lists were not a determining factor.

When the DCR matching mechanism was restored, however, the average fitness score typically leveled out at about 1.25 instead of evolving to the mutual cooperation payoff level 1.4, the outcome for the first batch of buyer-seller experiments. The explanation for this difference appears to lie in the changed nature of the refusal payoffs.

In the first batch of buyer-seller experiments, the acceptance quota (24) was large relative to the offer quota (1). In these circumstances, tradebots are generally refused by other tradebots only if the latter find them to be intolerable because

[9]Recall that each tradebot generation participates in only 150 trade cycles, and each tradebot can direct at most one trade offer to any particular tradebot during a trade cycle. Consequently, the maximum number of trades between any two tradebots in a given generation is equal to 300, which is attained only if each tradebot submits an accepted trade offer to the other tradebot in each of the 150 trade cycles in which they participate. Also, the trade strategies for the initial generation in each TNG experiment are randomly generated. A steady decline to mutual defection over the initial 50 generations is, therefore, not in conflict with the evolution of mutual cooperation observed for the IPD with random or round-robin matching by previous researchers when much longer player interaction lengths were permitted between players in each generation or when the initial strategy population was sufficiently seeded with cooperatively inclined strategies such as Tit-for-Tat.

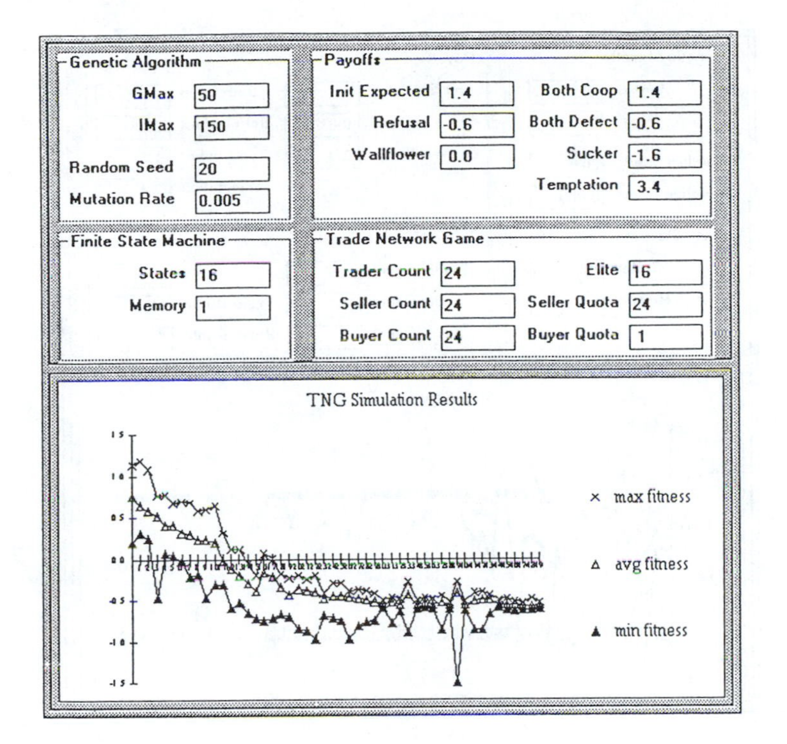

FIGURE 4 Buyer-seller average fitness with random matching and seller quotas equal to 24.

of past defections. Negative refusal payoffs received in response to defections should rightly count against the fitness of the trade strategies generating the defections, for this induces changes in these strategies in the generation cycle that tend to lead to higher future fitness scores. In the second batch of buyer-seller experiments, however, the acceptance quota (1) was much smaller in relation to the offer quota (1), implying that many more received trade offers had to be refused regardless of their desirability. In these circumstances, tradebots tend to accumulate large numbers of negative refusal payoffs purely as a consequence of the relatively small acceptance quota and the nature of the DCR mechanism, regardless of their trade strategies. Since the quotas and the DCR mechanism are not evolved in the current implementation of the TNG, penalizing the tradebots for these quota and DCR effects by including refusal payoffs in their individual fitness scores tends to lower their current average fitness score without inducing a higher average fitness score in the future.

As expected, the average fitness scores attained by the tradebots in the second batch of buyer-seller experiments markedly improved when refusal payoffs were

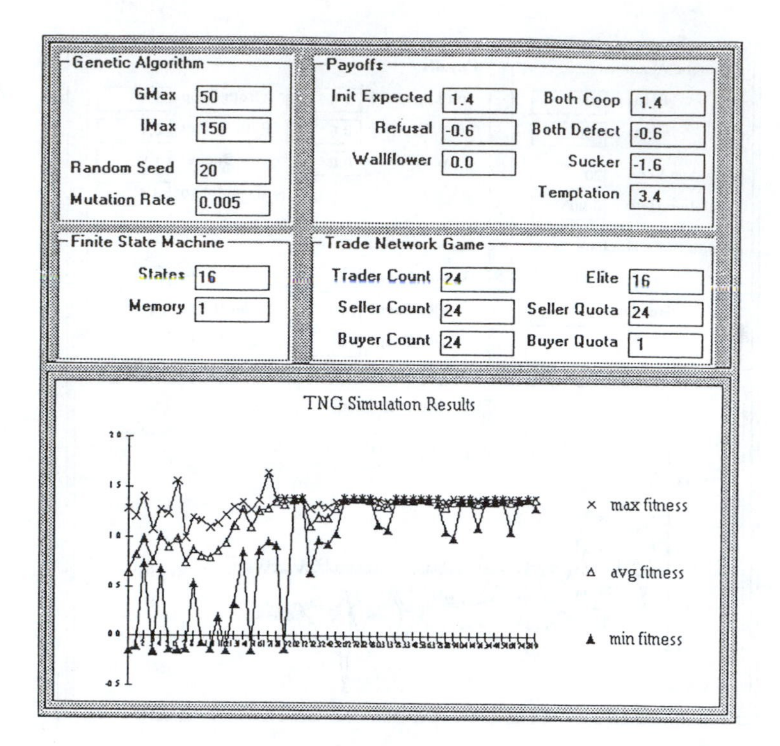

FIGURE 5 Buyer-seller average fitness with DCR matching and seller quotas equal to 24.

removed from the calculation of the tradebots' individual fitness scores; see Figure 6. Improvement continued to occur when, in addition, the refusal payoffs were reduced in magnitude from −0.60 to −0.30; but a further reduction in magnitude to −0.06 and then to 0 resulted in increasingly volatile maximum and minimum average fitness scores with no discernible improvement in average fitness scores.

The probable cause of this increased volatility is that tradebots receiving refusals during initial trade cycles have little incentive to direct their offers elsewhere in subsequent trade cycles when the magnitude of the negative refusal payoff is small. A negative refusal payoff guarantees that the continually updated expected payoff that a tradebot associates with another tradebot who repeatedly refuses him eventually falls below 0, the minimum tolerance level, at which point he ceases making offers to this other tradebot. Nevertheless, this learning process is slow when the magnitude of the negative refusal payoff is small, and it is nonexistent when the refusal payoff is simply set to 0.

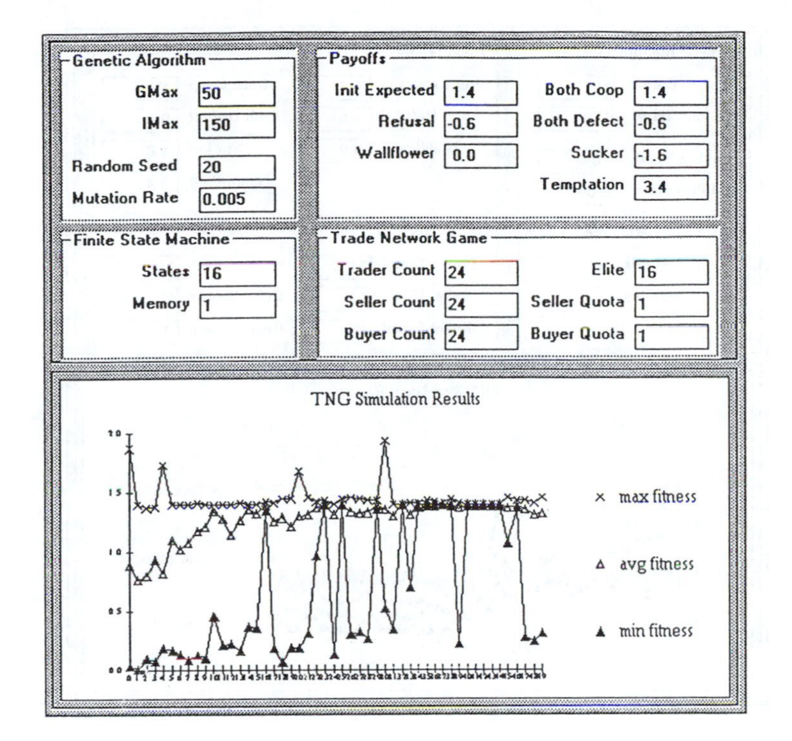

FIGURE 6 Buyer-seller average fitness with DCR matching, seller quotas equal to 1, and refusal payoffs omitted from fitness scores.

TWO-SIDED MARKETS. In each two-sided market experiment, the 24 tradebots were evenly divided into 12 pure buyers (makers of trade offers) and 12 pure sellers (receivers of trade offers).

In the first batch of experiments, the acceptance quota of each seller was set to 12 and the offer quota of each buyer was set at 1. Thus, sellers were effectively unconstrained regarding the number of trade offers they could have on their waiting lists at any one time. Experiments were first run with random partner matching in place of the DCR matching mechanism to obtain a benchmark for comparison. Interestingly, in contrast to buyer-seller experiments with nonbinding acceptance quotas and random matching, the average fitness score attained by the tradebots tended to fall to a level between -0.4 and the wallflower payoff 0 rather than dropping all the way down to the mutual defection payoff level -0.6; compare Figure 7 with Figure 4.

When the DCR matching mechanism was restored, the average fitness score of the tradebots typically evolved to about 1.2, a payoff level markedly below the

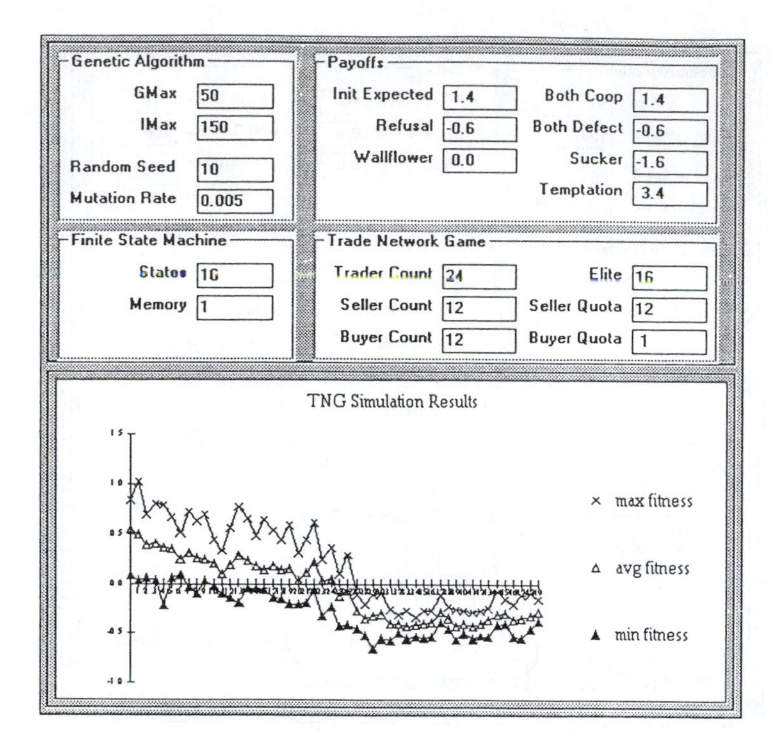

FIGURE 7 Two-sided market average fitness with random matching and seller quotas equal to 12.

mutual cooperation level 1.4 obtained in buyer-seller experiments with nonbinding acceptance quotas and DCR matching. Moreover, the maximum fitness score, the average fitness score, and the minimum fitness score attained by the successive tradebot generations persistently deviated from one another. Compare Figure 8 with Figure 5.

As established in Tesfatsion[48] (Proposition 4.3), the DCR mechanism is only guaranteed to be Pareto optimal for buyers, that is, for the active makers of trade offers. The effects of this bias on average fitness scores is hidden in buyer-seller markets, where each tradebot is both a buyer and a seller. However, for two-sided markets with buyer offer quotas set equal to 1 and with nonbinding seller acceptance quotas, the DCR mechanism appears to result in a "separating equilibrium" in which the buyers are generally achieving high fitness scores and the sellers are generally achieving low fitness scores. In particular, the extreme pickiness of buyers combined with the acceptance by sellers of all tolerable received trade offers appears to allow buyers to form long-run parasitic relations with sellers, i.e., relations characterized by successful defections within the limits permitted by the sellers' 0 minimum tolerance levels.

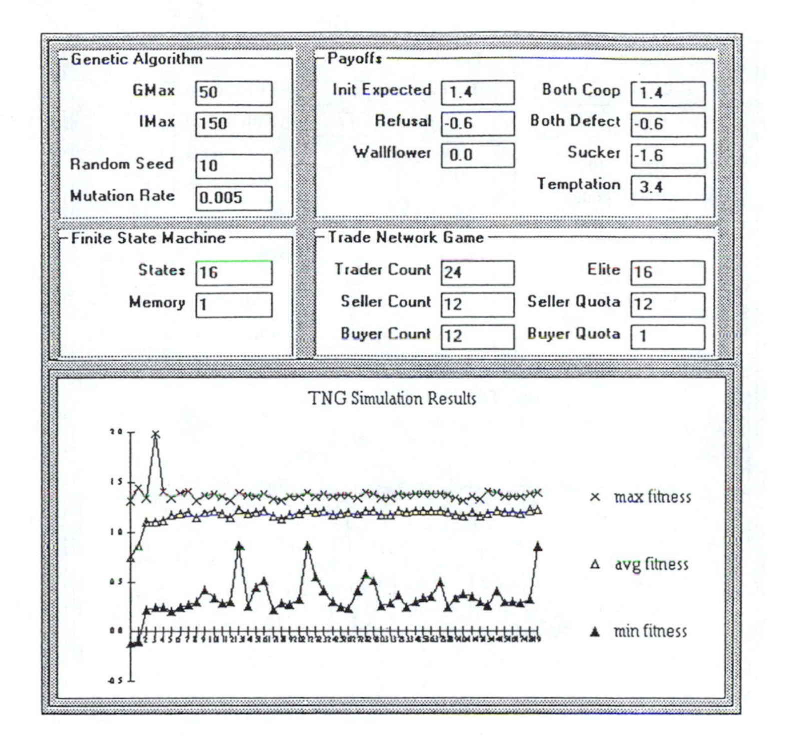

FIGURE 8 Two-sided market average fitness with DCR matching and seller quotas equal to 12.

In the second batch of two-sided market experiments, all parameter specifications were retained unchanged except that the seller acceptance quotas were decreased from 12 to 1. Thus, instead of accepting all tolerable received trade offers, the sellers now accepted at most one received trade offer per trade cycle.

When benchmark experiments were first run with random partner matching in place of the DCR mechanism, the typical outcome was the emergence of an average attained fitness score close to the mutual defection payoff, −0.6. This result obtained whether or not refusal payoffs were counted in the calculation of individual fitness scores. When the DCR matching mechanism was then restored, with refusal payoffs counted in the calculation of individual fitness scores, the accumulation of refusal payoffs tended to result in an average attained fitness score that was markedly below the mutual cooperation payoff level. When refusal payoffs were then omitted from the calculation of individual fitness scores, the average attained fitness score tended to evolve to the mutual cooperation level 1.4 and to be close to the maximum attained fitness scores; see Figure 9.

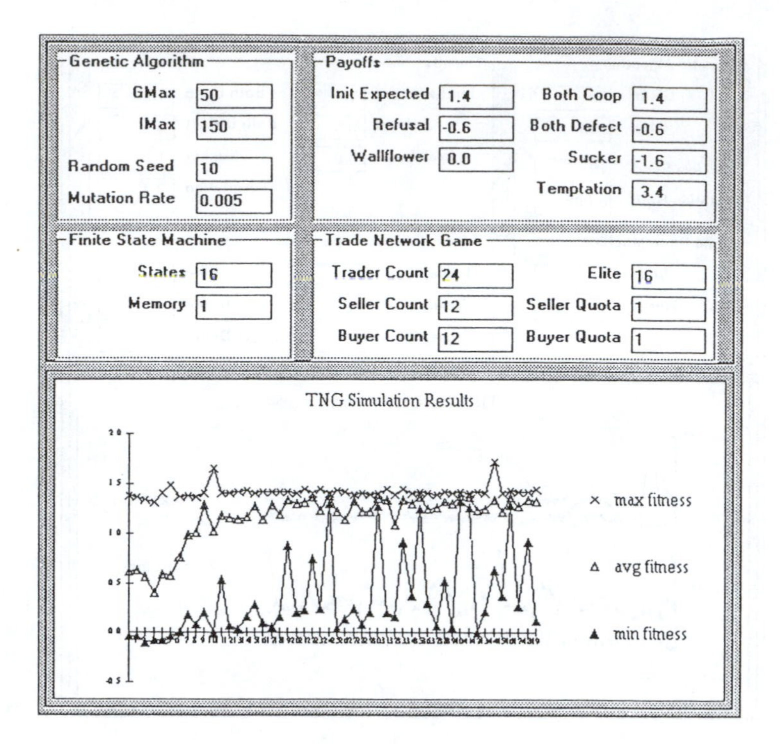

FIGURE 9 Two-sided market average fitness with DCR matching, seller quotas equal to 1, and refusal payoffs omitted from fitness scores.

Given the structuring of the two-sided market experiments currently under discussion, with equal numbers of buyers and sellers, common acceptance and offer quotas, and tradebots who are initially indifferent among their potential trade partners, the buyer offer quota roughly predicts the number of trade offers that each seller will receive in the initial few trade cycles. The larger this number is, the more chance there is that mutually cooperative matches between buyers and sellers will be quickly discovered. On the other hand, if the seller acceptance quota is small relative to the buyer offer quota, then many buyers will accumulate negative refusal payoffs unrelated to the nature of their trade strategies; and, if the seller acceptance quota is large relative to the buyer offer quota, opportunistic buyers have a greater chance to seek out and exploit sellers within the limits allowed by their minimum tolerance levels. Either circumstance could slow or even prevent the sustained emergence of mutually cooperative behavior.

It might therefore be conjectured that mutually cooperative behavior will best be induced in the current experimental setting when the seller acceptance quota and the buyer offer quota are equal and sufficiently large. Indeed, this turns out to be the case. In various two-sided market experiments with 12 pure buyers, 12 pure sellers, and equal seller and buyer quotas ranging from 3 to 12, the average fitness score attained by the tradebots tended to evolve to the mutual cooperation payoff level and to be close to the maximum attained fitness score even when refusal payoffs were included in the calculation of individual fitness scores.

4. CONCLUDING REMARKS

The hallmark of the alife approach to social and biological modeling is a bottom-up perspective, in the sense that global behavior is grounded in local agent interactions. The agent-based trade network game (TNG) illustrates how the alife paradigm can be specialized to an economic context. In particular, the analytical and simulation findings presented in the previous section illustrate how the TNG is currently being used to study the evolutionary implications of alternative market structures at three different levels: individual trade behavior; trade network formation; and social welfare as measured by average agent fitness.

As currently implemented, however, the TNG only partially achieves the goal of a bottom-up perspective. The TNG tradebots are surely more autonomous than agents in traditional economic models. For example, in order to determine their trade partners, the tradebots send messages back and forth to each other at event-triggered times. Nevertheless, they are still controlled by a main program that synchronizes the commencement of their trading activities and the evolution of their trade behavior. The advantage of imposing this synchronized dynamic structure is that it permits some analytical results to be obtained concerning the configuration, stability, uniqueness, and social optimality of the trade networks that emerge. The disadvantage is that these networks may not be robust to realistic relaxations of the imposed synchronizations.

As the TNG illustrates, then, the challenges to economists posed by alife modeling are great and the payoffs are yet to be fully determined. Using the alife approach, however, economists can at last begin to test seriously the self-organizing capabilities of decentralized market economies.

REFERENCES

1. Anderson, P. W., K. J. Arrow, and D. Pines, eds. *The Economy as an Evolving Complex System*, Proc. Vol. V, Santa Fe Institute Studies in the Sciences of Complexity. Redwood City, CA: Addison-Wesley, 1988.
2. Arifovic, J. "Genetic Algorithm Learning and the Cobweb Model." *J. Econ. Dynamics & Control* **18** (1994): 3–28.
3. Arthur, W. B. "On Designing Economic Agents that Behave Like Human Agents." *J. Evol. Econ.* **3** (1993): 1–22.
4. Ashlock, D., M. D. Smucker, E. A. Stanley, and L. Tesfatsion. "Preferential Partner Selection in an Evolutionary Study of Prisoner's Dilemma." *BioSystems* **37** (1996): 99–125.
5. Axelrod, R. *The Evolution of Cooperation*. New York, NY: Basic Books, 1984.
6. Axelrod, R. "The Evolution of Strategies in the Iterated Prisoner's Dilemma." In *Genetic Algorithms and Simulated Annealing*, edited by L. Davis. Los Altos, CA: Morgan Kaufmann, 1987.
7. Bell, A. M. "Locally Interdependent Preferences in a General Equilibrium Environment." Working Paper #97-W02, Department of Economics and Business Administration, Vanderbilt University, 1997.
8. Birchenhall, C. R. "Modular Technical Change and Genetic Algorithms." Working Paper No. 9508, University of Manchester, 1995.
9. Bosch, A., and S. Sunder. "Tracking the Invisible Hand: Convergence of Double Auctions to Competitive Equilibrium." Economics Working Paper # 91, Universitat Pompeu Fabra, 1994.
10. Brock, W., and S. N. Durlauf. "Discrete Choice with Social Interactions." Working Paper # 95-10-084, Santa Fe Institute, Santa Fe, NM, 1995.
11. Bullard, J., and J. Duffy. "A Model of Learning and Emulation with Artificial Adaptive Agents." Economics Working Paper, University of Pittsburgh, 1995.
12. De Vany, A. "The Emergence and Evolution of Self-Organized Coalitions." In *Computational Economic Systems: Models, Methods, and Econometrics*, edited by M. Gilli, 25–50. New York: Kluwer Scientific Publishers, 1996.
13. Durlauf, S. "Neighborhood Feedbacks, Endogenous Stratification, and Income Inequality." In *Disequilibrium Dynamics: Theory and Applications*, edited by W. A. Barnett, G. Gandolfo, and C. Hillinger. Cambridge: Cambridge University Press, 1996.
14. Epstein, J. M., and R. Axtell. *Growing Artificial Societies: Social Science from the Bottom Up*. Cambridge, MA: MIT Press/Brookings, 1996.
15. Fogel, D. B. "On the Relationship Between the Duration of an Encounter and the Evolution of Cooperation in the Iterated Prisoner's Dilemma." *Evol. Comp.* **3** (1995): 349–363.

16. Friedman, D. "Evolutionary Games in Economics." *Econometrica* **59** (1991): 637–666.
17. Gale, D., and L. Shapley. "College Admissions and the Stability of Marriage." *Am. Math. Monthly* **69** (1962): 9–15.
18. Goldberg, D. *Genetic Algorithms in Search, Optimization, and Machine Learning*. Reading, MA: Addison-Wesley, 1989.
19. Guriev, S., and M. Shakhova. "Self-Organization of Trade Networks in an Economy with Imperfect Infrastructure." In *Self-Organization of Complex Structures: From Individual to Collective Dynamics*, edited by F. Schweitzer. London: Gordon and Breach Scientific Publishers, 1996.
20. Hauk, E. "Leaving the Prison: A Discussion of the Iterated Prisoner's Dilemma under Preferential Partner Selection." Thesis, European University Institute, Florence, 1996.
21. Hirshleifer, D., and E. Rasmusen. "Cooperation in a Repeated Prisoners' Dilemma with Ostracism," *J. Econ. Behav. & Org.* **12** (1989): 87–106.
22. Holland, J. *Adaptation in Natural and Artificial Systems: An Introductory Analysis with Applications to Biology, Control, and Artificial Intelligence*, 2nd Ed. Cambridge, MA: MIT Press/Bradford Books, 1992.
23. Holland, J., and J. Miller. "Artificial Adaptive Agents in Economic Theory." *Am. Econ. Rev. Papers & Proc.* **81** (1991): 365–370.
24. Ioannides, Y. M. "Evolution of Trading Structure." This Volume.
25. Kirman, A. P. "Ants, Rationality, and Recruitment." *Quart. J. Econ.* **108** (1993): 137–156.
26. Kirman, A. P. "Economies with Interacting Agents." Working Paper # 94-05-030, Santa Fe Institute, Santa Fe, NM, 1994.
27. Kitcher, P. "The Evolution of Altruism." *J. Phil.* **90** (1993): 497–516.
28. Lane, D. "Artificial Worlds and Economics: Parts I & II." *J. Evol. Econ.* **3** (1993): 89–108; 177–197.
29. Langton, C., ed. *Artificial Life*. Santa Fe Institute Studies in the Sciences of Complexity, Proc. Vol. V., Redwood City, CA: Addison-Wesley, 1989.
30. Levy, S. *Artificial Life*. New York: Pantheon Books, 1992.
31. Lindgren, K., and M. G. Nordahl., "Cooperation and Community Structure in Artificial Ecosystems." *Artificial Life* **1** (1994): 15–37.
32. Mailath, G., L. Samuelson, and A. Shaked. "Evolution and Endogenous Interactions." SSRI Working Paper 9426, University of Wisconsin, Madison, WI, June, 1994.
33. Marimon, R., E. McGrattan, and T. J. Sargent. "Money as a Medium of Exchange in an Economy with Artificially Intelligent Agents." *J. Econ. Dynamics & Control* **14** (1990): 329–373.
34. Marks, R. E. "Breeding Hybrid Strategies: Optimal Behavior for Oligopolists." *J. Evol. Econ.* **2** (1992): 17–38.
35. McFadzean, D. *SimBioSys: A Class Framework for Evolutionary Simulations*. Master's Thesis, Department of Computer Science, University of Calgary, Alberta, Canada, 1995.

36. McFadzean, D., and L. Tesfatsion. "A C++ Platform for the Evolution of Trade Networks." *Comp. Econ.*, Special Issue on Programming Languages, (1997): to appear.
37. Miller, J. H. "The Coevolution of Automata in the Repeated Prisoner's Dilemma." *J. Econ. Behav. & Org.* **29** (1996): 87–112.
38. Mitchell, M., and S. Forrest. "Genetic Algorithms and Artificial Life." *Artificial Life* **1** (1994): 267–289.
39. Nelson, R. "Recent Evolutionary Theorizing About Economic Change." *J. Econ. Lit.* **33** (1995): 48–90.
40. Orbell, J. M., and R. M. Dawes. "Social Welfare, Cooperators' Advantage, and the Option of Not Playing the Game." *Am. Sociol. Rev.* **58** (1993): 787–800.
41. Routledge, B. "Artificial Selection: Genetic Algorithms and Learning in a Rational Expectations Model." Working Paper, Faculty of Commerce and Business Administration, UBC, 1994.
42. Sargent, T. *Bounded Rationality in Macroeconomics*. Oxford: Oxford University Press, 1993.
43. Sigmund, K. *Games of Life: Explorations in Ecology, Evolution, and Behavior*. Oxford: Oxford University Press, 1993.
44. Smucker, M. D., E. A. Stanley, and D. Ashlock. "Analyzing Social Network Structures in the Iterated Prisoner's Dilemma with Choice and Refusal." Department of Computer Sciences Technical Report CS-TR-94-1259, University of Wisconsin-Madison, WI, 1994.
45. Stanley, E. A., D. Ashlock, and L. Tesfatsion. "Iterated Prisoner's Dilemma with Choice and Refusal of Partners." In *Artificial Life III*, edited by C. Langton, 131–175. Santa Fe Institute Studies in the Sciences of Complexity, Proc. Vol. XVII, Reading, MA: Addison-Wesley, 1994.
46. Tesfatsion, L. "Direct Updating of Intertemporal Criterion Functions for a Class of Adaptive Control Problems." *IEEE Transactions on Systems, Man, and Cybernetics* **SMC-9** (1979): 143–151.
47. Tesfatsion, L. "How to Get Alife." *CSWEP Newsletter* American Economic Association, Winter Issue, (1995): 16–18.
48. Tesfatsion, L. "A Trade Network Game with Endogenous Partner Selection." ISU Economic Report No. 36, abbreviated version to appear in *Computational Approaches to Economic Problems*, edited by H. Amman et al. New York: Kluwer Academic, 1997.
49. Vriend, N. J. "Self-Organization of Markets: An Example of a Computational Approach." *Comp. Econ.* **8** (1995): 205–231.
50. Witt, U. *Evolutionary Economics*. London: Edward Elgar, 1993.
51. Young, P. "The Evolution of Conventions." *Econometrica* **61** (1993): 57–84.

Philip W. Anderson
Joseph Henry Laboratories of Physics, Badwin Hall, Princeton University, Princeton, NJ 08544

Some Thoughts about Distribution in Economics

The message of this talk is a philosophical one: scientists tend to place too much focus on averages, while the world is full of singularities.

In economics the oldest and best-founded singular distribution is the Pareto Law of incomes:

$$n(I > I_0) = \frac{\text{const}}{(I_0)^\alpha} .$$

For French towns in the nineteenth century, U.S. incomes in 1916, and a number of other populations, $\alpha \simeq 3/2$. But as we shall see, there are many societies in which α is different.

We can check that this is reasonably descriptive of the income statistics of the modern U.S. since we know minimum and maximum incomes—say \$5000 and \$$10^9$. Their ratio is indeed $(10^8)^{2/3}$, where 10^8 is the number of households.

Let us play with other aspects of this: in particular, integrate it down from high incomes and find the total *wealth* above a given level. This turns out to be

$$n = \frac{A}{I^{3/2}}$$

$$W(>I) = 3A\left(\frac{1}{I^{1/2}} - \frac{1}{I_{\max}^{1/2}}\right)$$

$$\left(\frac{1}{2} = \alpha - 1\right).$$

In terms of population N (P is total population)

$$W(N) = 3\text{hP}\left(\frac{N}{P}\right)^{1/3} \qquad \left(3 = \frac{\alpha}{1-\alpha}\right).$$

So to find half the wealth you need only 1/8 the people, etc. The *rich have all the money*! Policy implication: it pays to tax the rich, not the poor! The average income is $\alpha/\alpha - 1 \times$ minimum (a little low). (We have set α a little high for today, after the Reagan revolution: its effect was to increase unfairness, lower α.)

Is α a fixed constant of human nature? Pareto thought it was (he also thought *increasing* α made a more unfair economy, in which he was wrong).

In a truly sick society (Haiti, Zaire, etc.) $\alpha \to 1$ and $W(N)$ is all held by a few people. ($\alpha < 1$ is impossible—everyone starves.) α seems to be a measure of the health of the economy!

What causes Pareto? My instinct is that the economy is a system of competitive advantage, but that it contains, on the whole, positive sum, not zero-sum, games. A simple example of such a game: an elimination tournament in which, in the first round, the winning player wins 2, the lower keeps 1.

The loser is not allowed to play in the next round where the two outcomes are 2 (loser) and 4 (winner); etc. The number of players left after P plays is

$$n(p) = \frac{1}{2^p}$$

and the winnings:

$$W(p) = 2^p$$

so that we have a Pareto with $n = \frac{1}{W}$. In general, if the gain is G for the winner, $W = 1/n^{\ln(1+G)}$. An interesting sidelight for the scientists in the audience: if citations = wealth, they obey a moderately unfair Pareto. ($I_{\max} \simeq 10^4$, $I_{\min} \simeq 1$, 10^5 scientists.)

At the level of firms, however, the economy is more like Darwinian ecology than a sequence of games and, as in such an ecology extinctions, growth and decay, and competition are the rule (Dosi and Winter have suggested evolutionary models of the economy, for example).

In conclusion, I have tried to bring out one general ideas. Much of the real world is controlled as much by the "tails" of distributions as by means or averages: by the exceptional, not the mean; by the catastrophe, not the steady drip; by the very rich, not the "middle class." We need to free ourselves from "average" thinking.

Index

A

B

C

F

G

H

I

N

O

P

R

S

T

U

V

W